Theory of Accretion Disks

NATO ASI Series

Advanced Science Institutes Series

A Series presenting the results of activities sponsored by the NATO Science Committee, which aims at the dissemination of advanced scientific and technological knowledge, with a view to strengthening links between scientific communities.

The Series is published by an international board of publishers in conjunction with the NATO Scientific Affairs Division

A Life Sciences	Plenum Publishing Corporation
B Physics	London and New York
C Mathematical	Kluwer Academic Publishers
and Physical Sciences	Dordrecht, Boston and London
D Behavioural and Social Sciences	
E Applied Sciences	
F Computer and Systems Sciences	Springer-Verlag
G Ecological Sciences	Berlin, Heidelberg, New York, London,
H Cell Biology	Paris and Tokyo

Theory of Accretion Disks

edited by

Friedrich Meyer
Max-Planck-Institut für Physik und Astrophysik,
Institut für Astrophysik, Garching bei München, F.R.G.

Wolfgang J. Duschl
Institut für Theoretische Astrophysik,
Universität Heidelberg, Heidelberg, F.R.G.

Juhan Frank
Max-Planck-Institut für Physik und Astrophysik,
Institut für Astrophysik, Garching bei München, F.R.G.

and

Emmi Meyer-Hofmeister
Max-Planck-Institut für Physik und Astrophysik,
Institut für Astrophysik, Garching bei München, F.R.G.

Kluwer Academic Publishers

Dordrecht / Boston / London

Published in cooperation with NATO Scientific Affairs Division

Proceedings of the NATO Advanced Research Workshop on
Theory of Accretion Disks
Garching, F.R.G.
March 6–10, 1989

Library of Congress Cataloging In Publication Data

Theory of accretion disks / edited by Friedrich Meyer ... [et al.].
         p.   cm. -- (NATO ASI series. Series C, Mathematical and
physical sciences ; no. 290)
    "Proceedings of the NATO advanced research workshop Theory of
accretion disks, held at the Max-Planck-Institut für Astrophysik,
Garching, FRG, March 6-10, 1989, cosponsored by the Max-Planck
-Gesellschaft (MPG) and the European Space Agency (ESA)."
        ISBN-13:978-94-010-6958-8
    1. Accretion (Astrophysics)--Congresses. 2. Magnetic fields
(Cosmic physics)--Congresses. 3. Astrophysics--Congresses.
I. Meyer, Friedrich, 1928-   . II. Max-Planck-Gesellschaft zur
Förderung der Wissenschaften. III. European Space Agency.
IV. Series.
QB466.A25T47   1989
523.1--dc20   89-19912

ISBN-13:978-94-010-6958-8   e-ISBN-13:978-94-009-1037-9
DOI: 10.1007/978-94-009-1037-9

Published by Kluwer Academic Publishers,
P.O. Box 17, 3300 AA Dordrecht, The Netherlands.

Kluwer Academic Publishers incorporates the publishing programmes of
D. Reidel, Martinus Nijhoff, Dr W. Junk and MTP Press.

Sold and distributed in the U.S.A. and Canada
by Kluwer Academic Publishers,
101 Philip Drive, Norwell, MA 02061, U.S.A.

In all other countries, sold and distributed
by Kluwer Academic Publishers Group,
P.O. Box 322, 3300 AH Dordrecht, The Netherlands.

Printed on acid free paper

TABLE OF CONTENTS

vi

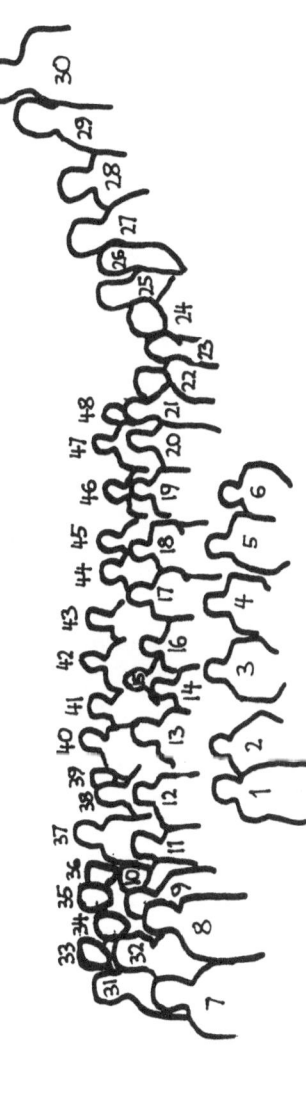

Abramowicz, 24; Adam, 45; Begelman, 8; Blandford, 41; Bodenheimer, 27; Camenzind, 39; Cannizzo, 21; Czerny, 14; Duschl, 35; Frank, 30; Glatzel, 23; Hawley, 40; Hessman, 4; Hubeny, 48; Hummer, 37; Innes, 29; Kato, 46; Kley, 5; Kumar, 20; Lasota, 32; Lin, 44; Livio, 2; Loska, 36; Malkan, 15; Matsuda, 38; Matsumoto, 11; Meyer, 1; Meyer-Hofmeister, 28; Mineshige, 31; Narayan, 9; Osaki, 6; Phinney, 3; Pringle, 42; Pudritz, 43; Rees, 12; Ritter, 7; Rozyczka, 33; Shaviv, 18; Shibata, 34; Siemiginowska, 13; Spruit, 25; Taam, 10; Tscharnuter, 47; Ulrich, 26; Wehrse, 19; White, 17; Whitehurst, 16; Zwitter, 22.

PREFACE

With the advent of space observatories and modern developments in ground-based astronomy and concurrent progress in the theoretical understanding of these observations it has become clear that accretion of material on to compact objects is an ubiquitous mechanism powering very diverse astrophysical sources ranging in size and luminosity by many orders of magnitude. A problem common to these systems is that the material accreted must in general get rid of its angular momentum and this leads to the formation of an *Accretion Disk* which allows angular momentum re-distribution and converts potential energy into radiation with an efficiency which can be higher than the nuclear burning yield.

These systems range in size from quasars and active galactic nuclei to accretion disks around forming stars and the early solar system and to compact binaries such as cataclysmic variables and low-mass X-ray binaries. Other objects that should be mentioned in this context are SS433, the black hole binary candidates, and possibly gamma-ray burst sources. Observations of these systems have provided important constraints for theoretical accretion disk models on widely differing scales, luminosities, mass-transfer rates and physical environments.

It was therefore appropriate to call together an expert meeting to discuss and evaluate the progress in these different research areas with a common theoretical paradigm: the *Accretion Disk*. This was the purpose of a NATO Advanced Research Workshop on *Theory of Accretion Disks* which took place at the Max-Planck-Institut für Astrophysik in Garching, from 6 to 10 March 1989. The participants constituted a significant fraction of the most active researchers in the field drawn from 10 nations, from different areas of expertise, mainly theoreticians but with a good representation of observers. The hope that recent advances in our understanding of different objects may lead to some cross-fertilization was in our opinion realized in the many open and informal discussions during the meeting.

We are deeply grateful to the NATO Science Committee for providing the main funding for this Workshop and to the European Space Agency and the Max-Planck-Gesellschaft for further support. We thank Prof. R. Kippenhahn for allowing the Workshop to be held at the Institute and extend our gratitude to the Max-Planck-Institut für Astrophysik and the Institut für Theoretische Astrophysik, Heidelberg, for use of facilities and support before, during and after the meeting. We are also grateful to BMW, the Cray Research Corporation, IBM, and the Tourist Office of the City of Munich for helping us to put together a nice registration folder. We thank all the participants for their lively contributions to the discussion during the sessions and for the prompt submission of their manuscripts. Finally we would like to thank Petra Berkemeyer and colleagues for organizatorial and secretarial help, and to all the members of our Institute who contributed towards a welcoming environment for our Workshop.

F. Meyer, W. Duschl, J. Frank, E. Meyer-Hofmeister

LIST OF PARTICIPANTS

Abramowicz, Marek A.	SISSA, Trieste, Italy
Adam, Johannes	Institut für Theoretische Astrophysik, Heidelberg, FRG
Begelman, Mitchell C.	JILA, Boulder, USA
Blandford, Roger D.	Caltech, Pasadena, USA
Bodenheimer, Peter	Lick Observatory, Santa Cruz, USA
Camenzind, Max	Landessternwarte, Heidelberg, FRG
Cannizzo, John K.	McMaster University, Hamilton, Canada
Czerny, Bozena	N. Copernicus Astronomical Center, Warszawa, Poland
Duschl, Wolfgang J.	Institut für Theoretische Astrophysik, Heidelberg, FRG
Frank, Juhan	MPI für Astrophysik, Garching, FRG
Glatzel, Wolfgang	MPI für Astrophysik, Garching, FRG
Hawley, John F.	Dept. of Astronomy, Charlottesville, USA
Hessman, Frederic V.	MPI für Astronomie, Heidelberg, FRG
Hubeny, Ivan	High Altitude Observatory, Boulder, USA
Hummer, David G.	JILA, Boulder, USA
Innes, Davina	Institut für Theoretische Astrophysik, Heidelberg, FRG
Kato, Shoji	Dept. of Astronomy, Kyoto, Japan
Kley, Willy	Universitätssternwarte, München, FRG
Kumar, Sanjiv	MPI für Astrophysik, Garching, FRG
Lasota, Jean-Pierre	Observatoire de Paris, Meudon, France
Lin, Douglas N.C.	Lick Observatory, Santa Cruz, USA
Livio, Mario	Technion, Haifa, Israel
Loska, Zbigniew	N. Copernicus Astronomical Center, Warszawa, Poland
Malkan, Matthew A.	Dept. of Astronomy, Los Angeles, USA
Matsuda, Takuya	Dept. of Aeronautical Engineering, Kyoto, Japan
Matsumoto, Ryoji	College of Arts & Sciences, Chiba, Japan
Meyer, Friedrich	MPI für Astrophysik, Garching, FRG

Meyer-Hofmeister, Emmi	MPI für Astrophysik, Garching, FRG
Mineshige, Shin	Dept. of Astronomy, Austin, USA
Narayan, Ramesh	Steward Observatory, Tucson, USA
Oegelmann, Hakki	MPI für Extraterrestrische Physik, Garching, FRG
Osaki, Yoji	Dept. of Astronomy, Tokyo, Japan
Phinney, E. Sterl	Caltech, Pasadena, USA
Pringle, James E.	Institute of Astronomy, Cambridge, UK
Pudritz, Ralph E.	McMaster University, Hamilton, Canada
Rees, Martin J.	Institute of Astronomy, Cambridge, UK
Ritter, Hans	MPI für Astrophysik, Garching, FRG
Rozyczka, Michał	Astronomical Observatory, Warszawa, Poland
Shaviv, Giora	Technion, Haifa, Israel
Shibata, Kazunari	Dept. of Earth Sciences, Aichi, Japan
Siemiginowska, Aneta	N. Copernicus Astronomical Center, Warszawa, Poland
Spruit, Henk	MPI für Astrophysik, Garching, FRG
Taam, Ronald	Northwestern University, Evanston, USA
Tscharnuter, Werner M.	Institut für Theoretische Astrophysik, Heidelberg, FRG
Ulrich, Marie-Helene	ESO, Garching, FRG
Verbunt, Frank	MPI für Extraterrestrische Physik, Garching, FRG
Wehrse, Rainer	Institut für Theoretische Astrophysik, Heidelberg, FRG
White, Nicholas E.	EXOSAT Observatory, SSD - ESTEC, Noordwijk, Netherlands
Whitehurst, Robert	Dept. of Astronomy, Leicester, UK
Zwitter, Tomaz	SISSA, Trieste, Italy

WELCOMING ADDRESS.

Rudolf KIPPENHAHN
Max-Planck-Institut für Physik und Astrophysik, Institut für Astrophysik,
Karl-Schwarzschild-Straße 1, D-8046 Garching bei München, FRG.

Ladies and Gentlemen,

I would like to welcome you all on behalf of our Institute.

The fact that I am standing here for the Max-Planck-Institut for Astrophysics is not completely unrelated to the topic of the conference. In order to explain this I want you to share with me a few of my reminiscences. I got my Ph.D. in pure mathematics in 1951 and then went into astronomy. In my first position I had to compare patrol plates in order to find new variable stars – a terribly boring business. At about the same time I saw a paper by Reimar Lüst on the evolution of rotating gas masses around a central body which contained the first solutions of the evolution of viscous accretion disks. This work was carried out under the guidance of Carl Friedrich von Weizsäcker while this Institute was still in Göttingen. The aim was to understand the formation of the planetary system. I was terribly depressed because there I saw that in astrophysics one could apply what I had learned in mathematics: differential equations and mathematical physics while I had to compare patrol plates. So I decided to get into close contact with the Max-Planck-Institute in Göttingen. This is how I started my work in theoretical astrophysics.

Now, in retrospect, if I think about my time with the patrol plates I see things in a less drastic way. Most of the theoretical work I had done meanwhile has been replaced or soon will be replaced by better papers. But the variables which I have found during that time are still variable today! And if one of them would stop being a variable this would only contribute positively to my reputation.

I remember that among the new variables I found there was an U Gem star – on almost all plates below the limit but once in a while popping up. Nothing was known about these stars, they were considered as some kind of a cheap edition of classical novae. I also found several new eclipsing variables. At that time I learned from papers of Kopal about semidetached systems.

When I had already joined the Max-Planck-Institute and was involved in stellar evolution the semidetached systems were considered as a proof that something was

1

F. Meyer et al. (eds.), Theory of Accretion Disks, 1–2.
© *1989 by Kluwer Academic Publishers.*

wrong with stellar evolution: There the less massive component turned out to be more advanced, in contradiction to what the stellar evolution people said.

It was then about 1960 when Donald Morton, Martin Schwarzschild's Ph.D. student, introduced the possibility of mass flow from one component to the other. Then – after one had learned to deal with the mathematics of stellar evolution – the problem was solved partially by work done in this institute and one learned that binary stars show the same basic evolutionary features as single stars, the only difference is the mass flow from the originally more massive component onto the other component.

At this time nobody cared too much about the hydro-dynamics of this flow. It was not so important because the receiving star was a main sequence object. The angular momentum of the overflowing material did not seem to become important before it hit the surface of the relatively large star and therefore no disks were expected. In 1956 Joy found that the minimum spectrum of SS Cyg is a composite one. In 1961 Kraft found more dwarf novae binaries and shortly afterwards Mumford detected the eclipses in U Gem. Since the receiving star was a white dwarf the problem of accretion disks should have occurred already to those who thought about mass overflow at that time. But the big step forward came later, when the UHURU data revealed the existence of X ray binaries. Then suddenly in the Soviet Union and in Cambridge the classical papers on accretion disks were written by Shakura and Sunyayev and by Lynden-Bell, Pringle and Rees. Much progress has been made and many problems have been solved since then.

There are still many open questions. I am sure you will answer them all during this conference. I wish you much success and I hope you will find some time to enjoy Munich.

OBSERVATIONAL EVIDENCE FOR ACCRETION DISKS IN GALACTIC NUCLEI.

Marie-Helene ULRICH
European Southern Observatory, Garching, FRG

ABSTRACT. The electromagnetic spectrum of quasars and Seyfert 1 galaxies shows an excess of energy at $\lambda > 4000$ Å over the extrapolation of the infrared/optical spectrum: This "blue bump" is generally interpreted as thermal radiation by gas at $T > 3 \cdot 10^4$ °K. But is it an accretion disk which emits this thermal radiation? To answer this question geometrical and kinematical information on the dense gas in the vicinity of the black hole is needed. We examine the most important observations in favor or against the presence of an accretion disk: the photoelectric X-ray absorption, the energy budget in the broad line region, and the profiles of the Balmer lines. The optical polarization and the occurrence of Lyman edge absorptions are also reviewed.

1. Introduction.

Theoreticians propose that accretion onto a massive object is the likely source of the energy emitted by quasars and Seyfert 1 galaxies. Because of the ubiquitous presence of angular momentum it is proposed further that the accreted gas is assembled in an accretion disk.

In the face of this attractive theoretical argument in favor of an accretion disk in quasars, one can ask the question as to whether there is observational evidence for such a disk. In favor of an accretion disk is the widespread "blue bump" which is generally interpreted as thermal radiation. The discovery of a soft X-ray excess is consistent with thermal radiation from an emitter with a range of temperatures $3 \cdot 10^4$ - 10^5 °K.

However establishing that the gas is or is not assembled in an accretion disk requires geometrical and kinematical information in addition to the continuous energy distribution between 4000 Å and 10 Å. Most important to this argument are the observations of the X-ray photoelectric absorption and fluorescence lines, the observations related to the energy budget in the broad emission line region and the emission line profiles. Further information comes from the optical polarization and the occurrence of Lyman edges in absorption. The data available at present are examined critically in the following.

F. Meyer et al. (eds.), Theory of Accretion Disks, 3–18.
© 1989 by Kluwer Academic Publishers.

2. A Powerful Argument in Favor of an Accretion Disk: The Properties of the Emission Lines.

2.1. THE ENERGY BUDGET IN THE BROAD EMISSION LINE REGION.

The properties of the low ionization lines and in particular the large amount of energy necessary to excite them (as compared to the high ionization lines) constitute one of the most convincing arguments for these lines being emitted for the most part by an accretion disk. The main properties of the broad emission lines (e.g. Collin-Souffrin and Lasota 1988) relevant to our discussion of accretion disks can be summarized as follows:

(i) The high ionization lines such as CIV, SiIV, NV have intensity ratios which can be satisfactorily modelled with clouds ionized by a hard UV ionizing continuum with $N_e \lesssim 10^{9.5}$ cm^{-3} and T $\sim 1.5 \cdot 10^4$ °K.

(ii) The low ionization lines - Balmer and Paschen lines, FeII, MgII - come from gas which is very optically thick as indicated by the observed intensity ratios Hα/Hβ = 3.5 and Lyα/H$\beta \sim 2$. The ratio between the total intensity of the two types of lines, I_T(low ionization lines)/I_T(high ionization lines) is $\sim$ 0.5 (Netzer 1985). This suggests that the clouds emitting most of the low ionization lines either have a larger cross section to the ionizing continuum than the clouds emitting the high ionization lines, or they have a very large optical depth, $N_H \gtrsim 10^{25}$ cm^{-2}, so as to be able to tap the energy in the range 10-500 keV.

(iii) There is a systematic difference between the velocities of the high ionization lines and the low ionization lines, the former being blueshifted by 500-1500 km s^{-1} with respect to the latter. This indicates that a large fraction of the high ionization line intensity does not come from the same gas clouds producing the low ionization lines.

These 3 properties taken together can be reproduced only in a two-component model (e.g. Netzer, 1987). Several such models are probably possible. A very attractive model is one in which the low ionization lines come from gas at the surface of a disk ($N_H \geq 10^{25}$ cm^{-2}) illuminated from above by the hard X-rays emitted near the black hole and scattered downwards in the direction of the disk surface.

Such a model has recently been investigated in detail by Collin- Souffrin and Dumont (1989a,b) and Dumont and Collin-Souffrin (1989) who have determined the structure at R $> 10^3$ R$_s$ of a stationary α-disk where the gas pressure dominates and the opacity is dominated by electron scattering, absorption by H, H^{-1} ions, molecules and dust depending on the radius. These authors calculate the total intensity of the low ionization lines and find it consistent with the observed value provided that 10 to 20% of the hard X-ray up to 500 keV is absorbed by the disk. This reproduces the basic property (ii) outlined above. By folding the emissivity law of such a disk with its dynamics they calculate the line profiles and verify the general property of the Balmer lines in Seyferts and quasars, that Hβ is broader than Hα (Shuder 1982). Moreover the profiles of the lines emitted by this disk model are found to be single peaked as is observed in general. Double peaked lines are emitted by the innermost section of a disk, R $\sim 10^3$ - 10^4 R$_s$, the outer regions where the velocity is lower filling the central section of the line profiles. Disks where the outer regions have low emissivity produce double peaked lines such as observed in Arp 102B for which Collin-Souffrin and Dumont find a satisfactory profile fit. The high ionization lines are emitted by gask flowing out in a cone with a net approaching velocity with respect to the Balmer lines reproducing property (iii) above.

2.2. OBSERVATIONS OF EMISSION LINE PROFILES.

Because of the possible presence of an accretion disk, one could expect a double peak to be present in the profiles of the emission lines. Indeed double peaked lines have been found, but only in a very small number of Seyfert 1 galaxies: Arp 102B, 3C 390.3, Akn 120. In NGC 5548 the temporal behavior of the Balmer line profile is sometimes consistent and sometimes inconsistent with these lines being emitted by a disk.

  - Arp 102B. This is the galaxy in which the profile of the Balmer lines has been analyzed in most details. Aside from the fitting achieved by Collin-Souffrin and Dumont (1989), another detailed fitting was done by Chen, Halpern and Filippenko (1989). These authors calculated the line profile of a Keplerian disk in the weak field approximation to first order in M/r. The variation of the surface emissivity with radius was chosen to be a single power law plus a factor simulating local broadening. This is a plausible but ad hoc assumption. A very good fit was obtained with a value of σ of 850 km s^{-1}. This value of σ can be produced by electron scattering in a photoionized atmosphere of the disk at $2 \cdot 10^4$ $^\circ$K and a mean number of scatterings of 4 (Shields and McKee, 1981).

  Electron scattering is interesting in a related point: When a disk is seen pole-on, the lines emitted by the disk have no contribution from the disk motion and have thus the profile of the local emission. The smallest observed value of the Balmer lines width is 1000 to 2000 km s^{-1} (Joly 1988; Wills and Browne 1986). Electron scattering provides an attractive explanation for the lack of Balmer lines narrower than about 1000 km s^{-1}.

  - 3C 390.3. In this broad line radio galaxy Perez et al. (1988) obtained an acceptable fit of the double peaked Hβ line profile with a relativistic accretion disk with an intermediate inclination to the line of sight. An independent information on the disk inclination comes from the asymmetry of the jet on VLBI maps (Linfield 1981; Preuss et al. 1981). This asymmetry when interpreted as differential Doppler brightening of an intrinsically symmetrical relativistic jet indicates an angle of $\lesssim 60^\circ$ between the jet direction and the line of sight (Preuss et al. 1980) consistent with the model reproducing the Balmer line profile.

- Akn 120 provides another example of a double peaked profile (Peterson 1983; Alloin, Boisson and Pelat 1988). The Hα line in the difference spectrum obtained by subtracting two spectra taken when Akn 120 was in bright and low states shows two peaks separated by 4000 km s^{-1} of same intensity and width ($\sim$ 2400 km s^{-1}). This is interpreted by Alloin et al. as the presence of a variable component emitted by a disk and they attribute the width of the peak to turbulence.

- NGC 5548 is a particularly interesting case. Stirpe, de Bruyn and van Groningen (1988) observed this galaxy spectroscopically at high wavelength resolution and with high S/N at several epochs when the nucleus was in low and high states. The profiles of the variable components of Hα and Hβ obtained by making difference spectra show the following: in one case there are two peaks of uneven height (ratio larger than 2); in another case there is only one large blue peak; in a third case one peak is positive and the other negative (Stirpe et al. 1988, figure). The situation is clearly different from the two peaks of comparable height expected in the case of a simple variable disk component. The disk model can still be saved if one assumes large local changes in the flux illuminating the disk (van Groningen 1989).

In summary, a few examples of double peaked lines have been found which are consistent with the presence of an accretion disk - NGC 5548 possibly being a counter example. In the majority of quasars and Seyfert 1's the lines appear single peaked. This could be due to the fact that the lines come in general from an extended disk in which the outer regions with low velocity make an important contribution to the line central part.

2.3. STATISTICAL STUDIES OF ORIENTATION.

If an accretion disk is present in quasars a clue to the inclination of the disk to the line of sight is provided, in the case of radio loud objects, by the intensity ratio R of the core radio component to the extended radio component. This is based on the assumption that the extended component emits isotropically, while the core component is beamed perpendicular to the disk and its observed intensity is therefore a function of its orientation. Wills and Browne (1986) analyzed a sample of 79 radio sources for which good Hβ profile and radio maps were available. The two variables, Hβ-width and R, are related in the sense that there are no objects with large Hβ width which also have a large value of R, i.e. for which the radio core is beamed nearly to the observer. This is consistent with the beaming hypothesis and the emission of Hβ by a disk.

On the other hand, it could be that the quasars with strong core sources have intrinsically narrow Hβ lines, independent of the orientation.

3. The X-Ray Spectrum.

3.1. THE PHOTOELECTRIC ABSORPTION AND ITS RELEVANCE TO THE AC-CRETION DISK AND TO THE UV/X BUMP.

The observations of photoelectric absorption are critically important to establish the location of the gas in the vicinity of the black hole. They give the frequency of occurrence of gas clouds along the line of sight and thus the covering angle of the X-ray source by these clouds if they are isotropically distributed.

A powerful, albeit indirect, argument for a disk has been that although 10% to 20% of the 2 to 500 keV flux must be intercepted by dense gas clouds with $N_H > 10^{25}$ cm^{-2} in order to produce low ionization lines with the total observed intensity, no evidence for photoelectric absorption by such large column density gas clouds had been found. Such clouds therefore could not be stributed isotropically. Thus the idea that they are in a disk.

New developments have recently taken place.

Before reporting the most recent relevant X-ray observations it must be pointed out that the past X-ray missions capable of observing a large number of quasars have covered the X-ray range up to $\sim$ 10 keV, and thus were practically insensitive to partial covering of the X-ray source by clouds with $N_H > 10^{24}$ cm^{-2} located along the line of sight to the source. Total covering of the X-ray source by clouds with $N_H > 10^{24}$ cm^{-2} and located along the line of sight would make a quasar or a Seyfert 1 galaxy appear very weak in the X-ray range while having normal properties in the other energy ranges. Such cases have not been found.

The currently flying X-ray satellite GINGA covers the range 2-30 keV. It has observed a number of Seyfert 1 galaxies and the quasar 3C 273. The results of these observations are known today for three of the Seyfert 1's, NGC 4051 and MCG 6-30-15 (Matsuoka, Yamauchi, Piro and Murakami 1989) and NGC 7469 (Piro 1989). In all three Seyfert galaxies the GINGA spectrum is consistent with partial absorption (30% to 60% of the X-ray source covered by clouds) with N_H in the range 3-$6 \cdot 10^{24}$ cm^{-2}. In addition the K-shell Fe fluorescence line at 6.4 keV is present in all three Seyferts with equivalent widths of $\sim$ 200 eV. (This line had already been detected by previous missions in a number of Seyfert 1 galaxies).

What is the origin of this absorption?

Are these absorbing clouds representative of a henceforth undetected ensemble of isotropically distributed dense clouds which would be capable of emitting the low ionization lines, thus weakening the argument outlined in Section 2.1 in favor of an accretion disk? Or could this absorption be caused by the edge of an accretion disk? Further observations are necessary to answer these questions.

We note that the observed value of N_H (a few 10^{24} cm^{-2}) is is barely sufficient to absorb the hard X-rays. Furthermore, such a large value has not been detected in intrinsically bright nuclei such as 3C 273.

Further observations will provide better statistics on the occurrence of large and partial covering absorption in low and high intrinsic luminosity and in radio quiet and radio loud objects. Another important statistical result to be expected is the one on absorption variability. In Seyfert 1's the X-ray flux is rapidly variable: a factor 2 in one day is not uncommon and implies a dimension of $3 \cdot 10^{15}$ cm. A cloud moving with a velocity of 2000 km s^{-1} will pass in front of the X-ray source in less than a year causing a large variation of the absorption column on this time scale.

3.2. FREE-FLYING DENSE CLOUDS IN THE VICINITY OF THE BLACK HOLE: A SOURCE OF UV/X ENERGY.

Recently Guilbert and Rees (1988) and Lightman and White (1988) independently showed that a collection of very dense and high optical depth gas clouds ($n_e \sim 10^{11}$-10^{15} cm^{-3}, $N_H > 10^{24}$ cm^{-2}) submitted to the non-thermal X-ray radiation emitted in the vicinity of a black hole could modify and reprocess the non-thermal radiation which would then re-emerge with an essentially thermal spectrum, with UV and soft X-ray properties not different from what is observed in Seyfert 1's and quasars. A blue bump therefore is not necessarily the signature of an accretion disk. The clouds also modify the X-ray energy distribution at higher energies resulting in an X-ray spectrum with bumps and troughs resembling that of the best observed objects NGC 4151, 3C 273, Cen A (Lightman and White 1988).

Even if an accretion disk is present, it is likely that very dense gas clouds are also present in the immediate vicinity of the black hole (20 R_s). The blue bump is then the sum of the disk thermal radiation and of the reprocessed non-thermal radiation due to the very dense clouds.

The dense material in the accretion disk and/or the dense free flying clouds in the vicinity of the black hole also produce iron - K shell features by fluorescence (Guilbert and Rees 1988; Lightman and White 1988). It is therefore particularly interesting that EXOSAT and GINGA observations show these features to be a general characteristic of the X-ray spectrum of Seyfert 1 galaxies and of the few quasars observed so far. The energy, width and equivalent width of these features provide constraints on the temperature, kinematics and location of the clouds emitting them (Fabian et al. 1989). The constraints will become very powerful with the much improved energy resolution offered by several X-ray missions of the 1990's.

3.3. THE SOFT X-RAY EXCESS.

The soft X-ray excess is observed in the range 0.1-2 keV above the extrapolation of the medium energy component.

The soft excess itself does not help answer the question as to whether there is an accretion disk or not, but it is important in the evaluation of the total energy emitted by the blue bump and in the modelling of a possibly present accretion disk, because it is quite likely the high energy tail of the UV bump[1]

The detection of the soft excess, with the present X-ray instruments, is not always straightforward because of the multiple spectral components contributing to the emission in the range 0.1-20 keV:

- First there is the medium energy component 2-20 keV with a photon index around $\alpha \sim 1.7$. In addition to a possible small real dispersion in α (Turner and Pounds 1989) it is now established for 6 sources that α varies with the flux level in the sense that the 2-10 keV component becomes softer when the flux increases. The pivot point of the medium energy component is around 30-50 keV (Perola et al. 1986; Matsuoka et al. 1989). Another complexity arises from the fact that in radio quasars the 2-10 keV component is harder and stronger (as compared to the optical luminosity) than in radio quiet quasars suggesting the presence of an additional flat spectrum component related to the radio emission. Some strong fluorescence lines also appear in this energy range.

- Second there are absorption components: the galactic absorption to which is always attached some uncertainty because of small scale angular variations in galactic N_H and one or several intrinsic absorption components which can cover the source uniformly or partially.

- Third there is the soft X-ray excess itself which is present in the range 0.15-2 keV range in a number of sources. The best evidence for this component comes from the observations of different time scales of the flux variations in the low energy and medium energy ranges (LE and ME).

Markarian 335, a typical Seyfert 1, is the best example so far of distinctive short term variability in the soft X-ray excess and the hard X-ray component. In particular, a 10 hour EXOSAT observation showed the 2-6 keV flux varying on time scales of 1 to 2 hours with amplitude by a factor 2 while the LE flux underwent slower variations (6 hours) with larger amplitudes.

There is also a correlation on large scale between the soft and hard X-ray components: both components brightened by a factor ~ 6 between November 83 and December 84. This could be attributed to changes in the accretion rate (Turner and Pounds 1988). In contrast with the X-ray flux, the UV flux shows very little variation. Spectacular examples of distinctive short term variability

---

[1] [with a known exception: NGC 4151 where the soft excess is constant with time and has been interpreted as emission by the intercloud medium of the inner part of the narrow line region; Pounds et al. 1986; Perola et al. 1986.]

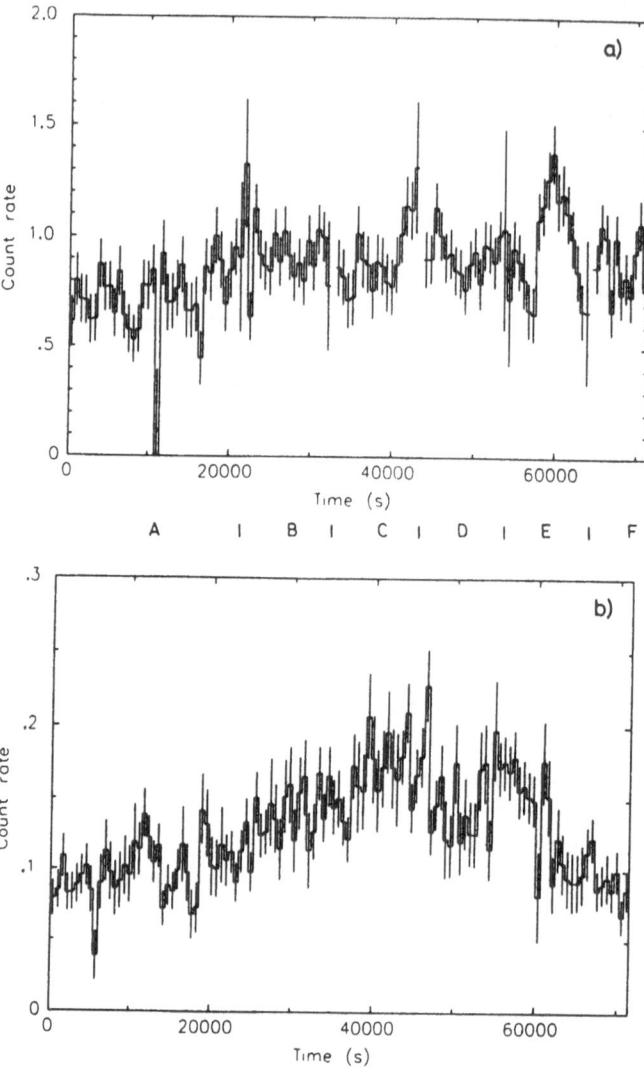

Figure 1: EXOSAT observations of Mkn 335 showing that the short term variability of the soft excess (bottom) and of the harder X-ray component (top) are clearly distinct. From Turner and Pounds, 1988.

in the LE and ME ranges are E1615+061 (Piro et al. 1988) and 1821+643 (Warwick, Barstow and Yaqoob 1989).

All the components contributing to the X-ray spectrum in the range 0.1-20 keV vary with time (except of course the galactic absorption) and, moreover, the absorption may not be the same for the different emission components. Because of these multiple components the result of the fitting and its meaning in terms of the

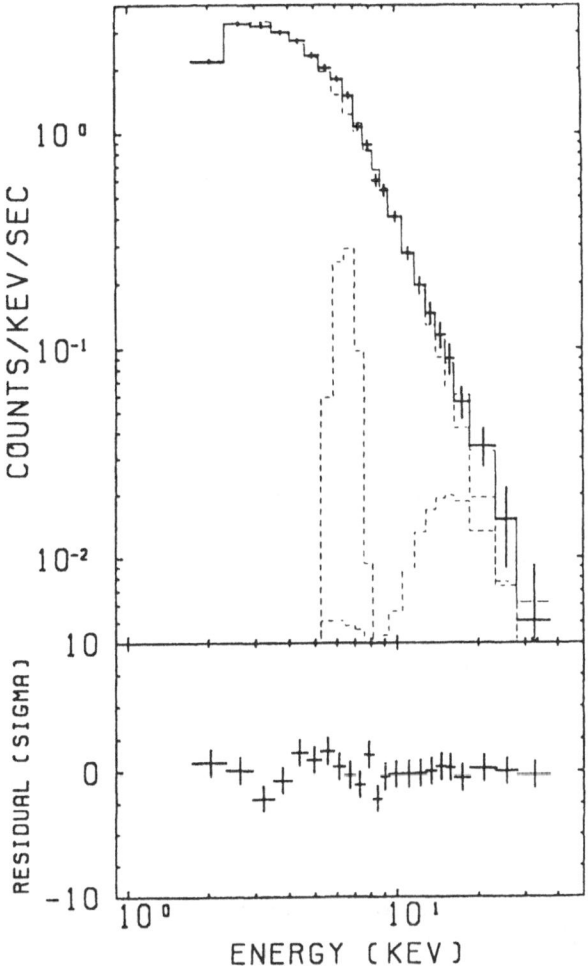

Figure 2: GINGA observations of MCG 6-30-15 (thick line) fitted with a power law of index 1.7 plus a covering of 50% of the source by a cloud with $N_H = 6 \cdot 10^{24}$ cm^{-2} and normal abundances. From Matsuoka et al., 1989.

soft X-ray excess is somewhat uncertain when the sources are complex and fairly faint. In other words establishing the existence of a soft X-ray component in the spectrum of a quasar or a Seyfert 1 is not often a straightforward matter.

3.4. FREQUENCY OF OCCURRENCE OF THE SOFT EXCESS.

Turner and Pounds (1989) in their EXOSAT survey of AGN and Nandra et al. (1989) estimate that half of the Seyfert galaxies observed by EXOSAT have a soft excess.

Its presence is certain or fairly certain in the 8 low luminosity objects - Mark

335, Mark 841, MCG 6-30-15, F9, NGC 4151, Mark 509, NGC 7469 and E1615+061 (Piro et al. 1988) - to which must be added the high luminosity objects PG 1211+143 (Elvis et al. 1989), 1821+643 (Warwick et al. 1989) and with less certainty 3C 273 (Courvoisier et al. 1987). Wilkes and Elvis (1987) estimated the frequency of occurrence from a sample of 33 quasars with well determined X-ray slopes in the Einstein IPC range 0.2-3.5 keV. The fit of the quasar spectra with a single power law component of energy index α and a value of N_H leads to values of N_H which are in 26 cases out of 33 lower than the galactic value. This is interpreted by Wilkes and Elvis (1987) as the presence of an "ultrasoft excess". A sample of 11 BL Lac objects with similar S/N analyzed in the same way yields values of N_H in excellent agreement with the galactic values, which shows that the "ultrasoft excess" (in the terminology of Wilkes and Elvis) found in quasars is not a spurious instrumental effect, and also confirms that BL Lac objects do not have soft X-ray excess (it is well known that they do not have a UV excess). The quasars analyzed by Wilkes and Elvis (1987) were selected for their good signal to noise in the range 0.2-2.5 keV. One can ask oneself whether the proportion of "ultrasoft" excess found would be the same in a sample selected at higher energies (e.g. 5-10 keV). Wilkes and Elvis offer reassuring comments about the absence of selection effects but these do not allay all doubts (at least not to this writer).

3.5. PROPERTIES AND MODELLING OF THE UV/X BUMP.

After the first suggestion of Shields (1978) that the blue bump in 3C 273 was due to thermal emission by an accretion disk, and the measurements of a fair number of UV spectra of Seyfert 1 and quasars with IUE, much effort has been devoted to the fitting of the blue bump, notably by Malkan and Sargent (1982) and Malkan (1983). Then came the suggestion by Arnaud et al. (1986) that the soft X-ray excess in Mark 841 was the high energy end of the blue bump. Then the effort shifted to the fitting of the whole UV/X bump by models of accretion disks.

It is pointed out that if accretion disks are indeed present in quasars, some free flying clouds can also contribute to the UV/X bump and their contribution should be first estimated. Moreover, the structure of thick disks is presently not known and therefore the theoretical predictions of the energy distribution of the UV/X bump are uncertain.

The temperatures which best represent the blue/UV part of the UV/X bump are all between 20000-35000 °K. There appear to be no systematic variations of this temperature with the absolute luminosity of the Seyferts and quasars. In individual objects however such as 3C 273 the temperature appears to be lower when the flux is lower (Ulrich, Courvoisier and Wamsteker 1988).

There is no relation between the fractional luminosity of the blue bump and the bolometric luminosity. Both intrinsically faint and weak objects show a blue bump (e.g. NGC 4151, 3C 273).

McDowell et al. (1989) find a positive correlation between the intensity of the

UV bump and the intensity of the soft X-ray excess which strongly suggests that the two have a common origin.

Several authors have remarked that the slope of the sum of black bodies representing satisfactorily the UV/X bump is steeper than the observed slope of the high energy tail. Several mechanisms have been invoked to obtain a slope harder than a black body. Pounds et al. (1988) use real atmospheres with the relevant effective temperatures and surface gravity but even there they cannot fit properly the slope of the soft X-ray excess in Mark 335. They consider Compton scattering of extreme UV photons by energetic electrons. The observed slope of 3.1 can be fitted with $\tau = 0.4$ and $kT = 50$ keV or $\tau = 0.2$ and $kT = 100$ keV. This is rather similar to the hot electron cloud with $\tau = 0.05$ and $kT = 80$ keV which fits the soft X-ray excess in PG 1211+143 (Bechtold et al. 1987).

Another treatment of this problem is the one by Laor and Netzer (1989) who include all significant sources of opacity, in particular the bound-free absorption, in a model of a massive geometrically thin accretion disk. This model is self-consistent in the sense that the requirement of a geometrically thin disk forces a limit on the accretion rate of $L < 0.3\ L_{EDD}$. The general result is that an acceptable fit of the big blue bump requires a large value of $\dot{m}$. The most luminous quasars have the most massive disks, but the massive disks are cool and their high energy tails do not reproduce the soft X-ray excess. Comptonization in a hot corona could reproduce the soft X-ray flux as suggested by Czerny and Elvis (1987) and Pounds et al. (1988). On the other hand the rapid variations of the soft excess in NGC 4051 (factor 2 in 100 seconds) and in PG 1211+143 (Elvis et al. 1989) seem to rule out a corona as the origin of the high energy tail of the soft excess.

A simple and attractive alternative for the origin of the high energy tail of the soft X-ray excess is as follows: In the part of the disk ($R > 10^2\ R_s$) producing the UV/X bump radiation pressure dominates by a large factor (typically 1000) the gas pressure. Differential acceleration can produce motions of 1000-2000 km s^{-1}, much larger than the sound speed. This produces shocks and thus X-ray emission by thermal bremsstrahlung, which can explain the high energy tail of the soft X-ray excess (Rees 1989). This was proposed by Lucy and White (1980) for the X-ray emission of OB stars.

4. Signatures of an Accretion Disk in the Ultraviolet Spectrum.

Models of thin accretion disks predict significant polarization of the UV flux (inclination dependent).Lyman edges in absorption are predicted in the models of Sun and Malkan (1988) and of Kolykhalov and Sunyaev (1984) but the edges appear in emission in the more recent models of Laor and Netzer (1989). Observations reveal only weak continuum polarization and show that strong Lyman edges are not a general property of the UV/X bump. They are summarized below.

4.1. IS THE RADIATION FROM THE BLUE BUMP POLARIZED?

Of interest here are the low polarization, low variability quasars where we see directly the nucleus. That is, the optical and UV emission we receive from them is neither beamed like in BL Lac nor obscured or scattered like in Seyfert 2's. The radio quasars are the most interesting ones because the radio axis gives some indication on the orientation of the object in space. Stockman, Angel and Miley (1979) discovered that in these quasars the polarization is ~ 1% and is in general aligned along the radio axis. More recently Miller and Antonucci (1988) measured spectropolarization in three of the quasars observed by Stockman et al. (Ton 202, 4C 34.47, 3C 232.1) and find the percentage polarization to decrease towards the blue, consistent with the weakly polarized continuum being diluted by the higher order Balmer lines and an unpolarized blue bump continuum. If confirmed by spectropolarimetry of a larger number of quasars this will implicitly show that the blue bump is even less polarized than the continuum at say 6000 Å. The commonly found 1% polarization aligned along the radio axis is thus a property of the visible continuum (thermal or non-thermal) and the disk emission is essentially unpolarized.

4.2. SPECTROSCOPIC OBSERVATIONS AT THE LYMAN EDGE.

The Lyman edge of thick disks has not yet been calculated. If present in absorption it is expected to be smaller than in thin disks because of a small net effective gravity g_{eff}, and possibly because the hotter parts of the disk will produce the edge in emission, and the cooler parts in absorption. As emphasized by Kolykhalov and Sunyaev (1984) the entire vertical disk structure must be considered in assessing g_{eff}. A Lyα absorption line of width and shift similar to the edge is expected to accompany the edge and its presence or absence provides a strong observational constraint.

At present there is no prediction of the Lyman edge in thick disks with which the observations can be compared. And in the spectra of Seyfert 1's and quasars there is no observational evidence for Lyman edges which can be unequivocally attributed to a disk.

The observational situation is as follows: The 912 Å region has been observed spectroscopically a) from the ground in quasars with $z > 2.5$ (Baldwin and Smith 1983; Antonucci, Kinney and Ford 1988), and b) with the satellite IUE in quasars of intermediate redshifts, $0.3 < z < 0.8$ (Kinney et al. 1985, 1987). To this, one must add the unique case of 3C 273 ($z = 0.158$) observed by Voyager (Reichert et al. 1988).

A Lyman discontinuity, partial or complete, is a rather common feature of high redshift quasars. In the sample of Smith et al. (1981) 14 out of 32 quasars show strong Lyman discontinuity. A detailed study of 8 of these 14 quasars by Baldwin and Smith (1983) show that all except 1 have strong black Lyα lines indicating the presence of an intervening absorber, or show the Lyman discontinuity to be significantly blueshifted with respect to the quasar again indicating that the Lyman discontinuity is not intrinsic to the quasar nor due to the broad line region.

Among the 11 high redshift quasars observed by Antonucci et al. (1988) only 2 have partial edges near zero redshift and the Lyα absorption lines in these quasars are narrow. So if the Lyα ption absorption lines in these 2 quasars are associated with the Lyman edge (they have enough column density), the edge cannot be due to the disk. On the other hand, the Lyα lines associated with the disk may be so broad as to be undetectable, but there is no evidence for this.

Among the 15 quasars observed by Kinney et al. at 912 Å with IUE only 3 show a Lyman discontinuity. Two of them have also MgII λ2800 in absorption and the discontinuity is attributed to an intervening absorber. There remains only 1 ambiguous case for which the Lyα line falls on a reseau mark of the IUE detector.

An alternative possibility, that the Lyα line corresponding to the edge is made up of a large number of weak lines, each individually undetectable, has been explored in the case of 3C 273 by Reichert et al. (1988) who find it unlikely.

5. Concluding Remarks.

One of the most important steps toward understanding the electromagnetic spectrum of quasars and Seyfert 1's has been the identification of the UV bump with thermal emission produced by accretion onto a massive object (e.g. Shields 1978). The recent observations of a soft X-ray excess are entirely consistent with this model even though it is not clear at present whether the soft X-ray excess is radiation emitted by the inner parts of the disk, by a hot corona surrounding the disk or by shocks at the disk surface.

The next important step is to establish that the thermal emission comes from an accretion disk. This requires geometrical and kinematical information on the cold gas clouds (T in the range $3 \cdot 10^4$ to 10^5 °K) emitting the UV/X bump. A powerful, albeit somewhat indirect argument in favor of a disk is the necessity of a two component model to explain the properties of the broad emission lines. In particular cold gas clouds with a high column density ($N_H > 10^{25}$ cm^{-2}) intercepting a substantial fraction of the 2 to 500 keV spectrum are required in order to produce the large amount of energy present in the low ionization lines. Since X-ray absorption with $N_H > 10^{25}$ cm^{-2} is never observed, the dense gas clouds are not isotropically distributed and are thus in a disk.

The recent observations by the X-ray satellite GINGA provide evidence in 3 nearby Seyfert 1's for partial absorption by clouds with N_H of a few 10^{24} cm^{-2}. This does not really weaken the argument in favor of a disk because the gas clouds seen by GINGA do not have enough column density to absorb hard X-rays and produce the low ionization lines. But it remains to be seen from additional observations what is the real frequency of occurrence of large N_H absorption and how large N_H can be in high luminosity quasars as well as in nearby Seyfert 1's. The forthcoming X-ray missions (such as SAX) with which a large energy range can be observed simultaneously, and which are therefore capable of detecting partial absorption with $N_H > 10^{25}$ cm^{-2}, will bring much needed data.

It is not clear at present whether the absorption seen by GINGA is caused by the edge of an absorption disk or by cold dense free flying clouds, at $R \lesssim 50$ R_s, such as those investigated by Guilbert and Rees (1988) and Lightman and White (1988). In fact it is plausible that some free flying clouds could add their thermal radiation to that of the disk. By what amount they can do this must be determined before modelling the disk. The wavelength equivalent width, and detailed profile of the optical and UV emission lines are entirely consistent with the presence of an accretion disk. Similarly the K-shell fluorescence lines will provide strong constraints on the location and kinematics of the dense gas clouds, and therefore on the accretion disk's existence and properties.

The observations of two other potentially important properties of the UV continuum - the Lyman edge and the polarization - are not decisive. The very weak polarization observed and the general absence of Lyman edges are inconsistent with thin accretion disk models but such models are very probably not applicable to the inner regions ($R \lesssim 10^{2.5}$ R_s) which emit the UV continuum. [The accretion disk, if present, is likely to be thin further out, where the emission lines are emitted.]

Even if a thick disk model produces a Lyman edge, the Lyman edge could be "scrambled" beyond detection by a scattering medium. The same scattering medium could also scramble and in effect reduce to essentially zero the polarization of the light emitted by the disk. A possible test proposed by Antonucci et al. 1988 is to see whether the few quasars which may have an intrinsic Lyman edge also have larger polarization than the others. This would indicate that in these special objects scattering is less effective.

ACKNOWLEDGEMENTS. I would like to thank the many colleagues who sent me results before publication and Suzy Collin-Souffrin and Martin Rees for stimulating discussions.

REFERENCES.

Alloin, D., Boisson, C., Pelat, D.: 1988, Astron. Astrophys. 200, 17.
Antonucci, R.R.J., Kinney, A.L., Ford, H.C.: 1988, STScI preprint 328.
Arnaud, K.A., et al.: 1985, Monthly Notices Roy. Astron. Soc. 217, 105.
Baldwin, J.A., Smith, M.G.: 1983, Monthly Notices Roy. Astron. Soc. 204, 331.
Bechtold, J., et al.: 1987, Astrophys. J. 314, 699.
Chen, K., Halpern, J.P., Filippenko, A.V.: 1988, preprint.
Collin-Souffrin, S., Lasota, J.-P.: 1988, Pub. Astron. Soc. Pacific 100, 1041.
Collin-Souffrin, S., Dumont, A.-M.: 1989a,b, preprints.
Courvoisier, T.J.-L., et al.: 1987, Astron. Astrophys. 176, 197.
Czerny, B., Elvis, M.: 1987, Astrophys. J. 321, 305.
Czerny, M., King, A.R.: 1989, Monthly Notices Roy. Astron. Soc. 236, 843.
Dumont, A.-M., Collin-Souffrin, S.: 1989, preprint.
Elvis, M., Giommi, P., McDowell, J.C., Wilkes, B.J.: 1989, BAAS p.1024.

Fabian, A.C., Rees, M.J., Stella, L. and White, N.E.: 1989, preprint.

Guilbert, P.W., Rees, M.J.: 1988, Monthly Notices Roy. Astron. Soc. 233, 475.

Joly, M.: 1987, in "Emission Lines in Active Galactic Nuclei", ed. P.M. Gondhalekar, RAL, Chilton OX11 OQX, England, p.166.

Kinney, A.L., Huggins, P.J., Bregman, J.N., Glassgold, A.E.: 1985, Astrophys. J. 291, 128.

Kinney, A.L., et al.: 1988, in "A Decade of UV Astronomy with IUE", ESA SP-281, p.277.

Kolykhalov, P.I., Sunyaev, R.A.: 1984, Adv. Space Res. 3, 249.

Laor, A., Netzer, H., 1989, preprint.

Lightman, A.P., White, T.R.: 1988, Astrophys. J. 335, 57.

Linfield, R.: 1981, Astrophys. J. 244, 441.

Lucy, L.B., White, R.L.: 1980, Astrophys. J. 241, 300.

Malkan, M.A.: 1983, Astrophys. J. 268, 582.

Malkan, M.A., Sargent, W.L.W.: 1982, Astrophys. J. 254, 22.

Matsuoka, M., Yamauchi, M., Piro, L., Murakami, T.: 1989, preprint.

McDowell, J.C., Elvis, M., Wilkes, B.J., Polomski, E.F., Oey, M.S.: 1989, BAAS p.967.

Miller, J.S., Antonucci, R.R.J.: 1988, result quoted by Antonucci in "Supermassive Black Holes", ed. M. Kafatos, Cambridge University Press, p.30.

Nandra, K., et al.: 1989, Monthly Notices Roy. Astron. Soc. 236, 39p.

Netzer, H.: 1985, Astrophys. J. 289, 451.

Netzer, H.: 1987, Monthly Notices Roy. Astron. Soc. 225,55.

Perez, E., et al.: 1988, Monthly Notices Roy. Astron. Soc. 230, 353.

Perola, G.C., et al.: 1986, Astrophys. J. 306, 508.

Peterson, B., et al.: 1983, Astron. J. 88, 926.

Piro, L.: 1989, private communication.

Piro, L., Massaro, E., Perola, G.C., Molteni, D.: 1988, Astrophys. J. 325, L25.

Pounds, K.A., Turner, T.J., Warwick, R.S.: 1986, Monthly Notices Roy. Astron. Soc. 221, 7p.

Pounds, K.A., et al.: 1986, Monthly Notices Roy. Astron. Soc. 224, 443.

Pounds, K.A., Nandra, K., Stewart, G.C., Leighly, K.: 1989, preprint.

Preuss, E., et al.: 1980, Astrophys. J. 240, L7.

Rees, M.: 1989, private communication.

Reichert, G.A., Polidan, R.S., Wu, C.-C., and Carone, T.E.: 1988, Astrophys. J. 325, 671.

Shields, G.A., McKee, C.F.: 1981, Astrophys. J. 246, L57.

Shields, G.A.: 1978, Nature 272, 706.

Shuder, J.M.: 1982, Astrophys. J. 259, 48.

Smith, M.G., et al.: 1981, Monthly Notices Roy. Astron. Soc. 195, 437.

Smith, P.S., et al.: 1988, Astrophys. J. 326, L39.

Stirpe, G.M., de Bruyn, A.G., van Groningen, E.: 1988, Astron. Astrophys. 200, 9.

Stockman, H.S., Angel, J.R.P., Miley, G.K.: 1979, Astrophys. J. 227, L55.

Sun, W.-H., Malkan, M.A.: 1988, ESA SP-281, p.283

Turner, T.J., Pounds, K.A.: 1988, Monthly Notices Roy. Astron. Soc. 232, 463.

Turner, T.J., Pounds, K.A.: 1989, preprint.

Ulrich, M.-H.: 1981, Space Sci. Rev. 28, 89.

Ulrich, M.-H., Courvoisier, T.J.-L., Wamsteker, W.: 1988, Astron. Astrophys. 204, 21.

van Groningen, E.: 1989, private communication.

Warwick, R.S., Pounds, K.A., Turner, T.J.: 1988, Monthly Notices Roy. Astron. Soc. 231, 1145.

Warwick, R.S., Barstow, M.A., Yaqoob, T.: 1989, preprint.

Wilkes, B.J., Elvis, M.: 1987, Astrophys. J. 323, 243.

Wills, B.J., Browne, I.W.A.: 1986, Astrophys. J. 302, 56.

Yaqoob, T., Warwick, R.S., Pounds, K.A.: 1989, Monthly Notices Roy. Astron. Soc. 236, 153.

York, D.G.: 1984, Astrophys. J. 276, 92.

THE LOW-FREQUENCY SPECTRA OF
ACCRETION DISKS IN ACTIVE GALACTIC NUCLEI:
THE DARK SIDE OF THE BIG BLUE BUMP.

Matthew MALKAN
Department of Astronomy, UCLA, Los Angeles, CA, USA

1. Previous Observation and Interpretation of the Big Blue Bump.

The flux density of most Seyfert 1 galaxy and quasar continua has an average slope of $\sim$-1 (*i.e.*, $f_\nu \propto \nu^{-1}$) from the visual to a far-infrared turnover (around 80 μm). Relative to this red "power law", there is generally excess flux in the blue and ultraviolet, a.k.a. the "Big Blue Bump" (BBB). Much effort has gone into measuring the shape of the high-frequency side of this BBB. (It is unfortunate that comparable effort has not been made to find a better name for it). These observations are difficult because of our lack of very powerful instrumentation in the ultraviolet, and absorption from gas along the line-of-sight, in our Galaxy and in intergalactic clouds (see Malkan, Alloin and Shore 1987 for a review). But fortunately the BBB component usually dominates the observed ultraviolet continuum, and may well extend out to the soft X-rays. The very steep soft X-ray continuum below 1 keV–*i.e.*the "Soft X-ray Excess" frequently detected by Exosat (Pounds and Turner 1989) and indirectly by the Einstein IPC (Wilkes and Elvis 1988)– appears to be the high-frequency tail of the BBB (Czerny and Elvis 1987). Observational progress is being made on closing the spectral gap between the far-UV and soft X-rays. Reimers et al. (1989) recently detected a quasar at z=2.75 down to the short-wavelength limit covered by IUE, corresponding to a rest wavelength of 330 A. Unfortunately, the observed far-UV fluxes must be multiplied by a large, but highly uncertain factor (6) to correct for absorption by intergalactic HI clouds along the line-of-sight. Detailed spectroscopy by HST with high resolution and high signal/noise ratio will be needed to determine column densities for the absorbing clouds, for accurate de-blanketing of the continuum.

An attractive interpretation of the BBB's high-frequency turnover is that it is the Wien cut-off of thermal emission from optically thick accreting gas (*e.g.*, Shields 1978; Malkan and Sargent 1982). If the BBB is indeed optically thick

F. Meyer et al. (eds.), Theory of Accretion Disks, 19–28.

thermal emission, its spectrum is far broader than that of a single-temperature blackbody. Models of emission from an optically thick, geometrically thin accretion disk naturally produce a superposition of thermal flux from a wide range of radii and temperatures, which successfully fit the observations of the BBB (*e.g.*, Malkan 1983). The accretion models continue to grow in number and sophistication. Many of these new complexities are driven by the need to explain new observations. In particular, producing the Soft X-ray Excess in the inner part of the accretion flow requires considering the effects of electron scattering (Czerny and Elvis 1987), Comptonization (Wandel and Petrosian 1987), geometry (Madau 1988), and relativity (Sun and Malkan 1989).

2. Observing the Long-Wavelength Side of the BBB.

While observational and theoretical work on the short-wavelength side of the BBB continues, it might be easier in theory to understand the outer parts of the accretion flow, where the temperatures are lower, and the complications listed above are negligible. Another advantage of testing models for the long-wavelength side of the BBB is that the near-infrared and optical data are generally quite complete and very accurate. *The disadvantage is that the BBB is no longer the only strong component present at longer wavelengths.*

In the less luminous Seyfert galaxies, starlight from the host galaxy contaminates the observed red and near-infrared continuum. Accurately isolating the nonstellar nuclear continuum is difficult when it produces less than half the total observed flux (*e.g.*, Malkan and Filippenko 1983). For this, spectrophotometry with excellent spatial resolution is needed, probably with HST. For the present, we can avoid this problem by confining our attention to luminous Seyfert 1 nuclei and quasars, where no more than a quarter of the observed near-infrared flux is starlight.

In most luminous active galactic nuclei, a substantial *nonstellar* infrared continuum component has a flux density rising roughly linearly with wavelength. Previous studies such as Malkan and Sargent (1982) and Edelson and Malkan (1986, EM hereafter) used the simplest description of this long-wavelength component, a power law: $f_\nu \propto \nu^\alpha$). This two-parameter expression is not a bad approximation to many observed infrared spectra, and does not contain any characteristic frequency. Because the power law slope α is always somewhat steeper than -1, and the flux density of the disk is constant or rising with frequency, the sum of these two components always has a clear minimum in νf_ν.

3. The 1 μm Miniumum in νf_ν and the Curvature of the Infrared Continuum.

The multi-wavelength energy distributions of more than 60 quasars have now been fitted with the power-law + accretion disk model (*e.g.*, Sun and Malkan, 1989). Recently Sanders *et al.* (1989) have discussed multi-wavelength spectra of the PG

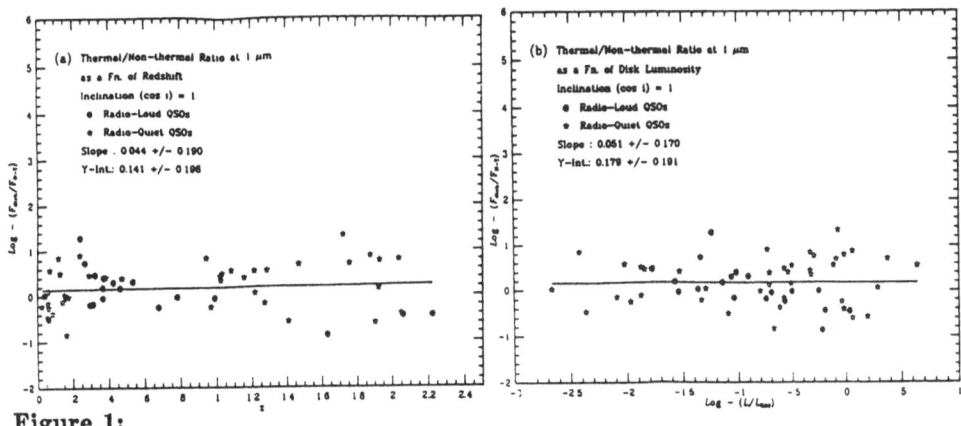

Figure 1:

Bright Quasars, which overlap the Sun and Malkan sample substantially. Sanders *et al.* point out that the quasar energy distributions very often have a minimum in νf_ν in the far-red or near-infrared. Examination of the Sun and Malkan data confirms that most of the quasars have minima at $log\nu_{min} = 14.55 \pm 0.2$. Sanders *et al.* further argue that a narrow range in ν_{min} requires a narrow range in the ratio of disk to power law fluxes. However, Sun and Malkan found disk/power law ratios (estimated at a reference wavelength of 1 μm) which range over two orders of magnitude, from 0.1 to > 10. As illustrated in Figure 1a and b from Sun and Malkan (1989), there is no clear dependence of this ratio on other quasar properties such as luminosity. This paradox arises because the simple power law *is not an adequate description of many of the observed near-IR continuum.*

About a third of the Seyfert 1 nuclei studied by EM had straight near- and mid-infrared continua, with no indication of downward curvature. In these spectra, the simple power law, with an exponential *low-frequency* cut-off around 80 μm, is an accurate description of the data. In the other two-thirds of the sample, however, a single power law does not fit all the infrared data. Mathematically, the second derivative of the observed logarithmic flux with respect to log frequency is significantly negative. EM distinguished two possible causes:

1) In Seyfert nuclei which were already known to be reddened, thermal emission from warm dust grains produces a steep infrared continuum, with an average slope of $\alpha < -1.5$. The continuum actually curves down more steeply in the near-IR (1-5 μm), because of the Wien cut-off of the hottest dust grains present. This thermal infrared continuum shape is the same as in the Seyfert 2 nuclei, where silicate absorption has established unambiguously that dust grains produce most of the infrared emission.

2) In contrast, in luminous Seyfert 1 nuclei and most quasars, the apparent lack of reddening does not suggest the presence of dust grains. The average infrared slopes are flatter, with $\alpha > -1.4$ (see examples in Figures 2–4). The downward curvature in the spectrum looks more like a "bump", centered on 5 μm. Thus EM proposed that the continua in these objects consist of a power law (presumed to be

nonthermal), and a separate near-infrared component, which could be described by either a parabola or a Gaussian. Since the high-frequency side of this near-IR bump occurs at 1–2 μm, the *local* continuum slope in that region is *steeper* than the underlying power law (which has a typical $\alpha \sim -1.2$).

The "magic" value of ν_{min} is caused by the fact that the infrared continuum component often has a sharp downward steepening above $log\nu = 14$. Thus a minimum in νf_ν occurs where this high-frequency cut-off merges into the BBB component. The small range in ν_{min} reflects mostly the small range of near-IR turnover frequencies, and is not very sensitive to the disk/power law ratio. This explanation has two implications:

(a) The luminosities of the infrared and ultraviolet continuum components in AGNs are *not* extremely closely linked. The fact that their ratios range over a factor of at least 100 lends support to the view that they are produced by different physical mechanisms, and perhaps in different locations in the central engine.

(b) Unlike a simple power law, the infrared component does appear to show some spectral structure (a steepening) at a roughly "characteristic" frequency. This could shed light on the nature of the emission, and influence our estimate of the low-frequency shape of the BBB.

4. Modeling Curvature in the Red "Power Law": Does it Have an Exponential Cut-off?

Since the IR curvature in the spectra of reddened AGNs is understood reasonably well, more effort focuses on understanding the curvature in the IR spectra of luminous blue quasars. For example, EM tried modifying the underlying power law in an unsuccessful attempt to fit the entire infrared continuum without this extra NIR bump component. The energy distributions of the highly polarized "blazars", which are clearly nonthermal, have been described as downward-curving parabolas (see Urry 1988 for a review of the subject). However, replacing the power law with a parabola fails to produce the sharp excess flux around 5 μm observed in normal (low-polarization) AGNs. Even using a "broken" power-law, with a sudden large steepening of slope around 3–5 μm, EM could not produce a sufficiently concentrated near-IR peak in νf_ν.

Recent experimenting has revealed an empirical scheme which *can* fit the full infrared spectrum without an explicit 5 μm bump component. Figures 2–4a show fits using an infrared power law with an exponential high-frequency cut-off for Markarian 335, 509 and 3C 273 . These are typical AGN multiwavelength energy distributions discussed in Sun and Malkan (1989) and Band and Malkan (1989). This does not imply there is anything special about these observations. The particular AGNs were chosen simply because they are bright and have been well studied. (Note that the energy distribution is of 3C 273 in its quiescent state, when its "blazar" component makes no significant contribution to the continuum–Impey *et al.* 1989). Due to the remarkable uniformity of quasar/Seyfert 1 spectra, similar

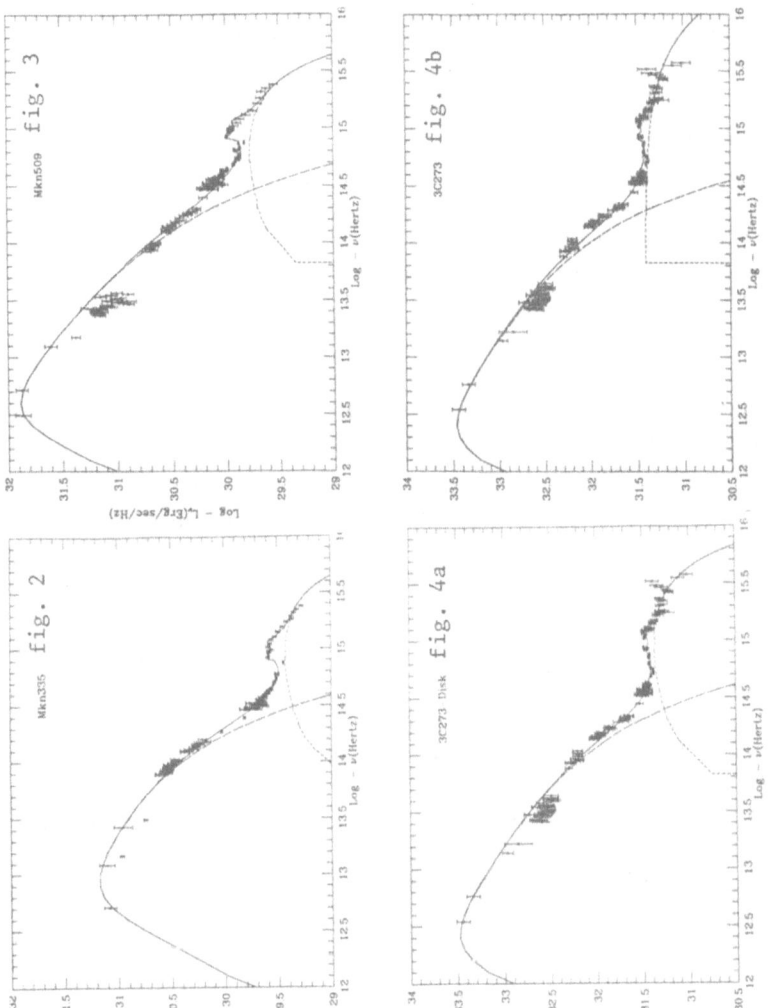

Figure 2-4:

conclusions would have been drawn from fitting any of the other $\gtrsim$ hundred AGNs with good multi-wavelength observations.

The only model component in the infrared is now a power law which is cut-off at long and short wavelengths: $f_\nu \propto \nu^\alpha \times e^{\nu/\nu_1} \times e^{-\nu/\nu_2}$. By manipulating the two cut-offs, equally good fits to the data can be obtained with a range of α, from the usual -1.0, to flatter slopes of -0.6 or -0.8. The latter values are characteristic of the observed slopes from 5–80 μm, significantly shallower than in the near-IR. It might not be a coincidence that these are the same as the typical observed X-ray slopes (*e.g.*the nonthermal models of Band and Grindlay 1986).

The fits to all 3 spectra are acceptable. Markarian 335 and 3C 273 had two of the strongest near-IR bumps measured by EM. There is a subtle hint that the exponential cut-off is not quite sharp enough to match their near-IR curvature. The models always have 3–10 μm slopes which are slightly too steep, and 1–3 μm slopes which are slightly too shallow, compared to the observations. This small flaw suggests that perhaps the near-IR bump really is better described by a separate component, a la EM. However, given the uncertainties of the currently available data, the exponential cut-off model cannot be ruled out.

EM found that the near-IR bump has the same central wavelength in all the Seyfert 1's they studied. Therefore it is not surprising that our new fits always require a high-frequency cut-off in the power law at $1.1 \pm 0.2 \times 10^{14} Hz$. The observed spectrum would be extremely steep at wavelengths shorter than 1 μm, were it not for the entrance of the BBB. What is the physical interpretation of this high-frequency turnover? If the power law were nonthermal, the cut-off would reflect some maximum energy of relativistic particles, possibly set by the energy-dependence of the particle acceleration or deceleration mechanisms. However, it would be surprising if this frequency were always the same from one AGN to another. A more likely physical explanation of the cut-off is available if the infrared emission is predominantly thermal. In this case, the high-frequency turnover is produced by the Wien peak from the hottest dust grains which can survive before evaporating, at $T_{max,dust} \sim 1000 - -1500K$.

5. Thermal vs. Nonthermal Origins for the Infrared Continuum.

The infrared data currently do not distinguish between the two possibilities (*i.e.*, nonthermal power law + separate NIR bump versus thermal exponentially cut-off "power law"). Other observations of radio-quiet quasars to resolve the thermal/nonthermal question have also yielded ambiguous results:

(a) The low polarization of the near-IR continuum is cited as evidence for a thermal origin. On the other hand, synchrotron emission would be strongly polarized only if the magnetic field is well-ordered and de-polarizing mechanisms are negligible.

(b) The far-infrared emission is not detectably variable on timescales of $\sim$a week and six months (Edelson and Malkan 1987). Since a nonthermal infrared source would have a shorter light-crossing time, such stability might not be expected. However, clear near-IR variability has been detected in many quasars (*e.g.*Neugebauer *et al.* 1989), implying the presence of some nonthermal emission at the shorter wavelengths, at least.

(c) The slope of the sub-millimeter continuum should probably not be much steeper than +2.5 if it is self-absorbed synchrotron emission (but see deKool and Begelman 1989); the expected slope from thermal dust grain emission is ~+3 to +4. Observations of the Seyfert 1 galaxy NGC 4151 (Edelson *et al.* 1988) and some reddened quasars has revealed slopes steeper than +2.5. However, millimeter detections of some bluer quasars (which are, a priori, more likely to be dust-free) have found several with slopes flatter than +2.5 (Chini *et al.* 1989).

d) Sensitive infrared spectroscopy has failed to detect any of the usual dust emission or absorption features, (*e.g.*, from silicate grains at 10 μm, or very small dust grains at 3.3 μm). Further, recent soft X-ray spectra of bright Seyfert 1 nuclei and quasars rule out the possibility of even tiny amounts of dust *along our line-of-sight to the nucleus.* Thus if the infrared continuum is thermal, it must be produced by unusual dust grains in a *special geometry* designed not to intersect our view to the nucleus. For example, Sanders *et al.* (1989) have proposed that the dust lies in a warped disk around the nucleus.

Given the ability each hypothesis has shown to wriggle out of possible observational difficulties, we should consider the implications both have for the BBB. In the new model only a few per cent of the visual flux comes from the (exponentially cut off, *i.e.*"thermal") red component; in the previous fits of Sun and Malkan (1989) and Band and Malkan (1989) the nonthermal power law contributed nearly a third of the total visual light. Therefore the new model requires some increase in the amount of emission from the accretion disk at low frequencies. In the fits shown in Figures 2–4a, the black hole mass is about twice its value in the previously published fits. Since the disk luminosity is constrained tightly by the UV data, the best-fitting accretion rate went up only slightly. Thus the new model would push the disk parameters up to a factor of two further below the Eddington limit. These fits show that even a dramatically different description of the NIR continuum would not have altered Sun and Malkan's conclusions substantially.

6. Does the BBB Flux Turn Drop at Low Frequencies?

At long wavelengths, the standard disk spectrum is roughly $f_\nu \propto \nu^{+1/3}$. With the exponential cut-off in the red "power law", is it possible to fit the BBB with a component which is flat at low frequencies? Figure 4b shows the result of a fit, in which the disk spectrum has been replaced by $f_\nu = Const \, e^{\frac{-h\nu}{kT}}$. This functional form (which approximates the spectrum of optically thin free-free emission) has the same high-frequency steepening that disk models do. But since it does not have flux density rising with frequency in the red, it is somewhat brighter at low frequencies (log $\nu \lesssim 14.5$). The exponentially cut-off IR power law + "free-free" model gives an acceptable fit to the spectrum of 3C 273. A stricter test is presented in Figure 5, where the error bars show observations of the quasar 1004+130, which has an unusually strong BBB. The solid-line fit, which is good, is the previous disk + power law model. The dashed-line fit, which is systematically worse, is the new free-free + exponentially truncated power law model. Admittedly, the differences are subtle: data of even higher quality would be required to *decisively* rule out the new model.

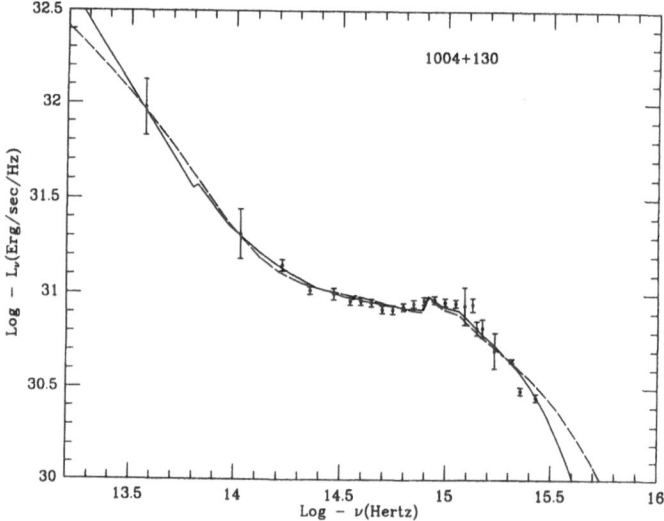

Figure 5:

The electron temperature for the free-free component, T_{ff}, is constrained to better than $\pm20\%$ by the downward curvature of the ultraviolet fluxes. The best-fitting temperatures range from 100,000 to 200,000 °K, except for 3C 273, which has T_{ff} slightly hotter than 300,000°K. Roughly, we expect T_{ff} should be similar to the maximum effective temperature in the face-on disk model which best fits the same spectrum.

7. Discussion.

Even if the "free-free" model did describe the observations of the BBB very well, its physical interpretation would be highly problematic. If the emission mechanism really were bremsstrahlung, the specific luminosity in 3C 273, 2.6×10^{31} erg/sec/Hz, would require an $n_e^2 V$ of $\sim 10^{71} cm^{-3}$. Plasma of such a large emission measure would probably emit detectable line radiation. The gas would have to be sufficiently compact that (a) its flux could vary on ~ 1 month timescales, and (b) it would not smother the compact VLB radio-emitting core. Yet the plasma could not be too optically thick, or it would not emit the flat bremsstrhalung spectrum. The electron temperature of 100,000–200,000 (needed to fit the high-frequency part of the BBB of a typical quasar) is not an obvious equilibrium for optically thin gas.

The "free-free" model for the BBB certainly does not match the observations *unless* the infrared component always "knows" to turn over sharply above a frequency of $\sim 10^{14}$ Hz. Thus the free-free model is probably incompatible with any nonthermal power law interpretation of the IR continuum. Nonetheless, the exercise serves to demonstrate that a major change in our model of the IR continuum requires only a small modification to models for the BBB. It also indicates that *observational* determination of the low-frequency shape of the BBB is not going to be easy. Perhaps because of this disappointment, I'll point out that *theories* of the low-frequency emission from AGN accretion disks also suffer from several ambiguities:

(a) Depending on the geometry of the disk, its outer parts may intercept more luminosity from the inner regions than they generate locally by viscous heating. This irradiation of the outer disk could dramatically decrease the its radial temperature gradient, and invert its vertical temperature gradient.

(b) at low temperatures (of 1,000–10,000°K), spectral lines become more important in the disk atmosphere. Accurate models must include the possibility that the outer disk could produce strong absorption or emission lines, possibly even from molecules (*e.g.*, Collin-Souffrin 1987).

(c) the outer radius of the disk is a completely unknown free parameter (*e.g.*, Czerny and Elvis 1987). Beyond several hundred Schwarzschild radii from the black hole, many accretion disk models predict that self-gravity becomes dominant. If the disk extends beyond this point, then theories must explain how it merges into the Broad Emission-Line Region.

Realistic disk models should address all three of these questions simultaneously. For observers these theoretical complications are either sour grapes, or encouragement that better data would provide interesting new information on the physics of accretion disks.

REFERENCES.

Synthesis of Accretion Disk and Nonthermal Source Models for Active Galactic Nuclei; D. Band and M. A. Malkan 1989, *Ap. J.*, Oct. 15 issue.

Chini, R., Kreysa, E, and Biermann, P. *et al.* 1989, *Astr. Ap.*, in press.

Collin-Souffrin, S. 1987; Astr. Ap.; 179; 60.

Collin-Souffrin, S. *et al.* 1986; Astron. Ap.; 166; 27.

Czerny, B. and Elvis, M. 1987; Ap. J.; 321; 305.

Davidson, K., and Netzer, H. 1979; Rev. Mod. Phys.; 51; 715.

Edelson, R. A. 1986; Ap. J. Lett.; 309; L69.

Edelson, R. A. 1987; Ap. J.; 313; 651.

Edelson, R. A. and Malkan, M. A. 1986; Ap. J.; 308; 59.

Edelson, R. A., Gear, W. K., Malkan, M. A., Robson, E. I. 1988; Nature; 336; 749.

Impey, C., Malkan, A. and Tapia, S. 1989, *Ap. J.*, in press.

de Kool, M. and Begelman, M. 1989; Nature; 338; 484.

Madau, P. 1988; Ap. J.; 327; 116.

Malkan, M. A. and Filippenko, A. V. 1983; Ap. J.; 275; 477.

IUE Observations of Markarian 3 and 6: Reddening and the Nonstellar Continuum; M. A. Malkan and J. B. Oke 1983; Ap. J.; 265; 92.

Malkan, M. A. and Sargent, W. L. W. 1982; Ap. J.; 254; 22.

Malkan, M. A. 1983; Ap. J.; 268; 582.

IUE Observations of Active Galactic Nuclei; M. A. Malkan, D. Alloin, and S. Shore 1987, *Scientific Accomplishments of the IUE*, ed. Y. Kondo, (Reidel), p. 655.

Mushotzky, R. F. 1984, Adv. Space Res., Vol. 3, No. 10-13, p. 157.

Neugebauer, G. *et al.* 1989; Astron. J.; 97; 823.

Pounds, K. A. and Turner, T. J. 1989, *Space Sci. Rev.*, in press.

Reimers, D. *et al.* 1989; Astr. and Ap., in press.

Sanders, D. *et al.* 1989, *Ap. J.*, submitted.

Shields, G. A. 1978; Nature; 272; 706.

Sun, W. H. and Malkan, M. A. 1989, *Ap. J.*, Nov. 1 issue.

Urry, C. M. *et al.* 1986, *in Variability of Galactic and Extragalactic X-Ray Sources*, (Milan: Assoc. Adv. Astr.) ed. A Treves, p.15..

Wandel, A. and Petrosian, V. 1988; Ap. J. Lett.; 329; L11.

Wilkes, B. and Elvis, M. 1987; Ap. J.; 323; 243.

Wills, B. J., Netzer, H., and Wills, D. 1985; Ap. J.; 288; 94.

OBSERVATIONAL CONSTRAINTS ON VISCOSITY IN THE AGN ACCRETION DISCS.

A. SIEMIGINOWSKA and B. CZERNY
Nicolaus Copernicus Astronomical Center,
Bartycka 18, 00–716 Warsaw, Poland

ABSTRACT. The optical/UV/soft X–ray big bump can be modelled as a thermal radiation from an accretion disc. Assuming that the observed UV variability in AGN spectra is caused by accretion disc instabilities we can set constraints on the viscosity. Simple bimodal disc approximation was used — the outer regions of the disc radiate locally as a black body whilst the inner regions locally become optically thin in a thermal time scale giving a negligible contribution to the UV radiation. Calculation were made for several models with $L/L_{Edd} = 1$ and $L/L_{Edd} = 0.01$ and various central masses. Comparison between thermal and observed variability time scales in 10 Seyfert galaxies and 16 QSOs indicate the parameter α value of the order of 0.01.

1. Model of variability of accretion disc spectrum.

The optical/UV/soft X–ray big bumps observed in some AGNs can be interpreted as a thermal radiation from accretion discs around supermassive black holes (Shields 1978, Malkan & Sargent 1982, Bechtold et al. 1987). The standard α–viscosity disc models are thermally unstable (Pringle et al. 1973) in the inner part dominated by radiation pressure. However, there are also stabilizing mechanisms, such as the existence of a hot corona (Ionson & Kuperus 1984), irradiation of the disc (Czerny et al. 1986) by the observed non–thermal component of the spectrum or the advection term (Abramowicz et al. 1988). Therefore this instability may develop only over a limited time scale, and only in some parts of the disc. It would cause temporary changes in the disc spectrum, mostly seen as a variability in the UV band.

We assume a very simplistic approximation of instability which develops in the radiative regions of the geometrically thin Keplerian accretion disc. In this model the outer disc regions are optically thick and radiate locally as a blackbody contributing to the optical and UV band, whilst the inner regions are hot, optically thin and radiate mostly in X–rays (Shapiro et al. 1976). Transitions between the two states proceed in the thermal time scale. Therefore, for every disc model we can identify the amplitude of the change in radiation flux with wavelength, the

F. Meyer et al. (eds.), Theory of Accretion Disks, 29–33.

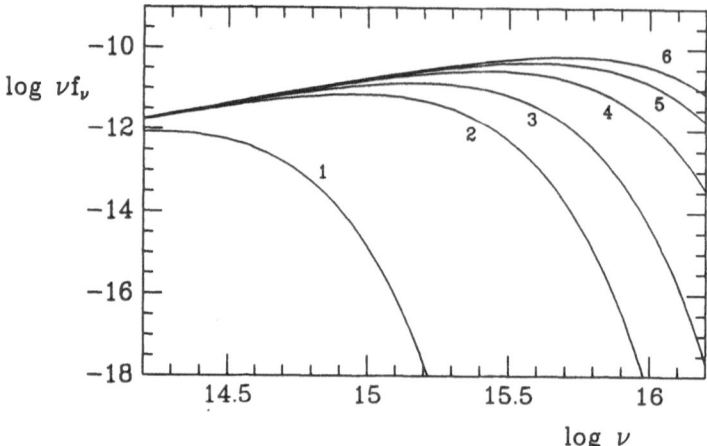

Figure 1: A plot of $\log \nu f_\nu$ vs. $\log \nu$ for accretion disc model around a black hole with $M = 10^8 M_\odot$ and $L/L_{Edd} = 1$. Shown is the resulting spectrum with different values of the distance from which the emitted flux is described as a blackbody radiation (r_{th} is equal $3r_g, 10r_g, 20r_g, 50r_g, 100r_g, 1000r_g$ for curves 6–1 respectively). As r_{th} increases, the radiation flux decreases at any given frequency. The characteristic time scale τ_{th} of such a process is given by the thermal time scale at the largest radius affected.

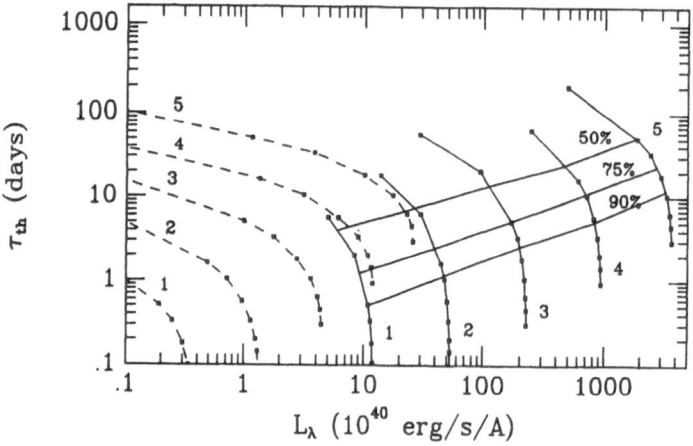

Figure 2: A plot of τ_{th} [in days units] vs. L_λ [in $10^{40} erg/s/\mathring{A}$ units] for $\lambda = 1740\mathring{A}$ and $\alpha = 0.1$. Shown are relations for disc models with different central masses ($1 - 10^7 M_\odot$; $2 - 10^{7.5} M_\odot$; $3 - 10^8 M_\odot$; $4 - 10^{8.5} M_\odot$; $5 - 10^9 M_\odot$) and accretion rates (the solid curves for $L/L_{Edd} = 1$ and dashed curves for $L/L_{Edd} = 0.01$). Points along the curves correspond to different values of parameter r_{th} equal $3r_g, 4r_g, 5r_g, 7r_g, 10r_g, 15r_g, 20r_g, 50r_g$ for curves number 3, 4, 5; $4r_g, 5r_g, 7r_g, 10r_g, 15r_g, 20r_g$, for 2. curves; $5r_g, 7r_g, 10r_g, 15r_g, 20r_g, 50r_g$ for solid 1. curve and $10r_g, 15r_g, 20r_g$ for dashed 1. curve respectively. The lines marked 90%, 75% and 50% indicate the amplitude of flux change.

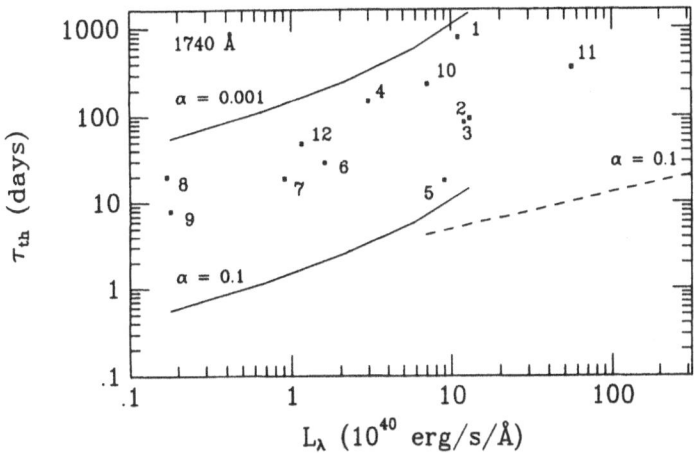

Figure 3: Two solid lines represent the two-folding time scales at $\lambda = 1740\text{Å}$ for $L/L_{Edd} = 0.01$ and two values of the viscosity parameter α. The dashed line corresponds to $L/L_{Edd} = 1$ and $\alpha = 0.1$. Data points lie above the line for $\alpha = 0.1$ and below the line for $\alpha = 0.001$, and this requires that the viscosity parameter α should be greater than 0.001 but may be smaller than 0.1. Observational points: 1–III–Zw2, 2–F9, 3–ESO141–G55, 4–Mkn279, 5–Mkn509, 6–NGC7469, 7–NGC3783, 8–NGC4593, 9–NGC4151, 10–PG1351+64, 11–PG1211+143, 12–3C120 (Chapman *et al.* 1985, Tanzi *et al.* 1986, Maraschi *et al.* 1986)

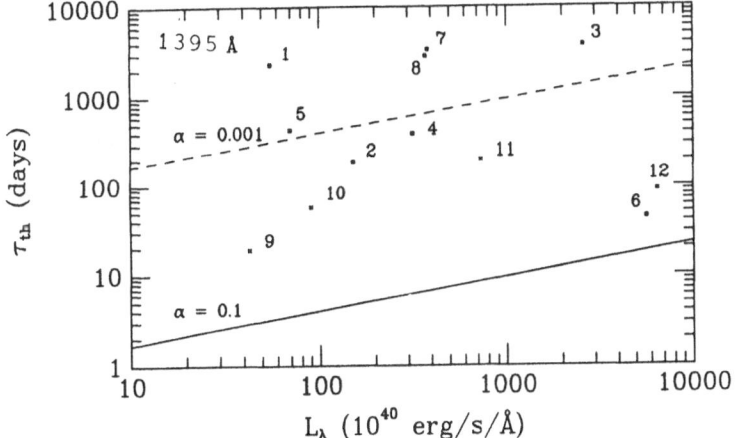

Figure 4: Two-folding time scale at $\lambda = 1060\text{Å}$ for $L/L_{Edd} = 1$ and two values of α: 0.1 and 0.001. Observational points: 1–Q0026+129, 2–Q0312–77, 3–Q0405–123, 4–Q0955+326, 5–Q1004+13, 6–Q1011+25, 7–Q1302–102, 8–Q2128–123, 9–Q0736+017, 10–Q1510–089, 11–Q1641+399, 12–Q2223–051 (O'Brien *et al.* 1988). Objects (9,10,11,12) marked with a star are blazars The data for five of the objects is consistent with a viscosity parameter α smaller than 0.001. The time spacing of observations of those objects, however, is not particularly good for detecting the time scales of interest in this paper.

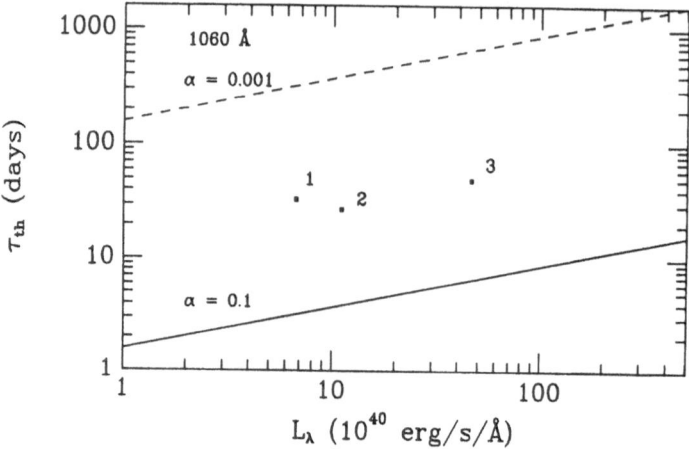

Figure 5: Two-folding time scale at $\lambda = 1395$Å for $L/L_{Edd} = 1$ and two values of α: 0.1 and 0.001. Observational points (from O'Brien *et al.* 1988): 1–Q1202+281, 2–Q1219+750, 3–PG1351+64 (also on Fig. 2).

parameter r_{th} (the distance from which the disc radiate locally as a black body) and thermal time scale τ_{th} corresponding to that distance from the central black hole. In particular, we obtain a two–folding time scale. By comparing this time scale with the two-folding time scale of the observed variability in the UV band we can estimate the viscosity parameter α (Shakura & Sunyaev 1973). This method gives much better accuracy than any estimates of viscosity from the shape of an accretion disc spectrum. Simplified analytical formulae for the two–folding time scale at a given wavelength corresponding to the assumed value of α were presented by Czerny & Czerny (1986). In this poster we present the results of more detailed analysis of the changes in the shape of the spectrum, and compare the predicted time scales with the observational data for 26 objects.

2. Results.

We consider the accretion disc models around black holes with masses in the range $10^7 M_\odot$–$10^9 M_\odot$ and sub–Eddington accretion rates such that $L/L_{Edd} = 1$ and $L/L_{Edd} = 0.01$.

3. Conclusions.

From our numerical calculations and the comparison of the data points with our estimation of thermal time scales we conclude that:
1) numerical calculations give more accurate results than the early analytic formula derived by Czerny & Czerny (1986),
2) the relation for the two–folding time scale is still approximately linear for

$L/L_{Edd} = 1$ but quasi–parabolic for $L/L_{Edd} = 0.01$ (Fig. 3),

3) viscosity parameter α in AGN accretion discs is required to be greater than 0.001 for most cases but may be smaller than 0.1,

4) the values of α are consistent with the assumed interpretation of the observed variability,

5) the α in AGNs and CVs are the same. If we adopt the interpretation of viscosity as a strong turbulence we can understand it. However the sources of turbulence may be different in each case.

REFERENCES.

Abramowicz, M.A. *et al.*, 1988. *Ap. J.* (in press)

Bechtold, J. *et al.*, 1987. *Ap. J.*, **314**, 699

Chapman, G. N. F., 1985. *Ap. J.*, **297**, 151

Czerny, B. & Czerny, M., 1986. in: *New Insights in Astrophysics: 8 Years of UV Astronomy with IUE*, Proc. Join NASA/ESA/SERC Conference, University College London

Czerny, B. *et al.*, 1986. *Ap. J.*, **311**, 241

Ionson, J. A., & Kuperus, M., 1984. *Ap. J.*, **284**, 389

Malkan, M. A. & Sargent, W. L. W., 1982. *Ap. J.*, **254**, 22

Maraschi, L. *et al.*, 1986. in: *New Insights in Astrophysics: 8 Years of UV Astronomy with IUE*, Proc. Join NASA/ESA/SERC Conference, University College London

O'Brien, P. T. *et al.*, 1988. *M.N.R.A.S.*, **233**, 845

Pringle, J. E. *et al.*, 1973. *Astr. Ap.*, **29**, 179

Shakura, N. I. & Sunyaev, R. A., 1973. *Astr. Ap.*, **24**, 337

Shields, G., 1978. *Nature*, **272**, 706

Shapiro, S. I. *et al.*, 1976. *Ap. J.*, **204**, 187

Tanzi, E. G. *et al.* 1986. in: *New Insights in Astrophysics: 8 Years of UV Astronomy with IUE*, Proc. Join NASA/ESA/SERC Conference, University College London

ACCRETION DISK MAGNETOHYDRODYNAMICS AND THE ORIGIN OF JETS.

Roger D. BLANDFORD

Harvard-Smithsonian Center for Astrophysics, 60 Garden St.,
Cambridge, MA 02138, USA.
Theoretical Astrophysics, 130-33 Caltech, Pasadena, CA 91125, USA.

ABSTRACT. Accretion disks are believed to be present in active galactic nuclei, binary X-ray sources, cataclysmic variables and protostars. Disks in all four types of source have been associated with collimated jets or bipolar outflows. It is argued that magnetic extraction of disk angular momentum provides a natural method for launching these jets and that poloidal field that passes through the disk is more likely to be effective than dynamo-generated field. Electromagnetic and hydromagnetic models for jet production are reviewed, emphasising their shortcomings and the dependence on physical conditions in the disk. Magnetic torques may also extract rotational energy from a central black hole. It is conjectured that, in practice, a black hole magnetosphere will be magnetically dominated and practically force-free and that no extra restrictions are imposed on its structure by the application of relativistic MHD theory. It is also argued that, despite recent concerns, the hole will be able to communicate its angular velocity to the magnetosphere and so drive a unipolar inductor.

1. Introduction.

The interplay of gravitation, rotation and magnetic field, long studied in the classical theory of star formation (e.g.Mestel 1961, Mouschovias, 1979), is notoriously subtle and resistant to simple arguments (e.g.Parker 1977). The initial concern that centrifugal force and magnetic pressure would prevent gravitational collapse has given way to the realisation that hydromagnetic torque provides a convenient (though not unique) mechanism for extracting angular momentum and that magnetic flux can subsequently escape from the gas either through ambipolar diffusion or buoyancy.

Accretion disks provide some special opportunities for investigating this interaction at work, distinguished, as they are, from our best local laboratory, the solar wind, by the dominance of angular momentum and the promise of observable, relativistic effects. Most accretion disk theory is essentially gas dynamical rather than magnetohydrodynamical and magnetic field is often sublimated as an α viscosity

F. Meyer et al. (eds.), Theory of Accretion Disks, 35–57.

or relegated to an inner boundary condition. By contrast, experience with space plasmas and the mounting evidence that collimated, bipolar outflow is a generic property of an accretion disk, suggest that magnetic fields are of primary importance and that their dynamical effects should be explicitly included from the start.

In this review, I shall briefly summarise observations of AGN, stellar and protostellar accretion disks which hint that they are endowed with large scale magnetic field that exerts a decelerating torque on the orbiting gas and uses the liberated angular momentum to launch a pair of jets (or bipolar outflows) normal to the disk. I shall not discuss the equally important interaction of a disk with the field of a central neutron star which is believed to dictate period changes in pulsating X-ray sources and perhaps also modulate the emission from QPOs. However, I shall review recent developments in understanding the interaction of a central black hole with an accretion disk field.

Recent reviews of magnetic accretion disks and related topics include Blandford (1985), Coroniti(1984), Frank, King and Raine (1986), Zel'dovich, Ruzmaikin and Sokoloff (1983) and Asseo and Gresillon(1987). A forthcoming conference proceedings, Belvedere(1989) is devoted to this general subject.

2. Observational Background.

Four different types of accretion disk are discussed. In every case, large scale magnetic fields have been invoked to extract angular momentum and propel an outflow.

2.1 ACTIVE GALACTIC NUCLEI.

The direct evidence for accretion disks in AGN is summarised by Ulrich and Malkan in these proceedings. Extensive reviews of jets can be found in Bridle and Perley(1984), Begelman, Blandford and Rees(1984) and the conference proceedings edited by Kundt(1986). To summarise, both powerful and weak radio galaxies and quasars and Seyferts exhibit collimated outflow emerging in anti-parallel directions from the galactic nucleus. Sometimes the resulting jets subtend angles $\lesssim 1°$ at their sources. Superluminal sources furnish strong evidence for relativistic expansion speeds and, by extension, a relativistically deep potential well. Several jets are apparently one-sided. In many sources, this is attributable to relativistic beaming of an intrinsically symmetric source. However, a few sources (*e.g.*M87, Reid *et al.*1988) may create only one jet at a time. A possible explanation is advanced in §3.1.

Perhaps the most compelling observational evidence that strong, ordered magnetic field exists in AGN comes from our own Galactic center where Yusef-Zadeh and Morris(1988) have discovered long linear features variously called arcs, filaments and threads, some of which exhibit linear polarisation consistent with synchrotron emission in a longitudinal magnetic field. These features are reminiscent of familiar magnetic structures, solar coronal loops and prominences.

Magnetically-dominated regions above accretion disks have been independently invoked by Rees (1987) to account for the confining pressures of broad emission line clouds.

2.2 X-RAY BINARIES.

A large fraction of high mass X-ray binaries (*e.g.*White 1989) exhibit regular pulsations attributed to accretion onto the poles of a spinning magnetized neutron star. In some systems, mass is believed to be transferred from the companion star via an accretion disk; in others a stellar wind is implicated. During episodes of rapid accretion in disk-driven systems, the X-ray period shortens. This is attributable to magnetic coupling of the star to the disk. Field lines that thread the disk within the corotation radius exert a stronger torque than those that attach to more slowly orbiting gas and the star is spun up. Many high luminosity, low mass X-ray binaries exhibit quasi-periodic oscillation (*e.g.*Haslinger 1989). These irregular $\sim$1-50Hz pulsations have been attributed to a changing beat frequency between the stellar rotational frequency and the orbital frequency in an accretion disk at the magnetospheric boundary. The interaction between poloidal field and orbiting disk has been widely studied for both types of object (*e.g.*Ghosh and Lamb 1979, Shibizaki and Lamb 1987 and references therein) and is similar to the interaction that I will argue is ultimately responsible for launching jets.

The best evidence for jets is found in SS433 (*e.g.*Margon 1984). The kinematics are, by now, well understood. Two anti-parallel jets with speed $0.26c$ precess on a cone with opening angle $\sim$ 20° every 164d. They originate from a binary X-ray source with orbital period 13.1d that appears to contain an accretion disk. Velocity measurements imply that the compact object has a high mass and is therefore a black hole (*e.g.*Zwitter and Calvani, these proceedings). There is no accepted explanation for the origin of the jets.

Further evidence for jets is provided by Sco X-1, which is a triple radio source, strongly reminiscent of extragalactic radio sources (Geldzahler and Fomalont 1986). Another low mass X-ray binary and prominent radio source, Cyg X-3, has been shown to be double (Strom *et al.*1988).

2.3 CATACLYSMIC VARIABLES.

Although they contain the best-studied accretion disks, cataclysmic variables, (low mass, white dwarf binaries), have so far produced the weakest evidence for bipolar outflow. They seem to lose appreciable amounts of mass through high velocity ($\sim$ 5000km s^{-1}) winds, accelerated well away from the white dwarf (*e.g.*Drew 1987, Mauche and Raymond 1987). UV line profiles are consistent with bipolar outflow but are also attributable to bipolar illumination of a spherical outflow. Some cataclysmic variables have been shown to contain variable radio features (Chanmugan 1987, Bookbinder and Lamb 1988). It is clearly of interest to see if these exhibit a jet morphology.

2.4 PROTO-STELLAR DISKS.

Young stellar objects appear to evolve from deeply embedded infra-red sources to "classical" T-Tauri stars and finally to "naked" T-Tauri stars. These stars have much lower specific angular momenta than the gas out of which they form by gravitational collapse. Indeed they only have surface rotation speeds typically $\lesssim$ 0.1 times the Keplerian value at their equators (Bouvier *et al.*1986, Hartmann *et al.*1986). Although it has long been suspected that gas accretes onto the proto-star via a circumstellar disk (Lynden-Bell and Pringle 1974) and that hydromagnetic torques are partly responsible for the angular momentum transport, the actual way by which this occurs remains controversial(*cf.*Bodenheimer, Lin, Pringle, these proceedings).

The best observational evidence for a circumstellar disk is probably the disk of size $\sim 6 \times 10^{16}$cm and mass $\sim 0.1 M_\odot$ reported around HL Tau (Sargent and Beckwith 1987)and seemingly supported centrifugally. Additional evidence for disks has come from infra-red spectra (Adams, Lada and Shu 1987) which require the presence of disks extending right down to the stellar surface, and observation of blue-shifted forbidden lines which point to an occulting disk (*e.g.*Appenzeller *et al.*1984, Edwards *et al.*1987). Mass accretion rates for T Tauri stars are typically $10^{-7} M_\odot yr^{-1}$ (Kenyon and Hartmann 1987, Bertout *et al.*1988). A subset of young stellar objects, the FU Orionis stars, undergo outbursts reminiscent of cataclysmic variables in white dwarf systems and accrete through the disk at a higher rate, up to $10^{-4} M_\odot yr^{-1}$. A significant fraction of the proto-stellar mass is probably accreted during FU Orionis phases (Kenyon and Hartmann 1988).

Many young stellar objects are associated with bi-polar outflows seen as expanding shells of molecular gas, winds of atomic hydrogen and small scale, collimated jets of ionised gas moving with speeds $\sim 200 - 400$km s^{-1}. These outflows, which carry discharges that can be several percent of the mass accretion rate, have too large a thrust to be caused by either radiation or gas pressure (*e.g.*Lada 1985, Mundt 1985 and references therein) and it is widely suspected that they are driven magnetically. In addition, the lobe/jet axes are correlated with the direction of the ambient magnetic field and the inferred disk angular momentum, implying that all three quantities are physically related (Strom *et al.*1986).

3. Self-Magnetised Disks.

3.1 ORIGIN OF DISK MAGNETIC FIELD.

The gas in most accretion disks is generally supposed to have a very high magnetic

Reynolds' number (the ratio of the Ohmic dissipation timescale to the convection timescale) so that MHD should be appropriate. Poloidal magnetic field extending perpendicular to the disk will be advected inward by the accreting gas and build up to a level where concentration and amplification balance escape through large scale reconnection and topological dissipation (Parker 1979). In most magnetic extraction models, it is proposed that a dynamically significant field strength can be built up in this manner.

In a dissenting view, Van Ballegooijen(1989) ($cf.$Lovelace, Wang and Sulkanen 1987) has pointed out that if the magnetic Prandtl number, (the ratio of the effective electrical diffusivity η_e to the effective kinematic viscosity ν_V), $P_m \sim 1$, as frequently assumed, then the time scale for inflow and convective amplification of the magnetic field, $\sim r^2/\nu_V$ exceeds the timescale for reconnection of the radial field across the disk $\sim rh/\eta_e$, (where $h << r$ is the disk thickness). This would imply that large scale poloidal field cannot be built up by advection. However, it is not certain that $P_m \sim 1$ because the turbulence in a strongly differentially rotating accretion disk must be highly anisotropic and the eddies responsible for a turbulent viscosity (should this be the main agency for effecting angular momentum transport), differ from those invoked to facilitate magnetic reconnection.

There is an alternative source of flux. Accretion disks are strongly differentially rotating and although they are not always convective, meridional circulation is hard to avoid. These conditions are propitious for dynamo action ($e.g.$Parker 1979, Zel'dovich, Ruzmaikin and Sokoloff 1983). In the usual treatment of turbulent dynamos, it is supposed that the small-scale velocity field exhibits a non-zero helicity measured by the quantity $\alpha = - < v \cdot \omega > \tau/3$ where ω is the vorticity and τ is a correlation timescale. Under these conditions, a small component of electric field along the magnetic field, $E = -\alpha B/c$ will develop and field growth will follow. In so-called $\alpha - \omega$ dynamos, driven by a combination of helicity and differential rotation, it can be shown that it is not possible to sustain even solutions in which the z-component of magnetic field has the same sign on both sides of the disk. Odd, ($e.g.$quadrupolar) modes can, however, be sustained. Unfortunately, an odd magnetic field can escape the disk quite easily and is probably less effective at exerting a torque (Lovelace $et\ al.$1989). A further complication is that the turbulent magnetic field may have to exceed the mean field by a large factor. However, the total magnetic energy density is limited by buoyancy to the gas pressure (Coroniti 1984). Therefore, the large scale field strength may be too small to be interesting dynamically. It is not possible to quantify this restriction at present.

An intriguing possibility is that an odd field created by a disk dynamo might co-exist with an even field convected inward by the accreting gas. There would then be a significant difference between the strengths of the poloidal field on the two sides of the disk and consequently a large difference in the powers of the two jets, which might then appear one-sided ($cf.$§2.1).

3.2 MAGNETIC VISCOSITY.

Conventional accretion disk models rely on either turbulent or local magnetic viscosity to transport angular momentum outward. In either case, it has become customary to suppose that the torque is directly proportional to the pressure, with constant of proportionality $\alpha \sim 0.1$ (Shakura and Sunyaev 1973). Although there is still no adequate justification of this prescription, it remains the best guess. Differential rotation probably builds up the toroidal component of the disk field until either magnetic reconnection (e.g.Eardley and Lightman 1975) or buoyant escape (e.g.Galeev, Rosner and Vaiana 1979) both limit the magnitude of the field and create a radial component so that there is a net torque $\propto < B_r B_\phi >$. Even in the context of 'α-models', it is important to decide whether or not the pressure is the total pressure or just the gas pressure. These two choices lead to quite different structures of the inner, radiation-dominated parts of accretion disks and, in particular, to differing conclusions about stability. Sakimoto and Coroniti (1981) and Coroniti (1984) have argued that in a radiation-dominated accretion disk, the magnetic pressure is limited, by buoyancy, to a fraction of the gas pressure.

By comparison with the solar photosphere, a supersonically rotating and rapidly shearing accretion disk that is being heavily irradiated with UV and X-ray photons, must be a very active environment. It is therefore quite likely that the disk is sandwiched between two hot, dense coronae (e.g.Galeev, Rosner, and Vaiana 1979, Aly 1984). Magnetic buoyancy within the disk will drive loops out into the corona where they are twisted by the differential rotation of their foot-points. In other words, field-aligned currents will flow which will supposedly dissipate in the corona. Indeed, it is conceivable that most of the dissipation associated with angular momentum transport in the disk actually occur in the corona (e.g.Mészaros, Meyer and Pringle 1977.) Coronal dissipation will create a plasma out of thermodynamic equilibrium with the disk and is quite possibly responsible for much of the non-thermal emission observable from all four types of disk.

3.3 ELECTROMAGNETIC MODELS.

Internal (e.g.viscous) torques are not the only possibility. The simplest type of external torque involves large scale magnetic stress acting on the surfaces of the disk. The first models were purely electromagnetic.

When a conducting disk of radius r is threaded by magnetic flux Φ, and dragged with angular velocity Ω, a radial electric field will be induced in the non-rotating frame giving a total electrical potential difference $\sim \int d\Phi\Omega/2\pi c$. (If, as generally assumed, the magnetosphere is stationary and axisymmetric, then the toroidal electric field must vanish.) In the absence of mobile charges, most of this potential difference would be made available along magnetospheric field lines and would be able to accelerate test charges to impressively large energies, $\sim 10^{11}$ GeV in the case of AGN (Lovelace 1976).

However, plasma is extremely mobile and should be able to flow freely along the magnetic field lines to keep them at nearly constant electric potential. Drawing the analogy with pulsar models, Blandford(1976) supposed that the charges and currents were arranged so that the electromagnetic force density on the plasma vanished, $i.e. \rho E + j \times B/c = 0$. A self-similar solution capable of extracting *all* the energy and angular momentum from a Keplerian disk was derived. (If the disk is a good conductor, then an element of magnetic torque dG must do work at a rate ΩdG and the ratio of energy to angular momentum release, Ω, is just the same as the net energy release by a ring of cold gas slowly giving up its orbital angular momentum.) A combination of magnetic pressure and "hoop" stress exerted by the toroidal field cause the poloidal field lines to adopt a paraboloidal shape thereby collimating the Poynting flux released by the disk into two anti-parallel jets. Energy is released in a form suitable for conversion into relativistic electrons as observations of AGN seem to require.

This model had two serious shortcomings. Firstly, as it was electromagnetic, it completely ignored the inertia of the plasma. Secondly, being self-similar, it failed to model conditions at small radii where most of the energy is liberated and at large radii where most of the angular momentum is released. In an attempt to address the first of these deficiencies, hydromagnetic disk magnetosphere models, that assume the existence of a fluid velocity v satisfying $E + v \times B/c = 0$, were developed. Numerical simulation is required to address the second shortcoming.

3.4 HYDROMAGNETIC MODELS.

If the disk is threaded by vertical magnetic field that extend through a dense corona, the magnetosphere above it will become filled with plasma which can be flung out centrifugally by the magnetic field lines. The inertia of the plasma will create toroidal field which will decelerate the disk and may react back on the corona to limit the flow. Hydromagnetic models differ from electromagnetic models because, although the field lines remain equipotential, poloidal current must now cross the field to create a $j \times B/c$ force density to accelerate the plasma. Furthermore, the plasma moves with a speed comparable with the Keplerian velocity rather than the speed of light.

Blandford and Payne(1982), Lovelace *et al.*(1986) and Camenzind(1987) have constructed hydromagnetic models of disk-driven centrifugal winds. In these models, flux Φ threading a radius r of the disk accelerates plasma to roughly the Alfvén speed $B/(4\pi\rho)^{1/2}$ at the Alfvén radius r_A, where the field lines are bent backward through an angle $\sim 45°$. The torque exerted on the disk is $\propto \Phi^2/r_A$. If it is to extract all the angular momentum liberated at that radius by steadily inflowing gas, the torque must scale $\propto r^{1/2}$ implying that the disk field scale as $B \propto r^{-5/4}$. The associated discharge in the wind is a fraction $(r_A/r)^{-2}$, typically 0.01 of the mass accretion rate in the disk. The asymptotic wind speed is usually comparable with the rotational velocity at the Alfvén radius, Ωr_A.

The wind can be launched centrifugally from a Keplerian accretion disk if the poloidal field direction is inclined at an angle of less than 60° to the radial direction. Stationarity and axisymmetry then require that the gas flow outwards along the rotating field lines like beads on a wire, conserving energy (electromagnetic plus mechanical) angular momentum, particles and the angular frequency of rotation of the field line Ω (equal to the angular frequency of the disk at its foot point). The flow must pass through a critical Alfvén surface where the flow speed normal to the surface equals the speed of an Alfvén wave travelling in the same direction. At this point, it can be shown that the specific angular momentum (mechanical plus magnetic) ℓ must be given by

$$\ell = \Omega r_A^2, \tag{1}$$

(e.g.Weber and Davis 1967). These flows are magnetically dominated and so the Alfvén speed exceeds the sound speed. There will therefore be a second critical surface much closer to the disk where the flow speed equals the sound speed. There will usually be a third critical surface beyond the Alfvén surface where the fast magnetosonic Mach number equals unity.

In order to devise self-consistent models in which there is force balance across the magnetic field, some version of the Grad-Shafranov equation, familiar from studies of Tokamak plasmas, must be solved (e.g.Lovelace 1987). As in the electrodynamic models, a combination of magnetic pressure and tension collimates the outflowing plasma. Asymptotically, the magnetic (as opposed to mechanical) contributions usually account for the majority of the energy and angular momentum fluxes (e.g.Blandford and Payne 1982, Lovelace, Mobarry and Contopoulos 1989). In some solutions, the jets were focussed onto the axis. In others, they were collimated to form cylindrical jets in which the electromagnetic energy flux was twice the mechanical energy flux corresponding to a fast magnetosonic Mach number of unity.

It is by now clear that even with the simplifications of axisymmetry, stationarity and perfect MHD, a consistent treatment of an accretion disk magnetosphere requires extensive numerical simulation. (Recent advances in the development of finite difference codes by Evans and Hawley 1989 and Zachary and Colella 1989 make this a more practical proposition.) Sakurai(1985,1987) has published such simulations starting from a "split monopole" field, both centered on the origin, and centered on a point on the axis and below a disk as a simple way of describing a magnetised disk. He is able to find stationary solutions that generalise the one dimensional Weber and Davies(1967) solutions and verifies that the outflow is indeed collimated toward the symmetry axis by the toroidal field. In another impressive series of numerical simulations, Shibata and Uchida(1985,1986, this volume) have studied impulsive energy release of gravitational energy by magnetic fields that are either rapidly spun up or compressed by collapsing gas clouds. They include the possibility that the ambient external medium be non-uniform. As with purely hydrodynamic explosions (*e.g.*Kompaneets 1960), density gradients can accentuate the focusing of the outflow.

3.5 APPLICATION TO ACCRETION DISKS.

All four types of accretion disk have been associated with hydromagnetic, disk-driven winds. Characteristic field strengths are given in Table 1. The first application of these ideas was to jets in AGN and was motivated by the failure of alternatives. In particular, an early model for radio jets involved a pair of nozzles that were excavated along the rotation axis of a spinning gas cloud allowing outflowing plasma to undergo a supersonic transition. However, the discovery that collimation could be established within ~ 1 pc of the galactic centre, where the gas would cool too rapidly and produce excessive X-ray emission, ruled out this mechanism for most sources. An alternative proposal, that jets be radiatively driven in the funnels of thick accretion disks, fails because radiation drag limits the outflow to mildly relativistic speeds, inadequate to account for superluminal expansion and, besides, there is no indication that an Eddington-limited, radiation-supported torus is present in most radio galaxies.

Disk-driven hydromagnetic jet models need not suffer these difficulties. They have the advantage of extracting angular momentum directly from accretion disks without requiring it to be continuously pushed out to larger and larger radii. In fact, for a Keplerian disk, the angular momentum, which is responsible for the collimation, will be mostly extracted from the outer radius, whereas the energy, which is responsible for the jet power is derived from the innermost parts. Unfortunately, in view of our ignorance of the conditions within AGN disks (or even their extent) it is very difficult to improve upon the above guesses as to the field strength and structure.

The best two examples of jets associated with X-ray sources are Sco X-1 and SS433 (§2.2). In Sco X-1, the minimum pressure in the radio components was too large to be confined by the ram pressure of the interstellar medium. Therefore, Achterberg, Blandford and Goldreich (1983) proposed that the jets were magnetically confined. Colgate (private communication) has argued that the same is true for SS433.

The application of these ideas to cataclysmic variable winds (§2.3) has been recently developed by Koen(1986) and Cannizzo and Pudritz(1988). They have argued that the outflows are hydromagnetically driven and that disk-driven extraction of orbital angular momentum is responsible for the observed period distribution.

3.6 MAGNETICALLY-MEDIATED ACCRETION OF NEUTRAL GAS.

This general mechanism must be modified when the accreting gas is largely neutral

CentralObject	r_{min}	$B(r_{min})$	r_{max}	$B(r_{max})$
Massive Black Hole	10^{14}cm	10^3G	3pc	10^{-3}G
Magnetised Neutron Star	10^8cm	10^5G	10^{12}cm	1G
Stellar Black Hole	10^6cm	10^8G	10^{12}cm	1G
White Dwarf	10^9cm	10^3G	10^{11}cm	1G
Protostar	10^{12}cm	100G	10^{17}cm	10^{-3}G

Table 1. Order of magnitude strength of magnetic field powerful enough to exert a dynamically significant external torque in different types of accretion disk.

as is the case with a disk orbiting a young stellar object. Pudritz and Norman(1983), Pudritz 1985, Pudritz and Norman (1986) envisaged that a disk of molecular gas of mass $\sim 100M_\odot$ and size $r_D \sim 10^{17}$cm be threaded by a field of strength ~ 0.01G. Although the gas may be predominately neutral, the ions are tied to the magnetic field and the ion-neutral friction has to be sufficiently great to ensure that the neutral gas is also efficiently extracted (*e.g.*Laetano 1988, Natta *et al.*1988). (This does not have to be the case at all radii.) Removal of a fraction of the disk as a hydromagnetic wind, with a discharge originally estimated to be $\dot{M}_W \sim 10^{-4}M_\odot yr^{-1}$ and speed $v_W \sim 50 kms^{-1}$, enables the remainder of the gas to accrete. The gas is heated by the protostar and partially ionised. It is accelerated to large speeds to and form a fast-moving core to the jet with speed $\sim 200 kms^{-1}$ and discharge $\sim 10^{-6}M_\odot yr^{-1}$. Criticism of the earlier version of this theory (*e.g.*Pringle 1988), namely that by operating in the outer parts of the disk, there is insufficient energy to account for the power in the winds, is ameliorated if the torques act throughout the disk extracting energy from close to the protostar and the wind has a more reasonable discharge of $\sim 10^{-6}$M$_\odot$yr^{-1}.

A more detailed model of the energy extraction has been developed by Königl (1989) who is able to combine a model of the steady disk field distribution, involving a self-consistent balance between ambipolar diffusion of the magnetic flux outward and convection inward by the inflowing gas,with a self-similar MHD wind.

An alternative model has been developed by Shu *et al.*(1988). They propose that the jets derive from the protostar (cf Hartmann and MacGregor 1982) rather than the disk and that the mass is flung out centrifugally from the stellar equator through an X-type neutral point. Collimation of the outflow toward the poles is then expected to follow. They cite evidence for fields of the requisite strength (~ 100G) being generated by dynamo action (*e.g.*Lago 1984). However, it is not clear why the field is confined to the rotational equator of the star instead of diffusing out through the disk and presumably removing some of its gas before it gets to the star. Of course this mechanism cannot explain the subsequent rotational deceleration of the star (*cf.*§2.4). (Fields of strength $\sim 10^4$G are necessary to allow magnetic torques exerted by the disk to decelerate the star by analogy with pulsating X-ray sources like Her X-1.)

If, alternatively, the protostar is rotating well below its maximum rate and is not strongly magnetised then the accretion disk should extend down to its surface where a boundary layer is formed. In principle, as much energy will be released in the boundary layer as in the whole of the accretion disk and this presumably is a is the source of the photo-ionising UV. Pringle(1989) has proposed that efficient dynamo action also occur within the boundary layer and that loops of toroidal flux be spun off to collimate the outflows. Unfortunately, both signs of toroidal field will be involved, and so there will not necessarily be any net collimating magnetic tension to counteract the de-collimating magnetic pressure. A similar difficulty may afflict the model of Shu *et al.*(1988).

3.7 THE LINK TO JETS.

It is apparent from these numerical simulations and general considerations that the degree of focusing of a hydromagnetic jet depends upon the distributions of the four conserved quantities, mass, energy, angular momentum and angular velocity on the field lines at their foot points in the accretion disk. In an important recent study, Haeverts and Norman(1989) have argued that axisymmetric perfect MHD winds with no closed field lines *must* collimate asymptotically, The degree of collimation depends upon the limiting value of the poloidal current flowing through a hoop of cylindrical radius r. If this decreases to zero as $r \to 0$, then it is possible for the field lines to lie on paraboloidal surfaces. Otherwise, they must asymptote to (right) cylindrical surfaces.

One concern about these results is that they make the assumption that there is an asymptotic flow in which the curvature on the poloidal field decreases asymptotically to zero. This need not be the case. To see this just follow Chan and Henriksen (1980) and Achterberg, Blandford and Goldreich (1985) and consider a cylindrically symmetric similarity solution. The radius of the jet satisfies a second order dynamical equation and behaves like a particle in a potential well with centrifugal force and magnetic pressure gradient acting as repulsive forces and magnetic tension acting as an attractive force. In the absence of dissipation, a stationary jet can oscillate in radius for ever. Nevertheless, this analysis does provide good evidence that jet collimation by magnetic field threading an orbiting disk is generic rather than just a feature of idealised solutions. It would be of considerable interest (though extremely difficult) to generalise these results to non-axisymmetric flows.

The supply of toroidal field to jets is very important in maintaining their collimation well away from their sources. In particular, toroidal field can produce a central jet pressure much larger than the ambient gas pressure in the surrounding medium as observations of extragalactic jets indicate is sometimes the case. To see this, consider a central jet core confined by toroidal field varying as $B_\phi \propto r^{-1}$ for $r_{min} < r < r_{max}$, with surface currents flowing along the cylindrical walls of the magnetic sheath at r_{min} and r_{max}. The external pressure required to maintain static balance in this example is only a fraction $(r_{min}/r_{max})^2$ of the pressure in the jet core. Many extragalactic jets exhibit perpendicular linear polarisation implying that the predominant field is aligned along the jet rather than toroidally around it. This field is almost certainly not unidirectional. It corresponds to far too much flux to be confined in the galactic nucleus. Instead it is probably generated by velocity shear in the jet.

4. Black Hole Electrodynamics.

4.1 BACKGROUND.

We have argued that twisting magnetic field can extract energy and angular momentum from accretion disks. We also know that they can decelerate stars, either electromagnetically, as in the case of radio pulsars, or hydromagnetically in the case of main sequence dwarfs. It is therefore natural to inquire if the spin energy of a rotating black hole can also be tapped in this way. Such a mechanism would allow a black hole to be active when not currently accreting much gas. This might provide an explanation for radio galaxies - active nuclei which can create high power, non-thermal jets in the absence of large UV emission. It might also operate in normally accreting black holes.

Blandford and Znajek (1977) explicitly developed the pulsar analogy and argued that the spin energy could be plausibly extracted with an efficiency of typically 50per cent using large scale magnetic field threading the horizon. They exhibited two solutions for the electromagnetic field demonstrating energy extraction from slowly rotating holes. However, this analysis avoided two important issues. Firstly, by (once again) only solving for mutually consistent currents and fields, (specifically by requiring that they be force-free), it ignored inertial effects associated with the charged particles carrying the currents. Secondly, it did not demonstrate that the electromagnetic field would evolve from generic initial conditions to the energy-extracting steady state solutions. In this section, we describe progress on the first of these issues and report on some doubts that have been raised concerning the second. Although this topic does not strictly come under the heading of magnetic accretion disks, it is important for understanding the inner boundary conditions for a magnetised disk and raises physical questions that are relevant to a complete understanding of the behaviour of a non-relativistic disk magnetosphere.

4.2 SOME RELATIVISTIC PRELIMINARIES.

A Kerr black hole obeys the so called "no hair" theorem which implies that it is completely described by its mass m and angular velocity $\Omega_H < 1/2m$, ($G = c = 1$) henceforth. (It can also carry an electric charge, though under astrophysical conditions this is gravitationally insignificant.) Information about the mass of the hole is contained in the g_{00} component of the metric tensor, and the angular velocity is coded into the $g_{0\phi}$ component. The charge is measured by the radial electric field in the usual manner. When a particle, or equivalently, electromagnetic Poynting flux, is directed toward the horizon, the mass, angular velocity and charge of the hole will change accordingly. All other attributes of the particle will be redshifted away as it approaches the horizon (which it crosses in a finite proper time by its clock but an infinite time according to a clock at infinity).

The horizon is a one-way membrane and an observer hovering just outside it (and necessarily orbiting with angular velocity Ω_H), can only see energy entering the hole. Nevertheless, energy can be released from a spinning hole to infinity by classical processes and the hole mass measured from infinity can decrease. This is because the observer can register a change in the angular momentum of the hole. When we Lorentz boost to the non-rotating frame, these changes in energy (dE) and angular momentum (dS) must be combined in such a way, $(\propto dE - \Omega_H dS)$ that the result can be negative. A limited quantity of energy can therefore be extracted without contradicting the usual absorbing property of black holes. In fact, we can define an irreducible or rest mass m_i of a spinning hole which is related to the true mass m through the formula $m = m_i(1 - a\Omega_H)^{-1/2}$, where the specific angular momentum of the hole is $a = 4m_i^2\Omega_H$. The maximum mass that can be released from a spinning hole is $m - m_i$.

It is straightforward in principle (although difficult in practice) to solve Maxwell's equations in a curved spacetime and impose the condition that the fields either be force-free or satisfy the equations of ideal MHD. The boundary condition at the black hole horizon (Znajek 1977) is basically derived by requiring that the electromagnetic field remain finite when measured by an infalling observer.

In order to elucidate the interaction between a black hole event horizon and electromagnetic field, MacDonald and Thorne (1982) developed an attractive analogy introduced by Damour(1978) and Znajek(1978) into an equivalent, quasi-Newtonian model in which spacetime is subjected to a "3+1 split" and the horizon is regarded as a spherical surface. This model - the so-called Membrane Paradigm - has been clearly described in Thorne, Price and MacDonald (1986). In this approach, it is argued that the electromagnetic properties of a black hole event horizon are equivalent to those of a spherical conductor endowed with a surface resistivity of $4\pi c$ in Gaussian units, equal to 377Ω in SI units. In particular, as far as a distant observer is concerned, it makes no difference if the currents complete well within the horizon or on the fictitious membrane.

4.3 BLACK HOLE MAGNETOSPHERE.

There are some important differences between a pulsar magnetosphere and a black hole magnetosphere. In the former case, charges of either sign can (at least in principle) be extracted from the stellar surface. However, in a black hole magnetosphere, charged particles must flow inwards across the horizon and (at least when the magnetosphere is matched to a wind), outwards at large radius. They must therefore be created, probably in the form of electron-positron pairs, within the magnetosphere. Current, however, can flow continuously from the horizon to infinity. (This only requires that the number of the electrons ultimately falling into the hole differ from the number of positrons.) Fortunately, sufficient pair production will almost certainly occur through γ-ray pair production within an active galactic nucleus to create enough electron-positron pairs to carry the current. For an electromagnetic power equal to the Eddington limit, this requires a particle density $n \sim 10^{-2}(M/10^8 M_\odot)^{-3/2} \text{cm}^{-3}$ which is much less than the density at which the particles become dynamically significant, $\sim 10^{14}\gamma^{-1}(M/10^8 M_\odot)^{-1} \text{cm}^{-3}$. Black hole magnetospheres should be essentially force-free.

A second difference is that a pulsar is an excellent electrical conductor and the field is supported by currents anchored in the star for an extremely long time. However a black hole magnetosphere must rely on poloidal field dragged inward by the infalling gas which may be able to escape through the disk by means of interchange instabilities. Although, it is possible, in principle, for the energy in the magnetic field to approach the total energy density in a ring of orbiting gas and for the electromagnetic power extracted to exceed the luminosity of the accreting gas, it is not understood how much flux will accumulate in practice. An indirect consequence of having the currents flow in the disk is that all the magnetic field lines are either open and extend to infinity, or intersect the disk and are frozen into it (Camenzind 1989). We are mostly concerned with the open field lines.

A third difference is that in a pulsar magnetosphere, where all field lines rotate with the angular velocity of the star, there is a well-defined light cylinder, beyond which particles attached to the field lines have to move outward. In a black hole magnetosphere, the open field lines rotate with an angular velocity roughly half that of the hole. There are then two *light surfaces*, bounding an intermediate zone where particles are created and can co-rotate with the field. Beyond the outer light surface, particles must move outward; within the inner light surface, particles must approach the horizon.

In a stationary, axisymmetric, force-free magnetosphere, the magnetic field lines can be thought of as wires, along which the electrical current is constrained to flow, rotating with a constant angular velocity Ω. (This is the angular velocity with which an observer must orbit the hole in order to transform away the electric field.) This permits a simple circuit analysis (*e.g.*Thorne, Price and MacDonald 1986). Consider two nearby magnetic surfaces, enclosing magnetic flux $\Delta\Phi$, connected across the horizon with a thin annular ring of resistance $\Delta R_H = \Delta\Phi/\pi g_{\phi\phi}B_P$ where B_P is the poloidal field and connected at infinity by a resistive load ΔR_L (given by essentially the same expression if the outflow is matched onto a magnetically-dominated, relativistic wind as we discuss below) orbiting with angular velocity Ω_L (equal to zero for a wind, though not for a disk).

The potential difference induced across the ring at the horizon is $\Delta V_H =$

$(\Omega_H - \Omega)\Delta\Phi/2\pi$. Similarly the potential difference across the load at infinity is $\Delta V_L = (\Omega - \Omega_L)\Delta\Phi/2\pi$. Now current conservation requires that $I = \Delta V_H/\Delta R_H = \Delta V_L/\Delta R_L$ or

$$\Omega = \frac{\Omega_H\Delta R_L + \Omega_L\Delta R_H}{\Delta R_L + \Delta R_H} \qquad (2)$$

The magnetic field acts as a sort of clutch that couples the hole to the load and rotates with a compromise angular velocity. When the load is on open circuit, the field is "frozen" into the horizon; when the load is short-circuited, the field rotates with the load. The field transmits a torque $\Delta G = I\Delta\Phi/2\pi = I^2(\Delta R_H + \Delta R_L)/(\Omega_H - \Omega_L)$ which does work at a rate $-\Delta G\Omega_H$ on the hole and $\Delta G\Omega_L$ on the load. The sum of these two powers, which must be negative represents the dissipation $I^2\Delta R_H$ in the hole plus $I^2\Delta R_L$ in the load. Field lines that attach to a relativistic wind will be roughly impedance-matched to the horizon so that the efficiency of energy extraction will be about 50per cent. Field lines that attach to a disk can transmit angular momentum of either sign, dependent upon the relative angular velocity.

A more complete theory can be derived if, following Phinney (1983) (to whom reference should be made for further details), Goldreich and Julian(1976), Michel(1982), Kennel, Fujimura and Okamoto(1983), Bekenstein and Eichler(1985), and Camenzind(1986ab, 1987), we describe the plasma using relativistic MHD. Just as with non-relativistic MHD, the fluid should pass through three critical points as it flows away from the region where the pairs are created and three more as it flows into the horizon. Under conditions of ignorable particle inertia, however, the electromagnetic field and efficiency of energy extraction are essentially those given by the force-free formalism.

We can illustrate this by considering the Alfvén critical surface in a relativistic MHD wind. For simplicity, we ignore the effects of gravity. This surface is defined by requiring that the relevant wave mode (in this case an Alfvén mode) propagate with zero frequency perpendicular to the surface. (This prescription is unchanged by Lorentz boosts parallel to the surface.) It is the one place in the flow where transients left over from initial conditions cannot decay away. Now for an axisymmetric magnetosphere, we can Lorentz transform into the frame moving with speed Ωr where the electric field vanishes. In this frame, the radial magnetic field is $B_r' = B_r(1 - \Omega^2 r^2)^{1/2}$, where B_r is the radial field measured in the non-rotating frame. The fluid proper velocity perpendicular to the surface is unchanged by the transformation and must equal the proper Alfvén speed, $B_r'/(4\pi n\mu)^{1/2}$, where n is the proper particle density and μ is the enthalpy per particle. In terms of quantities measured in the non-rotating frame, the condition for the critical surface is then that

$$M_I^2 = \frac{4\pi n\mu u_r^2}{B_r^2} = 1 - \frac{\Omega^2 r^2}{c^2} \qquad (3)$$

Now, when the flow is magnetically dominated, ($M_I^2 \ll 1$), the Alfvén critical surface will lie just within the light surface (at $r = c/\Omega$).

Now if the equations of perfect, relativistic MHD apply, (and there is no guarantee that this will be the case in a black hole magnetosphere), and we ignore gravity, then the conserved energy and angular momentum along the field lines will be

$$e = \mu\gamma - \frac{\Omega r B_r B_\phi}{4\pi n u_r}$$

$$\ell = \mu r u_\phi - \frac{r B_r B_\phi}{4\pi n u_r} \tag{4}$$

respectively. (See Phinney 1983, Camenzind 1987a, 1989 for the more complex, general relativistic expressions.) As the flow crosses the outgoing critical surface, the generalisation of condition (1) is

$$\ell = \Omega r^2 e \tag{5}$$

which is *automatically* satisfied provided that the particles move along the field lines, specifically that they satisfy the kinematic condition

$$u_\phi = u_r B_\phi / B_r + \Omega r \gamma \tag{6}$$

No extra condition is imposed on the electromagnetic field structure at the Alfvén critical surface. The slow mode critical surfaces do impose conditions on the fluid but as long as the magnetic field is dominant, these conditions only necessitate minor corrections to the electromagnetic solution. The fast mode critical surface is located well beyond the light surface. However, if the flow is still magnetically dominant at this point, then the fast mode speed is essentially that of light. Therefore, to good approximation, the requirement that the flow pass through a fast mode critical point translates into an electromagnetic boundary condition that the relativistic invariant $B^2 - E^2$ asymptotically approach zero from above.

If $B^2 - E^2 < 0$, beyond the light surface, then it is no longer possible to satisfy the kinematic condition (6). This is a circumstance that arises in attempts to solve the corresponding pulsar magnetosphere problem, (*e.g.*Mestel *et al.*1985), where it is found that it is not possible to satisfy the equations of perfect MHD globally. Large parallel electric fields are produced and strong dissipation ensues. However, in a black hole magnetosphere, we expect that field lines are either open and power a relativistic wind or are frozen into the disk, possess no outer light surface and extract little energy. Therefore, with the exception of isolated neutral lines, there should exist hydromagnetic solutions satisfying the kinematic condition (6) and these should be approximated by force-free solutions.

Similar considerations apply within the inner light surface. An electromagnetic solution that is continuous at the ingoing light surface is close to a magnetically dominated MHD solution with an ingoing Alfvén surface. At the horizon, the boundary condition of Znajek (1977) automatically implies that the fast speed be exceeded and so there has to be a fast mode surface. However, this is in practice very close to the horizon and requiring that an MHD solution pass through an ingoing fast mode surface is almost the same as requiring that an electromagnetic solution satisfy the Znajek condition.

In order to proceed further and compute the overall efficiency of energy extraction, (given by $< \Omega/\Omega_H >$), we need to know the shape of the magnetic surfaces produced by currents flowing both in the disk and in the magnetosphere. This requires solution of the relativistic Grad-Shafranov equation, a feat that has been accomplished using finite element methods by Camenzind(1987b) in his study of a spinar magnetosphere.

For the above reasons, I argue that a black hole magnetosphere, if it exists at all, probably contains enough charged particles to make the open field lines equipotential, though insufficient for the particles to contribute much to the energy or angular momentum fluxes. Under these conditions, I assert that solutions of the comparatively simple force-free equations with the boundary condition that $B^2 - E^2 \to 0$ at infinity and on the horizon are excellent approximations to relativistic hydromagnetic wind solutions at least for the open magnetic field. (This is in contrast to a disk magnetosphere where the electromagnetic and non-relativistic hydromagnetic solutions are qualitatively different.) A fully self-consistent MHD treatment is necessary to verify or refute this conjecture and for this, we await the publication of Phinney (in preparation).

4.4 THE QUESTION OF CAUSALITY.

The Membrane Paradigm has recently been criticised in an important analysis by Punsley and Coroniti(1989a). They draw a distinction between situations where the black hole behaves passively and accepts matter and Poynting flux just like radial infinity in the case of a supersonic and super-Alfvénic stellar wind and situations when energy is extracted and therefore reacts back on the magnetosphere. In the former case, they corroborate the results obtained using the membrane approach; however, in the latter case, Punsley and Coroniti argue that the existence of inflowing sonic points prevents the horizon from influencing the global current flow and consequently the energy flux at large radii.

In a subsequent preprint, Punsley and Coroniti(1989b) offer a critique of the black hole energy extraction mechanism and highlight some important differences between unipolar induction in black holes and in radio pulsars. They argue that as, in the MHD description, the inflowing plasma must pass through all three sonic points, the horizon is out of causal contact with the outflowing plasma, unlike in the pulsar case where MHD waves can be reflected back from the stellar surface and communicate the value of the stellar angular frequency to the magnetosphere. Therefore, they argue, it is not possible to establish the poloidal current flow essential to the energy extraction mechanism and it is not permissible to determine the field line angular velocity by simply equating the outgoing energy flux to the energy flux at the horizon (*cf.*§4.3). Instead, they assert that the global current flow must be determined by conditions in the particle injection region. This would probably ensure a low value for the field line angular velocity $\Omega \ll \Omega_H$ and a consequent low efficiency of extraction of energy from the hole.

Now, in an MHD treatment the two characteristics that convey temporal changes in the magnetic field from the vicinity of the horizon to the outer magnetosphere travel at the Alfvén speed and fast mode speed. The former carries information primarily about the poloidal field distribution and is not effective at propagating disturbances radially outward within the inner light surface. In other words, the poloidal field structure near the horizon is dictated by the currents flowing in the disk and between the two light surfaces and not *vice versa*. However, information about the toroidal field is carried by the fast mode and can be communicated outward from the inflowing fast mode critical surface, which is located quite close to the horizon in a magnetically-dominated magnetosphere. It is the strength of the toroidal field which is related to the poloidal current flow and the ultimate efficiency of energy extraction. Now, we remarked in §4.2, a black hole does communicate its angular velocity to infinity through the $g_{0\phi}$ component of the metric tensor. In fact a physical oberver moving on a timelike geodesic just outside the fast mode critical surface (though close to the horizon) must rotate at almost the hole angular velocity Ω_H. All structure in the electromagnetic field closer to the horizon except the total electrostatic charge, must be redshifted away. Therefore in a magnetically-dominated magnetosphere, there is causal contact with observers moving azimuthally on time-like geodesics with an angular velocity very close to the hole angular velocity. The small freedom in the angular velocity of these observers corresponds loosely to the freedom that exists in locating the fast critical point in outgoing hydromagnetic wind solutions. It is not necessary to have causal contact with the event horizon for an efficient unipolar inductor to operate. Time-dependent, force-free calculations in a simple spacetime embodying both lapse and shift seem to substantiate this interpretation, though more detailed computations are necessary to settle the matter.

For these reasons, I argue that, as long as a black hole magnetosphere is magnetically-dominated, it should evolve naturally in a few light crossing times to the type of stationary state described in the previous section in which the efficiency of energy extraction is of order 50 per cent. However if the magnetospheric plasma density becomes large enough to make the Alfvén and fast mode speeds non-relativisitic, then the field line angular velocity will indeed be fixed by the conditions well removed from the horizon and it will probably be small with consequent low energy extraction efficiency, in agreement with Punsley and Coroniti. The utility of the Membrane Paradigm is clearly a matter of taste. (This reviewer finds it useful.) However as long as it is applied to influences (such as electromagnetic wave modes and gravitational perturbations) that propagate along characteristics with near light speed, it ought not to be misleading in its conclusions.

5. Conclusion.

In this review, I have tried to give a unified discussion of hydromagnetic aspects of the theory of accretion disks emphasising the growing similarities in the observations of disks around protostars, white dwarfs, neutron stars and massive black holes. All four types of disk have been associated with collimated outflows and it is conjectured that the connection is through large scale poloidal magnetic field that threads the disk. (Only the ordered component of field is likely to be effective; small loops of field my however cause considerable dissipation in a disk corona.) Jet generation can then be seen as a natural mechanism for disposing of the angular momentum of the disk gas.

However, there are some important differences between the different types of disk. Protostellar jets are mostly neutral and a centrifugal outflow can only occur if there is strong ion-neutral coupling. Cataclysmic variable accretion disks extend through less than three decades of radius and will presumably create less well collimated outflows than X-ray binary disks. Outflows from AGN propagate out through perhaps eighteen decades of steadily diminishing pressure which must contribute to the uniquely fine collimation of radio jets. (Although the pressure drop between stellar objects and the interstellar medium can be even larger, most of this will usually occur fairly abruptly. Perhaps Sco X-1 is an exception.) Magnetised neutron stars are magnetically coupled to their inner accretion disks. Unmagnetised stars have boundary layer interfaces. Black holes are probably separated from their disks by substantial gaps through which matter trickles inwards.

The application of a surface Maxwell stress is just one of several possible types of torque that have been proposed in accretion disks and discussed at this workshop. Internal viscous torques, associated with hydrodynamic turbulence, small scale magnetic loops or large density irregularities created through Jeans' instability in self- gravitating disks are likely to be effective under appropriate conditions (*e.g.*Lin, Pringle, Livio, these proceedings). Internal global instabilities (*e.g.*Frank, Narayan, Spruit, this volume) may also be important. There are even alternative external torques that have been proposed (*e.g.*Gunn 1978, Begelman *et al.*1983, Ostriker 1983). It would clearly be of great importance to have ways of deciding if and when these different types of torque operate. The most basic distinction between internal and external torques is that the former require the disk to spread in order to conserve total angular momentum. The discovery that protostellar disks extend to radii where the specific angular momentum exceeds that in the ambient cloud might be one way to detect this. Alternatively a clearer understanding of the response of a dwarf nova disk to outburst could provide unambiguous evidence for viscosity, although we are some way from doing this at present. Perhaps the

best direct evidence that jets are magnetically collimated and carry away significant quantities of angular momentum would be the discovery of Faraday rotation of
opposite signs on opposite sides of a jet. (Indeed if, as we have argued, the poloidal
flux has even symmetry across the disk, then this polarity should be reversed for
the two jets, just like a BMW logo.) Differences in Doppler shift might also be
detectable though, in practice, these will be quite small at distances at which the
jet can be resolved. CO and emission line observations of bipolar outflows from
protostars may be the best hope for accomplishing this.

There are some important issues of theoretical principle to be resolved, that
will probably require numerical simulation. The first is to determine if magnetised
stellar winds from a star (or black hole) alone, will be collimated into anti-parallel
jets or if the field lines must also thread an extensive disk. Secondly, it is important to compute finite, hydromagnetic disk magnetospheres that are balanced by
small thermal pressure at large cylindrical radius. Thirdly, the relationship between
stationary solutions of MHD winds and true time-dependent evolution, briefly discussed in the context of black hole magnetospheres, should be clarified. Finally,
(and hardest of all) non-axisymmetric flows should be calculated. It is by no means
guaranteed that axisymmetric outflows are stable.

ACKNOWLEDGEMENTS. I acknowledge informative discussion with Fred Adams, Lee
Hartman, Arieh Königl, Colin Norman, David Payne, Aad van Ballegooijen and several
workshop participants. That I am indebted to Roman Znajek, Max Camenzind, Ferd
Coroniti, Sterl Phinney, Brian Punsley and Kip Thorne for invaluable debates on the
contents of §4.3, 4.4 does not imply that they agree with my conclusions.

I thank Harvard Smithsonian Center for Astrophysics and the University of Arizona for
hospitality during the preparation of this review. Support by the Smithsonian Institution
and Guggenheim Foundation, the National Science Foundation under grant AST86-15325
and NASA under grant NAGW-1301 is gratefully acknowledged.

REFERENCES.

Achterberg, A., Blandford, R. D. and Goldreich, P. 1983. *Nature*,**304**,607.
Adams, F. C., Lada, C. J. and Shu, F. H. 1987. *Astrophys. J.*,**326**,865.
Aly, J. J., 1984. *Magnetospheric Phenomena in Astrophysics*, ed. R. Epstein and P.
    Feldman. American Institute of Physics
Appenzeller, I., Jankovics, I., and Ostreicher, R. 1984. *Astr. Astrophys.*,**141**,108.
Asseo, E. and Gresillon, D. 1987 (ed). *Magnetic Fields in Extragalactic Objects*. Edi-
    tions de Physique, Les Vils Cedex, France.
Begelman, M. C., Blandford, R. D. and Rees, M. J. 1984. *Rev. Mod. Phys.*,**56**,265.
Begelman, M. C., McKee, C. F., Shields, G. 1983. *Astrophys. J.*,**271**,70.
Belvedere, G., ed. 1989. *Magnetised Accretion Disks*, in press.
Bertout, C., Basri, G., and Bouvier, J. 1988. *Astrophys. J.*,**330**,350.
Blandford, R. D. 1976. *Mon. Not. R. astr. Soc.*,**176**,465.
Blandford, R. D. 1982. *Radiation Hydrodynamics* Proc IAU Colloquium 115. Reidel,
    Dordrecht, Holland.
Blandford, R. D. and Payne, D. G., 1982. *Mon. Not. R. astr. Soc.*,**199**,883.
Blandford, R. D. and Znajek, R. L. 1977. *Mon. Not. R. astr. Soc.*,**179**,433.

Bookbinder, J. A. and Lamb, D. Q. 1988. *Astrophys. J. (Lett.)*,**323**,L131.

Bouvier, J., Bertout, C., Benz, W. and Mayor, M.. *Astr. Astrophys.*,**165**,110.

Bridle, A. H. and Perley, R. A. 1984. *Ann. Rev. Astr. Astrophys.*,**22**,319.

Camenzind, M., 1986a. *Astr. Astrophys.*,**156**,137.

Camenzind, M., 1986b. *Astr. Astrophys.*,**162**,32.

Camenzind, M., 1987. *Astr. Astrophys.*,**184**,341.

Camenzind, M. 1989. *Magnetised Accretion Disks.* ed G. Belvedere. in press.

Cannizzo, J. K. and Pudritz, R. E. 1988. *Astrophys. J.*,**327**,840.

Chan, K. L. and Henriksen, R. N. 1982. *Astrophys. J.*,,.

Chanmugan, G. 1987. *Astrophys. Sp. Sci.*,**130**,53.

Coroniti, F. V. 1985.*Unsteady Current Systems in Astrophysics, Proc. IAU Symposium No.???. ed. G. D. Holman and M. R. Kundu.* Reidel, Dordrecht, Holland.

Damour, T. 1978. *Phys. Rev. D.*,**18**,3598.

Drew, J. E. 1987. *Mon. Not. R. astr. Soc.*,**224**,595.

Eardley, D. M. and Lightman, A. P. 1975. *Astrophys. J.*,**200**,187.

Edwards, S., Cabrit, S., Strom, S. E., Heyer, I. and Strom, K. M., 1987. *Astrophys.
J.*,**321**,473.

Evans, C. R. and Hawley, J. F. 1989. *Phys. Rev. D.*,,in press..

Frank, J. H., King, A. R. and Raine, D. J. 1986. *Accretion Power in Astrophysics*,
Cambridge University Press, Cambridge.

Galeev, A., Rosner, R. and Vaiana, G. S. 1979. *Astrophys. J.*,**229**,318.

Geldzahler, B. J. and Fomalont, E. 1986. *Astrophys. J.*,**311**,805.

Ghosh, P. and Lamb, F. K., 1979. *Astrophys. J.*,**234**,296.

Goldreich, P. and Julian, W. H., 1976. *Astrophys. J.*,**160**,971.

Gunn, J. E. 1978. *Active Galactic Nuclei*, ed. C. Hazard and S. Mitton. Cambridge
University Press, Cambridge.

Haeverts, J. and Norman, C. A., 1989, preprint.

Hartmann, L., Hewitt, R., Stahler, S. E., and Mathieu, R. D. 1986. *Astrophys. J.*,**309**,275.

Hartmann, L. and Kenyon, S. J. 1985. *Astrophys. J.*,**299**,462.

Hartmann, L. and MacGregor, K. B. 1982. *Astrophys. J.*,**259**,180.

Hasinger, G. 1989. *The Physics of Neutron Stars and Black Holes.*, ed. Y. Tanaka,
ISAS, Tokyo, Japan.

Kennel, C. F., Fujimura, F. S. and Okamoto, I. 1983. *Geophys. and Astrophys. Fluid.
Dyn.*,**26**,147.

Kenyon, S. and Hartmann, L. 1987. *Astrophys. J.*,**299**,462.

Kenyon, S. and Hartmann, L. 1988. *Workshop on Pulsation and Mass Loss*, Trieste.

Koen, C. 1986. *Mon. Not. R. astr. Soc.*,**223**,529.

Kompaneets, A. S., 1960. *Sov. Phys. Doklady*,**5**,46.

Königl, A. 1989.*Astrophys. J.*,(inpress).

Kundt, W. (ed.)*Astrophysical Jets and Their Engines.* Nature Advanced Science Institute 208, Reidel, Dordrecht, Holland.

Lada, C. J., 1985. *Ann. Rev. Astr. Astrophys.*,**23**,267.

Lago, M. T. V. T., 1984. *Mon. Not. R. astr. Soc.*,**210**,323.

Lovelace, R. V. E. 1976. *Nature*,**262**,649.

Lovelace, R. V. E. 1987. *Magnetic Fields in Extragalactic Objects*, ed. E. Asseo and
D. Gresillon. Editions de Physique, Les Vils Cedex, France.

Lovelace, R. V. E., Mehanian, C., Mobarry, C. M. and Sulkanen, M. E.,. *Astrophys. J.
Suppl.*,**62**,1.

Lovelace, R. V. E., Mobarry, C. M. and Contopoulos, J. 1989. *Accretion Disks in
Astrophysics* ed. G. Belvedere, (in press).

Lovelace, R. V. E., Wang, J. C. L. and Sulkanen, M. E. 1987. *Astrophys. J.*,**315**,504.

Lynden-Bell, D. and Pringle, J. E. 1984. *Mon. Not. R. astr. Soc.,***168**,603.

MacDonald, D. and Thorne, K. S. 1982. *Mon. Not. R. astr. Soc.,***188**,345.

Margon, B., 1984. *Ann. Rev. Astr. Astrophys.,***22**,507.

Mauche, C. W. and Raymond, J. C., 1987. *Astrophys. J.,***323**,690.

Mestel, L. 1961. *Mon. Not. R. astr. Soc.,***122**,473.

Mestel, L., Robertson, J. A., Wang, Y. M. and Westfold, K. C. 1985.. *Mon. Not. R. astr. Soc.,***217**,443.

Mészáros, P., Meyer, F. and Pringle, J. E. 1977. *Nature,***268**,420.

Michel, F. C. 1982. *Rev. Mod. Phys.,***54**,1.

Mouschovias, T. Ch.,1979. *Astrophys. J.,***228**,475.

Mundt, R., 1985. *Protostars and Planets II,*,ed D. C. Black and M. S. Mathews, Tucson: University of Arizona Press.

Natta, A., Giovanardi, C., Palla, F.,and Evans, N. .J. 1988. *Astrophys. J.,***346**,168.

Ostriker, J. P. 1983. *Astrophys. J.,***273**,99.

Parker, E. N., 1979.*Cosmical Magnetic Fields*, Oxford University Press, Oxford, England.

Phinney, E. S. 1983. *A Theory of Radio Sources*, unpublished thesis, University of Cambridge.

Pringle, J. E. 1989. *Mon. Not. R. astr. Soc.,***236**,107.

Pudritz, R. E., 1981a. *Mon. Not. R. astr. Soc.,***195**,881.

Pudritz, R. E., 1981b. *Mon. Not. R. astr. Soc.,***195**,897.

Pudritz, R. E., 1985. *Astrophys. J.,***293**,216.

Pudritz, R. E. and Norman, C. A., 1983. *Astrophys. J.,***274**,677.

Pudritz, R. E. and Norman, C. A., 1986. *Astrophys. J.,***301**,571.

Punsley, B. and Coroniti, F. V., 1989a. *prd*, (in press).

Punsley, B. and Coroniti, F. V., 1989b.*Astrophys. J.* (in press).

Rees, M. J., 1987. *Mon. Not. R. astr. Soc.,***228**,47p.

Reid, M. J., Biretta, J. A., Junor, W., Muxlow, T. W. B. and Spencer, R. E., 1989. *Astrophys. J.,***336**,112.

Sakimoto, P. and Coroniti, F. V., 1981. *Astrophys. J.,***247**,19.

Sakurai, T., 1985. *Astr. Astrophys.,***152**,121.

Sakurai, T., 1987. *Publ. astr. Soc. Japan,***39**,821.

Sargent, A. I. and Beckworth, S., 1987. *Astrophys. J.,***323**,294.

Shakura, N. I. and Sunyaev, R. A. 1973. *Astr. Astrophys.,***24**,337.

Shibata, K. and Uchida, Y.,1985. *Publ. astr. Soc. Japan,***37**,31.

Shibata, K. and Uchida, Y.,1986. *Publ. astr. Soc. Japan,***38**,631.

Shibizaki, N. and Lamb, F. K. 1987. *Astrophys. J.,***318**,867.

Shu, F. H., Adams, F. and Lizano, S. 1987. *Ann. Rev. Astr. Astrophys.,***25**,23.

Shu, F. H., Lizano, S.Ruden, S. P. and Najita, J., 1988. *Astrophys. J. (Lett.),***328**,L19.

Strom, K. M., Strom, S. E., Wolff, S. C., Morgan, J., Wenz, M. 1986. *Astrophys. J. Suppl.,***62**,39.

Strom, R. G., van Paradijs, J. and van der Klis, M. 1988. *Nature,***337**,234.

Thorne, K. S., Price, R. M. and MacDonald, D. (ed.) 1986. *Black Holes. The Membrane Paradigm,*,Yale University Press, New Haven.

van Ballegooijen, A. 1989, preprint.

Weber, E. and Davies, L., 1967. *Astrophys. J.,***148**,217.

White, N. E. 1989. *Astron. Astrophys. Rev.,***27**,in press.

Yusef-Zadeh, F. and Morris, M. 1988. *Astrophys. J.,***329**,729.

Zachary, A. L. and Colella, P. 1989. *Jour. Comp. Phys.* (in press).

Zel'dovich, Ya. B., Ruzmaikin, A. A. and Sokoloff, D. D., 1983. *Magnetic Fields in Astrophysics*, Pergamon Press, Oxford, England.

Znajek, R. L. 1977. *Mon. Not. R. astr. Soc.*,**179**,457.
Znajek, R. L. 1978. *Mon. Not. R. astr. Soc.*,**185**,833.

FORMATION OF RELATIVISTIC MHD JETS
IN THE MAGNETOSPHERE OF ACCRETION DISKS.

Max CAMENZIND
Landessternwarte Königstuhl, Heidelberg, FRG.

1. Magnetized Jets in Quasars.

Quasars and Radio Galaxies are most probably powered by accretion onto super-massive black holes. These objects are however also sources of strong plasma out-flows which are driven from a region of a fraction of a parsec into the kiloparsec-domain (for this see e.g. the inner M87 jet, Reid et al. 1988, and the VLBI jets of superluminal sources, Porcas 1987). Accretion and jet outflows must therefore exist simultaneously in one and the same source. The properties of these jets in Quasars and other radio sources cannot be explained without the inclusion of mag-netic fields. The most natural way to carry away magnetic flux is through the formation of a magnetosphere of the accretion disk itself.

Blandford and Payne (1982) presented a self-similar solution of the Newtonian MHD equation for this particular case. We are, however, forced to use *relativistic MHD* for two basic reasons. First of all, the gravitational background of a rotating black hole is completely different from a Newtonian approximation, since the inner part of the disk, where most of the energy is produced, is too close to the event horizon. The second reason is even more compelling; the magnetic field lines em-anating from this inner part of the accretion disk have a high rotation speed with the result that the corresponding light cylinder $R_L = c/\Omega^F$ is extremely close to the injection point for the plasma flow. Newtonian MHD is, on the other hand, the limit of relativistic MHD only when all characterisitic radii including the jet radius itself are far inside the light cylinder. This is definitely not true for relativistic jets.

The problem of jet formation and jet propagation must be attacked therefore in terms of a fully relativistic treatment. There are essentially two ways of dealing with this problem. Method number one consists in looking for time-dependent solu-tions of at least 2D relativistic MHD. This is an intriguing challenge for the future. Method number two is somewhat easier and consists in searching for equilibrium solutions for *axisymmetric* configurations. This method has been successfully ap-

F. Meyer et al. (eds.), Theory of Accretion Disks, 59–63.
© *1989 by Kluwer Academic Publishers.*

plied in plasma physics. In the following we present a few results on this second way, which are intimately related to the physics of the accretion disk itself.

2. Relativistic Axisymmetric MHD.

The MHD approach is well justified for the description of plasma flows in jets, since the plasma density is many orders of magnitude above the Goldreich-Julian density for rotating magnetospheres. For this we consider systems consisting of an accretion disk surrounding a rapidly rotating supermassive Kerr black hole (see e.g. Fig. 1 in Camenzind 1989). We assume that the disk carries its own magnetic structure produced either by magnetic fields convected inwards in a disk of finite conductivity (Lovelace et al. 1987) or by a kind of dynamo action (Meyer and Meyer-Hofmeister 1983). The treatment of this problem is beyond the scope of the present investigation mainly due to a lack of a suitable understanding of the conductivity in the plasma of the disk. We assume therefore that the surface of the disk is covered by a particular distribution of magnetic flux. Part of the inner field lines are then also dragged onto the hole, and the inner accretion occurs in a magnetically structured boundary layer (see also Fig. 1 in Camenzind 1989). A corona above the disk initiates then a plasma flow along this global magnetosphere, which could extract the entire angular momentum from the disk (see Blandford, these Proceedings). Since geometrically thin disks have a well defined axisymmetric structure, axisymmetry is also a natural ingredient for the treatment of the disk outflow.

In the last years, we developed a new formalism suitable to treat axisymmetric and stationary plasma flows on arbitrary stationary and axisymmetric space-times represented in terms of the general metric for rapidly rotating objects (Carter 1973)

$$ds^2 = e^{2\nu}\, dt^2 - e^{2\psi}\, (d\phi - \omega\, dt)^2 - e^{2\mu}\, d\theta^2 - e^{2\lambda}\, dr^2 \,. \tag{1}$$

These objects include Kerr black holes or rapid rotators and rapidly rotating neutron stars. The details of this formalism can be found in Camenzind (1986a,b, 1987, 1989). A key ingredient is the concept of magnetic surfaces for axisymmetric configurations (see e.g. Okamoto 1978). The corresponding magnetic field can always be represented in terms of a magnetic flux function $\Psi = RA_\phi$ and the toroidal magnetic field B^ϕ

$$B = \frac{1}{R}\, (\nabla\Psi \wedge e_T) + B^\phi e_T \,. \tag{2}$$

In ideal MHD, the plasma is then streaming along these magnetic surfaces, and the entire *jet consists of a family of nested magnetic surfaces.*

Once we know the structure of a magnetic surface, we can solve the equations of motion for the plasma flow along that surface. By using general conservation laws (Bekenstein and Oron 1978), this equation can be brought into the form of a polynomial of degree 16 for the relativistic poloidal velocity u_p (polytropic index $\Gamma = 5/3$)

$$\sum_{n=0}^{16} A_n(\frac{R}{R_L}, E, L, T_*; \frac{\Phi}{\sigma_*}; g_{tt}, g_{t\phi}, g_{\phi\phi}) u_p^{n/3} = 0, \qquad (3)$$

or to a polynomial of degree 14 for electron-positron flows (Camenzind and Endler 1989). In this equation, R is the cylinder radius along the flux surface, $R_L = c/\Omega^F(\Psi)$ the light cylinder of the flux surface, E and L are the total energy and angular momentum carried away, T_* the initial temperature at the injection point R_*, and the g's are the metric coefficients of the background space-time. Two quantities turned out to be crucial for the question of the final Lorentz factor γ_∞ which can be achieved in these types of outflows. The first quantity is Michel's magnetization parameter σ_*, roughly defined as

$$\sigma_* \simeq \frac{\Phi_*^2}{2\dot{N}mcR_L^2} \simeq 10\,\Phi_{*,34}^2\,\dot{M}_{-1}^{-1}\,R_{L,16}^{-2} \qquad (4)$$

for a plasma flow of $\dot{M} \simeq 0.1\,M_\odot\,yr^{-1}$. The maximal magnetic flux carried by Quasar jets is estimated to be of order $\Phi_* \simeq 10^{34}$ Gauss cm^2, and the light cylinder R_L of the inner accretion disk around a supermassive black hole of 10^9 M$_\odot$ is roughly 10 Schwarzschild radii. It is well-known that in a split-monopole geometry, the asymptotic Lorentz factor γ_∞ follows from σ_*, $\gamma_\infty = \sqrt{1 + \sigma_*^{2/3}}$. The above hot wind equation on the background of a Kerr geometry with $\Phi = const$ and $\sigma_* = 20.0$ has various solution branches. We assume that the injection of plasma occurs near the radius of maximum energy production in the disk, and the field lines are assumed to rotate with the angular frequency of their foot points. The physical wind solution starts near the slow magnetosonic point, crosses an unphysical branch at the Alfvén point near the light cylinder and would cross it a second time at the fast magnetosonic point which is for cold flows practically at infinity. Despite a high initial magnetization, only moderate Lorentz factors are achieved beyond the light cylinder.

A split-monopole must be, however, a very poor approximation for a realistic situation. For this reason, one has to calculate self-consistently the structure of the magnetic surfaces according to the Grad-Schlüter-Shafranov equation for the magnetic flux function Ψ (here only given for a flat space-time)

$$R\nabla \cdot \left\{ \frac{1}{R} \nabla \Psi \right\} = \frac{4\pi}{c} j_\phi. \qquad (5)$$

The form of the toroidal current density follows thereby from a general force-balance perpendicular to the magnetic surfaces (Camenzind 1987, $x = R/R_L$)

$$\frac{1}{c} B_p j_\phi (1 - M^2 - x^2) = -\frac{RB_\phi}{4\pi R^2} + \rho_e E_n$$
$$- \mu\gamma n(\nabla_n \gamma - \Omega\nabla_n l) - \nabla_n P + \mu\eta B_p^2 \nabla_n(\eta/n). \qquad (6)$$

The first two terms on the right hand side determine the force-free part of this current, while the other terms are due to the finite inertia of the plasma. In particular, electric forces cannot be neglected on rapidly rotating magnetospheres. The force-free limit of this equation has first been derived by Scharlemann and Wagoner (1973) (see also Michel (1982) for applications in pulsar theory). The above equation shows in particular that inertia represented by the Lorentz factor γ and the specific angular momentum l plays the same role as pressure forces.

Parts of the above current can be absorbed into the left hand side of the Grad-Shafranov operator. In this way we obtain the relativisitic version of the Grad-Schlüter-Shafranov (GSS) equation

$$R\nabla \cdot \left\{ \frac{1 - M^2 - x^2}{R^2} \nabla \Psi \right\} = -\frac{4\pi}{c} j \qquad (7)$$

with a reduced current density j containing force-free terms and terms from the finite inertia. The Alfvén surface appears explicitly as a critical surface of the GSS equation, though the equation remains elliptic around this Alfvén surface. This inner boundary of the GSS equation gives in general a quite complicated restriction for the vertical component of the magnetic field, which is different from the condition used by Sakurai (1985, 1987) in the Newtonian limit.

3. Magnetized Outflows from Accretion Disks.

In the last years we developed a numerical scheme to solve iteratively the GSS equation simultaneously with the hot wind equation (3) (Camenzind 1987). The inner boundary conditions consist in the prescription of a flux distribution along the surface of the disk, a particular injection of plasma from the disk-corona and a finite temperature for this corona. As a result of the high value for the magnetization, the problem is nearly force-free, but not exactly force-free (the Alfvén radius is typically at 0.98 light cylinder radii). In particular, the boundary conditions at the Alfvén surface squeeze considerably the magnetosphere inside the light cylinder with a corresponding rapid opening just beyond the light cylinder (Camenzind 1989).

This particular behaviour of the magnetic surfaces has a great influence on the flowing plasma. The winding up of the toroidal field outside of the light cylinder produces a strong post-acceleration for the plasma flow and at the same time a strong reduction of the magnetic energy (Camenzind and Endler 1989). For the special range of parameters found in Quasar jets, we obtain rougly for the terminal Lorentz factor $\gamma_\infty \leq \sigma_*$. This means that we obtain much higher Lorentz factors and much lower Poynting fluxes in a realistic geometry than in a split-monopole for the same injection parameters. We also find that the force-free condition is strongly violated around the light cylinder, where the plasma decouples from the magnetic structure.

MHD jets produced in rapidly rotating magnetospheres of accretion disks can explain the range of Lorentz factors found in superluminal sources, $1.2 < \gamma_{jet} < 20$, with the parameters of typical accretion disks around supermassive black holes. The strong collimation observed in these sources is a natural consequence of the pressure in the ambient medium of galactic nuclei, and the corresponding jet radii are typically $10R_L < R_{jet} < 100R_L$. *The light cylinder of the magnetosphere is the natural scale of the problem.* Jets formed in this way also carry an appreciable amount of magnetic energy, which is an interesting input for the physics of the hot spots formed at the head of these jets (Appl and Camenzind 1988).

As already discussed above, VLBI observations show that Quasar jets have a diameter of a fraction of a light year. This is consistent with the self-absorbtion frequency of about 50 GHz measured e.g. for the VLBI jet of 3C 273 (Camenzind and Courvoisier 1984). Better resolution could then improve our understanding of the emission features seen in these jets. The various components in the spectrum of this Quasar also find a natural interpretation in terms of these magnetically confined outflows (Courvoisier and Camenzind 1989).

ACKNOWLEDGEMENTS. The research reported in this paper is partly supported by the Sonderforschungsbereich 328 in Heidelberg.

REFERENCES.

Appl, S., Camenzind, M.: 1988, *Astron. Astrophys.* **206**, 258
Bekenstein, J.D., Oron, E.: 1978, *Phys. Rev.* **D18**, 1809
Blandford, R.D., Payne, D.G.: 1982, *Mon. Not. Roy. Astron. Soc.* **199**, 883
Camenzind, M.: 1986a, *Astron. Astrophys.* **156**, 137
Camenzind, M.: 1986b, *Astron. Astrophys.* **162**, 32
Camenzind, M.: 1987, *Astron. Astrophys.* **184**, 341
Camenzind, M.: 1989, in *Accretion Disks and Magnetic Fields in Astrophysics*, ed. G. Belvedere, (Kluwer), in press
Camenzind, M., Courvoisier, T.J.-L.: 1984, *Astron. Astrophys.* **140**, 341
Camenzind, M., Endler, M.: 1989, to appear in *Astron. Astrophys.*
Carter, B.: 1973, in *Black Holes*, ed. C. DeWitt and B. DeWitt (Gordon and Breach, New York)
Courvoisier, J.L.-T., Camenzind, M.: 1989, *Astron. Astrophys.*, in press
Kennel, C.F., Fujimura, F.S., Okamoto, I.: 1983, *J. Ap. Geophys. Fluid Dyn.* **26**, 147
Lovelace, R.V.E., Wang, J.C.L., Sulkanen, M.E.: 1987, *Astrophys. J.* **315**, 504
Meyer, F., Meyer-Hofmeister, E.: 1983, *Astron. Astrophys.* **128**, 420
Michel, F.C.: 1982, *Rev. Mod. Phys.* **54**, 1
Okamoto, I.: 1978, *Mon. Not. Roy. Astron. Soc.* **185**, 69
Porcas, R.W.: 1987, in *Superluminal Radio Sources*, ed. J.A. Zensus and T.J. Pearson, (Cambridge University Press, Cambridge), p. 12
Reid, M.J., Biretta, J.A., Junor, W., Muxlow, T.W.B., Spencer, R.E.: 1988, *Astrophys. J.* **336**, 112
Sakurai, T.: 1985, *Astron. Astrophys.* **152**, 121
Sakurai, T.: 1987, *Publ. Astron. Soc. Japan* **39**, 821
Scharleman, E.T., Wagoner, R.V.: 1973, *Astrophys. J.* **182**, 951

NONSTEADY MHD JETS FROM MAGNETIZED ACCRETION DISKS
– SWEEPING-MAGNETIC-TWIST MECHANISM –

Kazunari SHIBATA[1] and Yutaka UCHIDA[2]

[1]Department of Earth Sciences, Aichi University of Education,
 Kariya, Aichi 448, Japan
[2]Department of Astronomy, University of Tokyo,
 Bunkyo-ku, Tokyo 113, Japan

ABSTRACT. By using two-dimensional axisymmetric MHD numerical simulations, we present an MHD mechanism (called *sweeping-magnetic-twist* mechanism) for the formation of astrophysical jets in which the directed flow are ejected along the rotation axis of an accretion disk. The acceleration of jet is due to the $\mathbf{J} \times \mathbf{B}$ force in the relaxing magnetic twist, which is produced by the interaction of the differentially rotating accretion disk with poloidal magnetic fields. It is shown that the accelerated gas is collimated by the $\mathbf{J} \times \mathbf{B}$ force and forms a supersonic bipolar jet which has a hollow cylindrical shell structure with helical motion in it. Interaction of the MHD jet with an interstellar condensation is also examined. Application to bipolar molecular flows in star forming regions is briefly discussed.

1. Introduction.

It has been revealed that there are collimated high-speed bipolar jets (sometimes called "astrophysical jets") having widely different scales in the universe, such as radio jets associated with active galactic nuclei, relativistic jets from SS433, bipolar molecular flows and optical jets in star forming regions. One of the most conspicuous characteristics common to these astrophysical jets is their configurations: the collimated supersonic flows are accelerated in the bipolar direction along the rotation axis of accretion disks found at the origin of these jets (e.g. Begelman *et al.* 1984). We have proposed a possible mechanism which may explain astrophysical jets in widely different scales (Uchida and Shibata 1985, 1986; Uchida *et al.* 1985; Shibata and Uchida 1986, 1987). In our mechanism, the acceleration and collimation of jets is due to the $\mathbf{J} \times \mathbf{B}$ force in a propagating magnetic twist which is generated by the winding-up effect of poloidal magnetic field lines by the differentially rotating accretion disk. Since the nonsteady relaxation and propagation of magnetic twist is essential in this mechanism, we called it the *sweeping-magnetic-*

65

F. Meyer et al. (eds.), Theory of Accretion Disks, 65–69.

66

twist mechanism. This mechanism is different from that of the centerifugal wind models (Blandford and Payne 1982; Pudritz and Norman 1983; Lovelace *et al.* 1987), but the centrifugal effect is automatically built-in.

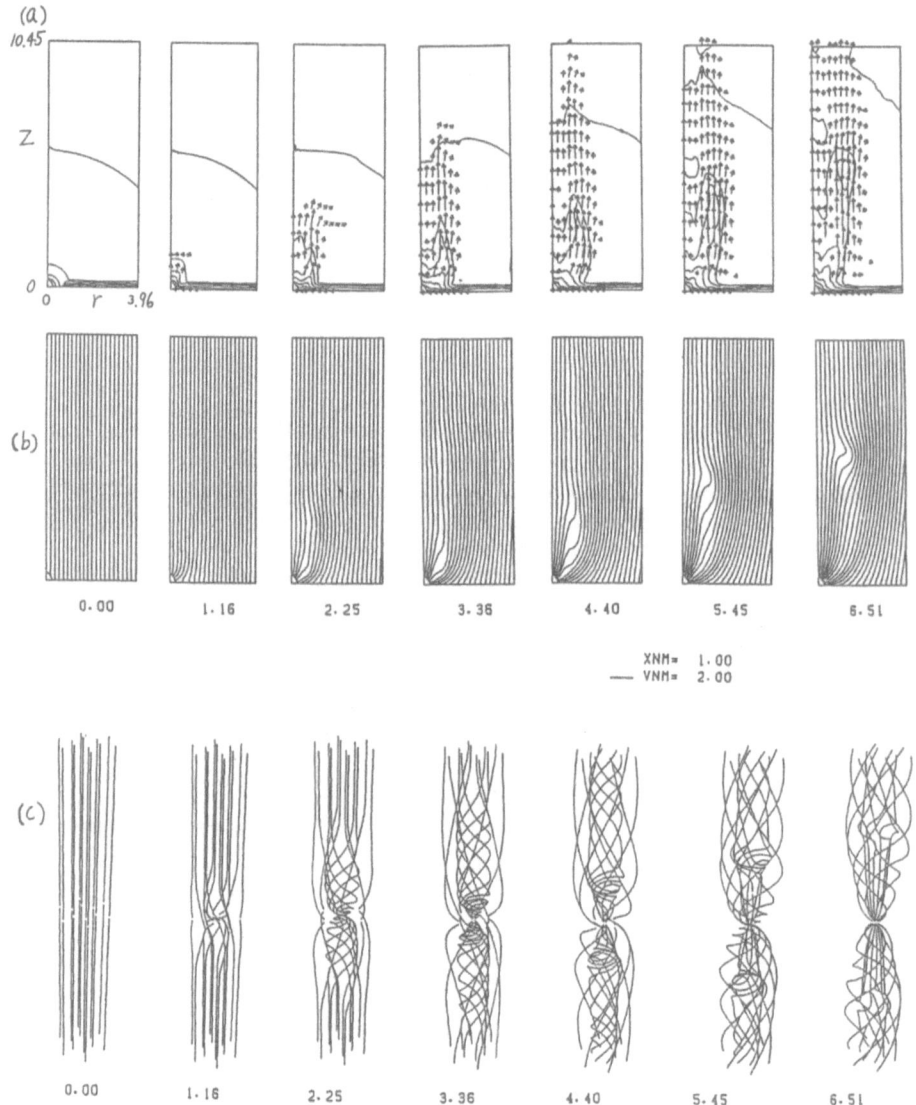

Figure 1: Typical model jet in the sweeping-magnetic-twist mechanism.
(a) The density contours plus velocity vectors
(b) The poloidal magnetic field lines
(c) Three dimensional configurations of magnetic field lines

2. Typical Model Jet in the Sweeping-Magnetic-Twist Mechanism.

The nonsteady MHD numerical simulations (see Shibata and Uchida [1986, 1989] for details of numerical methods) are performed on the assumption of (1) two-dimensional axisymmetry with respect to the rotation axis of the disk, (2) ideal MHD (adiabatic , frozen), (3) in a gravitational potential due to a point mass (self-gravity of the disk is not included). The initial magnetic field is assumed to be uniform and parallel to the rotation axis of the disk whose rotation velocity is equal to or smaller than the Keplerian value. Outside the disk, a hot tenuous corona is assumed to exist. The formulation was made in a non-dimensional scale-free representation, and hence similarity is expected if the initial situation in the relative coordinate $\mathbf{r}/r_d$ is similar, and if the set of non-dimensional coefficients of the equation, $R_1 = (V_s/V_k)^2, R_2 = (V_A/V_k)^2, R_3 = (V_\varphi/V_k)^2$ and the ratio of the coronal and disk temperatures ($= R_4$) are the same, where V_s, V_A, V_k and V_φ are initial sound, Alfven, Kepler, and rotational velocities at $r = r_d$ the inner edge of the initial disk.

Figure 1 shows the time evolution of a typical model jet with parameters ($R_1 = 3 \times 10^{-3}, R_2 = 7.2 \times 10^{-3}, R_3 = 0.64, R_4 = 400$), suitable for bipolar flows in star forming regions, though the model is not to the exact scaling to be compared directly with the observed molecular bipolar flows. Times are in unit of the free fall time at r_d. As the disk contracts toward the center, magnetic field lines (Fig. 1c) are pulled toward the inner region as well as to the azimuthal direction to produce helically twisted field lines. When the magnetic twist becomes sufficiently large, the gas in the surface layer of the disk starts to be accelerated upward due to magnetic pressure gradient ($\nabla B_\varphi{}^2/8\pi$) in a magnetic twist, generating supersonic bipolar jet which has a hollow cylindrical shell structure with helical motion in it. The maximum velocity of the jet is $\sim 1.3V_k$, comparable to the local Alfven speed in the jet, but is much larger than the local sound speed ($\sim 0.07V_k$). In the later phase, we find a low density flow preceding the dense jet. This flow is accelerated by the $\mathbf{J} \times \mathbf{B}$ force associated with the propagation of torsional Alfven wave.

The kinetic energy of a jet derives itself from the gravitational potential energy released in the contraction of the disk through the action of the helically twisted magnetic field, and the rate of conversion is about 10 percent. Since the angular momentum loss of the disk becomes more significant due to the production of the jet, the disk can continue to contract toward the center.

3. Interaction of the Jet with Interstellar Condensations.

Recent observations of detailed structure of L1551 CO bipolar flow (Uchida *et al.* 1987) revealed that the flow has a hollow cylindrical structure with a helical motion in it, both of which are consistent with the characteristics of the sweeping-magnetic-twist model discussed above. Moreover, the observations showed that

68

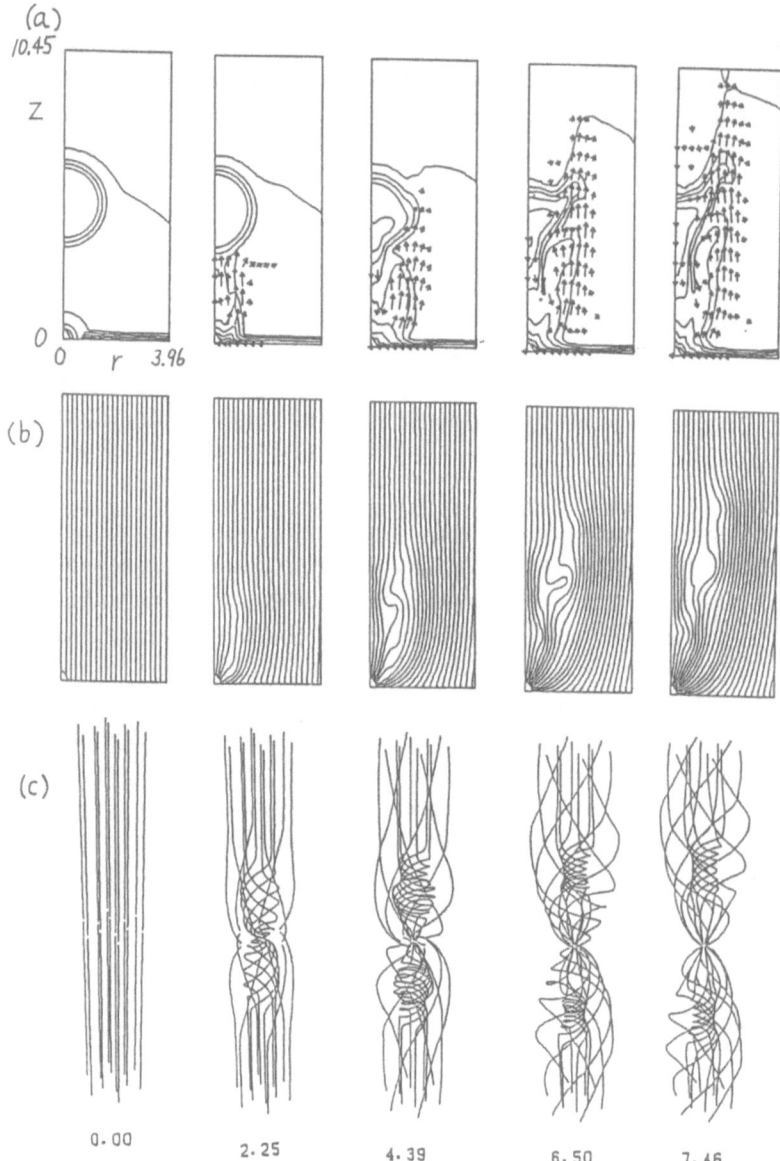

Figure 2: Interaction of the magnetic-twist jet with a spherical dense cloud.
(a) The density contours plus velocity vectors
(b) The poloidal magnetic field lines
(c) Three dimensional configurations of magnetic field lines

there is a dense blob structure in the flow, suggesting the interaction of the flow with pre-existed dense condensations in molecular cloud or interstellar space.

Motivated by this observation, we studied the interaction of the magnetic-twist jet with a dense cloud using the same model discussed in the previous section (Shibata and Uchida 1989). The only difference from the above model is that a spherical dense cloud ($\rho_{cloud}/\rho_{ambient} = 90$) is initially placed ahead of the jet on the z-axis as shown in Figure 2. Since the Alfven speed in the cloud is small, the magnetic twist is accumulated between the jet and the cloud, and finally the helical twist pinches and compresses the lower half of the cloud. On the other hand, the ourter part of the cloud is stripped off by the magnetic pressure gradient in the accumulated magnetic twist.

These nonlinear effects explain some of the features observed in L1551 flow (Shibata and Uchida 1989). Furthermore, if the cloud mass is nearly critical for the gravitational contraction , the cloud may collapse due to the increase in the self-gravity in the compression by the magnetic pinch effect. This jet-induced star formation may be the origin of the observed chain-like distribution of young stars (Uchida *et al.* 1989).

REFERENCES.

Begelman, M. C., Blandford, R. D. and Rees, M. J. 1984, *Rev. Mod. Phys.*, **56**, 255

Blandford, R. D. and Payne, D. G. 1982, *Mon. Not. R. Astr. Soc.*, **199**, 883

Lovelace, R. V. E., Wang, J. C. L. and Sulkanen, M. E. 1987, *Ap. J.*, **315**, 504

Pudritz, R. E. and Norman, C. 1983, *Ap. J.*, **274**, 677

Shibata, K. and Uchida, Y. 1986, *Publ. Astron. Soc. Japan*, **38**, 631

Shibata, K. and Uchida, Y. 1987, *Publ. Astron. Soc. Japan*, **39**, 559

Shibata, K. and Uchida, Y. 1989, submitted to *Publ. Astron. Soc. Japan*

Uchida, Y. and Shibata, K. 1985, *Publ. Astron. Soc. Japan*, **37**, 515

Uchida, Y., Shibata, K. and Sofue, Y. 1985, *Nature*, **317**, 699

Uchida, Y. and Shibata, K. 1986, *Can. J. Phys.*, **64**, 507

Uchida, Y., Kaifu, N., Shibata, K., Hasegawa, T., Hayashi, S. S., and Hamatake, H. 1987, *Publ. Astron. Soc. Japan*, **39**, 907

Uchida, Y., Mizuno, R. and Fukui, Y. 1989, submitted to *Publ. Astron. Soc. Japan*

STATISTICAL MODELLING OF IR/UV SPECTRA IN AGN.

Z. LOSKA and B. CZERNY
Nicolaus Copernicus Astronomical Center,
Bartycka 18, 00–716 Warsaw, Poland

ABSTRACT. We model IR/UV spectra of AGN assuming that the radiation originates in accretion discs, relativistic electrons, and dust. Within a broad range of parameters, the properties of the theoretical sample are similar to the properties of the Palomar–Green survey. The presence of a 'break' frequency at $log(\nu) = 14.5$ is particularly well reproduced. Better agreement with the data is for accretion discs radiating at their Eddington limits.

1. Continuum Radiation of AGN.

The variety of AGN spectral shapes and their complex variability patterns suggest a multicomponent nature of AGN. QSO and Seyfert galaxy spectra in the IR/UV band have been interpreted as a superposition of the following elements: an IR/UV power law, a small (or blue) bump, a big bump, and a microwave bump. The radiation is supposed to come from relativistic electrons, irradiated cool gas, accretion discs and dust (Zdziarski 1986, Wills, Netzer & Wills 1985, Malkan 1983, Edelson & Malkan 1986).

Data for the IR/UV spectra for a number of quasars have been obtained by Neugebaucr et al. (1987). The contribution from lines and a small bump were subtracted. The spectra were fitted by two power laws, and diagrams of both slopes and crossing points were presented. The distribution of the crossing points showed a surprisingly small dispersion (~ 0.1) around the median value $log(\nu) = 14.5$ (Figure 6 in Neugebauer et al. 1987).

Here we present our attempt to model the statistical sample of Neugebauer et al. (1987) using 'reasonable' ranges of parameters of accretion discs, relativistic electrons, and dust.

2. Model Sample.

We assume that an accretion disc radiates locally as a black body. We adopt val-

F. Meyer et al. (eds.), Theory of Accretion Disks, 71–74.

ues: 0.01,0.1 and 1.0 for accretion rate in critical units ($\dot{M}_{crit} = 64\pi GMm_p/c\sigma_T$); $10^7, 10^8, 10^9$ and $10^{10}M_\odot$ for the mass of the central black hole; 0.1 - 1.0 with step 0.1 for cosine of the inclination angle.

We parametrize radiation of relativistic electrons by a single power law of the energy index α and a ratio ϵ_{PL} of the power law luminosity (between 13 and 17 in logarithmic frequency) to the bolometric luminosity of the disc. We adopt: α from the range -1.2 to -2.0 with step 0.2, and ϵ_{PL} from 0.1 to 0.5 with the step 0.1.

We assume that the dust radiates as a black body at the temperature 1500 K. For the ratio ϵ_D of dust luminosity to the sum of luminosities of the accretion disc and the power law component, we adopt values from the range 0.02 - 0.1, with the step 0.02.

We calculate the resulting radiation spectrum for every set of parameters M, $\dot{m}$, $cos(i)$, α_{PL}, ϵ_{PL}, ϵ_D from the ranges stated above. Uniform distributions of the parameters (i.e. linear or uniform steps) were assumed because of our convenience. Next we repeat the analysis of the observational data on our theoretical spectra, i.e. we fit the resulting spectra by two power laws and obtain both slopes α_1 (at lower frequencies) and α_2 (at higher frequencies) and the crossing point $log(\nu_{cr})$.

3. Comparison of the Theoretical and Observational Sample.

Histograms constructed for our standard sample of theoretical spectra are shown in Figure 1a. They clearly show the presence of a 'break' at about 14.5 in logarithmic frequency. Also the distribution of high energy slopes dominated by the contribution from accretion discs is fairly well reproduced.

In order to discuss the influence of adopted ranges of parameters on resulting histograms we also show limited samples with particular values for some parameters (Figures 1b and 1c).

Formally, we could construct a broader sample allowing the dust to contribute more than 10% to the total luminosity in the IR/UV range. This would help to obtain a much steeper low energy slope. However, in such case the self–consistent approach would require taking into account the effect of absorption by the dust on the UV part of the spectrum.

4. Conclusions.

Statistical properties of the Palomar–Green survey (Neugebauer et al. 1987) can be approximately reproduced assuming that the spectrum in the IR/UV band originates in an accretion disc, relativistic plasma, and dust. Our analysis indicates that:

1) the almost fixed position of the crossing points of two power law fits to the IR/UV data arises naturally for broad range of parameters,
2) all three components of the spectrum are necessary to explain its overall shape,

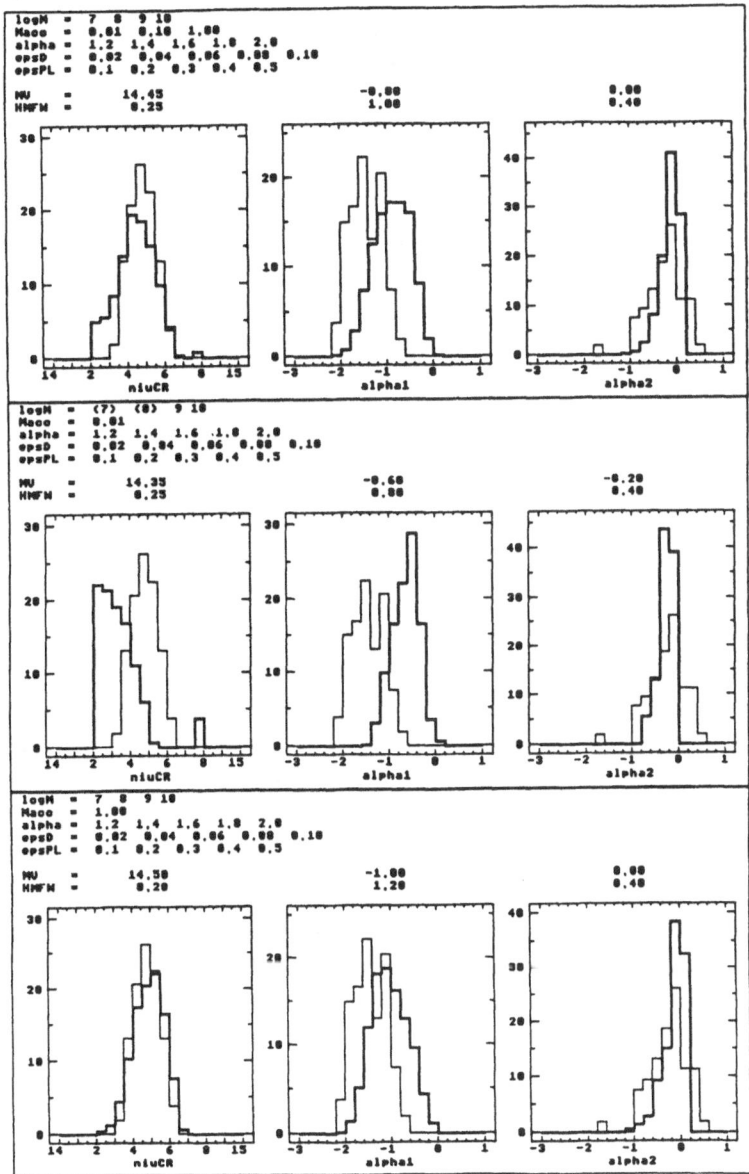

Figure 1: Histograms of the distributions of the low energy slopes α_1, high energy slopes α_2 and positions of crossing points in logarithmic frequency $log(\nu_{CR})$. Ranges of parameters are shown. Also given are median values and half-maximum full widths.

3) values of accretion rates $\dot{m} = 1$ (i.e. Eddington luminosity for the disc) are strongly favored,

4) bigger contributions from the dust and relativistic electrons are somewhat favored.

REFERENCES.

Edelson, R., & Malkan, M.A., 1986. *Ap. J.*, **308**, 59

Malkan, M.A., 1983. *Ap. J.*, **268**, 582

Neugebauer,G., Green, R.F., Matthews, K., Schmidt, M. Soifer, B.T., & Bennet, J., 1985.
   Ap. J. Suppl., **63**, 615

Wills, B.J., Netzer, H., & Wills, D., 1986. *Ap. J.*, **288**, 94

Zdziarski, A.A., 1986. *Ap. J.*, **303**, 94

FORMATION AND EVOLUTION OF THE SOLAR NEBULA.

Peter BODENHEIMER
Lick Observatory, Board of Studies in Astronomy and Astrophysics,
University of California, Santa Cruz, USA.

ABSTRACT. The protostellar phase of stellar evolution is of considerable importance in determining whether a solar nebula forms from the collapse of an interstellar cloud, what the physical properties of the nebula are at the onset of its evolution, what the dominant mechanisms for angular momentum transport will be during the subsequent evolution, and whether conditions are favorable for the formation of planets. The initial mass distribution and angular momentum distribution in the core of a molecular cloud determine whether a binary system or a single star is formed. A relatively slowly rotating and centrally condensed cloud is likely to collapse to a disk-like structure out of which planets can form. The above parameters then determine the temperature and density structure of the disk and the characteristics of the resulting planetary system.
    There has been considerable recent interest in two- and three-dimensional numerical hydrodynamical calculations with radiative transfer, applied to the inner regions of collapsing, rotating protostellar clouds of about 1 $M_\odot$. The calculations start at a density that is high enough so that the gas is decoupled from the magnetic field. Three-dimensional calculations show amplification of initial non-axisymmetric perturbations during collapse. If such perturbations are relatively small, angular momentum transport by gravitational torques is slow enough so that an axisymmetric approximation is sufficiently accurate to give useful results. Under the further assumption that angular momentum transport by turbulent viscosity is not important on a collapse time, calculations can be performed under the assumption of conservation of angular momentum of each mass element. Once the disk forms, however, transport processes must be included. This paper concentrates on the formation phase and its influence on the later evolutionary phases.
    With a suitable choice of initial angular momentum, the size of the disk is similar to that of our planetary system. The disk forms as a relatively thick, warm equilibrium structure, with a shock wave separating it from the surrounding infalling gas. The calculations give temperature and density distributions throughout the infalling cloud as a function of time. From these, frequency-dependent radiative transfer calculations produce infrared spectra and isophote maps at selected viewing angles. The theoretical spectra may be compared with observations of suspected protostellar sources, under the hypothesis that the observed objects actually represent precursors to "solar" nebulae.

1. Observational Constraints on the Properties of the Initial Solar Nebula.

From observations of physical and cosmochemical properties of the solar system

F. Meyer et al. (eds.), Theory of Accretion Disks, 75–87.
© 1989 by Kluwer Academic Publishers.

and from astronomical observations of star-forming regions and young stars certain constraints can be placed on the processes of formation and evolution of the solar nebula.

a) Low-mass stars form by the collapse of initially cold (10 K), dense (10^5 particles cm^{-3}) cores of molecular clouds. The close physical proximity of such cores with T Tauri stars, with imbedded infrared sources which presumably are protostars, and with sources with bipolar outflows, presumably coming from stars in a very early stage of their evolution, lends support to this hypothesis (Myers 1987).

b) The specific angular momenta (j) of the cores, where observed, fall in the range 10^{20} - 10^{21} cm^2 s^{-1} (Goldsmith and Arquilla 1985, Heyer 1988). In the lower end of this range, the angular momenta are consistent with the properties of our solar system: for example, Jupiter's orbital motion has $j \approx 10^{20}$ cm^2 s^{-1}. In the upper end of the range, collapse with conservation of angular momentum would lead to halt of collapse as a consequence of rotational effects at an outer radius of $\sim$ 500 AU, somewhat too large to account for the planetary orbits. In fact, hydrodynamical calculations suggest that collapse in this case is likely to lead to fragmentation into a binary or multiple system.

c) The infrared radiation detected in young stars indicates the presence of disks around these objects (reviewed by Hartmann and Kenyon 1988). A particularly good example, where orbital motions have been observed, is HL Tau (Sargent and Beckwith 1987). The deduced masses of the disk and star are 0.1 $M_\odot$ and 1.0 $M_\odot$, respectively. The radius of the disk is about 2000 AU. Roughly half of all young pre-main-sequence stars are deduced to have disks, mostly unresolved, with masses in the range 0.01-0.1 $M_\odot$ and sizes from 10 to 100 AU (Strom, Edwards, and Strom 1989).

d) The rotational velocities of T Tauri stars are small, typically 20 km s^{-1} or less (Hartmann et al. 1986). The distribution of angular momentum in the system consisting of such a star and disk is quite different from that in the core of a molecular cloud, which is generally assumed to be uniformly rotating with a power-law density distribution. Substantial angular momentum transport, from the central regions to the outer regions, must take place early in the evolution. The required transport is unlikely to occur during collapse; therefore it must occur during the disk evolution phase before the star emerges as a visible object.

e) The lifetime of the pre-main-sequence disks is difficult to determine from observation, but the phase during which infrared excess is present probably does not exceed 10^7 yr (Strom, Edwards, and Strom 1989). The mechanisms for angular momentum transport, which deplete disk mass by allowing it to fall into the star, must have time scales consistent with these observations as well as with time scales necessary to form gaseous giant planets.

f) The temperature conditions in the early solar nebula can be roughly estimated from the distribution of mass and chemical composition of the planets and the satellites (Lewis 1974). The general requirements are that the temperature be high enough in the inner regions to vaporize most solid material, and that it be low enough at the orbit of Jupiter and beyond to allow the condensation of ices. Theoretical models of viscous disks produce the correct temperature range, as do collapse models of disk formation with shock heating.

g) The evidence from meteorites is difficult to interpret in terms of standard nebular models. First, there is evidence for the presence of magnetic fields, and second, the condensates indicate the occurrence of rapid and substantial thermal fluctuations. Suggestions for explaining this latter effect include turbulent transport of material (Morfill *et al.* 1985), non-axisymmetric structure (density waves) in the disk (Boss 1989), or flare activity in the nebular corona (Levy and Araki 1989).

h) The classical argument, of course, is that the coplanarity and circularity of the planets' orbits imply that they were produced in a disk.

i) A large fraction of stars are observed to be in binary and multiple systems (Abt 1983); the orbital values of j in the closer systems are comparable to those in our planetary system. Several suggestions have been made regarding the conditions under which single stars form rather than binaries. Safronov and Ruzmaikina (1985) assume that the initial cloud is slowly rotating ($j \approx 10^{19}$ cm^2 s^{-1}); in this case most of the material of the protostellar cloud falls onto the central star and, later, angular momentum transport results in the formation of a disk. If j is somewhat higher ($\approx 10^{20}$ cm^2 g^{-1}) a single star is likely to form if the initial density distribution is centrally condensed (Boss 1987). However, observed cloud cores typically are centrally condensed (Myers *et al.* 1987), so that the above suggestion would lead to the conclusion that all stars are single, contrary to observation. In a discussion of this problem, Pringle (1989) points out that if a star begins collapse after having undergone slow diffusion across the magnetic field, it will be centrally condensed and will therefore form a single star. If, however, the collapse is induced by external pressure disturbances or collisions, the outcome is likely to be a binary. Still another possibility (Miyama 1989) is that single star formation occurs in initial clouds with $j \approx 10^{21}$ cm^2 s^{-1}. After reaching a rotationally supported equilibrium that is stable to fragmentation, the cloud becomes unstable to nonaxisymmetric perturbations, resulting in angular momentum transport and collapse of the central regions.

2. The Physical Problem.

The above considerations illustrate several of the important questions relating to the formation of the solar nebula: What are the initial conditions for collapse of a protostar? At what density does the magnetic field decouple from the gas? What conditions lead to the formation of a single star with a disk rather than a double star? Can the embedded IRAS sources be identified with the stage of evolution just after disk formation? What is the dominant mechanism for angular momentum transport that produces the present distribution of angular momentum in the solar system? Are the disklike configurations deduced to exist around young stellar objects suitable sites for planet formation? The goal of numerical calculations is to investigate these questions by tracing the evolution of a protostar from its initial state as an ammonia core in a molecular cloud to the final quasi-equilibrium state of a central star, supported against gravity by the pressure gradient, plus a circumstellar disk, supported in the radial direction primarily by centrifugal effects. A further goal is to predict the observational properties of the system at various times during the collapse. A full treatment would include a large number of physical effects: the hydrodynamics, in three space dimensions, of a collapsing rotating cloud with a magnetic field, the equation of state of a dissociating and ionizing gas of solar composition, cooling from molecules and grains in optically thin regions, frequency-dependent radiative transfer in optically thick regions, molecular chemistry, the generation of turbulent motions as the disk and star approach hydrostatic equilibrium, and the properties of the radiating accretion shock which forms at the edge of the central star and on the surfaces of the disk (see review by Shu, Adams, and Lizano 1987).

The complexity of this problem makes a general solution intractable even on the fastest available computing machinery, except under the simplifying assumption of spherical symmetry. For example, the length scales range from 10^{17} cm, the typical dimension of the core of a molecular cloud, to 10^{11} cm, the size of the central star. The density of the material that reaches the star undergoes an increase of about 15 orders of magnitude from its original value of $\sim 10^{-19}$ g cm^{-3}. Also, the numerical treatment of the shock front must be done very carefully. The number of grid points required to resolve the entire structure is very large in two space dimensions; in three dimensions it is prohibitively large. Even if the detailed structure of the central object is neglected and the system is resolved down to a scale of 0.1 AU, the Courant-Friedrichs-Lewy condition in an explicit calculation requires that the time step be less than one-millionth of the collapse time of the cloud, if the initial density for collapse is actually 10^{-19} g cm^{-3}. If the initial density is higher, the computation time decreases; determination of the actual value of this density therefore is of considerable practical interest.

A number of physical approximations and restrictions have been made in all recent numerical calculations of nebular formation. For example, magnetic fields have not been included, on the grounds that the collapse starts only when the gas has become almost completely decoupled from the field because of the negligible degree of ionization at the relatively high densities and cold temperatures involved. Also, in some calculations, turbulence has been neglected during the collapse. Even if it is present, the time scale for transport of angular momentum by this process is expected to be much longer than the dynamical time. It turns out that angular momentum conservation of individual mass elements is a reasonable approximation during the collapse. Three- dimensional effects, such as angular momentum transport by gravitational torques, are of importance during the collapse only if initial non-axisymmetric perturbations are quite large. In general, they become important later, during the phase of nebular evolution. Therefore, the two-dimensional (axisymmetric) approximation is useful, particularly since it allows calculations to be carried further and with more detail than in the 3-D case. A further approximation involves isolating and resolving only specific regions of the protostar. In one-dimensional calculations (e. g. Stahler, Shu, and Taam 1980) it has been possible to resolve the high-density core as well as the low-density envelope of the protostar. However, in two space dimensions, proper resolution of the region where the nebular disk forms cannot be accomplished simultanously with the resolution of the central star. In several calculations the outer regions of the protostar are also not included, so that the best possible resolution can be obtained on the length scale 1-50 AU. Thus the goal outlined above, the calculation of the evolution of a rotating protostar all the way to its final stellar state, has not yet been fully realized.

The stages of evolution of a slowly rotating protostar of about 1 $M_\odot$ can be outlined as follows:
a) The frozen-in magnetic field transfers much of the angular momentum out of the core of the molecular cloud, on a timescale of 10^7 yr.
b) The gradual decoupling between the magnetic field and the matter allows the gas to begin to collapse, with conservation of angular momentum.
c) The initial configuration is centrally condensed. During collapse, the outer regions, with densities less than about 10^{-13} g cm^{-3} remain optically thin and collapse isothermally at 10 K. The gas that reaches higher densities becomes optically thick, most of the released energy is trapped, and heating occurs.
d) The dust grains, which provide most of the opacity in the protostellar envelope, evaporate when the temperature exceeds 1500-2000 K. An optically thin region is created interior to about 1 AU. Further, at temperatures above 2000 K, the molecular hydrogen dissociates, causing renewed instability to collapse.
e) The stellar core and disk form from the inner part of the cloud. The remaining infalling material passes though accretion shocks at the boundaries of the core and disk; most of the infall kinetic energy is converted into radiation behind the shock. The surrounding infalling material is optically thick, and the object radiates in the infrared, with a peak at around 60-100 μm. During this phase, the object shows a steep negative spectral index in the 1-10 μm range in a plot of log (νF_ν) versus log ν (Adams, Lada, and Shu 1987).
f) A stellar wind is generated in the stellar core, by a process that is not well understood. The wind breaks through the infalling gas at the rotational poles, where the density gradient is steepest and where most of the material has already fallen onto the core. This bipolar outflow phase lasts about 10^5 yr.
g) Infall stops because of the effects of the wind or simply because the material is exhausted. The stellar core emerges onto the Hertzsprung-Russell diagram as a T Tauri star, still with considerable infrared radiation coming from the disk. Now, however, the spectral index in the [log ν, log (νF_ν)] plane is positive or flat in the

1-10 μm region.

h) The disk evolves, driven by processes that transfer angular momentum, on a timescale of 10^6 - 10^7 yr. Angular momentum is transferred outwards through the disk while mass is transferred inwards. The rotation of the central object slows, possibly through magnetic braking in the stellar wind. Possible transport processes in the disk include turbulent (convectively driven) viscosity, magnetic fields, and gravitational torques driven by gravitational instability in the disk or by non-axisymmetric instabilities in the initially rapidly rotating central star. Detailed calculations of the evolution of thin disks, under the assumption that convective turbulence is the transport mechanism, have been presented by Ruden and Lin (1986) and by Cabot *et al.* (1987a,b).

The following sections describe numerical calculations of phases b) through e), from the time when magnetic effects become unimportant to the time when at least part of the infalling material is approaching equilibrium in a disk.

3. Review of Hydrodynamic Calculations of the Formation Phase.

Modern theoretical work on this problem goes back to the work of Cameron (1962, 1963), who discussed in an approximate way the collapse of a protostar to form a disk. In a later work, Cameron (1978) solved numerically the one-dimensional (radial) equations for the growth and evolution of a viscous accretion disk, taking into account the accretion of mass from an infalling protostellar cloud. The initial cloud was assumed to be uniformly rotating with uniform density. The hydrodynamics of the inflow was not calculated in detail; rather, infalling matter was assumed to join the disk at the location where its angular momentum matched that of the disk. A similar approach was taken by Cassen and Summers (1983) and Ruzmaikina and Maeva (1986), who, however, took into account the drag caused by the infalling material, which has angular momentum different from that of the disk at the arrival point. The latter authors discuss the turbulence that develops for the same reason (see also Safronov and Ruzmaikina 1985). This section concentrates on full two- and three-dimensional calculations of the collapsing cloud during the initial formation of the disk.

One approach to this problem (Tscharnuter 1981, Regev and Shaviv 1981, Morfill, Tscharnuter, and Völk 1985, Tscharnuter 1987) is based on the assumption that turbulent viscosity operates during the collapse. The resulting transport of angular momentum out of the inner parts of the cloud might be expected to suppress the fragmentation into a binary system and to reduce the angular momentum of the central star to the point where it is consistent with observations of T Tauri stars. The procedure is to assume a simple kinematic viscosity $\nu = 0.33\alpha c_s L$, where c_s is the sound speed, L is the length scale of the largest turbulent eddies, and α is a free parameter. Subsonic turbulence is generally assumed, so that α is less than unity. Once the collapse is well underway, the sound crossing time is much longer than the free fall time, so that angular momentum transport is actually relatively ineffective. Binary formation is probably suppressed, but the material that falls into the central object still has high angular momentum compared with that of a T Tauri star.

Nevertheless, the model presented by Morfill *et al.* (1985) provides interesting information regarding the initial solar nebula. In contrast to the earlier calculation of Regev and Shaviv (1981), which used the isothermal approximation, this two-dimensional numerical calculation included the full hydrodynamical equations, applicable in both the optically thin and optically thick regions. Radiation transport was included in the Eddington approximation. The calculations started with 3 $M_\odot$ at a uniform density of 10^{-20} g cm^{-3} and with uniform angular velocity. Two different values for j were tested, 10^{21} and 10^{20} cm^2 s^{-1}, with qualitatively similar results. The case with lower angular momentum is the one of most interest. The collapse proceeds and results in the formation of a central condensation surrounded by a disk. However, the core does not reach hydrostatic equilibrium but exhibits a series of dynamical oscillations. This calculation gave the first indication of such oscillations in protostellar evolution. The instability is triggered when the adiabatic exponent $\Gamma_1 = (\partial lnP/\partial ln\rho)_S$ falls below 4/3, and the source of the energy for the reexpansion is association of hydrogen atoms into molecules. The oscillations can be of large amplitude, particularly in the non-rotating case (Tscharnuter 1987).

So that the development of the disk could be studied, the computational procedure was modified to treat the region interior to 2×10^{12} cm as an unresolved core, and thereby to suppress the oscillations. Matter and angular momentum were allowed to flow into this central region but not out of it. The kinetic energy of infall was assumed to be converted into radiation at the same boundary. The calculation was continued until about 0.5 $M_\odot$ had accumulated in the core, and about 0.1 $M_\odot$ had collapsed into a nearly Keplerian disk, with radial extent of about 20 AU. The calculation was stopped at that point because of insufficient spatial resolution in the disk region, and because the ratio (β) of rotational kinetic energy to gravitational potential energy of the core exceeded 0.27, so that dynamical instability to non-axisymmetric perturbations would be likely (see review by Durisen and Tohline 1985). The development of a triaxial central object is likely to result in the transport of angular momentum by gravitational torques (Yuan and Cassen 1985, Durisen *et al.* 1986). Angular momentum would be transported from the central object to the disk, and the value of β for the core would be reduced below the critical value; however its remaining total angular momentum would be still too large to allow it to become a normal star. A further important feature of the calculation was its prediction of the temperatures that would be generated in the planet-forming region. Over a time scale of 3×10^4 yr the temperature of material with the same specific angular momentum as that of the orbit of Mercury ranged from 400-600 K. The predicted temperature for Jupiter remained fairly constant at 100 K, while that for Pluto approximated 15 K. In the inner region of the nebula these temperatures are slightly cooler than those generally thought to exist during planetary formation or those calculated in evolving models of a viscous solar nebula (Ruden and Lin 1986).

A further 2-D calculation was made by Tscharnuter (1987) with similar physics but a different initial condition. A somewhat centrally condensed and non-spherical cloud of 1.2 $M_\odot$ starts collapse from a radius of 4×10^{15} cm, a mean density of 8 $\times 10^{-15}$ g cm^{-3}, and $j \approx 10^{20}$ cm^2 s^{-1}. In this case, the much higher starting density and the correspondingly higher mass inflow rate onto the core, as well as the effects of rotation, are sufficient to suppress the instability to a considerable extent. A few relatively minor oscillations, primarily in the direction of the rotational pole, damp quickly. The numerical procedure uses a grid that moves in the (spherical) radial direction and thus is able to resolve the central regions well, down to a scale of 10^{10} cm. This calculation is carried to the point where a fairly well-defined core of 0.07 $M_\odot$ has formed, which is still stable to non-axisymmetric perturbations ($\beta = 0.08$). A surrounding disk structure is beginning to form, out to a radius of about 1 AU. The density and temperature in the equatorial plane at that distance are about 3×10^{-9} g cm^{-3} and 2500 K, respectively. Further accretion of material into the core region is likely to increase the value of β. The calculation was not continued because of the large amount of computer time required.

A different approach to the problem of the two-dimensional collapse of the protostar has been considered by Adams and Shu (1986) and Adams, Lada, and Shu (1987). The aim is to obtain emergent spectra through frequency-dependent radiative transfer calculations. In order to bypass the difficulties of a full 2-D numerical calculation, they made several approximations. The initial condition is a "singular" isothermal sphere, in unstable equilibrium, with sound speed c_s, uniformly rotating with angular velocity Ω. In the initial state the density distribution is given by $\rho \propto R^{-2}$, where R is the distance to the origin, and the free-fall accretion rate onto a central object of mass M is given by $\dot{M} = 0.975\, c_s^3/G$. The hydrodynamical solution for the infalling envelope is taken to be that given by Terebey et al. (1984), a semianalytic solution under the approximation of slow rotation. The thermal structure and radiation transport through the envelope can be decoupled from the hydrodynamics (Stahler, Shu, and Taam 1980). The model at a given time consists of an (unresolved) core, a circumstellar disk, and a surrounding infalling dusty optically thick envelope. The radiation produced at the accretion shocks at the core and disk is reprocessed in the envelope, and emerges at the dust photosphere primarily in the mid-infrared. The thermal emission of the dust in the envelope is obtained by approximating the rotating structure as an equivalent spherical structure; however the absorption in the equation of transfer is calculated taking the full two-dimensional structure into account. The model is used to fit the observed infrared radiation from a number of suspected protostars, by variation of the parameters M, c_s, Ω, η_D, and η_*, where the last two quantities are the efficiencies with which the disk transfers matter onto the central star and with which it converts rotational energy into heat and radiation, respectively. These models provide good fits to the spectra of the observed sources for typical parameters M = 0.2 - 1.0 $M_\odot$, c_s = 0.2 - 0.35 km s^{-1}, Ω = 2 × 10^{-14} - 5 ×10^{-13} rad s^{-1}, η_D = 1 and η_* = 0.5. Of particular interest is the fact that in many cases the deduced values of Ω fall in the range $j \approx 10^{20}$ cm^2 s^{-1}, which is appropriate for "solar nebula" disks. The contribution from the disk broadens the spectral energy distribution and brings it into better agreement with the observations than does the non-rotating model. Recent observational studies of protostellar sources (Myers *et al.* 1987) also are consistent with the hypothesis that disks have formed within them.

Three-dimensional calculations of the collapse of a rotating protostar, with radiation transport, have been carried out by Boss (1985, 1989). The main purpose is to determine the degree to which the object becomes non-axisymmetric, so that the rate of angular momentum transport by gravitational torques can be determined. The earlier calculation involved a mass of 2 $M_\odot$, an initial uniform density of 10^{-18} g cm^{-3}, and $j = 10^{19}$ cm^2 s^{-1}. Non-axisymmetric perturbations with amplitudes of 5% in the even modes were introduced into the initial density distribution. This object did not fragment into a multiple system; instead, the perturbations amplified and a bar-like structure developed in the inner regions, on a scale of 100 AU. In the later calculations, the masses are 1 to 2 $M_\odot$, the initial densities 2 × 10^{-13} or 2 × 10^{-12} g cm^{-3}, and the specific angular momenta $j \approx 10^{18}$ - 10^{19} cm^2 s^{-1}. Initial non-axisymmetric density perturbations of ~ 1% are introduced. These perturbations grow into non- axisymmetric structures. The calculations are followed for about 1 free-fall time, and the deduced time scales for angular momentum transport by gravitational torques range from 10^3 to 10^6 yr for systems with a total mass of 1 $M_\odot$.

4. Recent 2-D Models with Hydrodynamics and Radiative Transport.

Full hydrodynamic calculations of the collapse, including frequency- dependent radiative transport, have recently been reported by Bodenheimer *et al.* (1988). The purpose of the calculations was to obtain the detailed structure of the solar nebula at a time just after its formation and to obtain spectra and isophotal contours of the system as a function of viewing angle and time. Because of the numerical difficulties discussed in Section 2, the protostar was resolved only on scales of 10^{13} - 10^{15} cm. These calculations have now been redone with the extension of the outer boundary of the grid to 5 ×10^{15} cm, with improvements in the radiative transport, and with a somewhat better spatial resolution, about 1 AU in the disk region (Bodenheimer *et al.* 1989).

The initial state, a cloud of 1 $M_\odot$ with a mean density of 4 ×10^{-15} g cm^{-3}, can be justified on the grounds that only above this value does the magnetic field decouple from the gas and allow a free-fall collapse, with conservation of angular momentum of each mass element, to start (Nakano 1984, Tscharnuter 1987). The initial density distribution is assumed to be a power law, the temperature is assumed to be isothermal at 20 K, and the angular velocity is taken to be uniform with a total angular momentum of 10^{53} g cm^2 s^{-1}. Because the cloud is already optically thick at the initial state, the temperature increases rapidly once the collapse starts. The inner region with R $\leq$ 1 AU is unresolved; the mass and angular momentum that flow into this core are calculated. At any given time, a crude model of this material is constructed, under the assumption that it forms a Maclaurin spheroid. From a calculation of its equatorial radius R_e, the accretion luminosity $L = GM\dot{M}/R_e$ is obtained. For each timestep Δt the accretion energy $L\Delta t$ is deposited in the inner zone as internal energy and is used as an inner boundary condition for the radiative transfer. Most of the energy radiated by the protostar is provided by this central source. During the hydrodynamic calculations radiative transfer is calculated according to the diffusion approximation, which is a satisfactory approximation for an optically thick system. Rosseland mean opacities are taken from the work of Pollack et $al.$ (1985). After the hydrodynamic calculations are completed, frequency-dependent radiative transfer is calculated for particular models according to the approach of Bertout and Yorke (1978), with their grain opacities which include graphite, ice, and silicates.

The results of the calculations show the formation of a rather thick disk, with increasing thickness as a function of distance from the central object. As a function of time the outer edge of the disk spreads from 1 AU to 60 AU, because of the accretion of material of higher angular momentum. The shock wave on the surface is evident, and the internal motions in the disk are relatively small compared with the collapse velocities. At the end of the calculation the mass of the disk is comparable to that of the central object, and it is not gravitationally unstable according to the axisymmetric local criterion of Toomre (1964). The central core of the protostar, inside 10^{13} cm, contains about 0.6 M$_\odot$ and sufficient angular momentum so that $\beta \approx$ 0.4. This region is almost certainly unstable to bar-like perturbations. Theoretical spectra show a peak in the infrared at about 40 μ; when viewed from the equator the wavelength of peak intensity shifts redward from that at the pole. A notable difference between equator and pole is evident in the isophotal contours. At 40 μm, for example, the peak intensity shifts spatially to points above and below the equatorial plane because of heavy obscuration there. This effect becomes more pronounced at shorter wavelengths. Maximum temperatures in the midplane of the disk reach 1500 K in the distance range 1-10 AU. At the end of the calculation, after an elapsed time of 2500 yr, these temperatures range from 700 K at 2 AU to 500 K at 10 AU and are decreasing with time.

5. Further Evolution of the System.

In the preceding example most of the infalling material joined the disk or central object on a short time scale, because of the high initial density. For a lower initial density, processes of angular momentum transport in the disk would begin before accretion was completed. The problem of the rapidly spinning central regions is apparently not solved by including angular momentum transport by turbulent viscosity during the collapse phase; furthermore no plausible physical mechanism for generating turbulence on the appropriate scale has been demonstrated. The central regions are unstable to nonaxisymmetric deformation because of rapid rotation ($\beta \geq .0.27$); the resulting angular momentum transport is likely to leave that region with values of β near 0.2 (Durisen *et al.* 1986). Further transport may occur through gravitational torques resulting from amplification of initial nonaxisymmetric perturbations (Boss 1985, 1989).

It is likely that some additional process is required to reduce j of the central object down to the value of 10^{17} characteristic of T Tauri stars. The approach of Safronov and Ruzmaikina (1985) is to assume that the initial cloud had an even smaller angular momentum ($j \approx 10^{19}$ cm^2 s^{-1}) than that assumed in most other calculations discussed here. The cloud would then collapse and form a disk with an equilibrium radius much less than that of Jupiter's orbit. Outward transport of angular momentum into a relatively small amount of mass is then required to produce the solar nebula. Magnetic transport could be important in the inner regions, which are warm and at least partially ionized (Ruzmaikina 1981). However, outside about 1 AU (Hayashi 1981) the magnetic field decays faster than it amplifies, and magnetic transport is ineffective. A supplementary mechanism must be available to continue the process; one possibility is the turbulence generated in the surface layers of the disk, caused by the shear between disk matter and infalling matter. Another possibility is that the initial cloud had higher j, and the rapidly spinning central object is braked through a centrifugally driven magnetic wind, which can remove the angular momentum relatively quickly (Shu *et al.* 1988).

As far as the evolution of the disk itself is concerned, other important mechanisms that have been suggested include (a) gravitational instability, (b) turbulent viscosity induced by convection, and (c) sound waves and shock dissipation. Process (a) can occur if the disk is relatively massive compared with the central star, or if the disk is relatively cold. Although it is still an open question whether this instability can result in the formation of a binary or preplanetary condensations, the most likely outcome is the spreading out of such condensations, because of the shear, into spiral density waves (Larson 1983). Lin and Pringle (1987) have estimated the transport time to be about 10 times the dynamical time. Processes (b) and (c) have time scales more in line with the probable lifetimes of nebular disks. Convective instability in the vertical direction (Lin and Papaloizou 1980), induced by the temperature dependence of the grain opacities, gives disk evolutionary times of about 10^6 yr (Ruden and Lin 1986). An alternate treatment of the convection (Cabot *et al.* 1987a,b) gives a time scale roughly a factor of 10 longer. Sound waves induced by various external perturbations give transport times in the range 10^6 - 10^7 yr (Larson 1989). A complete theory of how the disk evolves after the

86

immediate formation phase may involve several of the mechanisms just mentioned, for example, an initial rapid evolutionary phase driven by gravitational instability followed by a longer phase, suitable for planetary formation, during which convective instability dominates. These processes are discussed in detail in the chapter by D. LIN.

ACKNOWLEDGEMENTS. This work was supported in part by a special NASA theory program which provides funding for a joint Center for Star Formation Studies at NASA-Ames Research Center, UC Berkeley, and UC Santa Cruz. Further support was obtained from National Science Foundation grant AST-8521636.

REFERENCES.

Abt, H. A. 1983. *Ann. Rev. Astron. Astrophys.*, **21**, 343

Adams, F. C., Lada, C. J., and Shu, F. H. 1987. *Astrophys. J.*, **312**, 788

Adams, F. C., and Shu, F.H. 1986. *Astrophys. J.*, **308**, 836

Bertout, C., and Yorke, H. W. 1978. In *Protostars and Planets*, ed. T. Gehrels (Tucson: University of Arizona Press), p. 648

Bodenheimer, P., Różyczka, M., Yorke, H. W., and Tohline, J. E. 1988. In *Formation and Evolution of Low Mass Stars*, eds. A. K. Dupree and M. T. V. T. Lago (Dordrecht: Kluwer), p. 139

Bodenheimer, P., Yorke, H. W., Różyczka, M., and Tohline, J. E. 1989. In preparation

Boss, A. P. 1985. *Icarus*, **61**, 3

Boss, A. P. 1987. *Astrophys. J.*, **319**, 149

Boss, A. P. 1989. *Astrophys. J.*, submitted

Cabot, W., Canuto, V.M., Hubickyj, O., and Pollack, J. B. 1987a. *Icarus*, **69**, 387

Cabot, W., Canuto, V.M., Hubickyj, O., and Pollack, J. B. 1987b. *Icarus*, **69**, 423

Cameron, A. G. W. 1962. *Icarus*, **1**, 13

Cameron, A. G. W. 1963. *Icarus*, **1**, 339

Cameron, A. G. W. 1978. *Moon and Planets*, **18**, 5

Cassen, P., and Summers, A. 1983. *Icarus*, **53**, 26

Durisen, R. H., Gingold, R. A., Tohline, J. E., and Boss, A. P. 1986. *Astrophys. J.*, **305**, 281

Durisen, R. H., and Tohline, J. E. 1985. In *Protostars and Planets II*, eds. D. C. Black and M. S. Matthews (Tucson: University of Arizona Press), p. 534

Goldsmith, P. F., and Arquilla, R. 1985. In *Protostars and Planets II*, eds. D. C. Black and M. S. Matthews (Tucson: University of Arizona Press), p. 137

Hartmann, L., Hewitt, R., Stahler, S., and Mathieu, R. D. 1986. *Astrophys. J.*, **309**, 275

Hartmann, L., and Kenyon, S. 1988. In *Formation and Evolution of Low Mass Stars*, eds. A. K. Dupree and M. T. V. T. Lago (Dordrecht: Kluwer), p. 163

Hayashi, C. 1981. *Prog. Theor. Phys. Suppl.*, **70**, 35

Heyer, M. H. 1988. *Astrophys. J.*, **324**, 311

Larson, R. B. 1983. *Rev. Mexicana Astron. Astrof.*, **7**, 219

Larson, R. B. 1989. In *The Formation and Evolution of Planetary Systems*, eds. H. A. Weaver, F. Paresce, and L. Danly (Cambridge University Press), in press

Levy, E. H., and Araki, S. 1989. *Icarus*, in press

Lewis, J. S. 1974. *Science*, **186**, 440

Lin, D. N. C., and Papaloizou, J. 1980. *Mon. Not. R. astr. Soc.*, **191**, 37

Lin, D. N. C., and Pringle, J. E. 1987. *Mon. Not. R. astr. Soc.*, **225**, 607

Miyama, S. 1989. In *The Formation and Evolution of Planetary Systems*, eds. H. A. Weaver, F. Paresce, and L. Danly (Cambridge University Press), in press

Morfill, G. E., Tscharnuter, W., and Völk, H. J. 1985. In *Protostars and Planets II*, eds. D. C. Black and M. S. Matthews (Tucson: University of Arizona Press), p. 493

Myers, P. C. 1987. In *Star Forming Regions (IAU Symposium 115)*, eds. M. Peimbert and

J. Jugaku (Dordrecht: Reidel), p. 33

Myers, P. C., Fuller, G. A., Mathieu, R. D., Beichman, C. A., Benson, P. J., Schild, R. E. , and Emerson, J. P. 1987. *Astrophys. J.*, **319**, 340

Nakano, T. 1984. *Fund. Cosmic Phys.*, **9**, 139

Pollack, J. B., McKay, C., and Christofferson, B. 1985. *Icarus*, **64**, 471

Pringle, J. E. 1989. *Mon. Not. R. astr. Soc.*, in press

Regev, O., and Shaviv, G. 1981. *Astrophys. J.*, **245**, 934

Ruden, S. P., and Lin, D. N. C. 1986. *Astrophys. J.*, **308**, 883

Ruzmaikina, T. V. 1981. *Adv. Space Res.*, **1**, 49

Ruzmaikina, T. V. and Maeva, S. V. 1986. *Astron. Vestn.*, **20**, No. 3, 212

Safronov, V. S., and Ruzmaikina, T. V. 1985. In *Protostars and Planets II*, eds. D. C. Black and M. S. Matthews (Tucson: University of Arizona Press), p. 959

Sargent, A. I., and Beckwith, S. 1987. *Astrophys. J.*, **323**, 294

Shu, F. H., Adams, F. C., and Lizano, S. 1987. *Ann. Rev. Astron. Astrophys.*, **25**, 23

Shu, F. H., Lizano, S., Adams, F. C., and Ruden, S. P. 1988. In *Formation and Evolution of Low Mass Stars*, eds. A. K. Dupree and M. T. V. T. Lago (Dordrecht: Kluwer), p. 123

Stahler, S. W., Shu, F. H., and Taam, R. E. 1980. *Astrophys. J.*, **241**, 637

Strom, S. E., Edwards, S., and Strom, K. M. 1989. In *The Formation and Evolution of Planetary Systems*, eds. H. A. Weaver, F. Paresce, and L. Danly (Cambridge University Press), in press

Terebey, S., Shu, F. H., and Cassen, P. 1984. *Astrophys. J.*, **286**, 529

Toomre, A. 1964. *Astrophys. J.*, **139**, 1217

Tscharnuter, W. 1981. In *Fundamental Problems in the Theory of Stellar Evolution (IAU Symposium 93)*, eds. D. Sugimoto, D. Q. Lamb, and D. N. Schramm (Dordrecht: Reidel), p. 105

Tscharnuter, W. 1987. *Astron. Astrophys.*, **188**, 55

Yuan, C., and Cassen, P. 1985. *Icarus*, **64**, 435

ANGULAR MOMENTUM TRANSPORT IN PROTOSTELLAR DISKS.

Douglas N. C. LIN

Lick Observatory, University of California,

Santa Cruz, CA 95064, USA.

ABSTRACT. Analyses of angular momentum transport in protoplanetary disks are important for the investigation of star formation as well as the origin of the solar system. During the formation stages, the disk dynamics are regulated by mixing of infalling material and disk gas. In the outermost regions of the disk, self gravity may promote the growth of non axisymmetric perturbations and the associated tidal torque can provide an effective angular momentum transfer mechanism. After infall is switched off, convectively driven turbulence provides an effective angular momentum transfer which yields an evolutionary timescale of the order 10^{5-6} y. Convection in protoplanetary disks may eventually be stabilized by surface heating. When the grains in the disk settle into the midplane region, the disk can neither generate its own energy through viscous dissipation nor reflect radiation from the central star. Consequently, the infrared excess vanishes and the young stellar objects become "naked T Tauri stars." Protoplanetary formation modifies the structure and evolution of the disk when protogiant planets acquire sufficient mass to truncate the disk. In this case, a protoplanet's tidal torque induces the opening of a gap in the vicinity of the protoplanet's orbit. Gap formation also leads to the termination of protoplanetary growth by accretion. The condition for proto-Jupiter to acquire its present mass implies that the viscous evolution timescale for the disk to be comparable to the age of typical T Tauri stars with circumstellar protoplanetary disks.

1. Introduction.

The origin and evolution of the solar nebula and the formation of the planets provides a challenging problem for the theorist (Cameron 1988). Theories must explain extensive compilations of observational data, ranging from meteoritic abundance anomalies to the properties of outflows from young stars. Recent observations indicate that a significant fraction of pre-main-sequence stars are deduced to have disks with masses in the range 0.01-0.1 M$_\odot$ and sizes from 10 to 100 AU (Strom, Edwards, and Strom 1989). Evidence for protoplanetary accretion disks comes from infrared and UV excesses which are interpreted as radiation from the disk and the accretion disk-star boundary layer, respectively (Adams, Lada, and Shu 1987; Shu, Adams, and Lizano 1987; Kenyon and Hartmann 1987). These data are particularly useful

F. Meyer et al. (eds.), Theory of Accretion Disks, 89–104.

for constraining theoretical models, provided that we adopt the hypothesis that disks around young stellar objects are in fact protoplanetary disks.

Accretion disk theory is based on the conjecture that the source of energy is viscous dissipation of differentially rotating gas in the disk (Pringle 1981). Associated with energy dissipation is angular momentum transfer and mass diffusion. The rate of disk evolution is determined by the magnitude of an effective viscosity (Lynden-Bell and Pringle 1974). Molecular viscosity is generally too small to make a significant contribution. Disk models provide estimates on the timescale and physical mechanism for the viscous evolution. For example, the observed IR and UV signatures of protostellar disks are found in young stellar objects with ages less than 10^{6-7}y. If protostellar disks viscously evolve on this timescale, the magnitude of effective viscosity needed would correspond to $\alpha \sim 0.01 - 1$ in terms of an *ad hoc* α prescription (Shakura and Sunyaev 1973). In the discussion below, it is shown that accretion disks around protostars are intrinsically unstable against thermal convection in the direction normal to the plane of the disk and the turbulence generated from thermal convection may provide an effective $\alpha \sim 0.01$.

With an estimate of the magnitude of effective viscosity, the evolution of an accretion disk can be examined in detail. In the context of protostellar disks, the evolutionary disk model provides information on the environment out of which protostars and planets are formed. For example, giant planets are composed mostly of gas and therefore must have been formed in a gas-rich environment. But there is little of it left between the planets today. The determination of a characteristic timescale for gas depletion in the disk provides a constraint on the formation epoch of protogiant planets. This constraint also applies to prototerrestrial planet formation since protogiant planet formation proceeds through the formation of solid cores, which have masses comparable to or greater than those of the terrestrial planets (Bodenheimer and Pollack 1986). Detailed analysis of the protoplanet-disk interaction can also provide an important constraint on the magnitude of viscosity and the evolutionary timescale of protoplanetary disks (Lin and Papaloizou 1985).

2. Convectively Driven Turbulence in Protoplanetary Disks.

In order to rigorously demonstrate that an accretion disk is turbulent, we must show that a turbulence-free disk is intrinsically unstable. Up to now, there is no proof that an accretion disk is generally unstable (Pringle 1981). But in the context of protoplanetary disks, it can be shown that it is unstable against thermal convection in the direction normal to the plane of the disk (Lin and Papaloizou 1980). In a turbulence-free protoplanetary disk, in which rotation prevents gas from migrating in the radial direction, gas can contract in the vertical direction to establish local quasi hydrostatic equilibrium (Ruden 1987). After the disk has settled into a quasi hydrostatic equilibrium, slow contraction continues as thermal energy is lost from the disk's surface. In the absence of any dissipative source of energy, continuous adjustment leads to superadiabatic structure in the vertical direction. According to the standard Schwarzschild criterion, the superadiabatic gradient induces the disk to become convectively unstable. This tendency, which does not generally occur in accretion disks in other astrophysical contexts, is induced because the dominant grain opacity is an increasing function of temperature. In this case, the cooler surface region has a lower opacity and cools more efficiently (Lin and Papaloizou 1980, 1985; Lin 1981; Lin and Bodenheimer 1981).

Convective eddies in an accretion disk can induce mixing and angular momentum transport over a radial extent comparable to their own size. Convection also generates turbulence which causes dissipation of energy stored in differential rotation. Perhaps the simplest treatment for convection is to use the mixing length prescription in which the eddy viscosity is assumed to be the product of the convective speed and an effective radial mixing length which is determined by the epicyclic motion of the eddies (Lin and Papaloizou 1980). From such a treatment, we can build self-consistent models in which convection is responsible for 1) energy dissipation, 2) heat transport in the vertical direction, and 3) angular momentum transport in the radial direction. These models give an effective $\alpha \sim 0.01$.

The mixing length model, though informative and easy to use, is based on an *ad hoc* prescription of eddy viscosity. On the largest scales, which provide the dominant momentum transport, global effects such as rotation, anisotropy of convective motion, and radiative losses are important and therefore should be examined with a global analysis. In an attempt to carry out such an analysis, Cabot *et al.* (1987a,b) computed a vertically-averaged effective viscosity which is derived from integrating the linear growth rate through various distances above the midplane of the disk. They also assumed a non slip boundary condition at the mid plane, such that the convective velocity vanishes there. They deduced an effective $\alpha \sim 10^{-4}$. This method is inappropriate for evaluating the global properties of convective eddies because the linear growth rate, evaluated locally in the vertical direction, varies greatly as a function of distance from the mid plane and therefore cannot be attributed to a given eddy.

A more rigorous global treatment is to determine a unique growth rate and its associated eigenfunction for each characteristic convective mode (Ruden, Papaloizou, and Lin 1988). For computational simplicity, we considered linear axisymmetric perturbations on several equilibrium models. In the thin-disk limit, although the eigenfunctions extend over finite radial distances, the radial structure may be represented by a radial wavenumber through a WKBJ approximation, so that the perturbed equations are reduced to one dimensional ordinary differential equations in the vertical direction. These global linear stability analyses of axisymmetric perturbations indicate that: 1) rotation and compressibility tend to reduce the growth rate of the disturbances; 2) the growth rate is proportional to the square root of the radial wave number and is bounded by the maximum values of the Brunt-Vaisala frequency; 3) the maximum radial size of eddies scales as the square root of the superadiabaticity times the size of the convective region; 4) due to radiative losses, the short wavelength modes become overstable and only the fundamental and the first harmonic modes, where the wavelength is comparable to the thickness of the disk, can grow effectively; and 5) both even and odd modes exist in which a single eddy may either be confined to one side of or thread through the midplane and have a characteristic scale comparable to the thickness of the entire disk. The magnitude of the effective viscosity can be derived under the *ad hoc* assumption that gas, within a radial wavelength, mixes efficiently over the characteristic growth timescale. The effective α derived from this analysis is of the order 10^{-2}.

The axisymmetric normal mode calculation is the first step in the global analysis. Although it provides useful information on the growth rate and effective radial extent of convective elements, it does not really provide any rigorous information on angular momentum transport, because no non axisymmetric torque results from such a perturbation. Analyses on the growth of non axisymmetric perturbations are now underway. Eventually the perturbations become non linear, and dissipative processes become important. The determination of the torque associated with the growing non axisymmetric disturbances will provide a more rigorous estimate of the efficiency of angular momentum transport.

The results of the normal mode calculation can be readily applied to the studies of structure and evolution of protoplanetary disks. As the disks evolve from their formation to their depletion stages, several different boundary conditions need to be taken into account. For example, during the initial formation stage, shock dissipation near the region where the infalling material joins the disk induces an isothermal structure and therefore stabilizes against convection. When the infall rate is reduced, the superadiabatic gradient may be established as the surface heating becomes less important (Ruden 1987). At a later stage of disk evolution, surface heating, this time from the central object, may become important. Recently, Nakagawa, Watanabe, and Nakazawa (1988) have shown that surface heating due to radiation from the central star can reduce the temperature gradient in the vertical direction, a mechanism similar to shock dissipation associated with infall. This process may be particularly important during the late stage of protostellar disk evolution when a T Tauri star evolves into a "naked" phase.

3. Angular Momentum Transport by Wave Propagation.

An alternative scheme for angular momentum transport has been proposed by Shu (1975), Donner (1979), Spruit (1987), Spruit *et al.* 1987, Larson (1989), and Vishniac and Diamond (1989). In this scheme, wavelike disturbances induced by perturbations at very large disk radii may steepen into shock waves as they propagate inwards, and shock dissipation may induce effective angular momentum transport in the disk. Over an extended region, such a shock pattern may become self-similar so that it has a zero pattern speed and is stationary.

The scenario is particularly attractive because it does not depend on an *ad hoc* prescription of turbulence (Larson 1989). But it requires little or no dissipative damping of wave action as the wave propagates inwards in the disk, which is a questionable assumption. For example, refraction of waves in the vertical direction may be a possible source of dissipation of wave action. For the purpose of discussion, let us assume the wave front to be normal to the radius initially. When the z dependence of the disk is ignored, the well known local dispersion relation governing sound waves propagating in the radial direction is derived from the perturbation equations. From this it is apparent that for any wave to exist interior to the inner Lindblad resonance, its wavelength decreases to values comparable to or smaller than the vertical thickness of the disk. Thus, in the region interior to the inner Lindblad resonance, the disk acts like a wave guide for the propagating waves and ray tracing analysis can provide considerable insight. In a realistic disk model, the sound speed decreases with the distance from the midplane so that according to ray theory, the front must be retarded as the distance from the midplane increases. The wave front should tilt until it becomes normal to the vertical axis and the propagation becomes approximately vertical. Thus, vertical propagation is expected after the wave has propagated through on the order of a vertical scale height.

To examine wave propagation in a vertically stratified disk, we carried out a three dimensional numerical linear analysis (Lin, Papaloizou and Savonije 1989), in which we used a series of axisymmetric and vertically stratified, nevertheless thin, models for the unperturbed flow. It is computationally convenient to adopt a barotropic equation of state in which the variation in the sound speed can be simply adjusted with two model parameters. Provided the amplitude of the perturbations is small, wave propagation may be examined with a linear analysis. In the absence of any external forces and dissipation, the angular momentum of the propagating wave is conserved, *i.e.* there is no absorption or emission of angular momentum by the disk material as the wave propagates. However, flux of angular momentum can be transmitted in both the radial and vertical directions. For waves with large azimuthal wavenumbers, the wave energy may be readily lost through vertical propagation. However for waves with relatively small azimuthal wavenumbers, there may be reflection before densities sufficiently low for shock formation are attained. In both cases, waves are trapped in a limited radial extent. Thus, we conclude, angular momentum transport by wave propagation is unlikely to be efficient and effective over extended regions of the disk.

The most likely place for these waves to become non linear and be dissipated is in the low-density optically thin atmosphere above the disk. If the energy dissipation rate exceeds the cooling rate, the atmosphere of the disk may become thermally unstable. Consequently, a hot corona with a temperature greater than 10^{6-7} K may be formed (see chapter by G. Shaviv). Around protostars, the pressure scale height for such a hot temperature would be larger than the disk radius so that outflow from the corona would be expected (Murray, Bell, and Lin 1989).

4. Angular Momentum Transport by Gravitational Instability.

After the formation of the initial disk, it is likely to be relatively massive compared to the central object (Cassen and Moosman 1981). Gravitational instabilities are therefore possible, and they could result in a rapid early evolutionary phase, during which a considerable amount of mass is transferred to the central star. Observationally, the discovery of a massive disk around HL Tau (Sargent and Beckwith 1987) reveals the existence of disks with a relatively small gravitational stability parameter, Q, which is the ratio of the disk's self gravity, in the vertical direction, to that due to the central star. Local analyses indicate that when $Q \leq 1$, the disk becomes gravitationally unstable against axisymmetric perturbations (Safronov 1960; Toomre 1964).

The primary motivation for investigating gravitational instability is to establish an angular momentum transport mechanism in the outer regions of a protoplanetary disk where the effect of self gravity is relatively important. Toomre (1964) suggested that in a stellar disk, the effect of gravitational instability is to heat the disk locally until the stellar dispersion velocity becomes sufficiently large to render self gravity unimportant. A gas disk, however, can cool locally so that a balance between heating due to gravitational instability and cooling may be maintained (Paczynski 1978; Kozlowski, Wiita, and Paczynski 1979). In the context of protostellar disks, Larson (1984) has used the formula given by Lynden-Bell and Kalnajs (1972) for the torque produced by a particular spiral pattern and strength to estimate the torque which can be achieved. He concluded that such torques could be strong enough to dominate the angular momentum transfer in the self gravitating regions of the disk. However, he did not indicate how such a spiral pattern may be excited or the magnitude of the strength that could be attained. Recent theoretical investigations (Sellwood and Lin 1989, Papaloizou and Lin 1989, Papaloizou and Savonije 1989) showed that under certain circumstances, *e.g.* in disks with some surface density variations in the radial direction, non-axisymmetric instabilities can exist even when the disk is stable against axisymmetric perturbations.

The stability analyses of non-axisymmetric perturbations generally require a global analysis because non-axisymmetric perturbations induce torque which can transport angular momentum over extended regions of the disk. We analyzed the growth of non axisymmetric perturbations by introducing small-amplitude perturbations on equilibrium disk models, in which the flow is assumed to be inviscid with a given isentropic equation of state and an equilibrium surface density distribution. The unperturbed flow is restricted in azimuthal direction with a velocity which is determined by both gravity and a radial pressure gradient. In the geometrically thin disk limit, the vertical structure can be averaged. The linear growth rate of non axisymmetric perturbations can be determined as eigenvalues of normal mode oscillations. Using these methods, we obtained some initial results (Papaloizou and Lin 1989). For example, we have established the existence of various kinds of non axisymmetric instabilities. Each of the various instabilities can be understood in terms of behavior of waves or disturbances associated with the various resonances that occur in the problem. We intend to carry out an extensive analysis for a variety of equilibrium disk models and equations of state.

From these stability analyses, we can deduce an approximate prescription for the rate of angular momentum transport in terms of an effective kinematic viscosity (Lin and Pringle 1987). This approximation is based on the assumption that waves, induced by gravitational torque of growing non axisymmetric normal modes, do not propagate very far from the region where they are excited. A natural next step is to evaluate the effect of non axisymmetric gravitational instability on the evolution of the disk (see chapter by J.E. Pringle). These analyses provide the distributions of surface density and temperature for these outermost, gravitationally unstable regions of the disk, which are most likely to be resolvable observationally (Sargent and Beckwith 1987).

Self-gravitational effects will be particularly important during the early formation stage of the disk, when the mass of the central star is still relatively small (see chapter by Bodenheimer). During this stage, infall effects are likely to be important. If the infall flux onto the disk is larger than the rate at which mass is redistributed in the disk, we would expect the disk mass to increase during the infall stage. The energy release from the disk would be primarily determined by the energy dissipation rate of the infalling material rather than by viscous dissipation in the disk. The rate of infall per unit area, as a function of radius, depends on the specific angular momentum and the initial density distribution of the infalling cloud. Under some conditions most of the energy may be released in the far infrared wavelengths because more material infalls onto the extended outer regions of the disk. This scenario may provide one possible explanation for the strong far infrared excess observed in the spectrum of some T Tauri stars which cannot be explained by simple steady-state accretion disk models (Adams, Lada, and Shu 1987). An alternative mechanism for generating this excess has been proposed recently by Adams, Ruden, and Shu (1989) based on the excitation of the m=1 mode in massive self gravitating disks. In this scenario, infall may still be necessary to sustain a large amount of material in the outer regions, since dissipation tends to drive rapid inward diffusion of disk material. Thus our calculation may provide information on the conditions under which gravitationally unstable regions of the disk may be maintained.

5. Dynamical Evolution of Protoplanetary Disks.

We now turn our attention to the main evolutionary phase after infall onto the disk is switched off. The dynamics of the disk during this stage is essentially determined by the viscous diffusion process (Lüst 1952; Lynden-Bell and Pringle 1974) and can be analyzed by applying a simplified prescription for convectively induced turbulent viscosity (Lin and Papaloizou 1985) in the diffusion equation. In principle, the surface density distribution at the end of the infall stage may be used as the initial condition for the main evolution phases. The evolution of the disk can also be computed with arbitrary initial conditions.

In the context of solar system formation, an interesting initial condition is the "minimum mass" nebula model (Cameron 1973; Hayashi 1981). In this model, the surface density distribution is derived by augmenting gas to the present mass distribution in the solar system based on the assumption that the protoplanetary disk had a solar composition and planetary formation is totally efficient at retaining all the heavy elements in the disk. Applying the effective viscosity associated with convection into a mass diffusion equation, we find the evolution of the minimum-mass nebula with a physical dimension comparable to that of the present day solar nebula takes place on the timescale of $\sim 10^6$ years (Lin and Papaloizou 1985). The temperature distribution resembles that deduced from the condensation temperatures for various terrestrial planets and satellites (Lewis 1972). At radii interior to the orbit of Mercury, the midplane temperature exceeds 2,000K, so that grains would be mostly evaporated.

It is of interest to note that the mass transfer rate onto the accreting protosun, i.e., 10^{-8} to 10^{-7} $M_\odot$ y^{-1} is comparable to that deduced for the boundary layer regions, between protostellar disk and accreting stars, in typical T Tauri stars (Bertout, Basri, and Bouvier 1988). With this accretion rate and $\alpha \sim 0.01$, the boundary layer region is expected to be optically thin. Strong emission lines are observed for this region.

The disk becomes optically thin exterior to the orbit of Neptune and attains an isothermal vertical structure. Consequently, convection would not occur and dust would settle towards the midplane. Even if there are some other local instabilities, we do not expect turbulence to be sustained for an extended period. In the marginally optically thin region, the characteristic timescale for heat loss is comparable to or shorter than the orbital timescale of the disk. Energy transfer between eddies with different sizes occurs on the eddy turnover timescale which is comparable to or longer than the orbital timescale. Since thermal energy loss is faster than the kinetic energy of the eddies, the eddy motion becomes supersonic. Shocks during eddy mixing could cause turbulence to decay.

However, the absence of local turbulence may not imply the termination of disk evolution. Accompanying the decay of turbulence is the decline in mass diffusion and energy dissipation so that the disk temperature and optical depth decrease with the surface density. The effect of self gravity also increases with time. Eventually, the Q value decreases to order unity and gravitational instability becomes important in promoting the growth of non axisymmetric disturbances and in inducing angular momentum transfer (see chapter by J.E. Pringle). In this region, the evolutionary timescale is determined by the cooling timescale since cooling is essential in maintaining a sufficiently low value for Q. In a relatively extended massive disk, when the temperature decreases to ~ 10 K, the cooling timescale at 100 A.U. exceeds 10^{6-7}y. The disk temperature can not decrease below 10 K, which is the temperature of the protostellar cloud. Thus, when the disk surface density decreases to a sufficiently small value, the effect of self gravity can no longer be important and the evolution of the solar nebula must stop at that region. A natural outcome for this process is the development of an extended, circumstellar, optically-thin disk.

For the minimum-mass solar nebula model, the surface density of the disk is so low that self gravity becomes at best marginally important beyond the orbit of Neptune even when the disk temperature is 10 K. If the outer region is both optically thin and non self-gravitating, the disk terminates its evolution there. This may be the reason why there is no major planet beyond the orbit of Neptune. Inside the outermost region, the disk remains opaque and viscous evolution continues. A fraction of disk material would be deposited into the outermost region to carry the excess angular momentum. While gas may eventually be evaporated by the photodissociation process, dust particles would be left behind. In the absence of turbulence, dust particles descend towards the midplane. When the dust layer becomes sufficiently thin, gravitational instability would cause the dust to clump and form 10 kilometer size objects which may be the progenitors of planetesimals or comets (Goldreich and Ward 1973; Weidenschilling 1987).

6. Causes for Unsteady Flow and Outburst Phenomena.

In addition to the usual infrared signatures of disks, the prototype of young stellar objects, T Tau, has a variable light curve. Large magnitude luminosity variations, on the timescale of months to years, are found in other young stellar objects such as FU Ori and V1057 Cyg (Herbig 1977). It was suggested that these outbursts are associated with variation in mass transfer rate in accretion disks (Lin and Papaloizou 1985). There is observational evidence which supports this scenario (Hartmann and Kenyon 1987a,b). For example, the pre outburst spectrum of V1057 Cyg is that of a typical T Tau star and the accretion disk around a typical T Tau star has a mass transfer rate of 10^{-7} to $10^{-6} M_\odot$ y^{-1}. In outbursts, however, the estimated mass accretion rate ranges from $\sim 10^{-4} M_\odot$ yr^{-1} for FU Ori to $\sim 10^{-3} M_\odot$ y^{-1} (Kenyon, Hartmann, and Hewett 1988).

The rise timescale for the FU Orionis events is typically a few months. This rapid rise implies that the variations in accretion occurred in the inner regions of the disk close to the accreting star. However, during the rise, the inferred disk mass being accreted by the star is greater than $\sim 10^{-3} M_\odot$. Unless the entire disk is gravitationally unstable, we deduce that such a large mass concentration in the inner regions of the disk requires perturbations which can sweep all the disk material from 10^{14} cm onto the accreting star (Clarke, Lin, and Pringle 1989). Mass transfer variation may originate from a more extended region but it would require a more rapid propagation of disk response than that which can be provided by viscous diffusion. One possible mechanism for inducing a rapid rise is thermal instability in the disk analogous to that attributed to dwarf nova outbursts (Lin and Papaloizou 1985).

For accretion rates exceeding $\sim 10^{-5} M_\odot \, \mathrm{y}^{-1}$, the associated energy dissipation would partially ionize the gas in the inner regions of a protoplanetary disk. According to standard α models, partially ionized regions of accretion disks are thermally unstable because the opacity increases rapidly with temperature (Meyer and Meyer-Hofmeister 1981; Smak 1982; Cannizzo, Ghosh and Wheeler 1982; Faulkner, Lin, and Papaloizou 1983). A small temperature increase would cause a large increase in opacity and a decrease in the heat diffusion rate. Consequently, the disk would undergo an upward thermal transition. If the magnitude of viscosity is a function of temperature, as is the case in the α prescription, thermal instabilities lead to changes in the mass and angular momentum transfer rate. Thermal transition at one radius would cause a large local temperature gradient in the radial direction. The sudden increase in viscous stress across the transition front would cause the front to propagate throughout the disk at α times the sound speed, such that the entire disk enters an outburst state.

Most regions of protoplanetary disks are sufficiently cool that radiation transfer is determined by grain opacity so they are thermally stable. If the mass transfer rate into the inner regions of the disk is insufficient to sustain the rate of mass transfer in the disk during the outburst, the disk material would be depleted and the disk would return to quiescence. This type of thermal relaxation limit cycle has been examined in detail for dwarf nova outbursts In the case of protoplanetary disks, in order to cause a T Tauri type disk to become thermally unstable, large amplitude perturbations are needed (Clarke, Lin, and Pringle 1989).

However, unlike the situation in dwarf novae, such a large amplitude perturbation is unlikely to be caused by thermal instability alone. The major difference to dwarf nova outbursts is that FU Orionis has remained in a post outburst high state for more than thirty years. In V1057 Cyg, the decline rate is very slow. An important theoretical issue is how to maintain the high state. One difference is that protoplanetary disks are much larger than dwarf-nova disks so that the critical mass transfer rate for the onset of thermal instability is also larger (Clarke, Lin, and Papaloizou 1989). After an upward transition, the thickness is a large fraction of the radius of the disk. The advective transport of heat is much more important in the FU Orionis disks than in dwarf-nova disks. Over the temperature range near the transition front, advective heat transport can stabilize against thermal runaway and therefore prevent the propagation of the transition front throughout the disk. It is of interest to note that this stabilizing effect is only effective for disks with a mass transfer rate comparable to or larger than $10^{-4} M_\odot \, \mathrm{y}^{-1}$. Therefore, it may be very useful for the observers to determine whether there is any correlation between the light curves and mass accretion rate in the FU Orionis stars.

Another indication for the importance of non local effects may be found in the spectroscopic data (Kenyon, Hartmann, and Hewett 1988). These spectra can be decomposed into stellar and steady state disk components. Conspicuously absent is any break in the continuum which is often used as an indicator of boundary layer radiation. Since the energy released in the boundary layer is comparable to that from the entire disk, it would be detectable unless it is redistributed over an area larger than the accreting star's surface. In an outburst state, significant radiation transfer in the radial direction can occur since 1) the opacity near the midplane is small compared to that near the surface of the disk and 2) the thickness is comparable to the radius of the disk. Detailed radiative transfer calculations are underway (Bell, Clarke, and Lin 1989).

There are several possible mechanisms which may induce large amplitude perturbations to trigger thermal instability in the disk. For example passage of hypothetical companions through the inner regions of the disk may cause significant perturbations to the surface density and mass transfer rate in the disk (Clarke and Pringle 1989). Alternatively, rapidly accreting disks dissipate energy at a high rate. If the surface of the disk is exposed to the radiation that is emitted by the star or the boundary layer, convection may be stabilized. If convectively driven turbulence is the only source of viscosity, accretion flow may be quenched. The outer regions of the disk with such a large accretion rate are probably self gravitating. Large surface heating may sustain a relatively high disk temperature so that the disk may be stabilized against gravitational instability. In either case, a potential feedback mechanism may be induced if the surface heating flux is proportional to the mass transfer rate into the inner regions of the disk. Preliminary investigation indicates that such a feedback mechanism may lead to regular or chaotic limit cycles. Large amplitude variations in the mass transfer rate in these cycles provide a good model for FU Ori outbursts.

7. Clearing Processes.

On the evolutionary timescale of typical T Tauri stars, $\sim 10^6$ y, a large fraction of these young stellar objects become naked T Tauri, i.e., they lose any indication of the presence of a circumstellar disk (Walter 1986, 1987). The disappearance of these signatures may require that the entire disk has become optically thin since optically thick regions of the disk would be able to reprocess the stellar radiation into infrared radiation. The Optically thick disk could also induce turbulence from convective instability resulting in dissipation of differential rotation energy into infrared radiation.

One possible mechanism for the disk to become optically thin is through viscous diffusion. However, in this case, even with the assumption that some hypothetical transonic turbulence may be responsible for angular momentum transfer, the timescale for the inner regions, (within $\sim 0.1 - 1$ AU), of the disk to evolve into an optically thin state is larger than 10^7 y. Thus, the disappearance of IR and UV excess is unlikely to be due to viscous diffusion. An alternative mechanism for eliminating the UV and IR excess from the disk is through dust settling. In §2, we indicate that the surface heating effect can stabilize against convection when the surface heating is sufficiently large to induce a black body temperature comparable to the temperature near the midplane. This critical flux increases with the surface density of the disk since opacity, and consequently temperature, near the midplane also increase with the surface density. For a typical protoplanetary-disk model (Ruden and Lin 1986), the surface density reduces below the critical values for most regions of the disk, on a timescale $\sim 10^6$ y, such that convection may become stabilized. Thereafter, unless there are other instabilities, turbulence would decay and dust would settle toward the midplane on the relatively short timescale of 10^{3-4} years (Hayashi, Nakazawa, and Nakagawa 1985). As dust particles segregate from the gas they coagulate rapidly and form planetesimals either through gravitational instability (Goldreich and Ward 1973) or cohesive collisions (Weidenschilling 1987). Consequently, the protoplanetary disk would become transparent so that it would no longer reprocess radiation from the central star. The lack of turbulent viscosity would also quench viscous dissipation and mass transfer in the disk.

Through this mechanism, radiative flux from the disk is significantly reduced while most of the disk gas is retained. The disk gas may be eventually eliminated on a somewhat longer timescale by 1) stellar wind ablation, 2) wind generated by the dissociation or ionization of disk gas by the incident solar radiation, and 3) disk-protoplanet tidal interaction.

8. Protoplanet-Disk Tidal Interaction.

Through its tidal torque on the protoplanetary disk, a protogiant planet can excite waves which carry angular momentum and energy. In the regions of the disk interior to the protoplanet's orbit, these waves carry negative angular momentum flux with respect to the local fluid. When the wave is dissipated, it deposits negative angular momentum and thereby causes the disk material to drift inwards. If these waves can propagate deep into the interior region of the disk, the tidal influence of the protoplanet is distributed over an extended region, whereas if these waves are dissipated in the close vicinity of the protoplanet's orbit, the tidal effect is localized. Because the protoplanetary disk has a thermally stratified structure in the vertical direction, *i.e.* the temperature decreases with distance from the midplane, the propagation of the waves in the radial and vertical directions is closely linked. In §3 we indicated that thermal stratification causes refraction such that initially radially propagating waves are deflected in the vertical direction. For moderate temperature contrast between the disk's midplane and its surface, wave transmission into the tenuous upper atmosphere is allowed. For protoplanetary disks, these two effects inhibit wave propagation through large distances in the radial direction. Consequently, waves excited by tidal disturbances due to a protoplanet are not expected to reach the interior regions of the disk with a significant amplitude (Lin, Papaloizou, and Savonije 1989). Based on these results, the rate of energy dissipation can be evaluated as a function of the sound speed and the distance from where the waves are launched. From these rates, the structure of the disk in the neighborhood of the protoplanet can be computed with a two dimensional numerical hydrodynamic model in the limit that the disk is relatively thin (Papaloizou and Lin 1985; Lin and Papaloizou 1986 a,b). The results of these computations indicate that when the mass of the protoplanet is sufficiently large, its tidal torque may induce the formation of a gap in the vicinity of its orbit.

There are two conditions for gap formation. One criterion is a necessary condition: the Roche radius of the protoplanet must be comparable to or larger than the scale height of the protoplanetary disk. If the necessary condition is not satisfied, the moment a protoplanet tries to open up a gap, the pressure gradient, at the boundary of the disk near the protoplanet, would become sufficiently large to force the epicyclic frequency to become negative and therefore cause dynamical instability. The other criterion is that the rate of angular momentum transfer between the protoplanet and the disk, due to the tidal torque of the protoplanet, must exceed that within the disk due to viscous stress. In typical protoplanetary disk models, the second condition is automatically satisfied when the first criterion is satisfied provided the effective α is somewhat less than unity.

After a protoplanet opens up a gap in the vicinity of its orbit, mass growth of the protoplanet is essentially terminated. Thus, we can deduce some useful constraints on the dynamics of the primordial solar nebula from the present mass of Jupiter. For example, let us consider the scenario that the disk is not turbulent and viscous dissipation is ineffective. In this case, only the first criterion for gap formation applies. In order to avoid gap formation before Jupiter acquired its present mass, the nebula must have a relatively large vertical scale height. The nebular temperature necessary for such a large scale height is several times that which can be provided by the solar radiation. Thus, an additional heat source is required. Viscous dissipation in the disk can provide sufficiently high temperature but it requires an effective viscosity considerably larger than the molecular viscosity, which is inconsistent with the assumption that the disk is quiescent. Let us consider the other extreme scenario that the primordial solar nebula is massive and its evolution is regulated by torque induced by gravitational instabilities. In this case, the disk would be relatively hot and the viscosity relatively large, so that the disk truncation would be unlikely to occur until Jupiter had acquired a mass substantially larger than its present mass. Unless Jupiter could have lost most of its initial mass, this scenario also seems unlikely. From the above arguments, we deduce that in order for both criteria to be satisfied for the present mass of Jupiter, the viscous evolution time of the disk must have been of the order 10^5 - 10^6 y. It is of interest to note

102

that this is comparable to the typical age of the T Tauri stars.

After the protoplanet opens up a gap, it continues to tidally interact with the disk. Such interaction causes the orbit of the protoplanet to evolve, and the protoplanet continues to have a strong influence on the structure and evolution of the disk (Lin and Papaloizou 1986b). When gap formation first occurs, the amount of material cleared to either side of the gap is determined by the motion as well as the mass of the embedded protoplanet. The disk responds by modifying the local surface density gradient such that the angular momentum transfer rate between the protoplanet and the region interior to it is delicately balanced by that between the protoplanet and the disk exterior to it. If the mass of the protoplanet were relatively large compared with that of the disk, the protoplanet would not undergo significant orbital evolution. The disk interior to the protoplanet would be depleted on a relatively short timescale. The disk mass exterior to the protoplanet would be conserved; the surface density decreases slowly as it expands viscously. This process would lead to the formation of a hole in the central region of the disk which may be directly observable. In the limit that the protoplanet's mass is small compared with that of the disk, the protoplanet's orbital evolution would proceed with the disk on the viscous diffusion timescale of the disk. In this limit, if the protoplanet were formed at relatively small radii, it would migrate inward on a relatively short timescale, whereas if the protoplanet were formed at a region near the outer edge of the disk, it would migrate outward until the disk material interior to its orbit was somewhat depleted.

Based on these results, one can deduce the mass distribution of the protoplanetary disk from Jupiter's orbit. The disk mass exterior to the proto-Jupiter could not have exceeded that interior to its orbit by more than $0.1M_\odot$, otherwise proto-Jupiter would have migrated significantly toward the Sun. Similarly, it can also be argued that the mass of the disk interior to Jupiter could not have exceeded that exterior to its orbit by more than $0.1M_\odot$, otherwise the resonant asteroids would have escaped from their commensurable orbit with Jupiter.

9. Summary.

In this paper we have reviewed various angular momentum transport mechansims in the protoplanetary disk. During the formation of the protoplanetary disk, mixing of infalling material with the disk gas can lead to significant mass transfer. Rapidly rotating clouds can lead to the formation of extended disks. In the outer regions of the disk, self gravity of the disk can promote growth of non axisymmetric disturbances. The tidal torque associated with these growing unstable modes induces angular momentum transport and regulates mass transfer. After the infall is switched off, convectively driven turbulence provides an effective viscosity such that the typical evolutionary timescale of the disk is $\sim 10^6$ y. This timescale is consistent with the typical age of young stellar objects with signatures of circumstellar disks. It is also consistent with the timescale derived from the condition for tidal truncation of protoplanetary disks by proto-Jupiter.

ACKNOWLEDGEMENTS. Aspects of the work reported here are due to P.H. Bodenheimer, R. Bell, C. Clarke, L. Hartmann, S. Murray J.C.B. Papaloizou, J.E. Pringle, S. Ruden, G. Savonije, J.A.Sellwood, and F. Shu. Their contributions are greatly appreciated. This work is supported in part by grants AST-85-21636 and NAGW 1211 from NSF and NASA respectively. Part of this work has been conducted under the auspices of a special NASA astrophysics theory program that supports a Joint Center for Star Formation Studies at NASA-Ames Research Center, University of California, Berkeley and University of California, Santa Cruz.

REFERENCES.

Adams, F.C., Lada, C., and Shu, F.H. 1987, *Ap. J.*, **312**, 788

Adams, F. C., Ruden, S. P., and Shu, F. H. 1989. *Ap. J.*, submitted

Bell, R., Clarke, C., and Lin, D.N.C. 1989, in preparation

Bertout, C., Basri, G., and Bouvier, J. 1988, *Ap. J.*, **330**, 350

Bodenheimer, P.H., and Pollack, J. 1986, *Icarus*, **67**, 391

Cabot, W., Canuto, V.M., Hubickyj, O., and Pallock, J.B. 1987a, *Icarus*, **69**, 387

Cabot, W., Canuto, V.M., Hubickyj, O., and Pallock, J.B. 1987b, *Icarus*, **69**, 423

Cameron, A.G.W. 1973, *Icarus*, **18**, 407

Cameron, A.G.W. 1988, *Ann. Rev. Astr. Ap.*, **26**, 441

Cannizzo, J. K., Ghosh, P. and Wheeler, J. C. 1982. *Ap. J.* **260**, L83

Cassen, P., and Moosman, A. 1981. *Icarus* **48**, 353

Clarke, C., Lin, D.N.C, and Papaloizou, J.C.B. 1989, *M.N.R.A.S.*, in press

Clarke, C., Lin, D.N.C, and Pringle, J.E. 1989, in preparation

Clarke, C. and Pringle, J.E. 1989, in preparation

Donner, K. 1979, Ph.D. thesis, Cambridge University

Faulkner, J., Lin, D.N.C., and Papaloizou, J.C.B. 1983, *M.N.R.A.S.*, **205**, 359

Goldreich, P. and Ward, W.R. 1973, *Ap. J.*, **183**, 1051

Hartmann, L. and Kenyon, S.J. 1987a, *Ap. J.*, **312**, 243

Hartmann, L. and Kenyon, S.J. 1987b, *Ap. J.*, **322**, 393

Hayashi, C. 1981, *Prog. Theo. Phys. Suppl.*, **70**, 35

Hayashi, C., Nakazawa, K., and Nakagawa, Y., 1985 in *Protostars and Planets II*, eds. D. Black and M.S. Matthews, (University of Arizona Press:Tucson), p. 1100

Herbig, G.H. 1977, *Ap. J.*, **217**, 693

Kenyon, S.J. and Hartmann, L. 1987, *Ap. J.*, **323**, 714

Kenyon, S.J., Hartmann, L., and Hewett, R. 1988, *Ap. J.*, **325**, 231

Kozlowski, M., Wiita, P. J., and Paczynski, B. 1979. *Acta Astr.* **29**, 157

Larson, R. B. 1984. *M. N. R. A. S.* **206**, 197

Larson, R. 1989, in *The Formation and Evolution of Planetary Systems*, eds H.A. Weaver, F. Paresce, and L. Danly, (Cambridge University Press: Cambridge), in press

Lewis, J.S. 1972, *Icarus*, **16**, 241

Lin, D.N.C. 1981, *Ap. J.*, **242**, 780

Lin, D.N.C. and Bodenheimer, P.H. 1981 *Ap. J.*, **262**, 768

Lin, D.N.C. and Papaloizou, J.C.B. 1980, *M.N.R.A.S.*, **191**, 37

Lin, D.N.C. and Papaloizou, J.C.B. 1985, in *Protostars and Planets II*, eds. D. Black and M.S. Matthews, (University of Arizona Press:Tucson), 981

Lin, D.N.C. and Papaloizou, J.C.B. 1986a, *Ap. J.*, **307**, 395

Lin, D.N.C. and Papaloizou, J.C.B. 1986b, *Ap. J.*, **309**, 846

Lin, D.N.C., Papaloizou, J.C.B., and Savonije, G.J. 1989, in preparation

Lin, D.N.C. and Pringle, J.E. 1987, *Ap. J.*, **225**, 607

Lüst, R. 1952, *Z. Nat.*, **7a**, 87

Lynden-Bell, D., and Kalnajs, A. J. 1972. *M. N. R. A. S.* **157**, 1

Lynden-Bell, D., and Pringle, J. E. 1974. *M. N. R. A. S.* **168**, 603

Meyer, F., and Meyer-Hofmeister, E. 1981. *Astr. Ap.* **104**, L10

Murray, S., Bell, R., and Lin, D.N.C. 1989, in preparation

Nakagawa, Y., Watanabe, S., and Nakazawa, K. 1989, in *The Formation and Evolution of Planetary Systems*, eds H.A. Weaver, F. Paresce, and L. Danly, (Cambridge University Press: Cambridge), in press

Paczynski, B. 1978. *Acta Astr.* **28**, 91

Papaloizou, J.C.B. and Lin, D.N.C. 1985, *Ap. J.*, **285**, 818

Papaloizou, J.C.B. and Lin, D.N.C. 1989, *Ap. J.*, in press

Papaloizou, J.C.B. and Savonije, G.J. 1989, in preparation

Pringle, J.E. 1981, *Ann. Rev. Astr. Ap.*, **19**, 137

Ruden, S.P. 1987, Ph.D thesis, University of California, Santa Cruz

Ruden, S.P. and Lin, D.N.C. 1986, *Ap. J.*, **308**, 883

Ruden, S.P., Papaloizou, J.C.B., and Lin, D.N.C., 1988, *Ap. J.*, **329**, 739

Safronov, V.S. 1960, *Sov. Phys. Dokl.*, bf 5, 13

Sargent, A.I. and Beckwith, S. 1987, *Ap. J.*, **323**, 294

Sellwood, J., and Lin, D. N. C. 1989. In preparation

Shakura, N. I., and Sunyaev, R. A. 1973. *Astr. Ap.* **24**, 337

Shu, F.H. 1975, in *Structure and Evolution of Close Binary Stars*, eds. P.P. Eggleton, S.A. Mitton, and J.A.J. Whelan, (Dordrecht:Reidel), p. 251

Shu, F. H., Adams, F. C., and Lizano, S. 1987. *Ann. Rev. Astr. Ap.* **25**, 23

Smak, J. 1982. *Acta Astr.* **32**, 199

Spruit, H.C. 1987, *Astr. Ap.*, **277**, 312

Spruit, H. C., Matsuda, T., Inoue, M., and Sawada, K. 1987. *M. N. R. A. S.* **229**, 517

Strom, S. E., Edwards, S., and Strom, K. M. 1989. In *Formation and Evolution of Planetary Systems*, eds. H. A. Weaver, F. Paresce, and L. Danly (Cambridge University Press), in press

Toomre, A. 1964, *Ap. J.*, **139**, 1217

Vishniac, E.T. and Diamond, P. 1989, preprint

Walter, F. M. 1986. *Ap. J.* **306**, 573

Walter, F. M. 1987. *P. A. S. P.* **99**, 31

Weidenschilling, S.J. 1987, *Gerlands. Beitr. Geophys.*, **96**, 21

PROTOSTELLAR DISCS.

James E. PRINGLE
Institute of Astronomy, Madingley Road, Cambridge, CB3 0HA, England.

ABSTRACT. We argue on both observational and theoretical grounds that protostellar discs are often large and massive. We propose that for the short–lived stars there is evidence that such discs can exist through the mainsequence phase and become rejuvenated at the planetary nebula stage.

1. Introduction.

In the early half of the century ideas about star formation were dominated by interest in the formation of one particular star — the sun. It was well accepted that rotation played an important rôle in shaping the collapse. As Kippenhahn commented in his introductory address, it was from such considerations that the first papers on accretion discs by von Weizsäcker[1] and by Lüst[2] emerged. With the advent of computers, however, interest regressed to the study of spherically, symmetric star formation. What these studies have shown is that the collapse process is of such intricate complexity that modern computational techniques can barely resolve the necessary detail of the one–dimensional calculation, and have little chance of coping adequately with the required three–dimensional calculations[3,4].

I do not wish to review the large amount of work that has been done on rotational collapse[5], but will, rather, concentrate on what is expected to be the later stages of the collapse process, that is the development and evolution of an accretion disc. A simple picture of how such collapse might proceed is given by Cassen & Moosman[6]. Much of the work in this area has been carried out with the solar system in mind, with the result that discs considered to form are usually small (10–30 a.u.) and assumed to evolve rapidly (on a timescale of $\lesssim 10^6$ y) so that they fit with the minimum mass solar nebula of $\sim 10^{-2}\, M_\odot$[7]. However, if one looks at the observations of rotation in molecular clouds, one comes rapidly to the conclusion that the initial disc should often be made larger than this. The specific angular momentum in a Keplerian disc is

$$h_K = 4.5 \times 10^{19}\ R_{AU}^{\frac{1}{2}}(M/\, M_\odot)^{\frac{1}{2}}\ cm^2\ s^{-1}\, .$$

F. Meyer et al. (eds.), Theory of Accretion Disks, 105–111.

The specific angular momenta of rotating clouds tabulated by Goldsmith & Arquilla[8] range from $3 \times 10^{24}\ cm^2\ s^{-1}$ to $1.2 \times 10^{21}\ cm^2\ s^{-1}$, indicating discs (post–collapse) of radii extending upwards from ~ 700 a.u. Further if one assumes that binary star separations are indicators of the angular momentum contained in the initial collapsing cloud, then, again, the larger fraction of binaries with separations greater than ~ 30 a.u.[9] indicates that the initial discs can often be large.

With this in mind, Douglas Lin and I[10] have recently undertaken calculations similar to those of Cassen & Moosman[6] but considering the initial infall to discs of much greater radius. For such discs it is important to take account not just of the usual α–parametrized viscosity, but also to allow for the effect of the disc being self–gravitating. We take account of this effect in the manner suggested by Lin & Pringle[11] modified by an efficiency parameter η. Numerical computations of the effect of self gravity on discs[12] indicate that η might be in the region of $10^{-2} - 10^{-3}$. These calculations will be presented in detail elsewhere and I just mention here some relevant results. If we consider the collapse of a 1.5 $M_\odot$ cloud on a timescale of $t_f = 2.2 \times 10^5$ y and with sufficient rotation to give a disc radius of ~ 1700 a.u., then we find that if $\alpha = \eta = 1$, that is if the processes which redistribute angular momentum are at their most efficient, then the disc decays rapidly and by the time infall is complete most of the mass has accreted onto the central star. If however we take $\alpha = \eta = 0.01$, which are probably still on the high side of expectations, then the disc evolves much more slowly, and even after a time of 2×10^6 y the disc mass still exceeds the central stellar mass by a factor of three. The general conclusion from the calculations is that even some time after the infall is completed, the new–born star might have around it a large and massive disc. How long such a disc might survive is subject to considerable uncertainty, and rather than pursue the matter further by theoretical and speculative means, I would now like to consider what evidence there might be for such discs from the observational point of view.

2. Observations.

In the search for protostellar discs and their remnants there are two basic questions which need to be asked. First, is there evidence for a disc? Here the evidence is often compelling, but unfortunately often indirect (for example do jets or bipolar flows imply the presence of a disc?). Second, if the answer to the first is affirmative, can we learn anything about disc properties from it (e.g. disc mass, radius, evolution timescale). It is the second question which is of crucial importance to disc theorists, for little progress can be made in understanding disc processes until there is some interplay between theory and observations.

To consider what answers can be given to such questions I shall now consider some of the observational evidence. The examples I pick are intended to be indica-tive and not exhaustive.

2.1. PROTOSTARS.

The obvious place to look for evidence of discs is among the youngest stars. The best case for a recently formed massive disc is the disc observed by Sargent & Beckwith[13] around HL Tau. Using interferometric observations in the CO emission line to give a resolution of 6 arc sec (equivalent to ~ 1000 a.u.), they produced strong evidence of a disc around HL Tau. The disc appears to have a mass of $\sim 0.1\,M_\odot$ and to be rotating in Keplerian fashion around the $\sim 0.5\,M_\odot$ central star out to a radius of ~ 2000 a.u.

There are a few other examples of differentially rotating discs of much larger extent and around higher mass stars. Jackson, Ho & Haschick[14] demonstrate the presence of a differentially rotating disc or torus in NGC 6334I seen in the NH_3 emission. The central object appears to have a dynamical mass of $\sim 30\,M_\odot$, and deviations from Keplerian behaviour in the disc indicate that it has a comparable mass of $\sim 30\,M_\odot$, out to a radius of $\sim 0.3\,pc$. Observations of Vogel et al.[15] of IRc2 provide similar evidence of a differentially rotating flow around an object of mass $\sim 35\,M_\odot$, with the disc mass comparable to that of the central object out to a radius of $\sim 0.2\,pc$.

These observations hold out an exciting prospect for accretion disc theorists in that they indicate the future possibility of resolving such discs spatially. Such observations will make a tremendous impact on our understanding of the inner workings of accretion discs.

2.2. PRE MAIN SEQUENCE STARS.

2.2.1. T Tauri Stars. For the T Tauri stars the indications that discs are present is more indirect. One line of evidence is the observation[16,17] that the emission line profiles of forbidden lines (*e.g.* [OI] $\lambda6300$) show only the blue wing. If one assumes that the lines are formed in an outflowing wind then the natural assumption is that the geometry of the situation obscures the red wing. The presence of an accretion disc provides the required geometry. Then a calculation of the emitting volume in the forbidden line yields an estimate of how large the obscuring disc must be. Specifically it is found that the discs must have radii of at least 20–100 a.u.

Further indirect evidence for the presence of discs comes from the recent elaborations of the proposal by Lynden–Bell & Pringle[18] that the infrared and ultraviolet excesses in T Tauri stars could be explained by accretion through a disc. Using IRAS data, Cohen et al.[19] have suggested that if one takes a strong infrared excess (IR luminosity comparable to stellar luminosity) as evidence of a disc then about one third of all T Tauri stars have discs. A number of groups have been working on computing the spectra of such discs[20,21,22] and comparing them with the observations. The general conclusion appears to be that most of the discs are passive in the sense that most of the radiation they emit is not internally generated from accretion but merely irradiated stellar flux. This is bad news for disc theorists in that it means that the discs tell us little about the accretion process, although the flaring thought to be present in the outer parts of some such discs[20] may tell us about disc dynamics, and in particular about the dynamics of twisted discs. Calculations indicate that discs with accretion rates of $\dot{M} \lesssim 10^{-8} \, M_\odot \, yr^{-1}$ are passive. However some T Tauri stars — the 'continuum' stars with strong excesses do appear to have genuine accretion discs. Modelling of the disc and of the boundary layer can provide a good fit to the 2500 A° – 3.5μm region of the spectrum [22,23], although longer wavelengths are somewhat more problematic. The accretion rates required are in the region of $10^7 - 3 \times 10^7 \, M_\odot \, yr^{-1}$ and the disc radius is modelled out to ~ 14 a.u., although the region providing the best spectral fit has a radius of less than this. No estimate of disc mass is available directly from the observations, although if one estimates that such an accretion rate might continue for $\sim 10^6$ years (a typical T Tauri star age) then this implies a reservoir of matter in the outer disc regions of at least $\sim 0.1 \, M_\odot$.

2.2.2. FU Orionis stars. The estimate of disc mass made above is in line with recent work on the FU Ori stars[24,25]. These stars appear to be T Tauri stars whose accretion rate through the disc has suddenly increased to a value of $\sim 10^{-3} - 10^{-4} \, M_\odot \, yr^{-1}$. Continuum energy fits require disc radii of around 1 a.u. For these stars the evidence for a disc comes not just through spectral fits, but more directly from modelling of absorption line widths in the disc. Specifically, the absorption line widths are narrower in the infrared than in the optical demonstrating that the cooler radiating material rotates more slowly in line with accretion disc expectations[26]. Based on an outburst length of ~ 100 years and the observationally deduced accretion rate, one concludes that $\sim 10^{-2} \, M_\odot$ is accreted in an outburst. From estimates of the observed number of such stars it is concluded that each star undergoes several such outbursts in its lifetime. This in turn implies that a sizeable mass ($(\gtrsim 0.1 \, M_\odot)$ is stored at large disc radii, and further, that some mechanism is required to instigate the outbursts [25].

2.3. MAINSEQUENCE STARS.

We have seen above that premainsequence stars can have discs which are sizeable and massive, although getting a good handle on disc properties has proved somewhat elusive. The next obvious questions to ask are how long do such discs last and in what form, and the next obvious place to look for the answers to these questions is among the mainsequence stars.

The sun and the solar system demonstrate the existence of disc remnants around a solar mass star of age $\sim 4.5 \times 10^9$ y. The planets extend out to about 30 a.u. and the deduced initial disc mass (allowing for loss of H and He) is $\sim 10^{-2} \, M_\odot$[27]. It is by no means clear, however, that the radius of Neptune should be taken to demarcate the outer edge of the initial solar disc. For example, estimates of the mass of the comet cloud range up to $\sim 10^{-3} \, M_\odot$ [28] which implies an initial hydrogen and helium cloud (or disc) mass of $\sim 0.1 \, M_\odot$.

Dust discs have been found around α Lyr (A0V) and β Pic (A5V) which have ages $\sim 10^8$ y [29,30]. Although the discs have little residual mass, they have large radii extending for α Lyr to at least 85 a.u. These cases, however, are remnants of discs but no more than that. At some stage the disc has been dissipated, perhaps as much by radiation or wind from the central star as by accretion. It might be, therefore, that long–lived protostellar discs, are more readily found among the younger mainsequence stars, that is the O and B stars. There is of course, one obvious class of mainsequence stars which is modelled in terms of a surrounding disc — the Be stars. About 10 per cent of all B stars are Be stars with ages in the range $10^7 - 10^8$ y [31], and most models of the Be star phenomenon involve discs extending out to about 10 stellar radii. These discs are highly variable, and there appears to be a relationship between the disc, non–radial stellar pulsations and small outbursts. Most modellers have in mind (e.g. Ebbets[32]) that the star itself initiates an outburst (for no specified reason) and throws off a disc. To someone who works on accretion discs, it seems more plausible if it were the disc which initiated the outburst. If the disc is in fact a permanent feature, left over from the star formation process, and of large radius and mass, then one could envisage the inner edge occasionally dropping on to the star to produce the observed phenomena.

2.4. POST MAINSEQUENCE STARS.

Having perhaps strained credulity in asking you to consider Be stars as being the proud possessors of leftover protostellar discs, I now ask you to strain it still further and consider some post mainsequence stars.

Indeed the presence of discs around post mainsequence stars has been suggested by Balick & Preston[33] as a mechanism for shaping certain types of planetary nebulae in the 'bow tie' or 'butterfly' class, and these authors suggest that the discs could be remnants from the premainsequence phase. Further weight is given to these arguments by the detection of molecular hydrogen in such shaped planetary nebulae and the discovery that such planetary nebulae appear to be concentrated towards the galactic plane, indicating more massive and therefore shorter–lived, progenitors*[34].

It is also worth considering the object CRL2688 — the 'Egg Nebula'. This is thought to be an evolved star passing through the precursor stage to a planetary nebula. Observations using HCN emission[35] indicate that the HCN morphology is consistent with a toroidal density distribution. The HCN velocity maps show outflow at $\sim 20 \, km \; s^{-1}$, together with strong rotation, with rotation velocity $\sim 1 \, km \; s^{-1}$ at a radius of 4000 a.u. From the amount of angular momentum present in the flow it seems plausible that the velocity pattern is brought about by the interaction of a wind from the central star with a large and massive circumstellar disc left over from the period of star formation[36].

3. Discussion.

In conclusion, I would like to stress the following points:

- On theoretical grounds one expects that for many stars the formation process leaves behind a disc remnant which is large ($\sim 10^2 - 10^3$ a.u.) massive (disc mass comparable to stellar mass), and potentially long–lived.

- There is evidence for such discs around premainsequence stars.

- Disc remnants are seen around the older mainsequence stars.

- The Be star phenomenon may be the manifestation of such discs around the younger mainsequence stars.

- There is evidence that such discs might play a rôle in the shaping of planetary nebulae formed by the shorter–lived stars.

There is still much work to be done, both theoretical and observational on the nature, ubiquity and longevity of protostellar accretion discs.

REFERENCES.

1. von Weizsäcker, C.F. 1943, *Z. Astrophys.*, **22**, 319
2. Lüst, R. 1952, *Z. Naturforsch.*, **7a**, 87
3. Appenzeller, I. & Tscharnuter, W. 1975, *Astr. Astrophys.*, **40**, 397
4. Winkler, K.-H. 1978, *Moon & Planets*, **19**, 237
5. Bodenheimer, P. in *IAU Symposium No. 115*, eds. D Sugimoto *et al.*, Reidel, p.5
6. Cassen, P. & Moosman, A. 1981, *Icarus*, **48**, 353
7. Cassen, P., Shu, F.H. & Terebey, S. 1985, in *'Protostars and Planets II'*, eds. D.C. Black & M.S. Matthews, Univ. of Arizona Press, p.448
8. Goldsmith, P.F. & Arquilla, R. 1985, in *Protostars & Planets II'*, eds D.C. Black & M.S. Matthews, Univ. of Arizona Press, p. 137
9. Halbwachs, J.L. 1983, *Astr. Astrophys.*, **128**, 399
10. Lin, D.N.C. & Pringle, J.E., in preparation
11. Lin, D.N.C. & Pringle, J.E. 1987, *Mon. Not. R. astr. Soc.*, **225**, 607
12. Anthony, D.M. & Carlberg, R.G. 1988, *Astrophys. J.*, **332**, 637
13. Sargent, A.I. & Beckwith, S. 1987, *Astrophys. J.*, **323**, 294
14. Jackson, J.M., Ho., P.T.P. & Haschick, A.D. 1988, *Astrophys. J. Lett.*, **333**, L73
15. Vogel, S.N., Bieging, J.H., Plambeck, R.L., Welch, W.J. & Wright, M.C.H. 1985, *Astrophys. J.*, **296**, 600
16. Appenzeller, I., Jankovics, I. & Ostreicher, R. 1984, *Astr. Astrophys.*, **141**, 108

17. Edwards, S. Cabrit, S., Strom, S.E., Heyer, I., Strom, K.M. & Anderson, E. 1987, *Astrophys. J.*, **321**, 473
18. Lynden-Bell, D. & Pringle, J.E. 1974, *Mon. Not. R. astr. Soc.*, **168**, 603
19. Cohen, M., Emerson, J.P. & Beichman, C.A. 1989, *Astrophys. J.*, in press
20. Kenyon, S.J. & Hartmann, L. 1987, *Astrophys. J.*, **323**, 714
21. Adams, F.J., Lada, C.J. & Shu, F.H. 1988, *Astrophys. J.*. **326**, 865
22. Bertout, C., Basri, G. & Bouvier, J. 1988, *Astrophys. J.*, **330**, 350
23. Basri, G. & Bertout, C. 1989, *Astrophys. J.*, in press
24. Hartmann, L. & Kenyon, S.J. 1985, *Astrophys. J.*, **299**, 462
25. Hartmann, L. & Kenyon, S.J. 1987, *Astrophys. J.*, **312**, 243
26. Hartmann, L. & Kenyon, S.J. 1987, *Astrophys. J.*, **322**, 393
27. Hayashi, C. 1981, *Prog. Th. Phys. Suppl.*, **70**, 35
28. Bailey, M.E., McBreen, B.P. & Ray, T.P. 1984, *Mon. Not. R. astr. Soc.*, **209**, 881
29. Aumann, H.H., Gillett, F.C., Beichman, C.A., de Jong, T., Houck, J.R., Low, F.J., Neugebauer, G., Walker, R.G. & Wesselius, P.R. 1984, *Astrophys. J. Lett.*, **278**, L23
30. Hobbs, L.M., Vidal-Madjar, A., Ferlet, R., Albert, C.E. & Gry, C. 1985, *Astrophys. J. Lett.*, **293**, L29
31. Slettebak, A. 1982, in *IAU Symposium No. 98*, eds. M. Jaschek & H.-G. Groth, p.109
32. Ebbets, D. 1981, *Pub. Astr. Soc. Pacific*, **93**, 119
33. Balick, B. & Preston, H.L. 1988, *Astr. J.*, **94**, 958
34. Zuckerman, B. & Gatley, I. 1988, *Astrophys. J.*, **324**, 501
35. Bieging, J.H. & Nguyen-Quang-Rieu, 1988, *Astrophys. J.*, **324**, 516
36. Pringle, J.E. 1989, *Mon. Not. R. astr. Soc.*, in press

FORMATION OF VISCOUS PROTOSTELLAR ACCRETION DISKS.

Werner M. TSCHARNUTER
Institut für Theoretische Astrophysik, Universität Heidelberg, FRG

ABSTRACT. Recent calculations of the axisymmetric collapse of rotating interstellar clouds have shown that initially relatively slowly rotating, centrally condensed fragments are most promising candidates for modelling the formation of the (pre-)solar nebula. The gas density at which the collapse starts out is considered to be so high that the magnetic flux, which might have been able to stabilize the cloud fragment (also a substantial redistribution of angular momentum could have taken place) during the preceding evolutionary phase, has already been lost. Further transport of angular momentum is done by turbulent friction. Turbulence is accounted for as an α-disk-type viscosity with $0.1 \leq \alpha \leq 1$.

Angular momentum is efficiently removed from the first quasi-hydrostatic core, and the second collapse, triggered by dissociation of molecular hydrogen, is very similar to the well-known spherically symmetric case. A rapidly rotating, flattened stellar core forms. While gradually losing angular momentum this core tends to become increasingly more pressure supported, i.e., more spherical. But since the mass-weighted mean adiabatic exponent $\bar{\Gamma}_1 \approx 4/3$, which is the critical Γ_1 for dynamical instability of spherical stars, the core is always on the verge of being dynamically unstable. In addition, it is also vibrationally unstable, because the κ-mechanism is very efficient in the extended hydrogen ionization (and dissociation) zone. As a result, angular momentum transport *destabilizes* rotating protostellar cores. Depending on the initial and boundary conditions, the quasi stationary core formation and accumulation process evolves into a non-stationary phase during which core oscillations of large amplitudes take place. In the worst case (low initial density) the stellar core undergoes a series of complete disruptions and subsequent re-formations. The implications of such 'core activities' for the physical and chemical evolution of the ambient disk-like nebula remain to be discussed.

1. Introduction.

It is a well known fact that the specific angular momentum of dense molecular cloud fragments is still sufficiently high to inhibit single star formation, unless there is a possibility to redistribute angular momentum very effectively during contraction. Three principal mechanisms have been discussed which could be made responsible for the desired process that separates mass and angular momentum:

1. Turbulence, which gives rise to an effective turbulent eddy viscosity or 'turbulent friction'.

F. Meyer et al. (eds.), Theory of Accretion Disks, 113–123.
© 1989 by Kluwer Academic Publishers.

2. Magnetic fields, which can transport angular momentum in two different ways, namely,

   a) *within* the differentially rotating disk-like structure due to 'magnetic viscosity', so that the total amount of angular momentum is conserved,

   b) *away* from the system into the ambient interstellar medium, which will lead to a net loss of angular momentum ('magnetic braking').

3. Gravitational torques, if non-axisymmetric structures (e.g., trailing spirals) develop in the flattened system.

Since almost nothing is known about turbulence and magnetic fields in protostellar fragments, mechanisms 1 and 2a above (turbulent friction and magnetic viscosity) have been modelled in a very simple-minded fashion by using the so-called α-ansatz (Shakura and Sunyaev, 1973). Numerical α-disk models can be computed in axial symmetry, whereas mechanisms 2b and 3 require a fully three-dimensional (3-D) description. It is not surprising that numerical 3-D models are rare and suffer from poor spatial resolution and accuracy (cf. Dorfi, 1982; Boss, 1989).

As far as the importance of magnetic fields is concerned, above a certain critical density in between 10^{-14} and 10^{-12}g cm^{-3}, there should be substantial leakage of magnetic flux due to ambipolar diffusion and/or Ohmic dissipation (Nakano, 1984; Mestel, 1985). This process is highly welcome, because otherwise we would be faced with an inconvenient magnetic flux problem in addition to the already very troublesome angular momentum problem. The numbers for the critical density given above indicate that magnetic flux is lost either before or during the formation of the first, optically thick, core. This means that further evolution of the central high-density region of a rotating protostar is no longer dictated by magnetic forces. On the one hand there are still 10 orders of magnitude to be gained in order to arrive at stellar densities; on the other hand, however, there is presumably still so much angular momentum left in the first core that, without further efficient removal of angular momentum, e.g., by means of turbulent friction and/or gravitational torques, contraction or collapse of the first core to a *single* central stellar object is not possible.

Results discussed below refer to axially symmetric models with $\alpha = 0.1$ throughout. The kinematic turbulent viscosity ν_t is assumed to be

$$\nu_t = \alpha \cdot c_s \cdot \ell \tag{1}$$

where c_s denotes the local speed of sound and ℓ a length scale which is typically the local half-width of the disk. The full set of non-linear equations describing the physics of the collapse flow has been solved by virtue of an implicit (backward time differencing, in order to avoid the stringent Courant-Friedrichs-Lewy-condition for numerical stability) numerical scheme and a Newton-Raphson technique. I have used a pseudo-spectral method (finite differences in the radial direction, expansion into Legendre polynomials with respect to the polar angle) which takes the global conservation laws for mass, momentum, and internal energy strictly into account.

The numerical grid is self-adaptive in the radial direction, the mesh points are locally condensed, depending on the spatial variation of the physical quantities. The advection term is first order according to the simple upwind 'donor cell' differencing, but the effective accuracy obtained is locally equivalent to higher order schemes by virtue of the adaptive grid. This has been verified by test calculations which have been designed for the shock tube problem (slab symmetry) and for a (spherically symmetric) supernova explosion model (Dorfi and Drury, 1987).

Energy transfer by radiation is modelled using the Eddington approximation. Shock fronts are smoothed out by an artificial tensor viscosity (Tscharnuter and Winkler, 1979; Tscharnuter 1987b). Constitutive relations (equation of state, internal energy, opacities, ...) for a Pop. I mixture (X = 0.770, Y = 0.213, Z = 0.017) have been used. The most abundant species taken into account are hydrogen in the form of H_2, H°, H^+, helium as He°, He^+, He^{++}, and electrons e^-, heavier elements do not affect the thermodynamical equilibrium to any noticeable extent (cf. Wuchterl, 1989).

In the following Sections I shall discuss results and conclusions that have been derived from recent model calculations. Particular attention is drawn to the rôle of the initial conditions and symmetry assumptions for the stability behavior of protostellar cores. Furthermore, a sequence of collapse models referring to the formation of the solar nebula is briefly discussed. Finally, further arguments in favor of the physical reality and importance of core instabilities are presented on the basis of a simplified model.

2. Results.

The most important finding of earlier model calculations referring to the formation of the solar nebula was the onset of instabilities in the stellar core, i.e., the protosun. The core, containing only a few percent of a solar mass, was found to become dynamically unstable shortly after formation. The bounce and the ensuing expansion phase gives rise to a complete disruption of the core. Since the bulk of the matter is collapsing, the expansion in the central part of the cloud is eventually stopped and the overall collapse starts anew (Morfill et al., 1985). The whole cycle can be regarded as an oscillation of very large amplitude. Disruption and subsequent re-formation of the core taking place in a recurrent way have also been found in spherically symmetric collapse flows (Tscharnuter, 1987a).

2.1. HOMOGENEOUS STARTING CONDITIONS, LOW INITIAL DENSITY.

Two starting configurations were chosen for the calculations, namely, a $3M_\odot$ cloud fragment at uniform density of 10^{-20}g cm^{-3} and temperature of 8.3 K, with uniform angular velocity corresponding to two different values for the specific angular momentum, 10^{21} and 10^{20}cm^2 s^{-1}; $\alpha = 0.33$ was adopted for the turbulent viscosity parameter in eq. (1) above. Both model sequences exhibited strong core oscillations.

These results came as a surprise, and the question of the physical reality of this phenomenon — A. Boss named it protostellar core 'hiccups' — has been justly raised, because the numerical scheme adopted was not set up in conservation form. One major shortcoming of non-conservative schemes is that one cannot rule out a possibly inconsistent numerical treatment of the advection terms. As an example, mass and internal energy can be advected during one time step at different relative rates, in this way simulating a spurious, unphysical net energy transport. However, test calculations with the new code described briefly above, which does contain the strict conservation form for the finite differences, have quantitatively confirmed the previous results. Moreover, the adaptive grid method by Dorfi and Drury (1987) turned out to be so robust that artificial damping of the stellar core was not necessary for achieving numerical convergence. As a matter of fact, hiccups of the stellar core have been found to occur even in non-rotating, spherically symmetric collapse flows (Tscharnuter, 1987a). The crucial phase which commences just before the bounce takes place has been re-calculated by means of an explicit scheme (forward time differences) for comparison purposes (Boss and Tscharnuter, 1989). We followed the evolution of the core bounce for the classical $1M_\odot$ protostar with homogeneous initial density distribution at 10^{-19}g cm^{-3} and temperature at 10 K. Again, dynamical expansion ensued, and after several 10^5 explicit time steps we stopped the calculation, when the inner shock front that had been driven outward by the expanding core was about to catch up with the outer shock surrounding the first, optically thick core. We found *quantitative* agreement between the results which were obtained by entirely different numerical methods.

Of course, all these numerical findings cannot be considered as a *rigorous* proof that 'hiccups' are real physical phenomenon. It is therefore not surprising that some members of the scientific community who, for some reason or another, stick to the idea of quasi-stationary accretion, which makes the computational aspect of the stellar formation problem considerably less involved, and reject any other possibility as being 'unphysical' (e.g., F. Shu, private communication). Unfortunately, it is true that collapse calculations are very computer time consuming. So it is almost impossible to cover the whole interesting parameter space, which would be desirable to explore the dependence of the results on the physical (initial conditions, boundary conditions, approximations with regard to transport processes taken into account, constitutive relations, ...) and numerical (number of gridpoints, degree of local condensation of gridpoints, choice of timestep, numerical scheme, ...) model assumptions. However, there are some general, more qualitative results which might be relevant to the hiccup-problem.

According to test calculations, hiccups are *suppressed* if the accretion rate $\dot{M} \gtrsim 10^{-3}M_\odot\text{yr}^{-1}$ for a time long enough that the core is able to accumulate a certain minimum amount of mass (in between 0.05 and $0.1M_\odot$). A more rapid accretion rate implies the building up of a bigger core of low mean density. This is to be compared with more compact cores where a much larger fraction of hydrogen already starts ionizing, while dissociation of molecular hydrogen is not yet completed. A quantitative measure of the increasing 'delay' of *complete* H_2-dissociation with increasing density is the logarithmic derivative of the pressure P with respect to the density ρ at constant entropy S, i.e., the adiabatic exponent $\Gamma_1 = (\partial \ln P/\partial \ln \rho)_S$. For *spherically symmetric* hydrostatic layers there is a critical $\Gamma_1^{(crit)} = 4/3$. Γ_1 greater (or less) than $4/3$ indicates dynamical stability (or instability). So, the onset of global dynamical instability is expected if the bulk of the material satisfies the condition $\bar{\Gamma}_1 < 4/3$, where $\bar{\Gamma}_1$ is the mass-weighted mean of Γ_1. An adiabatically contracting core will then begin to collapse dynamically (second protostellar collapse!), because the rising pressure cannot compensate for the rising density, hence gravitation, if the flow is spherically symmetric. But after the core bounce ($\bar{\Gamma}_1 > 4/3$) the velocity field is directed outward and, if the flow were *strictly* adiabatic and $\bar{\Gamma}_1$ became smaller than $4/3$ again, the gravitational forces would not be able to catch up with the pressure forces, but would reinforce the expansion until $\bar{\Gamma}_1$ rises above $4/3$. Oscillations of large amplitude should then be observed. Non-adiabatic effects, such as radiation transport, generally tend to damp stellar oscillations, but the κ-mechanism is a good example of how energy of the radiative flow can be transferred into kinetic energy. Moreover, radiation losses from the inner protostellar core are almost negligible (Boss and Tscharnuter, 1989). Thus, protostellar core embryos have a chance to survive, only if they can grow very fast. A simple model which makes these lines of reasoning somewhat more precise, though it is *not* meant to be a *rigorous* mathematical proof of protostellar core 'hiccups', is given in Section 3 below.

2.2. SOLAR NEBULA FORMATION: INHOMOGENEOUS STARTING CONDITIONS, HIGH INITIAL MEAN DENSITY.

Let me now turn to the latest calculation (Tscharnuter, 1987b) that has been carried out with the particular aim to simulate the formation of the solar nebula by means of an axially symmetric collapse model including turbulent friction (the viscosity parameter $\alpha = 0.1$ throughout). A substantially precondensed, non-homogeneous cloud fragment of total mass M = $1.2 M_\odot$ and specific angular momentum $j = 1.26 \times 10^{20} \mathrm{cm}^2 \mathrm{\ s}^{-1}$ was chosen as the initial configuration. The central density and the mean density are 2.81×10^{-14} and $8.51 \times 10^{-15} \mathrm{g\ cm}^{-3}$, respectively, and the mean free-fall time is 721.3 yr accordingly. The fragment is almost rigidly rotating with an angular velocity of about $2.4 \times 10^{-11} \mathrm{s}^{-1}$, the temperature at the center is 47 K, the equivalent temperature of the external radiation field was assumed to be 10 K; the ratios of internal energy over gravitational energy and rotational energy over gravitational energy are 0.050 and 0.054, respectively.

Fig. 1. shows the evolution of the center in a *log* central density versus *log* central temperature plot. The path A–B–C–D is very similar to spherically symmetric models. It reflects the cooling phase (A–B) which should be compared with the isothermal evolution, if the cloud were chosen to be thermally adjusted to the external radiation field. Cooling is efficient if the Kelvin-Helmholtz time is smaller than the free-fall time. But the initial density is already so high that the collapse soon becomes adiabatic in the central parts of the cloud where the optical depth is rising quickly. Angular momentum is transported outward to such an extent that the second dynamical contraction phase (C–D) triggered by the destruction of molecular hydrogen is possible. The rising of Γ_1 slightly above $\Gamma_1^{(crit)}$ which is *lower* than 4/3 for flattened equilibria halts the contraction, and a phase of rapid mixing and redistribution of angular momentum within the stellar core ensues (D–E). Rotational energy is transferred into heat, the dissociation process completes and hydrogen is found only in the form of H° and H^+. During this evolution the core becomes closer and closer to a spherical shape on length scales of $6-8\times10^{11}$cm. $\bar{\Gamma}_1$ is found to vary always around 4/3; at point E a series of damped quasi-adiabatic oscillations starts in the partially ionized region in the absence of molecular hydrogen (E–F–G, G–H–I). The steepening of the adiabat just before point F indicates the narrow strip within which only atomic hydrogen is present and the adiabatic exponent Γ_1 approaches 5/3.

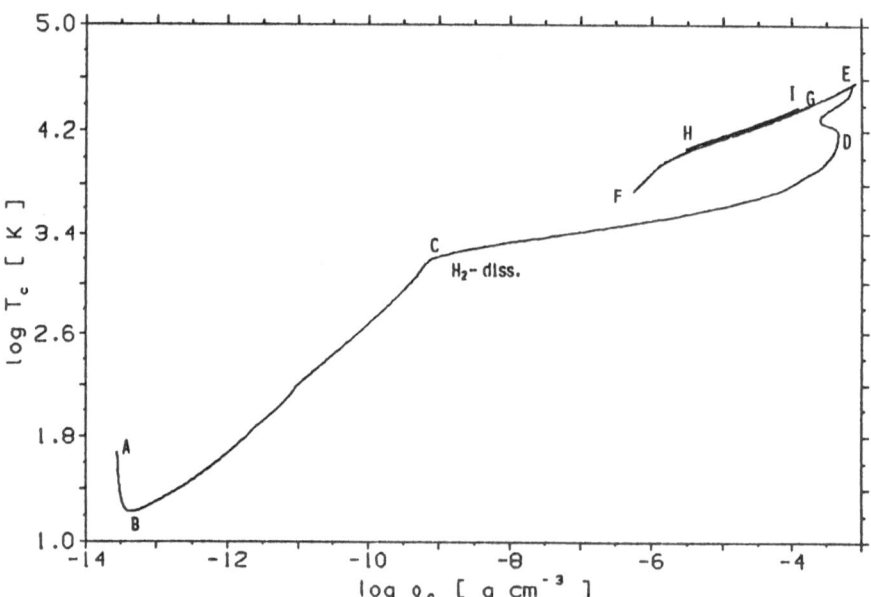

Figure 1: Central density ρ_c – central temperature T_c diagram. The labels "A" through "I" refer to characteristic stages of the evolution (for details see text).

The mass accretion rate turns out to be 10^{-3} to $10^{-2}M_\odot yr^{-1}$ which, according to spherically symmetric models, would be high enough to produce a stable stellar core (cf. Balluch, 1988).

As an illustration of the rather complex disk structure that has developed after 623 years (point "I" in Fig. 1), Fig. 2 and 3 show meridional cross sections with two different length scales. Plotted are equi-density, -temperature, and -angular velocity contours; the velocity field projected onto the meridional plane is indicated by arrows whose length are proportional to the absolute value of the (projected) velocity vector. The z-axis coincides with the rotational axis, symmetry is also assumed with respect to the equatorial plane ($z = 0$). One can distinguish between an outer disk extending to about the dimensions of our planetary system (Fig. 2) and an inner disk surrounding a spheroidal, pressure supported "protosun" (Fig. 3) within a distance of a few astronomical units (AU). The protosun contains only about $0.07 M_\odot$, more than 90% of the total mass is still far outside the — geometrically and optically — thick outer disk at distances larger than 100 AU or so. Material is continuously "raining" onto this disk which will grow to about $1 M_\odot$ within a few 10^3yr.

One might speculate that after the core oscillations have died out the outer disk will gradually evolve into a normal α-accretion disk, whereas the inner disk will probably degenerate into the boundary layer between the centrifugally supported disk and the pressure supported protosun. Since this evolution proceeds on a viscous time scale of 10^5 to 10^6yr which is much longer than the mean free-fall time of less than 10^3yr, we may conclude that the protoplanetary accretion disk is expected to be a massive one (cf. J. Pringle, this conference).

Unfortunately, it does not seem feasible at present to push the model calculations a major step forward and cover the viscous evolution of the protoplanetary disk together with the gradually changing internal constitution of the growing protosun. This difficult task could presumably be accomplished, only if a *fully adaptive, implicitly defined two-dimensional numerical grid* were available; without such an essential numerical tool, which has to be developed in the near future in order to be able to proceed, the computational problems are and will stay insurmountable, even though more powerful supercomputers of the new generations to come will be at our disposal.

3. Discussion.

The most salient feature of the early phases of viscous protostellar disk formation is the 'activity' of the young stellar core. Up to now I have reported only on numerical experiments which have led us *a posteriori* to the conjecture that the 'observed' disruptive core expansions or quasi-adiabatic core oscillations of large amplitudes relate to

1. a mass-weighted mean adiabatic exponent $\bar{\Gamma}_1 < 4/3$,
2. the geometrical shape of the core becoming spherical after angular momentum has been removed due to turbulent friction.

It should be noted in this context that the numerical scheme adopted is by no means

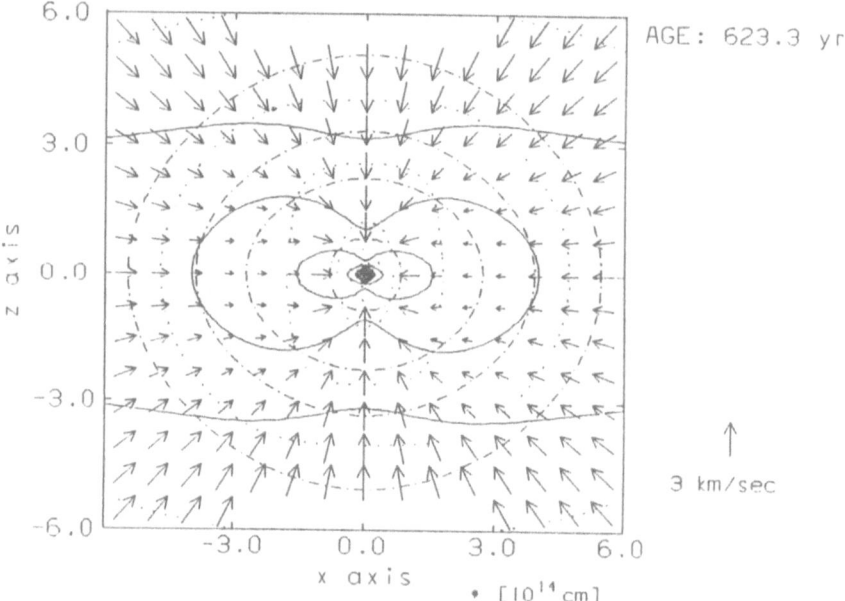

Figure 2: Meridional cross section of model "I" in Fig. 1 at a length scale of 40 AU ($\approx$ semi-major axis of Pluto's orbit) after 623.3 yr since the beginning of the collapse at point "A". The z-axis is the rotational axis. Plotted are equi-contours of the density ρ (full line), temperature T (dash-dotted line), and angular velocity Ω (dotted line). Between two successive contour lines the respective distances are: $\Delta \log \rho = 1.0$, $\Delta \log T = 0.2$, $\Delta \log \Omega = 0.5$. All quantities are increasing with decreasing radial distance from the center starting from outside with $\log \rho_1/\mathrm{g\,cm^{-3}} = -13.0$, $\log T_1/\mathrm{K} = 2.6$, $\log \Omega_1/\mathrm{s^{-1}} = -10.0$. Arrows indicate the velocity field projected onto the meridional plane. The scale (3 km/s) is given by the single arrow on the right hand side. The optical depth in the vertical direction at a distance of 40 AU from the axis is of the order of unity.

appropriate for studying stellar oscillations. First order backward time differences which allow large time steps are well known to introduce a substantial numerical damping. This is exactly why the scheme is so very powerful if applied to collapse flows. Variations on small time scales are effectively damped, unless there are real physical excitation mechanisms, e.g., the κ-mechanism, operating which *enforce* small time steps for the calculation. In other words, it would be rather unwise to use backward time differencing for investigating pulsations of cepheïds, since the physical excitation mechanism could easily be overwhelmed by the huge amount of numerical damping introduced by the code. But in cases where we do find oscillations in the calculation we must conclude that we should indeed look for a

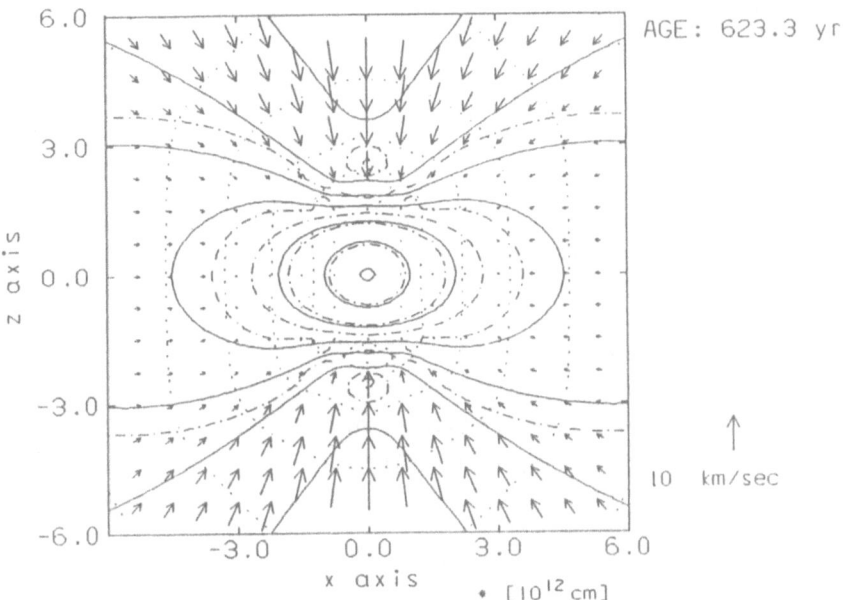

Figure 3: Meridional cross section of model "I" in Fig. 1 at a length scale of 0.4 AU ($\approx$ semi-major axis of Mercury's orbit). The equi-contour lines and the arrows have the same meaning as in Fig. 2. $\Delta \log \rho = 1.0$, $\Delta \log T = 0.2$, $\Delta \log \Omega = 0.2$. The distribution of density and angular momentum is monotonic, the temperature distribution shows a (certainly not well resolved) 'hot spot' produced by the polar accretion shock with a peak value of about 4000 K. $\log \rho_1/\mathrm{g\ cm}^{-3} = -10.0$, $\log T_1/\mathrm{K} = 3.4$, $\log \Omega_1/\mathrm{s}^{-1} = -6.6$. Note the equi-contour lines for Ω which turn out to be parallel to the rotation axis in "quiet" (low linear velocity) regions inside the "inner" (see text) disk.

strong driving mechanism. In the sequel, we shall briefly discuss the one-zone-model (OZM) by N. Baker (1966) which was developed originally for interpreting model calculations pertaining to cepheïd pulsations.

The basic idea of the OZM is immediately written down: Consider a "representative" optically thick, spherically symmetric layer and linearize the three basic structure equations (equations of continuity, motion, and energy balance). In doing so, one ends up with an algebraic equation of third order for the eigenvalues which yield three necessary conditions for stability. Let $\Gamma_1 := \left(\frac{\partial \ln P}{\partial \ln \rho} \right)_S$, $\nabla_{ad} := \left(\frac{\partial \ln T}{\partial \ln P} \right)_S$, $\rho_P := \left(\frac{\partial \ln \rho}{\partial \ln P} \right)_T$, $\rho_T := - \left(\frac{\partial \ln \rho}{\partial \ln T} \right)_P$, $\kappa_P := \left(\frac{\partial \ln \kappa}{\partial \ln P} \right)_T$, $\kappa_T := \left(\frac{\partial \ln \kappa}{\partial \ln T} \right)_P$, where ρ, P, T, S, and κ denote the density, pressure, temperature, entropy, and opacity, respectively. We then find necessary conditions for

$$\text{dynamical stability:} \quad \Gamma_1 > 4/3, \tag{2}$$

$$\text{secular stability:} \quad (4\rho_P - 3)(\kappa_T - 4) + 4\rho_T(\kappa_P + 1) > 0, \tag{3}$$

$$\text{vibrational stability:} \quad \nabla_{ad} - \frac{1}{4}(\nabla_{ad}\kappa_T + \kappa_P) - \frac{1}{3\Gamma_1} > 0. \tag{4}$$

In Section 2, I have already discussed the significance of dynamical stability. Putting in numbers for the material functions and the equation of state for typical densities and temperatures in protostellar cores we find that strong vibrational instability is indicated due to increasing opacity with increasing temperature (κ-mechanism) by H^- absorption and hydrogen ionization. Wuchterl (1989) found the vibrational instability to inhibit the direct formation of (spherically symmetric) Jupiter-like gaseous planets with solid cores on the basis of the so-called core-induced instability; most of the gas in the planet's gaseous envelope is lost after outward travelling shocks have triggered heavy mass loss. Here again an expansion phase ensues after some accumulation has already taken place.

I should like to emphasize the fact that a simple-minded and rather straightforward application of the OZM to protostellar core embryos must be treated with caution, since the quasi-hydrostatic protostellar cores are not necessarily in thermal equilibrium. Nevertheless, the OZM can be taken as a useful hint, though certainly not as a compelling proof, that protostellar core 'hiccups' are real physical phenomena and not merely numerical artefacts. First results of a much more involved linear stability analysis of protostellar accretion flows point toward a confirmation of the results obtained by applying the simple OZM. A more detailed comparison will be made in M. Balluch's forthcoming Ph.D. thesis (University of Heidelberg).

4. Conclusions.

There remain a few general statements to be made:

- We have seen that there is good evidence for protostellar collapse flows to be intrinsically non-stationary. If this is true then the question is legitimate: What do we really learn from quasi-stationary similarity solutions?
- Collapse calculations *excluding* the interaction of the central object with the ambient accretion disk do not seem to be very useful, because the accretion rate of mass and angular momentum, and hence the global properties of the (unresolved) central star, depend on the dimensions of the central "sink"; core activities are suppressed, but the periodical heating and cooling of the disk as a consequence of an oscillating protosun could have important consequences for the chemistry in the ambient solar nebula.
- Permanent cores will form, only if the mass accretion rate remains high enough for a sufficiently long time interval subsequent to the formation of the stellar core. This is achieved by choosing a much higher density for the initial fragment

than the Jeans criterion would indicate. This implies very short time scales for the formation of a massive protostellar accretion disk (at most a few 10^3 years) for rotating clouds, or otherwise, for spherical stars which would appear as luminous object on top of the Hayashi track.

To summarize, we may state that we have just begun to realize how complicated the star formation process really is. Both extended numerical experiments and thorough (semi-)analytical investigations are necessary to settle the intriguing issue of the stability of protostellar cores and accretion shocks.

ACKNOWLEDGEMENTS. I thank my collaborators M. Balluch and G. Wuchterl for many enlightening discussions about the stability behavior of accreting very low-mass stellar objects. I am indebted to Y. Osaki for focussing my attention to the weak points concerning the application of the OZM to protostellar cores during a most interesting breakfast conversation. I also thank Dr. Davina Innes for reading the manuscript carefully, her remarks and suggestions greatly helped to make the presentation much more readable and transparent.

REFERENCES.

Baker, N.: 1966, *Simplified Models for Cepheïd Instability*, in: *Stellar Evolution*, eds. R.F. Stein and A.G.W. Cameron, Plenum Press, New York, p. 333

Balluch, M.: 1988, *Astron. Astrophys.* **200**, 58

Boss, A.P.: 1989, *Evolution of the Solar Nebula. I. Nonaxisymmetric Structure During Nebula Formation, preprint*, to appear in *Astrophys. J. October 1st issue*

Boss, A.P, Tscharnuter, W.M.: 1989, *Protostellar Core Instabilities: Verification of Dynamical Hiccups, preprint*, to appear in *Astron. Astrophys. Letters*

Dorfi, E.A.: 1982, *Astron. Astrophys.* **114**, 151

Dorfi, E.A., Drury, L.O'C.: 1987, *J. Comput. Phys.* **69**, 175

Mestel, L.: 1985, *Physica Scripta* **T11**, 53

Morfill, G.E., Tscharnuter, W.M., Völk, H.J.: 1985, *Dynamical and Chemical Evolution of the Protoplanetary Nebula*, in: *Protostars and Planets II*, eds. D.C. Black,and M.S. Matthews, Univ. Arizona Press, Tucson, p. 493

Nakano, T.: 1984, *Fundamentals of Cosmic Physics* **9**, 139

Shakura, N.I., Sunyaev, R.A.: 1973, *Astron. Astrophys.* **24**, 337

Tscharnuter, W.M.: 1987a, *Models of Star Formation*, in: *Physical Processes in Comets, Stars, and Active Galaxies, Lecture Notes in Physics*, eds. E. Meyer-Hofmeister, H.C. Thomas, and W. Hillebrandt, Springer Verlag, p. 96

Tscharnuter, W.M.: 1987b, *Astron. Astrophys.* **188**, 55

Tscharnuter, W.M., Winkler, K.-H.A.: 1979, *Computer Physics Communications* **18**, 171

Wuchterl, G.: 1989, *Zur Entstehung der Gasplaneten: Kugelsymmetrische Gasströmungen auf Protoplaneten*, Dissertation, Univ. Wien

FORMATION AND STRUCTURE OF PROTOSTELLAR ACCRETION DISKS.

Ralph E. PUDRITZ

Department of Physics, McMaster University,
Hamilton, ON L8S 4M1, Canada

1. Introduction.

The physics of accretion disks revolves around the two issues of how disks form and how angular momentum is transported within them. In the case of cataclysmic variables, the formation mechanism is well understood to be a Roche lobe overflow in a low mass, binary stellar system. There is increasing evidence for the existence of accretion disks around young stellar objects. Both low and high mass star formation regions contain protostellar disks (eg.; S106 and HL Tau respectively) and these are at least a tenth the mass of the central star (and perhaps more). While the angular momentum transport mechanisms in all types of disks could be similar, their formation mechanisms are of course not. Disks in molecular clouds could be produced as a consequence of gravitational instability and collapse in strongly magnetized, dense, cold gas.

A plausible reason why disks form so freely in molecular clouds is that the latter have globally ordered magnetic fields whose energy density is comparable to gravity. If the fields were purely static, it is well known that rapid collapse along field lines will produce disks. There is one caveat-fields cannot be static because line widths tell us that supersonic motions abound in clouds. It is well known that MHD waves in a strong magnetic field can provide effective pressure support of molecular gas against collapse (Arons and Max 1975). However, MHD waves do not propagate on all scales in molecular clouds. At sufficiently high freqencies ω, wave motions must damp out because of friction effects in the cloud. Thus, there is some scale λ_{min} smaller than which waves cannot provide pressure against gravity. This scale is typically 10^{-1} pc.-which is comparable to the size of dense cores and extended flattened structure in molecular clouds. Since magnetic fields still thread gas on scales $\lambda < \lambda_{min}$- collapse into disks is likely to occur.

This paper summarizes two stages of work on disk formation and evolution:

1. Disk Formation: I present a a linear instability analysis of a two fluid (ions

125

F. Meyer et al. (eds.), Theory of Accretion Disks, 125–133.

and neutrals) self gravitating, partially ionized, warm, and magnetized cloud. The
new issue studied here is the effect that a background spectrum of MHD waves
has in the linear stability problem (Pudritz 1989). A spectrum of Alfvén waves
traversing a cloud produces a stress in the gas (Dewar 1970) which can in principle
stabilize it against collapse (Bonazzola et al 1987).

2. Disk Structure: I discuss recent observations of disks around young stellar
objects - in particular-HL Tau. We have made Infrared Images of these objects
using the French INSU CIRCUS camera (Monin, Pudritz, Lacombe, and Rouan
1989). These systems still have flattened, disk like structures around them. Between
$1 - 2\mu$m, these dusty disks scatter light from the central star. By using theoretical
scattering calculations, we attempt to reproduce these images in order to constrain
the disk geometry and density profile (Lazareff, Pudritz, and Monin 1989).

2. Disk Formation: Instabilities in Magnetized Clouds.

A molecular cloud can be treated as a two fluid medium; a neutral one, and an
ion-magnetic one (denoted by the suffixes n, i respectively). Each of the two fluids
obeys a continuity equation. In the absence of a wave field in the gas, the equations
of motion of the neutral and ion fluids are

$$\rho_n \frac{\partial v_n}{\partial t} + \rho_n(v_n.\nabla)v_n = -\rho_n\nabla\Phi - \nabla p_n - F_f$$

$$\rho_i \frac{\partial v_i}{\partial t} + \rho_i(v_i.\nabla)v_i = -\rho_i\nabla\Phi - \nabla p_i + \frac{(\nabla \times b) \times b}{4\pi} + F_f$$

where $p_{i,n}$ are the ion and neutral gas pressures, $\rho_{i,n}$ are the ion and neutral gas
densities and $v_{i,n}$ are their velocities, Φ is the gravitational potential, b is the
magnetic field, and F_f is the friction force which arises from neutral-ion collisions.
The response of the magnetic field to the ion motions is described by

$$\frac{\partial b}{\partial t} = \nabla \times (v_i \times b); \qquad \nabla.b = 0.$$

These equations are supplemented with Poisson's equation for a two fluid medium
and the equation of state of the fluids which are both taken to be isothermal.
    The friction force density is well known;

$$F_f = \rho_n.x\nu_n.(v_n - v_i)$$

where $x\nu_n$ is the collision frequency of a given neutral with the ions. For typical
conditions in molecular clouds, the ionization fraction is predominantly due to
metallic ions although this may break down in the densest inner portions of accretion
disks.

The gravitational response frequency in the neutral fluid is given by $\omega_{g,n}^2 = 4\pi G \rho_{n,o}$ so that the gravitational response time scale is typically

$$\omega_{g,n}^{-1} = 0.72 \times 10^6 n_3^{-1/2} \quad years = 1.63 t_{ff}$$

The fact that this number is of order 10^6 yrs, while molecular clouds survive for 10^7 years clearly shows the importance of identifing the support mechanism for molecular clouds. It is interesting to note that the neutral-ion collision time scales directly with this gravitational time scale since

$$\frac{x \nu_n}{\omega_{g,n}} = 3.57 (\zeta_{-16})^{1/2}.$$

for an ionization rate per H atom of ζ_H s^{-1}.

2.1 DISPERSION RELATIONS.

Observations show that magnetic fields in clouds cannot be static. It is nonetheless important to consider first the condition for local stability of the cloud in a uniform field. Following the Jeans analysis for local stability, the unperturbed state is taken to have constant density, uniform magnetic field, and zero velocities. The perturbations have the form $\Psi_1 \propto exp[i(k.x - \omega t)]$ where ω is the wave frequency of the perturbation and k the wave vector. Two geometries are possible for perturbations in static, ordered magnetic fields:

1. *Perturbed fluid motion parallel to field:* $\rightarrow$ magnetic field cannot support the gas. It is easy to show that when $k \parallel v_{i,1} \parallel b_o$, the magnetic force term in the dispersion relation vanishes.

2. *Perturbed fluid motion perpendicular to field:* $\rightarrow$ field can strongly retard the motion and help support the gas. In this case $v_{i,1} \perp b_o$, there is a net magnetic contribution to cloud support. In the remainder of sections 2.1 and 2.2, I *assume Case 2.*

The full dispersion relation for clouds with arbitrary ionization may be found in Pudritz 1989 (see also Langer 1978). This relation is a quartic in the perturbing frequence ω as is to be expected in any gravitational stability calculation. The presence of friction in general introduces damping, which formerly introduces a cubic term. Fortunately, this most general fourth order dispersion relation can be considerably simplified in the case of molecular clouds.

2.2. STABILITY OF STATIC CLOUDS.

Since molecular clouds have tiny ionization fractions ($x \simeq 10^{-7}$), one may safely take the limit $x \rightarrow 0$. Real clouds have magnetic energy densities comparable to gravity so that $\omega_n \ll \omega_{A,i}$ The dispersion relation for molecular clouds then reduces to the cubic

$$(\omega^2 - \omega_{A,i}^2)\omega + i\nu_n(\omega^2 - x\omega_{A,i}^2\Gamma) = 0. \tag{2.1}$$

where $\omega_{A,i} \equiv kV_{A,i}$, the Alfvén speed of waves in the ionized medium is

$$V_{A,i}^2 \equiv \frac{b_o^2}{4\pi\rho_{i,o}},$$

and where it is convenient to define an Alfven speed in the neutrals as $V_{A,n} \equiv x^{1/2}.V_{A,i}$. The function

$$-\omega_{A,n}^2\Gamma(k) \equiv \omega_{g,n}^2 - k^2(V_{A,n}^2 + c_n^2) \equiv \Omega_n^2 \tag{2.2}$$

measures the relative importance of magnetic and pressure support with respect to gravity at any wavenumber k. In a Jeans analysis, the sign of Ω_n^2 is crucial in deciding the stability of the gas, and this same feature is preserved as I will shortly demonstrate.

There is one final simplification of the relation (2.1) that is useful. In the *long wavelength limit*, we consider frequencies $\omega \ll \nu_n$. This regime corresponds to scales $> x^{1/2}\lambda_c \simeq 10^{-4}$ pc. In this case (2.1) reduces to a quadratic;

$$\omega^2 + i(\frac{\omega_{A,i}^2}{\nu_n})\omega - x\omega_{A,i}^2\Gamma = 0 \tag{2.3}$$

whose solutions are;

$$\omega_\pm = \frac{1}{2}[-\frac{\omega_{A,i}^2}{\nu_n}i \pm \sqrt{(-\frac{\omega_{A,i}^4}{\nu_n^2} + 4x\omega_{A,i}^2\Gamma(k))}] \tag{2.4}$$

In the case that gravity and gas pressure are negligible: then $\Gamma = 1$ and the relation reduces to that of MHD waves in a partially ionized gas first derived by Braginskii (1965) and Kulsrud and Pearce, 1969 (KP). These solutions correspond to two damping waves. One sees that there is a critical Alfven frequency

$$\omega_{A,n;c} = 2x\nu_n,$$

such that waves of higher frequency are completely damped by friction. Thus wave propagation for long wavelength modes occurs only if $\omega_{A,n} \equiv kV_{A,n} < \omega_{A,n;c}$. Equivalently, wave propagation is only possible for wavelengths longer than a critical length $\lambda > \lambda_c$. The ratio of this critical scale to the cloud size λ_o is

$$\lambda_c/\lambda_o = 3.5 \times 10^{-2}(B/B_c)\zeta_{-16}^{-1/2}$$

where (B/B_c) is the ratio of the magnetic field to an isotropic critical field B_c that could support the cloud.

The *stability* of the cloud (using 2.4) is determined by the following results:

$\Omega_n^2(k) < 0$ two oscillating solutions above critical wavelength; pure exponential damping at smaller wavelengths

$\Omega_n^2(k) > 0$ no oscillations; one exponentially growing and one damping mode.

The frequency which demarcates the transition from oscillating to growing or damping modes is found by setting the discriminant in (2.4) equal to zero. Thus the critical frequency is;

$$\omega_{A,c;\pm}^2 = 4x\nu_n^2\kappa[\frac{1}{2} \pm \frac{1}{2}\sqrt{(1 - (\frac{\omega_{g,n}}{x\nu_n\kappa})^2)}]$$

where the factor $\kappa \equiv 1 + (c_n/V_{A,n})^2$ takes finite gas pressure effects into account.

Propagation occurs for in the semi-infinite set of long wavelengths for which $\omega_{A,i} < \omega_{A,c}$. When gravity is included propagating waves are restricted to a band, ie

$$\omega_{A,c;-} < \omega_{A,i;propag} < \omega_{A,c;+}$$

where the upper and lower limits are the two roots. The reason for this is that at long enough wavelengths (small enough wavenumber and hence small enough $\omega_{A,i}$) the gas becomes gravitationally dominated and Jeans instability must ensue. Thus wave propagation is restricted by friction at the small wavelength end and by gravity at the long wavelength end of our regime. In between, the waves are slow enough that the fluids respond together as a fast magnetosonic mode, yet quick enough to beat the deadening gravitational response.

Instabilities develop in the regime $\Omega_n^2 > 0$. Thus, there is a critical wave number k_J such that gravitational instabilities occur for $k > k_J$, and is

$$k_J^2 = \frac{(V_{A,n}^2 + c_n^2)}{\omega_{g,n}^2}$$

as is to be expected for warm magnetized, self gravitating gas. The *growth rates* of these modes is however modulated by friction in an important way (see Figure 1).

2.3. STABILITY OF CLOUDS TRAVERSED BY MHD WAVES.

Clouds cannot have static fields and mode-mode coupling makes it likely that a spectrum of MHD waves will be present (McIvor 197). Consider a wave spectrum of the form

$$V_{A,n}(k) = V_{A,o}.(\frac{k_o}{k})^\beta. \tag{2.5}$$

where the power law index β for the MHD wave spectrum is taken to be arbitrary. The scale λ_o is taken to be the cloud size and $V_{A,o}$ is denotes the Alfvén speed associated with this largest possible scale. For a cloud of uniform density, the ratio of magnetic to gravitational energy scales as

$$\frac{U_{mag}}{U_{grav}} = \frac{V_{A,n}^2}{V_{ff}^2} \propto \lambda^{2(\beta-1)}.$$

This shows that waves support large scales against collapse with spectra $\beta > 1$ while small scales are supported for spectra $\beta < 1$. The most relevant issue for cloud stability therefore is how to self-consistently treat the stability of a gas which is itself traversed by waves.

The physics of such media has been examined by Dewar (1970). For an ionized fluid, the equation of motion is modified by the addition of the radiation stress of the wave.

$$-\nabla.P_w$$

to the ordinary gas pressure term $-\nabla p_i$. For Alfven modes, the pressure tensor is isotropic and Dewar (1970) has prescribed how to calculate it for a collection of waves. I consider stability to collapse in directions parallel to the field where the Jeans condition is now modified by a term given in equation (2.5).

Numerical Results. The exact growth rate for any mode in a given wave spectrum depends upon 2 parameters, namely the ratio of the sound to global Alfvén velocity,

$$c_n/V_{A,o} \equiv c$$

and the overall ratio of the gravitational to magnetic energy density on the largest cloud scales

$$a \equiv (\frac{\omega_{g,n}}{k_o V_{A,o}})^2$$

where the unperturbed cloud size is λ_o. In real clouds a is observed to be within a factor of two of unity (Myers and Goodman 1988).

It is convenient to scale the growth time of the unstable modes in terms of the gravitational frequency; $(t_{inst}\omega_{g,n})$. For convenience, the function Ω_n^2 normalized to the square of the gravitational frequency $\omega_{g,n}^2$. The unstable modes are those for which $\Omega_n^2 > 0$. All the calculations shown in the figures assume an ionization rate of $\zeta_{-17} = 1$.

The steeper the wave spectra develop the faster growing instabilities since intermediate cloud scales are effectively without magnetic support. Two important features about the results emerge. First, *wavelengths for fastest growth only occur for spectra with $\beta > 1$*. For power laws with $\beta < 1$, the growth rate is a monotonic function of the wavenumber in which the *largest* scales grow the fastest. The presence of a maximum when $\beta > 1$ is new. Thus a spectrum of hydromagnetic waves

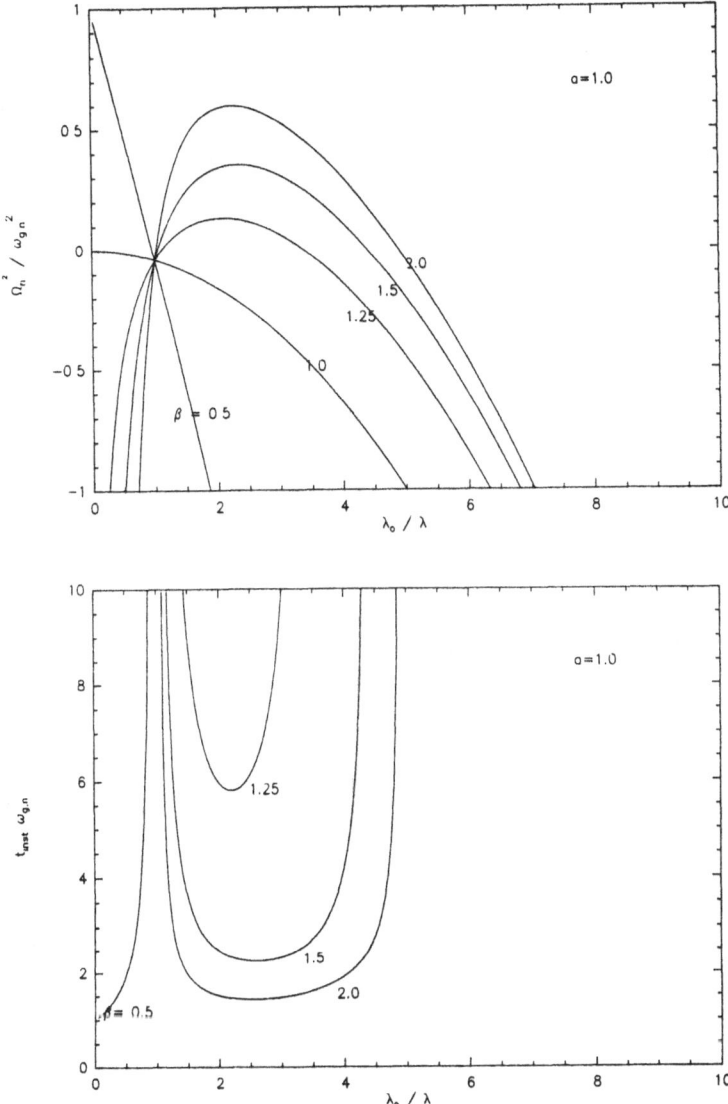

Figure 1: Dependence of instability wavelengths and growth times upon cloud magnetization (measured by $a \equiv (\omega_{g,n}^2/k_o V_{A,o})^2$). The figures show the function Ω_n^2 (1 a) and the instability growth time t_{inst} (1 b) as functions of the ratio λ_o/λ where the cloud size is (λ_o) and wavelength of the excitation (λ). The clouds have equal gravitational and magnetic energy densities $(a = 1.0)$. Unstable modes are all those for which $\Omega_n^2 > 0$. The plots show that for spectral indices $\beta > 1$, there is a most rapidly growing mode present (t_{inst} is a minimum at this wavelength). In all plots the ratio of the cloud sound speed to global Alfvén speed is c=0.2. The global gravitational response time of the cloud is $\omega_{g,n}^{-1} = 1.63 t_{ff}$.

with a sufficiently steep spectrum stabilizes the longest scales in a cloud against gravitational collapse.

The second point to note is that *the exact range of modes which goes unstable depends not only on the cloud magnetization a, but also on the perturbing wave spectrum β.* Thus in Figure 1a and b, one sees that for a spectrum with $\beta = 2$, there is a small band of unstable modes in spite of the fact that globally speaking, the cloud is well stabilized by the magnetic field. This result suggests that the problem of core formation cannot depend upon just one parameter such as the cloud magnetization a, but also upon the spectrum of hydromagnetic disturbances. Temperature effects also have an important effect on the instability scale (see Pudritz 1989).

The wavelength of the fastest growing mode $\equiv \lambda_m$ is tabulated in Pudritz 1989. As an example, it is found that for $\beta = 3/2$, $a = 1.0$, and $c = 0.1$, it is $\frac{\lambda_m}{\lambda_o} = 0.23$. The growth time for this mode is $t_m = 1.5\omega_{g,n}^{-1} = 2.45 t_{ff}$. In terms of Jeans lengths, the dominant mode develops at

$$\frac{\lambda_m}{\lambda_J} = 1.34$$

Thus the dominant instability scale is different than the Jeans length, the modification having to do with the presence of an MHD wave spectrum.

3. Disk Structure.

There is considerable evidence that many, if not most, young stellar objects are associated with disks. Millimetre observations are limited to scales of no smaller than 3". CO observation of HL Tau have detected a flattened, rotating structure (Sargent and Beckwith 1987). We observed this system using an infrared camera, achieving 0.8" seeing at CFHT. The images in the 4 IR bands are given in Monin et al (1989). Since scattering dominates the dust opacity in the $1 - 2\mu m$ band, our J,H, and K images are produced by the scattering of photons emitted from the central star off the surrounding, disk like structure. By imaging disks in the very earliest phases of protostellar evolution, we hope to gain valuable information about the nature of the disk formation mechanism, as well as constraints on disk models.

We have resolved the scattering region in HL Tau which has a radius $R_d = 500 A.U.$ Assuming that scattering off of grains is the dominant opacity, then the disk mass can be constrained. The ratio of scattered to "stellar" light at H band is $\simeq 0.15$ so $\tau_{scat} \geq 0.15$. For scattering at 2.2 μm from a mixture of graphite and silicate grains the scattering opacity is $\sigma = 20$ cm^2 per gram of gas, while for pure silicates it is $\sigma = 2$. The dust masses for the disk of radius 500 A.U. then ranges from $.7 \times 10^{-3} - .7 \times 10^{-2} M_\odot$. Since the gas to dust ratio in typical gas is a factor of 100, we have evidence for a mass in the HL Tau system of $.07 - .7 M_\odot$. The co-rotation radius for a centrifugally balanced disk is

$$r_{co-rot} = 500 \left(\frac{M_d}{0.15 M_\odot} \right) \cdot \left(\frac{v_\phi}{0.5 km \quad s^{-1}} \right)^{-2} \quad A.U.$$

We are currently calculating theoretical IR images to be expected from disks. We assume the disk density, scale height, etc. are power laws in radius. These disks are embedded in standard surrounding isothermal spheres (the core). Our *preliminary* calculations (Lazareff, Pudritz, and Monin 1989) suggest that the HL Tau disk flares geometrically and is substantially inclined. The fact that a bright star like image is always found in the centre of the highest intensity contours suggests a concave disk surface. A structure such as a thick disk with a narrow funnel is unlikely to produce such images.

It is a pleasure to thank the organizers for orchestrating such a stimulating workshop, and for providing me with financial support. I acknowledge support for this research through a grant from NSERC, Canada.

REFERENCES.

Arons, J., and Max, C.E. 1975, *Ap. J.*, **196**, L77
Bally, J., Langer, W.D., Stark, A.A., and Wilson, R.W. 1987, *Ap. J.*, **312**, L45
Bonazzola, S., Falgarone, E., Heyvaerts, J. Pérault. M. and Puget, J.L. 1987, *Astron. Ap.*, in press
Dewar, R.L. 1970, *Phys. Fluids*, **13**, 2710
Lazareff, B. Pudritz, R.E., and Monin, J-L 1989, in prep
Langer, W.D. 1978, *Ap. J.*, **225**, 95
Kulsrud, R., and Pearce, W.P. 1969, *Ap. J.* **156**, 445
Myers, P.C., and Goodman, A. 1988, *Ap. J. Lett*, **326**, L31
Monin, J-L, Pudritz, R.E., Lacombe, F., and Rouan, D 1989, *Astron. Ap. Letters*, in press
Pudritz, R.E. 1989, *Ap. J.*, submitted
Sargent, A.I. and Beckwith, S. 1987, *Ap. J.*, *323*, 294

MERGING WHITE DWARFS, DISK FORMATION AND TYPE I SUPERNOVAE.

Mario LIVIO
Department of Physics, Technion, Haifa 32000, Israel.

ABSTRACT. The evolutionary path leading to the formation of double white dwarf systems is briefly reviewed. The coalescence process of the two white dwarfs is described and its implications for Type Ia supernovae are discussed. It is shown that the fate of the merged configuration, depends crucially on the evolution of the heavy disk formed when the secondary white dwarf is totally destroyed following Roche lobe overflow.

1. Introduction.

Most models for Type Ia supernovae involve in one form or another an accreting carbon-oxygen white dwarf in a binary system. This is a consequence of the fact that exploding white dwarfs are capable of producing ejecta abundances, as well as expanding envelope dynamics, which can lead to synthetic spectra which agree quite well with observations (e.g. Nomoto et al., 1984, Woosley and Weaver 1986a, b, Branch et al. 1985, Wheeler, Harkness and Cappellaro 1987).

It is, however, very questionable whether a white dwarf can be driven to the Chandrasekhar mass by the accretion of hydrogen rich material (e.g. MacDonald 1984). This is a consequence of the fact that periodic mass ejections during nova erruptions, can cause even a decrease in the white dwarf's mass, if the accretion rate is below some critical value $\sim 10^{-8} M_\odot /yr$. The last conclusion seems to be supported by observations of nova ejecta which all show enrichment in material from the underlying white dwarf (e.g. Truran and Livio 1986, Livio and Truran 1987). Some fraction of Type Ia supernovae may still result from symbiotic stars which accrete at a rate above the critical value (e.g. Kenyon 1986, Livio, Prialnik and Regev 1989), or from recurrent novae, in which the accreting white dwarf is probably very close to the Chandrasekhar limit (Webbink et al. 1987). The difficulty in accounting for the statistics of Type I supernovae in terms of hydrogen accreting white dwarfs, has led to a scenario which involves the coalescence of two white dwarfs (Iben and Tutukov 1984, Webbink 1984), with a total mass exceeding the Chandrasekhar limit. In what follows, we shall describe briefly the evolutionary

135

path which can lead to the formation of double white dwarf systems (Section 2), we shall then discuss the merger process and its implications for Type I supernovae (Section 3).

2. Common Envelope Evolution and the Formation of Double White Dwarf Systems.

The evolutionary path that is supposed to lead to a close binary white dwarf system, starts either with two stars of comparable masses (in the range $4 - 6M_\odot$) at separations $10-100R_\odot$, or with two stars of disparate masses (in the range $5-9M_\odot$) at separations $70-1500R_\odot$. In the former case, the first mass transfer episode (the primary undergoing an early case B Roche lobe overflow), is expected to be approximately conservative and only when the original secondary fills its Roche lobe, a common envelope phase is expected to ensue. In the case of disparate masses, two (at least) common envelope phases can be expected to occur (the first mass transfer episode is a case C or late case B).

The term "common envlope" refers to a situation in which the binary components (or their cores) spiral-in inside a typically non-corotating extended envelope. The principal outcome of a common envelope phase is a reduction in the binary separation (due to the loss of orbital energy and angular momentum), which can be accompanied by the ejection of the common envelope (Ostriker 1975, Paczynski 1976). The formation of the common envelope is a consequence of a dynamical mass transfer event, which, in turn, results from the fact that mass is transferred from the more massive component, in a stage in which it possesses a convective envelope (giant or AGB phase). Under such circumstances, an unstable mass transfer process is obtained (e.g. Paczynski and Sienkiewicz 1972), since the mass losing primary is unable to contract as rapidly as its Roche lobe.

The most direct evidence for the reality of a common envelope phase is provided by the existence of planetary nebulae with close binary nuclei (see e.g. Bond and Livio 1989, for an updated list). The process which led to the ejection of the nebula in this case, must have involved a phase in which the core of the AGB star and the secondary, spiralled-in, embedded in a common envelope. Furthermore, the application of the "interacting winds" model (Kwok 1982, Kahn 1983) for the shaping of the planetary nebula in this case (Balick 1987, Soker and Livio 1989), resulted in morphologies which are consistent with the observed ones (Bond and Livio 1989).

The difficulty in calculating common envelope evolution, is associated mainly with the fact that this is a three-dimensional problem, in which processes occuring on a wide range of length scales and timescales take place. Nevertheless, two-dimensional calculations (assuming axial symmetry) were attempted by Bodenheimer and Taam (1984) and Taam and Bodenheimer (1989) and recently, a preliminary three-dimensional calculation was performed (Livio and Soker 1988).

These calculations have demonstrated that : (i) **a large fraction of the envelope can be ejected**, and (ii) **mass ejection occurs in a narrow angle around the orbital plane**.

Probably the most important physical quantity determining the fate of a binary undergoing common envelope evolution, is the efficiency with which orbital energy is deposited into the ejection of the common envelope. This can be expressed by the parameter α_{CE} defined by (Livio and Soker 1988, Tutukov and Yungelson 1979)

$$\alpha_{CE} \equiv \frac{\Delta E_{bind}}{\Delta E_{orb}}, \tag{1}$$

where ΔE_{bind} is the binding energy of the ejected material and ΔE_{orb} is the change in the binary orbital energy (between the initial and final configurations). Clearly the value of α_{CE} determines the final separation of the binary emerging from the common envelope.

The processes which can act to decrease the value of α_{CE} are:
(1) Efficient energy transfer, which can transport the energy generated by the drag to the surface, thus suppressing the deposition into mass motion.
(2) The ejection of a relatively small amount of mass around the orbital plane with velocities exceeding the escape velocity, rather than imparting exactly the energy needed for escape to the entire envelope. The calculations of Taam and Bodenheimer (1989) and Livio and Soker (1988) implied typical values of $\alpha_{CE} \sim 0.3 - 0.6$.

The value of α_{CE} can be increased (in principle at least) if additional energy sources are included (retaining, however, the formal definition in Eq. (1)). These possible sources include (see Livio 1989 for a discussion): the recombination energy in the ionization zones and mass ejection by the AGB star itself (like in the formation of planetary nebulae by single stars).

We shall now describe a typical evolutionary path which can lead to the formation of a double white dwarf system (see also Iben and Webbink 1988 for discussion). For this, we note that if approximate expressions are used for ΔE_{bind} and ΔE_{orb}, Eq. (1) can be expressed as (Iben and Tutukov 1984, 1985)

$$\frac{GM_1^2}{a_i} \simeq \alpha_{CE} \frac{GM_{1R}M_2}{a_f} \tag{2}$$

where M_1 and M_2 are the masses of the primary and secondary (respectively), M_{1R} is the mass of the primary's core and a_i, a_f are the initial and final separations. If we then, start with an initial binary of masses $M_1 = 9M_\odot$, $M_2 = 5M_\odot$ at a separation $a_i = 1200R_\odot$, then following two common envelope phases, a final pair of carbon-oxygen white dwarfs will be obtained, at a separation

$$a_f \simeq 0.47R_\odot\left(\frac{\alpha_{CE}^{(1)}}{0.4}\right)\left(\frac{\alpha_{CE}^{(2)}}{0.4}\right)\left(\frac{M_{1R}}{M_\odot}\right)^2\left(\frac{M_{2R}}{M_\odot}\right) \tag{3}$$

where $\alpha_{CE}^{(1),(2)}$ denote the values of α_{CE} in the first and second common envelope phases (respectively). Such two white dwarfs will be brought together by gravitational radiation on a timescale (M_1, M_2 denote here the white dwarf masses)

$$\tau_{GR} \simeq 6 \times 10^8 yr (\frac{M_1}{M_\odot})^{-1} (\frac{M_2}{M_\odot})^{-1} (\frac{M_1 + M_2}{M_\odot})^{-1} (\frac{a}{R_\odot})^4. \tag{4}$$

An important prediction of the proposed scenario is that a population of close binary white dwarf systems should exist. The existence of such a population has been recently established, with the discovery of Saffer, Liebert and Olszewski (1988) that L 870-2 is a double line, spectroscopic binary white dwarf system, with an orbital period of $P = 1.55578$ days. In addition, Ringwald (1988) found evidence that the white dwarfs WD 1919+145 and WD 2032+248 have close companions, with the latter system being potentially a double white dwarf binary. A few other candidates (in which variability was detected) were discovered by Tytler (1989). In this respect, the null results reported by Robinson and Shafter (1987), from a radial velocity search intended to detect pairs with orbital periods between 30 sec and 3 hr, in a sample of 44 white dwarfs, is somewhat disappointing.

In the next section we shall outline the evolution of a binary white dwarf system, following the point in which the separation between the two components has been sufficiently reduced, for the secondary to fill its Roche lobe.

3. The Coalescence Process and Type I Supernovae.

Once the lighter white dwarf fills its Roche lobe, mass transfer begins. For a primary of mass $M_1 \simeq 1 M_\odot$ a dynamical mass transfer event is obtained whenever $M_2 > 0.5 M_\odot$. Benz et al. (1988) and Benz, Cameron and Bowers (1988) have presented a three-dimensional simulation of the mass exchange process, using Smooth Partcle Hydrodynamics, for a $1.2 + 0.9 M_\odot$ binary. They found that in a little more than two orbital periods, the secondary is completely destroyed, forming a thick disk around the primary. Mochkovitch and Livio (1989a,b) constructed models of such primary - massive disk configurations, using the self consistent field method for differentially rotating, self gravitating systems (Ostriker and Mark 1968, Clement 1974). The important point to note, in relation to Type I supernovae, is the fact that **although the total mass of the configuration exceeds the Chandrasekhar mass, collapse does not immediately ensue.** The reason for this is that most of the disk is rotationally supported and in addition, thermal pressure plays a non-negligible part in its inner parts (and in the primary's outer parts). Mochkovitch and Livio (1989b) have shown that cooling alone is not sufficient to induce collapse, the central density still being considerably lower (in the cold configurations) than $\rho_{crit} \sim 10^9$ g cm^{-3}. The fate of the merged configuations thus relies mainly on the rate of removal of angular momentum from the disk, which in turn, determines the accreting rate. In particular, for accretion rates $\dot{M} < 10^{-6} M_\odot/yr$ central carbon ignition can ensue, leading (potentially) to a standard Type Ia supernova. For higher accretion rates, carbon may ignite off-centered, with the possible transformation of the entire white dwarf to O-Ne-Mg composition (Nomoto and Iben 1985).

Mochkovitch and Livio (1989a,b) considered viscous transport of angular momentum and obtained lower and upper limits to the viscous timescale. They have shown that if the disk remains laminar, with viscosity being that of degenerate electrons, then a viscous transport timescale of $\tau_{vis} \simeq a\ few \times 10^8\ yrs$ is obtained (when the disk-primary boundary layer is included). If the high shear boundary layer is excluded, the timescale is of order 10^{10} yrs. However, Mochkovich and Livio have shown that the Richardson criterion

$$R_i = \frac{N^2}{(r\frac{d\omega}{dr})^2} > \frac{1}{4} \tag{5}$$

is violated over a large part of the disk (here N is the Brunt-Vaisala frequency and ω is the angular velocity). This may imply that a large part of the disk is turbulent. When a turbulent viscosity $\eta_{turb} \simeq \rho v_t \ell_t / R_e^c$ is used (R_e^c is the critical Reynolds number), a lower limit for the timescale for viscous transport in the disk is obtained, of order $\tau_{vis} \sim 10$ days. Associated with the large turbulent viscosity is an extremely high rate of dissipation, $L_{Diss} \simeq 10^{44}\ ergs^{-1}$, which would lead to carbon ignition.

We thus find, that the final outcome of the coalescence of the two white dwarfs, depends crucially on the evolution of the massive disk that is formed following dynamic mass exchange. Since this is a self-gravitating disk, various instabilities (such as spiral ripples) may be expected to appear and dominate the subsequent evolution. Off-center ignition will certainly occur, both when the stream of mass transfer strikes the primary (e.g. Benz, Cameron and Bowers 1988) and in the disk. Depending on the ensuing accretion rate, the primary will expand and evolve in a way similar to that obtained in the calculations of Iben (1988) or Kawai, Saio and Nomoto (1988). The question whether this scenario will lead to a Type Ia supernova remains entirely open. However, due to the fact that the total mass exceeds the Chandrasekhar limit, a violent outcome seems inevitable.

REFERENCES.

Balick, B., 1987, Astron. J., **94**, 671.
Benz, W., Bowers, R.L., Cameron, A.G.W. and Press, W.H. 1988, preprint.
Benz, W., Cameron, A.G.W. and Bowers, R.L. 1988, preprint.
Bodenheimer, P. and Taam, R.E. 1984, Ap. J. **280**, 771.
Bond, H.E. and Livio, M. 1989, Ap. J., submitted.
Branch, D., Doggett, J.B., Nomoto, K. and Thielemann, F.K. 1985, Ap. J., **294**, 619.
Clement, M.J. 1974, Ap. J. **194**, 709.
Iben, I. Jr. 1988, Ap. J. **324**, 355.
Iben, I. Jr. and Tutukov, A.V. 1984, Ap. J. Suppl. **54**, 335.
Iben, I. Jr. and Tutukov, A.V. 1985, Ap. J. Suppl., **58**, 661.

Iben, I. Jr. and Webbink, R.F., 1988, preprint.

Kahn, F.D. 1983, in *Planetary Nebulae*, IAU Symp. 103, ed. D.R. Flower (Dordrecht: Reidel), p. 305.

Kaway, Y. Saio, H. and Nomoto, K. 1988, Ap. J. **328**, 207.

Kenyon, S.J. 1986, *The Symbiotic Stars*, Cambridge University Press.

Kwok, S. 1982, Ap. J. **258**, 280.

Livio, M. and Truran, J.W. 1987, Ap. J. **318**, 316.

Livio, M. and Soker, N. 1988, Ap. J., **329**, 764.

Livio, M., 1989, in IAU Coll. 107, *Algols*, Ed. A.H. Batten, in press.

Livio, M., Prialnik, D. and Regev O. 1989, Ap. J., in press.

MacDonald, J. 1984, Ap. J. **283**, 241.

Mochkovitch, R. and Livio, M. 1989a, Astron. Ap., **209**, 111.

Mochkovitch, R. and Livio, M. 1989b, in preparation.

Nomoto, K., Thielemann, F.K. and Yokoi, K. 1984, Ap. J. **286**, 644.

Nomoto, K. and Iben, I. Jr. 1985, Ap. J. **297**, 531.

Ostriker, J.P. 1975, paper presented at IAU Symp. 73, Cambridge, England.

Paczynski, B. 1976, in IAU Symp. 73, *Structure and Evolution of Close Binary Stars*, eds. P. Eggleton, S. Mitton, and J. Whelan (Dordrecht:Reidel). p. 75.

Paczynski, B. and Sienkiewicz, R. 1972, Acta Astr. **22**, 73.

Ringwald, F.A. 1988, preprint.

Robinson, E.L. and Shafter, A.W. 1987, Ap. J. **322**, 296.

Saffer, R.A., Liebert, J. and Olszewski, E.W. 1988, preprint.

Soker, N. and Livio, M. 1989, Ap. J., in press.

Taam, R.E. and Bodenheimer, P. 1989, Ap. J., in press.

Tytler, D. 1989, in *White Dwarfs*, ed. G. Wegner (Berlin: Springer Verlag), in press.

Truran, J.W. and Livio, M. 1986, Ap. J. **308**, 721.

Tutukov, A.V. and Yungelson, L.R. 1979, Acta Astr., **23**, 665.

Webbink, R.F. 1984, Ap. J., **277**, 555.

Webbink, R.F., Livio, M., Truran, J.W. and Orio, M. 1987, Ap. J. **314**, 653.

Wheeler, J.C., Harkness, R.P. and Cappellaro, E. 1987, in *13th Texas Symp. on Relativistic Astrophysics*, ed. M.P. Ulmer, (Singapore: World Scientific), p. 402.

Woosley, S.E. and Weaver, T.A. 1986a, Ann. Rev. Astron. Ap. **24**, 205.

Woosley, S.E. and Weaver, T.A. 1986b, in *Radiation Hydrodynamics in Stars and Compact Objects*, IAU Colloq. 89, ed. D. Mihalas and K.H. Winkler (Springer Verlag), p. 91.

VARIABILITY OF ACTIVE GALACTIC NUCLEI AND GALACTIC QPO SOURCES: A DIAGNOSIS.

Marek A. ABRAMOWICZ, Ewa SZUSZKIEWICZ[1], and
Frederik WALLINDER[2]
Scuola Internazionale Superiore di Studi Avanzati
Strada Costiera 11, Trieste 34014, Italy
[1] Present address: Max Planck Institute for Astrophysics, Munich, Germany
[2] Permanet address: Lund Observatory, Lund, Sweden

ABSTRACT. The similarities in variability patterns of Galactic X-ray sources and active galactic nuclei are quite apparent when one realises that the corresponding time scales must differ by a factor $\sim 10^8$ because of the mass difference. This indicates a very similar physical mechanism of variability operating in all these objects. We argue that the "normal branch" quasi periodic oscillations (2-5 Hz) observed by *Exosat* and *Ginga* in about twenty low mass X-ray binaries, slightly slower ($\sim 10^{-1}$ Hz) oscillations recorded by *Ginga* in two Galactic black hole candidates, and the much longer quasi periodic variability with timescales of a few hundreds days found in optical light curves of several active galactic nuclei are all related to a thermal instability of the innermost part of accretion disks around neutron stars and black holes.

We propose a specific mechanism in which the thermal instability and a stabilizing effect of a strong advective cooling, caused by the general relativistic effect of the Roche lobe overflow, produce together cyclic changes in the structure of the innermost part of the disk.

1. Symptons and Signs.

About a year ago we presented at the 20th Yamada Conference in Tokyo a long and very detailed review on both observations and theory of variability of active galactic nuclei (Abramowicz and Szuszkiewicz, 1988). We concluded, not very optimistically, that neither was there a general agreement concerning the interpretation of observational data, nor were the theoretical concepts close enough to reality. We listed several theoretical suggestions of how to explain fragmented and often confusing observations in terms of non-stationary accretion processes, varying supply rates, radiation transfer instabilities, or gravitational lenses. In the present paper we examine more closely the possibility connected with the accretion disk instabil-

F. Meyer et al. (eds.), Theory of Accretion Disks, 141–166.
© *1989 by Kluwer Academic Publishers.*

142

ities. Like the other ones also this particular possibility is described, at its present state of development, more in a qualitative than quantitative way. The best it can offer is a *diagnosis* rather than deep or consistent understanding.

According to *The Oxford Paperback Dictionary* a diagnosis is "a statement of the nature of a disease or other condition made after observing its signs and symptons", but we have used here this word in a slightly different sense, which cannot be better explained than by a short dialogue[1] between Doctor Faustus and a Professor who discovered an infallible method of diagnosing cancer at a very early stage.

"It is simple", explained the Professor, "One must carbonize the organism and pulverize it first, then one adds a quite cheap reagent which without question detects cancer viruses. I have made thousands of successful tests. The method works best when the carbon comes from the bones."

"But the organism itself", asked Faustus, "I suppose it is then not worth anything ?"

"This is a problem for a therapist. Let everybody mind his own business."

Our diagnosis is based on an attempt to unify variability properties of active galactic nuclei (AGN), low mass X-ray binaries (LMXB), and galactic black hole candidates (GBHC) which are thought to be connected with accretion disks around black holes or neutron stars.

In Figure 1.1 we present three "typical" examples of power spectra (power vs. frequency) of an LMXB object (Cyg X-2), a galactic black hole candidate (LMCX-1) and a hybrid from two active galactic nuclei (high frequencies: X-ray data for NGC 5506, low frequencies: optical data for ON 231). The power spectra look quite similar, being featureless at both low and high frequencies and showing a broad band bump at intermediate ones. The bump indicates quasi periodic behaviour. There are about twenty galactic X-ray sources which show similar behaviour as Cyg X-2. Most of them are accreting neutron stars (with very weak or no magnetic fields) in low mass X-ray binaries. There is one[2] example of another galactic black hole candidate (GX339-4) showing behaviour similar to that of LMCX-1. The quasi periodic oscillations (QPO) in the Galactic X-ray sources have periodicities either in the "slow" range of about $2-5$Hz or in the "fast" range of about $20-50$Hz. The slow QPO belong to the so-called normal branch and fast ones to the horizontal branch. The names refer to the X-ray "colour - colour" diagrams. We shall not discuss details, which may be found in a long and full review by Lewin, van Paradijs and van der Klis, 1988 or more concise reviews by Lamb, 1988 and Hasinger, 1988. (See also White 1989, this Workshop).

Quasi periodic variability with much longer time scales, typically $10^{-7}-10^{-8}$Hz was also reported in several active galactic nuclei and reviewed in our Yamada

---

[1] Taken from "Faust" by Leszek Kołakowski. Our translation from Polish original (in rhymes) is rather approximate.
[2] Most recently Tanaka (1989) reported QPO in still another black hole candidate GS2000+25. It has the period of 2.6 Hz.

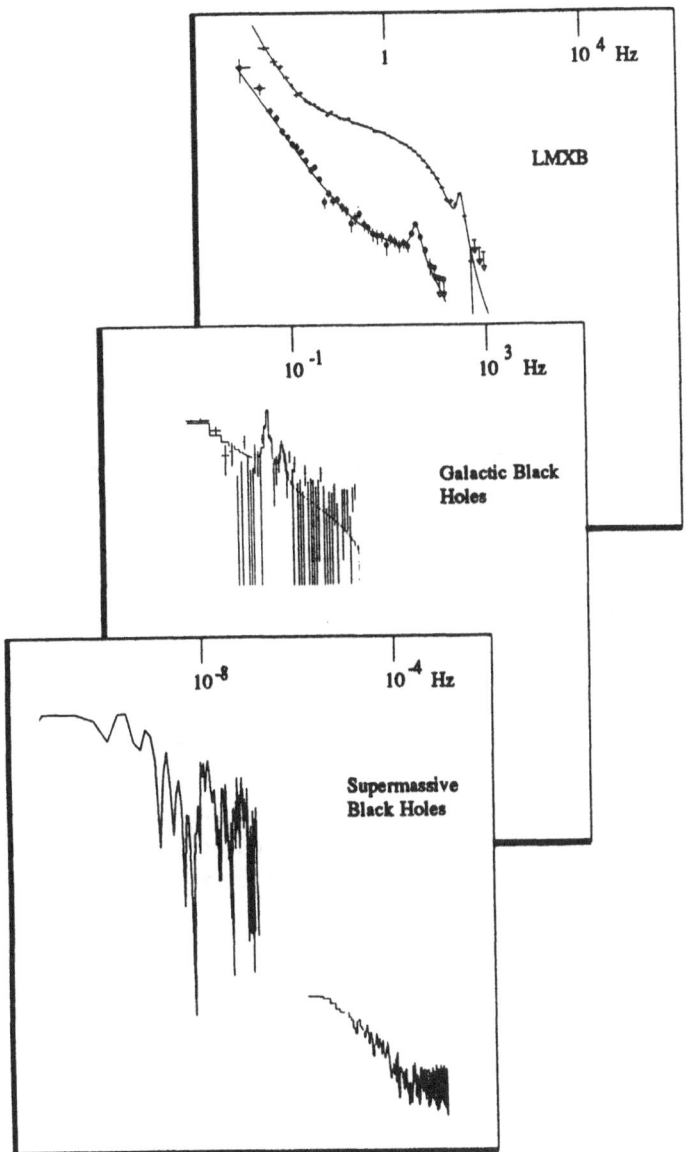

Figure 1.1: Power spectra of several sources showing quasi periodic variability.

article. Perhaps because of an old fashioned technique used to detect them (optical observations with the same instrument span over many years) the QPO in active galactic nuclei have gained almost no publicity. However, they may be just a scaled up version of QPO in low mass X-ray binaries. An observational support for such a suggestion is presented in Figure 1.2. The QPO frequencies in the sources connected with Galactic or extra galactic black holes have been scaled (according to the masses of the holes) by a factor shown in Figure 1.2 before the name of each object. In all but one case (NGC 1566) the QPO frequencies connected with black holes fall in the same, rather narrow, range. They are smaller by a factor of about 10 than the slow, normal branch, QPO frequencies connected with LMXB, which contain neutron stars.

2. Diagnosis.

If the similarity in variability patterns reflects (as we assume) similarity in basic physics of time dependent processes in LMXB, GBHC and AGN, then gravitational lensing is ruled out as a possible explanation for variability (LMXB and GBHC are not lensed). Varying supply rate is excluded if one adopts the theory of AGN fueling worked out by Begelman and Frank (1989, this Workshop). According to the theory the inner accretion disk is supplied with the interstellar gas coming from more distant parts of the host galaxy. The supply rate does not change in timescales much longer than those observed.

Thus, there are only two likely theoretical possibilities left: instabilities of accretion disk or some variability in transfer of radiation.

The second possibility was extensively discussed by several authors in terms of the inverse Compton scattering model. It has been suggested that the high energy X-rays from accreting compact objects are the result of high energy gain in the inverse Compton scattering of low energy photons emitted from the accretion disk by hot plasma clouds or corona. There are several instabilities of this process (some of them involve electron - positron pairs) which may, in principle, explain variability. However, they depend strongly on many unknown parameters and it is still not possible to estimate their characteristic timescales of variability with the same precision as in the case of the accretion disk instability model. In addition there are some observational indications that things may work differently than assumed in the inverse Compton scattering model. According to the model if there is a time variation in the initial low energy X-ray photons, the corresponding time variation of the high energy X-rays will be delayed. Miyamoto et al. (1988) have examined recently the delayed hard X-rays from Cygnus X-1 using the *Ginga* satellite and concluded that the observed delay time *cannot* be due to inverse Compton scattering but may be connected, for example, with the emission mechanism in the accretion disk. In our opinion, because of these uncertainties and difficulties, the inverse Compton scattering model (in its present form) cannot be used to explain the observed variability.

On the other hand the accretion disks instability model quite accurately predicts the characteristic variability timescales. They are very similar to those which are

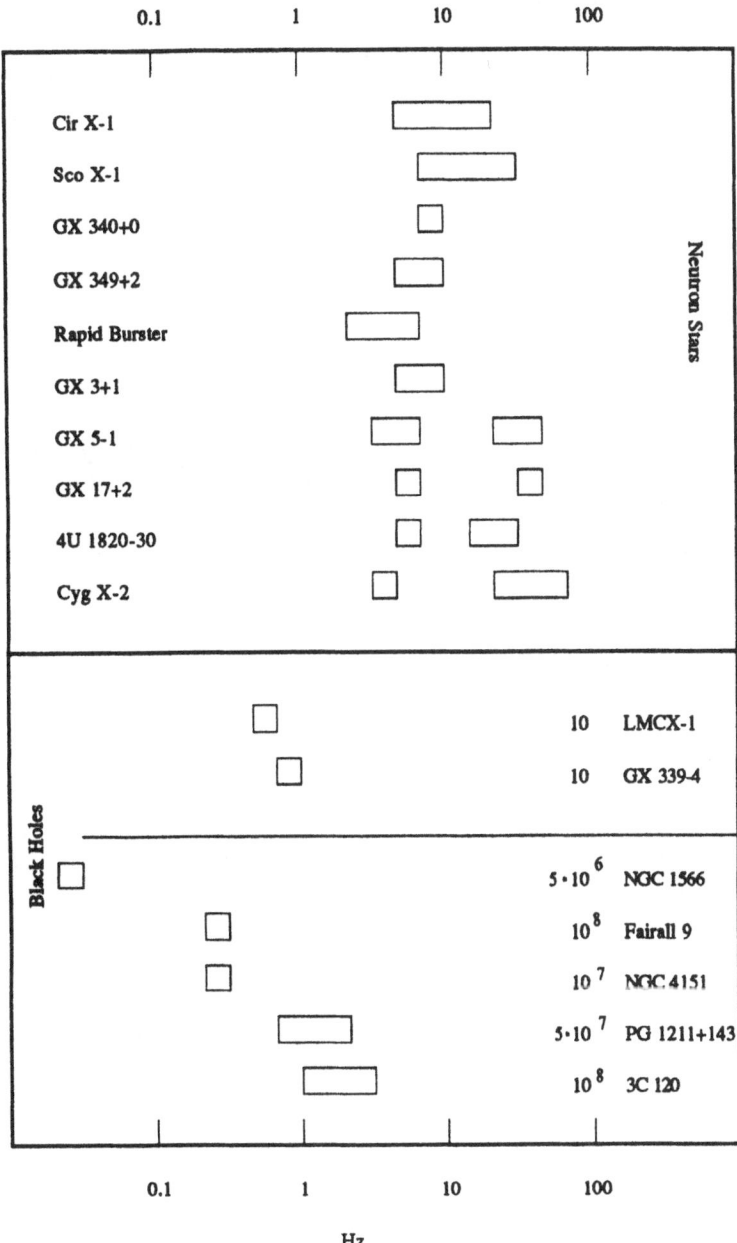

Figure 1.2: Quasi periodic oscillations in Galactic and extragalactic objects.

observed. There are three main parameters determining accretion disk structure, namely the mass of the central accreting body (black hole or neutron star) M, the viscosity parameter α and the dimensionless accretion rate $\dot{M}/\dot{M}_E$ (where $\dot{M}_E$ is the Eddington rate). Most of the luminosity comes from the innermost part of the disk, close to the radius $r \approx 5r_G$, with $r_G = 2GM/c^2$ being the gravitational radius of the central body. For this particular radius, and for disks which are radiation pressure supported and have opacity dominated by electron scattering, the dynamical timescale t_{dyn}, thermal timescale t_{th} and viscous timescale t_{vis} are respectively (in days):

$$t_{dyn} \approx 2 \times 10^{-8} \frac{M}{M_\odot}, \quad t_{th} \approx 2 \times 10^{-8} \alpha^{-1} \frac{M}{M_\odot}, \quad t_{vis} \approx 5 \times 10^{-8} \alpha^{-1} \frac{M}{M_\odot} \left(\frac{\dot{M}}{\dot{M}_E} \right)^{-2}.$$

$$(2.1)$$

The dynamical timescale is too short to be relevant as an explanation of quasi periodic variability shown in Figure 1.2. From observations, the estimated accretion rates in AGN are close to $0.1 - 1\dot{M}_E$ (Malkan 1989, this Workshop) and the estimated value of α is in the range $0.01 - 0.001$ (Siemiginowska and Czerny 1989, this Workshop). Therefore, the thermal and viscous timescales are comparable to each other and they both are very close to the *observed* periodicities. (Accretion rates in LMXB and GBHC are also in the same range. See *e.g.* White 1989, this Workshop)

We now formulate our diagnosis:

The "normal branch" quasi periodic oscillations observed in LMXB with timescales ~ 1 Hz are similar in nature to variability observed in GBHC (timescales $\sim 10^{-1}$ Hz) and to variability observed in AGN (timescales $\sim 10^{-7} - 10^{-8}$ Hz). The variability is due to thermal and viscous processes (most probably an instability) operating at the innermost $(r \approx 5r_G)$ part of accretion disks. It is important that the accretion rates for these objects fall (approximately) in the range $0.1 - 1\dot{M}_E$ and the viscosity parameter is small $(\alpha \approx 0.001 - 0.01)$.

In Section 4 we describe a thermal limit cycle mechanism which operates at the innermost part of an accretion disk around a compact object and may produce quasi periodic oscillations according to the above diagnosis. Its *qualitative* behaviour depends on very basic physics of accretion and is well understood in terms of the *slim accretion disk* model. However, the details of the mechanism have not yet been worked out in sufficient depth and therefore a *quantitative* comparison of theoretical light curves with observational ones is not yet possible.

3. Basic Equations for the Slim Disk Model.

Neither the standard "thin" accretion disk model of Shakura and Sunyaev nor its general relativistic version of Novikov and Thorne can correctly describe the innermost part of an accrertion disk close to a black hole. They both ignore several physical effects which dominate the *qualitative* behaviour of the accretion flow there. The ignored effects are in fact much bigger than the *quantitative* relativistic corrections studied by Novikov and Thorne.

These effects are connected with the transonic nature of the flow, deviations from the Keplerian rotation, and non-local cooling by advection. The region controlled by the effects ignored in the standard model increases with increasing accretion rate. When the accretion rate is highly super-Eddington the whole disk is affected by them and the "thick" accretion disk model is more appropriate. However, these effects are very important *sufficiently close* to the inner edge of the disk even for very small, sub-Eddington, accretion rates. This was first noticed by Paczyński and Bisnovatyi-Kogan (1981) who numerically constructed the model for transonic dissipative flow close to the black hole.

All these effects are present in the "slim" accretion disk model introduced by Abramowicz, Czerny, Lasota and Szuszkiewicz (1988) who followed the approach of Paczyński and Bisnovatyi-Kogan. The slim disks are characterised by $\dot{M} \approx \dot{M}_E$ and $\alpha \ll 1$. Therefore, they are in the range of parameters appropriate to LMXB, GBHC and AGN. They combine the good features of both thin and thick disk models while avoiding many of their problems. In particular, by adopting the procedure of vertical averaging of all the equations, the same as used for the thin disks, the slim disk models inherit the advantage of a one dimensional treatment. However, as in the thick disk models, they also include the effects of horizontal pressure gradient and horizontal heat flux, the dynamical importance of the accretion velocity and deviations away from strictly Keplerian rotation. The slim accretion disk models have been constructed so far in the pseudo-Newtonian potential

$$\Phi = -\frac{GM}{r - r_G} \tag{3.1}$$

which very accurately mimics all the relevant strong field effects of general relativity (at least in the gravitational field of a non-rotating black hole). Magnetic fields and self gravity of matter in the accretion flow have been neglected. The slim disks are dynamically stable, also with respect to Papaloizou and Pringle instability, due to the strong stabilizing effect of the steep angular momentum gradient, high accretion rate and large radial extent (see *e.g.* Nayaran 1989, this Workshop).

The time dependent state of the slim accretion disk is given by differential equations which describe the mass, angular momentum, radial momentum, and energy conservation:

$$\frac{\partial \Sigma}{\partial t} = -\frac{1}{r}\frac{\partial}{\partial r}\left(r\Sigma v\right), \tag{3.2a}$$

$$\left[\frac{\partial l}{\partial t}\right] = -v\frac{\partial l}{\partial r} + \frac{1}{2\pi r \Sigma}\frac{\partial g}{\partial r}, \tag{3.2b}$$

$$\left[\frac{\partial v}{\partial t}\right] = \frac{l^2 - l_K^2}{r^3} - \left[v\frac{\partial v}{\partial r}\right] - \left[\frac{1}{\rho}\frac{\partial P}{\partial r}\right], \tag{3.2c}$$

$$\Sigma T\frac{\partial S}{\partial t} = \Sigma g\frac{\partial}{\partial r}\left(lr^{-2}\right) - 2F - \left[v\Sigma T\frac{\partial S}{\partial r}\right]. \tag{3.2d}$$

Here v is the radial component of accretion velocity and Σ is the surface density, connected with the density of matter ρ and the vertical thickness of the disk H by an approximate formula

$$\Sigma = 2\rho H. \tag{3.3a}$$

The vertical thickness follows from the assumed hydrostatic equilibrium in the vertical direction

$$\frac{c_S^2}{H} = \frac{l_K^2}{r^2}\frac{H}{r^2} \tag{3.3b}$$

where c_S is the sound speed approximated as the square root of the ratio of pressure P and density ρ

$$c_S^2 = \frac{P}{\rho} \tag{3.3c}$$

and l_K is the Keplerian angular momentum which in the pseudo Newtonian potential (3.1) is given by

$$l_K^2 = \frac{GMr^3}{(r-r_G)^3}. \tag{3.3d}$$

It is not assumed that the angular momentum of the matter in the disk l follows the Keplerian distribution. Note, that $l_K(r)$ has a minimum for $r = 3r_G$, which indicates that the Keplerian orbits with $r < 3r_G$ are unstable, exactly like in the exact general relativistic case. The temperature and specific entropy are denoted by T and S. The thermodynamical quantities are connected by the equation of state for a mixture of perfect gas and radiation

$$P = P_{gas} + P_{rad}, \qquad P_{gas} = (\mathcal{R}\mu)\rho T, \qquad P_{rad} = \frac{a}{3}T^4 \tag{3.3e}$$

and the first law of thermodynamics

$$\frac{\rho T}{P}dS = (12 - \frac{21}{2}\beta)\frac{dT}{T} - (4 - 3\beta)\frac{d\rho}{\rho}, \qquad \beta = \frac{P_{gas}}{P}. \tag{3.3f}$$

The expression for the vertical radiation flux F follows from the assumption that the disk is optically thick in the vertical direction

$$F = \frac{acT^4}{3\kappa\rho H}, \tag{3.3g}$$

where κ is the opacity, assumed to be a known function of density and temperature

$$\kappa = \kappa(\rho, T). \tag{3.3h}$$

The last quantity which appears in the structure equations (3.2) is the viscous torque g connected with the (φ, r) component of the viscous stress τ_{ik} by

$$g = -4\pi r^2 H \tau_{\varphi r}. \tag{3.3i}$$

The standard α viscosity prescription assumes that

$$\tau_{\varphi r} = -\alpha P. \tag{3.3j}$$

The algebraical equations (3.3) are the same as in the standard thin disk model of Shakura and Sunyaev, however the differential equations (3.2) contain more terms than the standard ones. The terms absent in the Shakura-Sunyaev model are indicated by the square brackets.

Note, that in the Shakura-Sunyaev model only two time derivatives appear which means that this model admits only two modes of pulsation. These are thermal and viscous modes. In the slim accretion disk model there are four time derivatives, corresponding to thermal, viscous and two acoustic modes. The presence of acoustic modes suggests that the transonic part of the flow may be important for variability.

In the stationary case all the differential equations in the Shakura-Sunyaev model can be trivially integrated. This is basically because $l = l_K(r)$ is an explicitly known function. Thus, the stationary Shakura-Sunyaev disk model is totally described by purely algebraical equations. In addition, when opacity is given by Kramer's formula and either $P_{gas} \gg P_{rad}$ or $P_{rad} \gg P_{gas}$, all these equations are linear in their logarithmic version, so all the physical quantities can be explicitly given as products of powers of M, $\dot{M}$, α, r, and $l_K(r) - l_K(r_0)$, where r_0 is the radius at which $g(r_0) = 0$. These expressions are most often referred to as the Shakura-Sunyaev accretion disk model. They are very useful in astrophysical discussions because of their simplicity.

The stationary slim disk model does not offer such simple and useful formulae. This is because only equations of the mass conservation (3.2a) and angular momentum conservation (3.2b) can be explicitly integrated:

$$\dot{M} = 2\pi r \Sigma v, \tag{3.4}$$

$$\dot{M}(l - l_0) = 4\pi r^2 H \alpha P. \tag{3.5}$$

The two integration constants appearing here are accretion rate $\dot{M}$ (which may be assumed known) and the specific angular momentum of matter at the horizon of the black hole l_0, which is not known *a priori*. From the differential form of the mass and angular momentum conservation equations and still not integrated differential equations (3.2c,d) one derives

$$\frac{d\ln v}{d\ln r} = \frac{f(v, c_S, l, r)}{v^2 - c_S^2}, \tag{3.6}$$

where $f(v, c_S, l, r)$ is an explicitly known algebraical function. At the critical (or "sonic") point r_S it is $c_S = v$. Therefore, for a regular solution it should be $f(v, c_S, l, r_S) = 0$. This additional non trivial condition reduces the number of independent integration constants by one. Thus, one of these constants becomes an eigenvalue of the problem. Because l_0 appears in the equations and it is not known *a priori* it is convenient to treat this particular constant as the eigenvalue.

Note, that nothing was assumed about the value of the torque at the inner edge of the disk. The problem of what is the "correct" boundary condition for the viscous torque at the inner edge of the disk is often discussed in the context of the Shakura-Sunyaev model. However, the whole problem is totally artificial and due solely to an incorrect treatment of the inner edge in the Shakura-Sunyaev (and Novikov-Thorne) model. In the slim accretion disk model the inner edge is non singular and the value of the viscous torque is calculable (should not to be assumed *a priori*).

4. Slim Disks, S-Curves and the Thermal Limit Cycle.

We now describe stationary slim accretion disk models computed recently by Abramowicz, Czerny, Szuszkiewicz and Lasota (1988, "Trieste models") and later by Kato, Honma and Matsumoto (1989, "Kyoto models"). These models form a three parameter family, with dimensionless parameters $(\alpha, \dot{m}, m)$. Here $m = M/M_\odot$ and $\dot{m} = \dot{M}/\dot{M}_C$. The accretion rate is scaled in terms of $\dot{M}_C = 16\dot{M}_E$ rather than $\dot{M}_E$, because, for small accretion rates, 1/16 is the efficiency of accretion in the pseudo-Newtonian potential.

A sequence of equilibrium slim accretion disk models with fixed radius, mass of the central object and microphysical properties (equation of state, opacity, α viscosity) depends on only one parameter, $\dot{M}$ (accretion rate): all of the physical properties of the disk, including the surface density Σ, are functions of $\dot{M}$ only. The relation between the accretion rate $\dot{M}$ and the surface density Σ at the fixed radius r has a characteristic S-shape with the three branches (lower, middle and upper) defining three regimes of accretion. This is shown in Figure 4.1. The sequences of models shown in this Figure have been computed in Kyoto (solid lines) and Trieste (circles and triangles) for two particular radii and for the mass of the central object $M = 10M_\odot$ and Shakura-Sunyaev viscosity parameter $\alpha = 10^{-2}$. The slight difference for small accretion rates is due to different opacities used.

On the *lower branch* the gas pressure P_{gas} is greater than the radiation pressure P_{rad} and the opacity is dominated by electron scattering. The cooling is provided by the vertical radiative flux. Accretion is stable against local thermal and viscous perturbations - as indicated by the positive slope of $\dot{M} = \dot{M}(\Sigma)$. On the *middle branch* the opacity and cooling mechanism are the same as on the lower branch, but $P_{gas} \ll P_{rad}$ here. Accretion is thermally and viscously unstable - as indicated by the negative slope of $\dot{M} = \dot{M}(\Sigma)$. The thermal instability is due to an insufficient dependence of the rate of radiative cooling Q^- on the vertical thickness of the flow, H. For radiative cooling $Q^- \approx H$, while for viscous heating $Q^+ \approx H^2$. Thus, overheating causes expansion and expansion overheating and a thermal runaway instability arises (Pringle, Pacholczyk, and Rees, 1973). On the

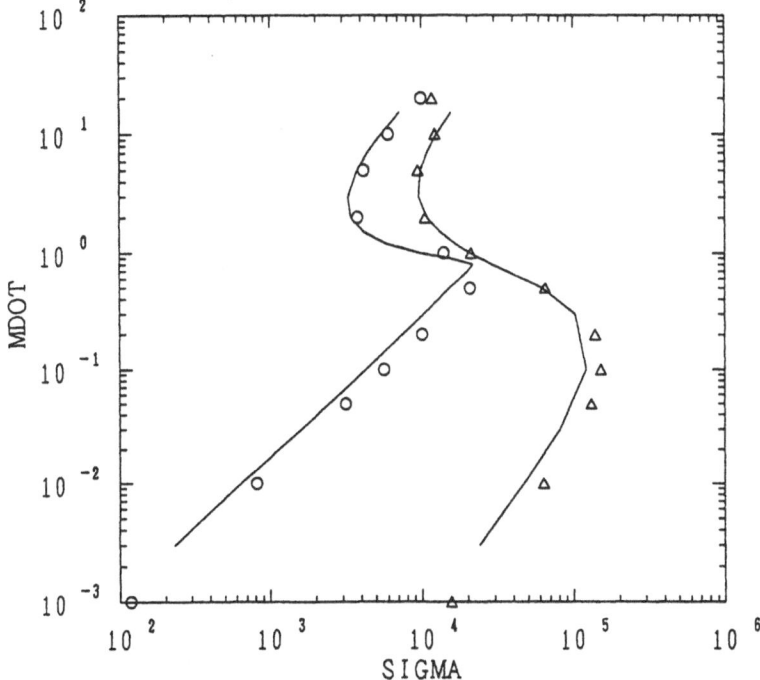

Figure 4.1: Sequences of slim accretion disk models for $r = 3r_G$ and $r = 5r_G$.

upper branch accretion flows cannot be described by the Shakura-Sunyaev model. Here $P_{gas} \ll P_{rad}$ and the cooling is provided by both vertical radiative flux and horizontal advection. We shall show later that for the strong advective cooling connected with the Roche lobe overflow it is $Q^- \approx H^3$. Thus, thermal runaway is avoided and the accretion flow is thermally stable (Abramowicz, 1981). This corresponds to the positive slope of the $\dot{M} = \dot{M}(\Sigma)$ curve.

We shall now discuss stability of the models in the three branches in a more detailed way. We start from the lower and middle branches, which may be accurately described (at the radii sufficiently greater than r_S) by the standard, Shakura-Sunyaev, thin accretion disk model.

In this case it is convenient to introduce (Piran 1978) the phenomenological parameters $\mathcal{K}$, $\mathcal{L}$, $\mathcal{M}$, $\mathcal{N}$ which describe dissipative processes:

$$\mathcal{K} = \left(\frac{\partial \ln Q^-}{\partial \ln H} \right)_\Sigma \qquad \mathcal{L} = \left(\frac{\partial \ln Q^-}{\partial \ln \Sigma} \right)_H$$

$$\dot{\mathcal{M}} = \left(\frac{\partial \ln Q^+}{\partial \ln H} \right)_\Sigma \qquad \mathcal{N} = \left(\frac{\partial \ln Q^+}{\partial \ln \Sigma} \right)_H \tag{4.1}$$

Here Q^+ and Q^- are the total heating and cooling rates respectively.

In the case of the standard Shakura-Sunyaev disk the slope of the $\dot{M}(\Sigma)$ curve is related to Piran's coefficients by

$$\frac{d\log\Sigma}{d\log\dot{M}} = \frac{1}{2}\frac{\mathcal{K} - \mathcal{M}}{\mathcal{K}\mathcal{N} - \mathcal{M}\mathcal{L}}$$

The general criteria for thermal and viscous stability found by Piran (1978) are also related to these coefficients:

$$\mathcal{K} - \mathcal{M} > 0, \tag{4.2}$$

$$\frac{\mathcal{K}\mathcal{N} - \mathcal{M}\mathcal{L}}{\mathcal{K} - \mathcal{M}} > 0. \tag{4.3}$$

Thus, the stability of the thin disk models is connected with the slope of the $\dot{M}(\Sigma)$ curve and the turning points of this curve correspond to changes in stability. One can show that the existence of the turning points is always connected with strong non-linearities in the physical functions which (implicitly) describe Piran's coefficients (Abramowicz, Lasota and Xu, 1986). This typically happens when a small change in some parameters of the flow causes a change in the physical mechanism of cooling or heating. For example, in the accretion disks relevant for dwarf novae the very strong dependence of opacity on temperature close to hydrogen ionization ($T \approx 10^4 K$) causes the $\dot{M}(\Sigma)$ curves (at all relevant radii) to bend twice and form the S-curve.

In the standard thin disk models the Piran coefficients are given in terms of β, (the ratio of the gas pressure to total pressure) by

$$\mathcal{K} = 4\frac{1+\beta}{4 - 3\beta}, \quad \mathcal{L} = -\frac{\beta}{4 - 3\beta}, \quad \mathcal{M} = 2, \quad \mathcal{N} = 1.$$

Together with the criteria (4.1), this implies that there will be thermal and viscous instabilities when $\beta < 2/5$,(i.e. for the radiation pressure supported standard thin accretion disks). Let us denote by $\tau = |Im(\omega)|^{-1}$ the growth time of an unstable mode with frequency ω. Unstable modes have $Im(\omega) < 0$. In the long wavelength limit $\lambda/H \gg 1$, the growth time of the unstable is independent of the wavelength:

$$\tau = \frac{1}{30}\frac{56 - 57\beta - 3\beta^2}{0.4 - \beta}t_{th}.$$

Here t_{th} is the time scale characteristic to thermal processes. This type of instability was discovered and studied by Pringle, Rees and Pacholczyk (1973), and Shakura and Sunyaev (1976). In the same limit, the growth time for the unstable viscous mode is:

$$\tau = \frac{3}{10}\frac{2 - 3\beta}{0.4 - \beta}\left(\frac{\lambda}{r}\right)^2 t_{vis}.$$

Here t_{vis} is characteristic time scale for viscous processes. This type of instability was found by Lightman and Eardley (1974) and Lightman (1974).

The stability properties of slim disks with low accretion rates are illustrated in Figure 4.2 which gives the local dispersion relation, *i.e.* the relation between Im(ω) and λ, for two models: one from the lower branch of an S-curve corresponding to $r = 5r_G$, and the other one from the middle branch. Im(ω) is given in units of $6\alpha\Omega_K$ and λ is given in terms of H/λ.

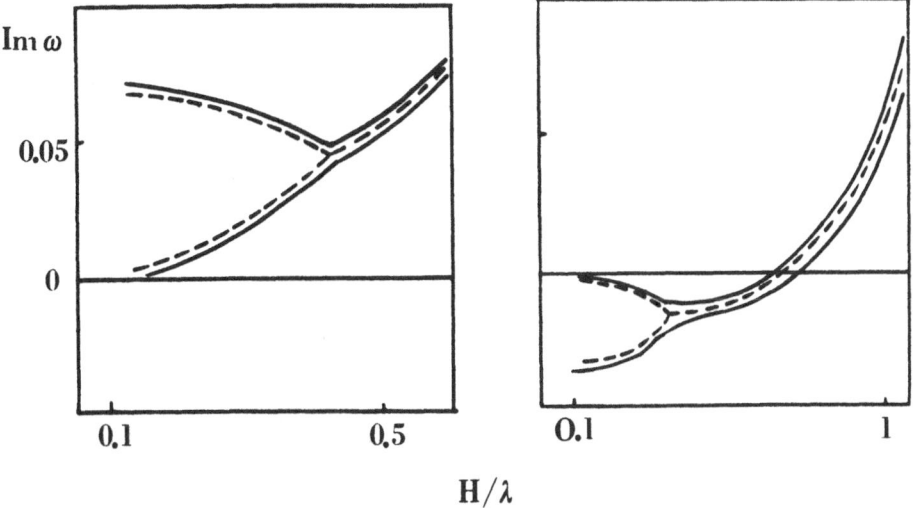

Figure 4.2: Dispersion relations for slim disk models at the lower and middle branch.

In both cases $m = 10$ and $\alpha = 0.001$, but for the model from the lower branch $\dot{m} = 0.1$, while for the model at the middle branch $\dot{m} = 1$. The Figure was prepared by Z. Loska, who used a numerical code for solving the dispersion relation for pulsations of slim disks developed by himself and B. Czerny (unpublished).

The dispersion relations for the Shakura-Sunyaev model are shown by broken lines. They have been obtained by linearization of equations (3.2) without terms in square brackets. Two separate modes, thermal and viscous, are present at long wavelengths and merge at shorter ones. The model from the lower branch is stable with respect to perturbations with all wavelengths, but the model from the middle branch is unstable at long wavelengths. Thermal and viscous modes for the slim disks (shown in solid lines) have been computed from the linearized version of (3.3) in which the two time derivatives in square brackets have been neglected, but all other terms included. (In this way the acoustic modes have been eliminated, but

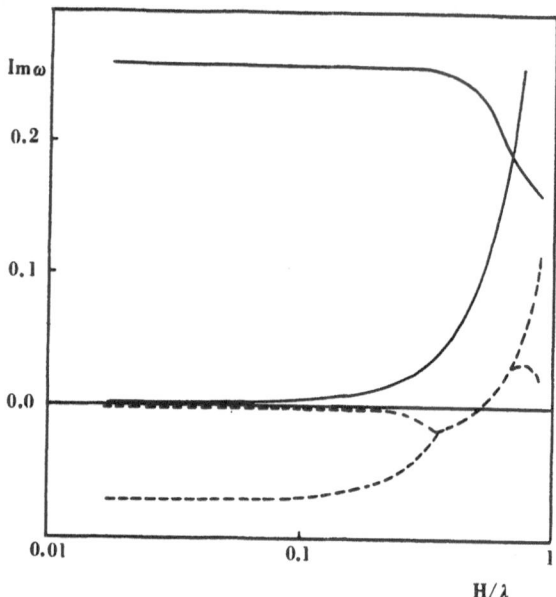

Figure 4.3: Dispersion relation for a slim disk model at the upper branch.

influence of the advection into thermal and viscous modes included.) There is only a little change in stability properties due to advection.

Figure 4.3 shows the dispersion relation, obtained for the same sequence as in Figure 4.2, but for $\dot{m} = 10$, *i.e.* for a model at the upper branch. The meaning of all the lines is the same as in the previous Figure. The slim disk model is stable despite the fact that the corresponding Shakura-Sunyaev model predicts instability. This suggests that exactly as in the Shakura-Sunyaev approach the positive slope of the $M(\Sigma)$ curve indicates *local* stability and that in particular all the models at the upper branch are locally stable. However, in order to reach a final conclusion concerning the stability at the upper branch one must discuss the *global* cooling and heating. This can be seen from the following argument:

Close to the sonic radius $r = r_S$ one of the equipotential surfaces crosses itself due to general relativistic effects (Figure 4.4). This self-crossing surface is called the *Roche lobe* and the place of crossing is called the *cusp*. If the surface of the disk overflows the Roche lobe, then hydrostatic equilibrium close to the cusp is not possible, and a dynamical mass loss takes place. The matter flowing out of the disk through the cusp carries away heat. Thus, the Roche lobe overflow is always connected with cooling. The cooling connected with the overflow is different that the advective cooling far away from the cusp.

The first one has a global, while the second one has a local effect on the heat balance.

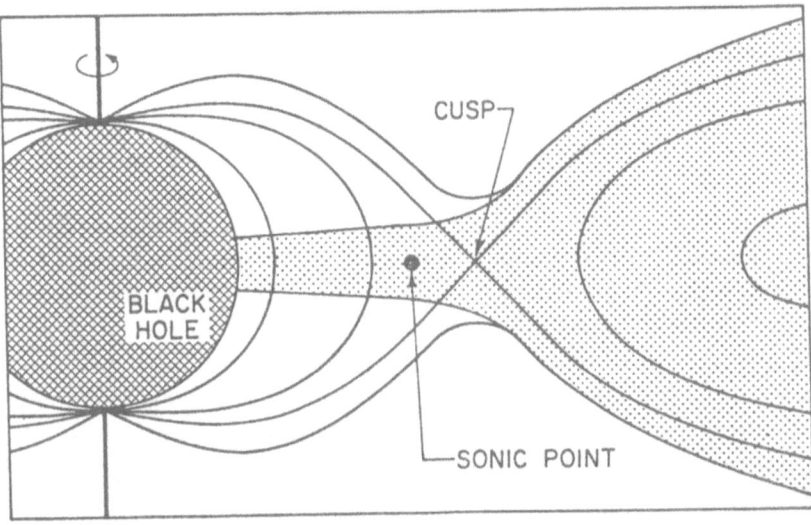

Figure 4.4: Stabilization by the Roche lobe overlow.

Far away from the cusp the advective heat transport may *locally* heat up or cool down the disk at some fixed place according to whether the local entropy gradient is positive or negative there. However, close enough to the cusp the advective flux always cools *globally* the whole region of the disk which is affected by the mass loss. The region of the disk cooled by the Roche lobe overflow mechanism is thermally stable because the "Roche cooling" goes like $Q^-_{Roche} \sim H^3(r_{cusp})$, *i.e.* stronger than viscous heating $Q^+ \sim H^2$, and therefore the thermal runaway is avoided (*cf.* criterion 4.2). For small accretion rates the stabilized region lies in the immediate vicinity of the inner edge, but for $\dot{M} > \dot{M}_C$ its size may be substantially greater. The fact, that the slim disks at the upper branch have comparable pressure heights in both vertical and radial direction (we shall denote them by h) may also affect global thermal stability. Kato (1988, unpublished) considered thermal equilibrium of a very slender, two dimensional, toroidal "disk". When the disk expands due to overheating its density decreases like h^{-2}. If the hydrostatic equilibrium is maintained, $dP/dz \sim \rho/h^2$, which means that the pressure decreases like h^{-3} during expansion. Because the viscous heating rate $Q^+ \sim \tau_{\varphi r}h \sim \alpha Ph$, one gets that $Q^+ \sim h^{-2}$ during expansion. The radiative cooling rate, on the other hand, is approximately given by $Q^- \sim P_{rad}/\Sigma \sim h^{-2}$. Thus, when only viscous heating and radiative cooling of the disk are considered $Q^+ \sim Q^- \sim h^{-2}$, *i.e.* the disk is in a neutral thermal equilibrium. Let us now consider what happens in a more general situation when $Q^+ \sim h^{-a}$, $Q^- \sim h^{-b}$, where a and b are some positive numbers (perhaps both close to 2) and in addition $a < b$. Clearly, the last condition introduces thermal instability: the disk will expand and heats itself up more and more. Hovever, when its expanded surface finally reaches the Roche lobe a new cooling mechanism turns on. Its efficiency scales with the overflow as Δh^3 and therefore quickly dominates the viscous heating. Thus, the disk will have its size almost exactly equal to the Roche lobe and it will be in a stable thermal state, providing that the mass supply from outside compensates the loss through the cusp. If it does not, then a stationary stable state is not possible and instead the disk will experience a *limit cycle*, expanding and contracting in a roughly thermal timescale. The expanded "hot" state corresponds to the upper branch of the S-curve, while the contracted "cool" state to the lower branch.

This situation resembles (formally) the very-well known limit cycle mechanism which gives quasi periodic dwarf novae outbursts. It also works due to: (1^0) the relation between accretion rate and surface density is characteristically S-shaped (2^0) the lower and upper branches of the S-curve correspond to stable, while the lower one to unstable stationary equilibria, (3^0) the accretion rate fixed by the outside conditions lies tn the unstable range (see *e.g.* Osaki 1989, this Workshop). As Bath and Pringle (1982) proved, these are the three most general conditions for a limit cycle to operate.

In general it would not be possible to accurately estimate the timescales and amplitudes connected with the *global* response of the disk to instability by considering a particular (fixed radius) S-curve. The problem of how different local cycles combine to produce one global cycle must be studied numerically. Unfortunately, this is much more difficult than in the case of dwarf novae: there are two more time derivatives and three more radial derivatives in the slim disk problem. So far there has been only one computation done in which the full set of time dependent equations (3.2) has been included (Matsumoto, Kato and Honma 1989, this Workshop). Only half of the cycle was followed because of computer time limitation.

However an accurate analytic estimation can be done when the instability region has a very small radial extent and all the local S-curves have approximately the same shape (the same difference in $\dot{M}$ between the lower and the upper branch).

Figure 4.5 shows the location of the lower and upper turning points and instability regions (dashed) for slim disk models corresponding to $M = 10 M_\odot$ and to the two different viscosity prescriptions. In the left panel the standard Shakura-Sunyaev prescription (3.3j) was used, while in the right one it was assumed that

$$\tau_{\varphi r} = -\alpha\sqrt{PP_{gas}}. \tag{4.4}$$

This particular prescription was introduced according to an argument that in general the viscosity coefficient should be proportional to the product of the turbulent velocity of *gas* elements U and the *total* height scale of the disk H. It is

$$U \sim c_S[gas] \sim P_{gas}^{1/2} \quad and \quad H \sim c_S[total] \sim P^{1/2}.$$

157

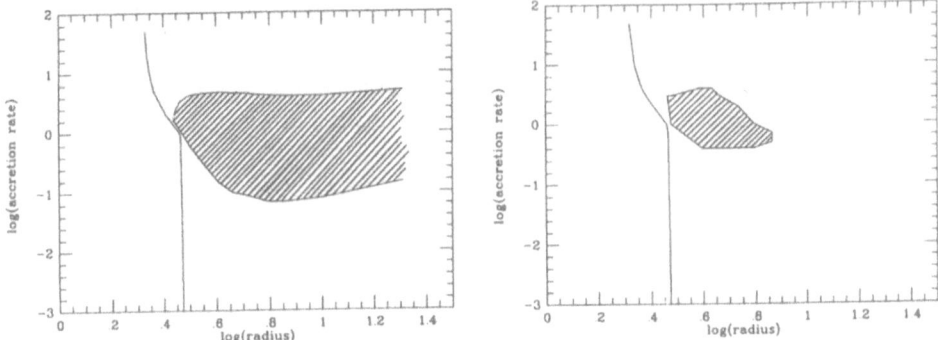

Figure 4.5: Instability regions for slim disk models.

The prescription (4.4) is no more arbitrary than the standard one.

The instability region corresponding to the standard viscosity case is very large. It is practically unbounded in the radial direction and corresponds to a rather large separation (in terms of $\dot{M}$) between upper and lower stability regions. For this reason the limit cycle based on the standard prescription would surely have too long a timescale and too large an amplitude. It cannot explain the observed variability.

The other viscosity prescription gives a *very small* instability region, so local arguments can be quite accurate in determining both the typical frequencies and amplitudes of the limit cycle. It follows from Figure 4.5 that the frequency of the cycle is roughly given by the thermal timescale at $r \approx 5r_G$, while the amplitude is less than about a factor of 10. These values are close enough to the observed ones in quasi periodic variability of LMXB, GBHC and AGN.

We propose to consider (in the full time dependent calculations) the viscosity prescription

$$\tau_{\varphi r} = -\alpha \sqrt{P^\mu P_{gas}^{2-\mu}}, \tag{4.5}$$

which becomes the standard prescription when $\mu = 2$ and the modified one when $\mu = 1$. In addition, when $\mu = 0$ viscosity is proportional to the gas pressure alone. The last prescription gives no unstable region at all.

There are two unknown viscosity parameters α and μ in our viscosity formula (4.5). By tuning them in the right way a better agreement with observations may be acheived. Two parameter viscosity tuning is also a standard procedure in the dwarf novae research, where one assumes not only α but also the change of it between the upper and lower branch $\Delta\alpha$.

We conclude, that the slim disk thermal instability and the Roche lobe overflow advective cooling provide a very probable explanation of the QPO phenomenon in the low mass X-ray binaries, Galactic black hole candidates and active galactic nuclei. The main arguments to support our conclusion have been already discussed, but we list them here again altogether:

1) The similarity in variability patterns of all these objects,
2) The constancy of the mass supply rate over many QPO periods,
3) The existence of S-shaped $\dot{M}(\Sigma)$ curves for slim disks,
4) The right range of accretion rate and viscosity parameter,
5) Preliminary numerical simulations of the limit cycle.

We shall now discuss two different, but related subjects: possibility of long timescale switching between "high" and "low" states, and high frequency oscillations with featureless power spectra. They both connect to the transonic part of the black hole accretion flow.

The left and right panels in Figure 4.6 compare the angular momentum distribution in the accretion flow with the Keplerian one and give examples of two physically different types of accretion. In the first type (left panel, and curve marked $\alpha = 0.001$ in the right panel) the angular momentum of matter equals to the Keplerian one in exactly two points, in the second type (curve marked $\alpha = 1$ in the right panel) the angular momentum of the flow is always sub-Keplerian. What type of accretion is realized for a flow with fixed accretion rate depends *both* on the values of angular momentum l_∞ and energy E_∞ far away from the center (outer boundary conditions) *and* on the value of viscosity α: the division between the two types in the full parameter space is given in general by

$$f(l_\infty, E_\infty, \alpha) = 0. \tag{4.6}$$

Abramowicz and Zurek (1981) were the first to recognize the existence of these two types of accretion. They called the first type "disk accretion" and the second type "Bondi accretion". They found a part of the boundary in the parameter space dividing the two types in a special case of $\alpha = 0$, but with general boundary conditions,

$$f(l_\infty, E_\infty, 0) = 0. \tag{4.7}$$

Thus, their condition had the form $E_\infty = E_\infty(l_\infty)$. Muchotrzeb (1983) studied the complementary case: she assumed *fixed* (Keplerian) outside boundary conditions, but a general α:

$$f(l_K, E_K, \alpha) = 0. \tag{4.8}$$

Because l_K and E_K were two given constant numbers, the condition found by Muchotrzeb by solving equation (4.8) was $\alpha = \alpha_{crit} = const$. She later interpreted this result assuming that there is a critical value of the viscosity parameter $\alpha_{crit} \approx$

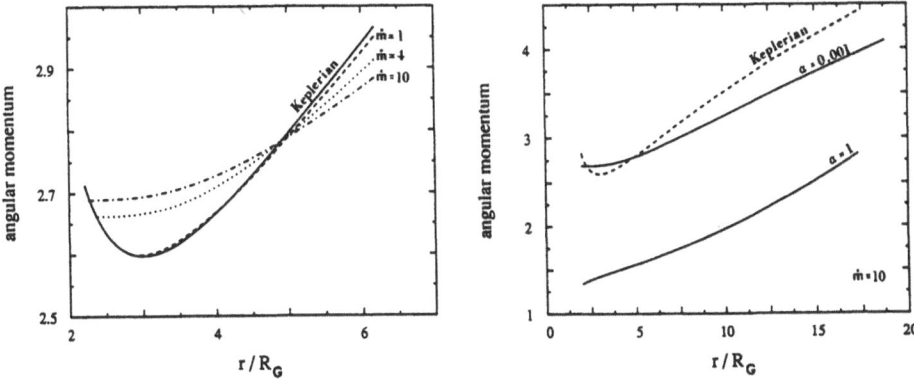

Figure 4.6: Angular momentum in slim disks.

0.01 which separates the two types of accretion. The issue of the critical α was discussed by Matsumoto, Kato, Fukue and Okazaki (1984), Paczyński (1987) and others. More recently Abramowicz and Kato (1989) unified the condition found by Abramowicz and Zurek with that found by Muchotrzeb by showing that they describe the same boundary (4.6), but in different sections of the parameter space.

Disk accretion The labels 1, 4, 10 in the left panel refer to the accretion rate (in terms of $\dot{m}$). The viscosity parameter is $\alpha = 0.001$, *smaller* than the critical one, and the mass of the central object is $M = 10 M_\odot$. There is an increasing deviation from Keplerian distribution when the accretion rate increases, but angular momentum in the disk is always Keplerian at the two points: at the "center" $r = r_C$ very close to the pressure maximum, and at the cusp $r = r_{cusp}$. These two particular radii, the radius of marginally stable circular orbit $r_{MS} = 3 r_G$ at which the Keplerian angular momentum has a minimum, and the sonic radius r_S satisfy, for the disk accretion:

$$r_G < r_S \lesssim r_{cusp} < r_{MS} < r_C < 5 r_G.$$

In this case the accretion in the innermost part of the flow goes because of the Roche lobe overflow: the energy of the flow is just slightly greater than the top of the centrifugal potential barrier located at the cusp.

Bondi accretion. In the examples shown in the right panel of Figure (4.6) the outer boundary condition was assumed to be Keplerian. It is obvious, that when one changes α between 0.001 and 1 there is a particular value for which the angular momentum in the disk is Keplerian in exactly one point. This is the "critical" α found by Muchotrzeb. Flows with $\alpha > \alpha_{crit}$ belong to the Bondi type of accretion. This type of accretion is characterized by the slight influence of the effects of rotation and general relativity, because the sonic point is located relatively far from the center ($r_S > r_{MS}$). Accretion goes close to the center just because there is no centrifugal potential barrier there, and therefore resembles spherical accretion.

The existence of the two types of accretion is important in the context of variability because it is possible that every black hole accretion flow is intrinsically non-stationary. This is due to the fact that the regularity condition at the sonic point (3.6) which fixes one integration constant is not compatible, in a general astrophysical situation, with the outer boundary conditions. As a result the flow will switch between the disk and Bondi states with a characteristic timescale, estimated by Abramowicz, Livio and Lu (1987),

$$t_{h-l} \sim t_{vis}(50r_G),$$

i.e. about weeks or months for the Galactic black holes and about 10^7 years (say) for extra galactic sources. Presence of a global shock structure does not change this conclusion. The characteristic time for switching is much longer than the quasi periodic oscillations, and therefore we can assume that the accretion rate is constant (on average) during many QPO periods corresponding to either Bondi or disk state.

Some black hole candidates in our Galaxy, including Cygnus X-1, show bimodial switching between "high" and "low" states with timescales of weeks or months. In active galactic nuclei a similar recurrent activity may also be present. This is indicated by the observation of roughly equispaced knots in the optical jet of M87 (Rees 1978) and in the helical radio structure of 4C 29.47 (Condon and Mitchell 1984). The derived outburst periods lie in the range 300 - 3×10^6 years and 3 × $10^7 - 10^9$ years respectively and agree with the scaled up (by a factor of about 10^8) version of bimodal switching between high and low states in the Cyg X-1 case. The observed periods of high state - low state switching agree roughly with the one estimated theoretically. One should stress, however, that the theoretical estimation is *very* approximate.

Thermal and viscous modes interact with acoustic ones in the transonic region. We shall shortly discuss two aspects of this which may be relevant for variability. They are connected with the existence of a non propagating acoustic mode at the sonic point, and with modes trapped in the region close to the maximum of the epicyclic frequency.

The standing mode. Since the propagation speed of acoustic modes is $u = v \pm c_S$, one of them does not propagate at the sonic radius r_S where $v = c_S$. The non propagating mode becomes unstable when the viscosity coefficient is high enough,

$$\alpha \Omega(r_S) > \left| \frac{dv}{dr} \right|_S.$$

(For the Keplerian outer boundary conditions this criterion demands $\alpha \lesssim 0.01 \approx \alpha_{crit}$ for stability.)

In the non linear regime the unstable mode saturates at a finite amplitude which is modulated by a slower thermal mode. This was studied recently in detail by Kyoto the group (see *e.g.* Matsumoto, Kato, Honma 1989, this Workshop).

We would like point out here some similarity in the computed, *very compli-cated*, variability patterns resulting from interaction of acoustic and thermal modes to the featureless power spectra of variability observed in several active galactic nuclei during *Exosat* long looks (McHardy and Czerny 1987). Both the computed and observed frequency range is roughly from minutes to days. Note, that a corresponding high frequency noise cannot be observed in Galactic sources because of too low a time resolution of the present X-ray detectors.

The trapped modes. The epicyclic frequency

$$\chi^2 = \frac{2\Omega}{r} \frac{dl}{dr}$$

has a maximum close to the transonic region. This is because $l \approx const$ for $r < r_S$, while $\Omega \approx (GM/r^3)^{1/2}$ for $r \gg r_S$. Thus, for both $r < r_S$ and $r \gg r_S$ one has $\chi \approx 0$ which proves that $\chi(r)$ has a maximum. Okazaki, Kato and Fukue (1987) considered a perturbation of an isothermal accretion disk with n nodes in the vertical direction. They found that the radial wave vector component k can be written as

$$k^2 = \frac{(\omega^2 - \chi^2)(\omega^2 - n\Omega^2)}{\omega^2 c_S^2}.$$

Because $k^2 > 0$ for radially propagating modes, the above expression shows that that local waves with $n = 0$ propagate in the regions where $\omega > \chi$. Knowing that $\Omega > \chi$ and denoting by χ_{max} the maximal value of the epicyclic frequency and by r_1 and r_2 the smaller and larger roots of the equation $\omega = \chi$, one concludes that waves with $n \neq 0$ can propagate in the region between r_1 and r_2 when $\omega < \chi_{max}$. This suggests that there will be global oscillation modes trapped in the region between r_1 and r_2. The pulsation period of these modes cover some range close to about one day for disks around a 10^8 black hole. Both r_1 and r_2 are located closely to the sonic radius. Therefore, it is likely that interaction of the trapped modes will produce local shocks and in consequence a chaotic, featureless power spectrum with no characteristic frequency.

We shall end this section by showing the thermal spectra of slim accretion disks (Figure 4.7). They are so-called *modified black body* spectra, which means that they include effects of electron scattering, but not Comptonization. They have been calculated (Szuszkiewicz 1988, unpublised Ph.D. thessis) in a similar way as that described by Malkan (1989, this Workshop). The effective optical depth in the vertical direction

$$\tau_{eff} = (\kappa_{es}\kappa_{ff})^{1/2}\Sigma$$

was greater than one for all the models. Here κ_{es} and κ_{ff} are electron scattering and free-free opacity coefficients. In the calculations the total opacity coefficient, $\kappa = \kappa(\rho, T)$, was taken from Cox and Steward (1970). Typical parameters for the

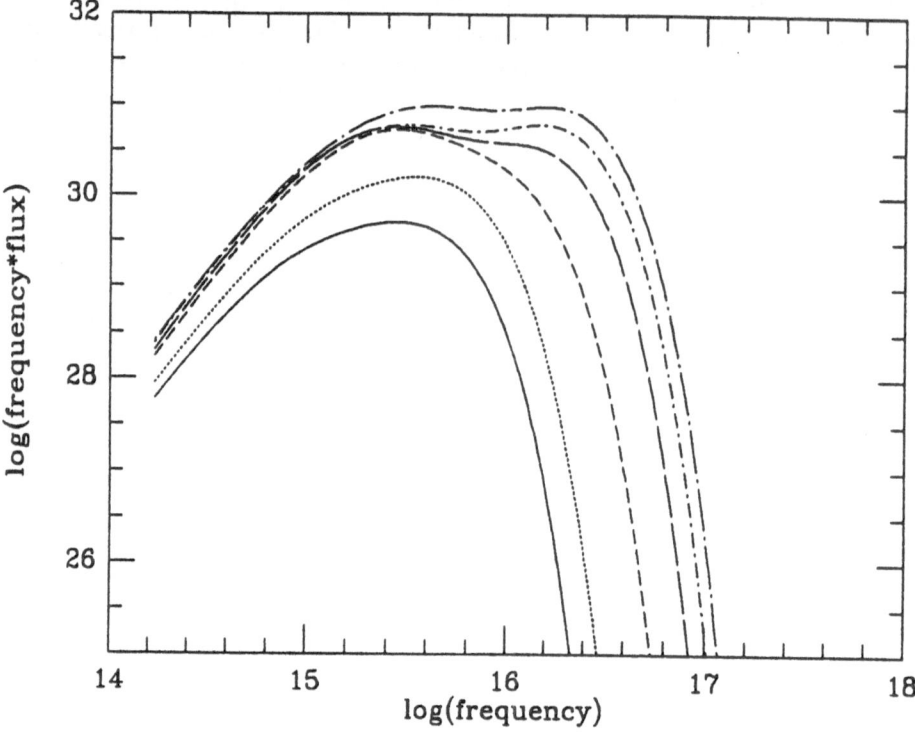

Figure 4.7: The spectra of slim accretion disks.

active galactic nuclei have been adopted: mass $M = 10^8 M_\odot$, viscosity coefficient $\alpha = 0.001$, and six different accretion rates

$$\dot{m} = 0.03, \ 0.01, \ 1.0, \ 2.0, \ 5.0, \ 50.$$

The accretion rate corresponding to the curve at the far left is the lowest. It increases continously to the right.

All the curves peak at the ultraviolet (UV) part of the spectrum. For the small accretion rates the frequency of the peak increases together with the intensity and with the spectral slope in the UV range. Such a behaviour is observed in Seyfert galaxies which have relatively small luminosities, *i.e.* small accretion rates. For higher accretion rates the slope in the UV range is approximately constant but the spectral hardness increases with increasing luminosity at the peak rates.

5. Problems for a Therapist.

We briefly comment now on the most important observational tests for our theoretical idea. According to it, the quasi periodic oscillations (normal branch type) should occur for accretion rates corresponding to the unstable middle branch of the S-curve in accretion disks around black holes and compact enough neutron stars*. Objects with higher or lower accretion rates should not experience this type of variability. Therefore, the first and the most direct observational test of the hypothesis should be a check whether there is a correlation between occurence of the QPO variability and the luminosity expressed in the Eddington units. To perform this check one needs to know the mass of the accreting object.

In the case of AGN a more convincing observational indication of the occurence of QPO with typical timescales of a few hundred days is also necessary. The existing indications are based mainly on the Fourier analysis of the optical light curves (see Abramowicz and Szuszkiewicz, 1989). It would be now rather difficult to make these indications more accurate, or repeat them for the UV or X-ray frequencies. The first difficulty is connected with the length of the period: to get interrupted data on just ten full cycles, one needs (say) thirty, fifty or even more years of observations. Surely, this is not yet possible for ultraviolet or X-ray data. The long time optical data exists on archival plates collected for some different purposes *e.g.* the search for supernovae outside the Galaxy. The problem with this data is the infrequent and irregular sampling of a given object. This makes the computation of the Fourier power spectra very difficult. The same difficulty is present in the case of ultraviolet data from *IUE* and X-ray data from *Exosat* (except for the *Exosat* long looks) and other X-ray satellites. The "true" time series $F(t)$ is present in the data, but it is observed by a "window" in time described by $W(t)$. The Fourier transform method produces a *convolution* $\hat{F} \circ \hat{W}$, rather than the desired $\hat{F}$, with a strong but practically unknown mixing occuring in a wide range of frequencies. Direct inversion of this convolution is very unstable numerically and therefore the method of inverse Fourier transform is not very accurate.

Edelson and Krolik (1988) discussed another method which may solve this problem and which is based on the fact that the variability power spectrum is just the Fourier transform of the autocorrelation function. They have studied the autocorelation functions of three AGN (optical data on Seyfert galaxy Akn 120 and quasar 0957+561, *IUE* data on Seyfert galaxy NGC 4151) and found that these objects have featureless continuous power spectra (no periodic behaviour) in the frequency

---

* The radius of the neutron star should be less than the radius of the marginally stable circular orbit. This is a typical situation for non-rotating neutron stars with "soft" equation of state (see Brown, 1988 for references). The discovery of the 1968.629 Hz periodicity of the pulsar in SN1987A seems to support the view in which equation of state is soft and the neutron stars are small enough. It is still not known, however, how rapid rotation changes this picture. In any case, the sometimes expressed opinion that rapidly rotating neutron stars must have equatorial radii greater than r_{MS} is based on an incorrect estimate based on the Kerr geometry.

range corresponding to 100-500 days. In collaboration with T. Courvoisier they plan to collect all available X-ray data since 1970 and all the archival *IUE* data for the CfA redshift survey Seyferts (Krolik, 1989, private information). Although the preleminary results of this research indicate no QPO in the three AGN studied, it may be that the accretion rates are outside the instability range, or that the periodicity is of the order of 500 days (or longer).

In this context an important study by McHardy (1988, see also Pounds and McHardy 1988) should be mentioned. He has examined the very low frequency power spectrum of a bright X-ray source NGC 5506 including data from *Ariel V, Einstein Observatory, HEAO-1,* and *Exosat.* He found a "knee" frequency $\sim 10^{-7}$Hz at which the power spectrum becomes and remains flat to about 10^{-9}Hz. No data is available at still lower frequencies (see Figure 5.1).

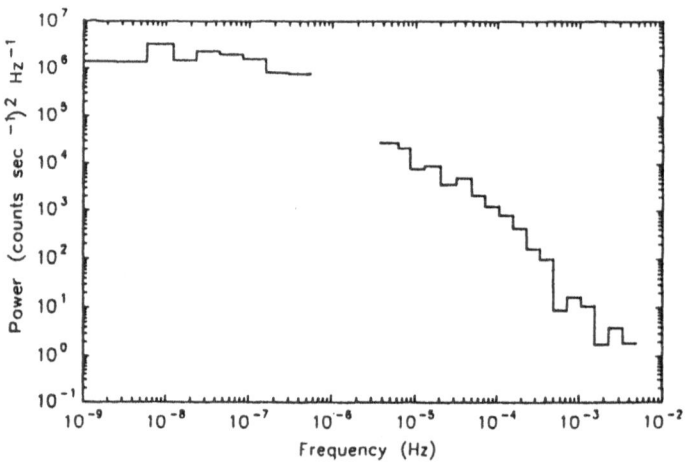

Figure 5.1: Power spectrum for NGC 5506 (from Pounds and McHardy 1988).

The low frequency power at the flat part of the spectrum should be taken, according to Pounds and McHardy (1988), only as an upper limit. Therefore, it is more likely that the true shape of the spectrum shows a *maximum* somewhere between 10^{-7}Hz and 10^{-9}Hz. In addition, the mass estimation for this source $\sim 5 \times 10^6 M_\odot$ gives luminosity close to the Eddington limit. Thus, we may have here the first indication of QPO in active galactic nuclei based on the X-ray data. Both the frequency of the oscillations and the accretion rate are quite consistent with our theory.

We agree with Pounds and McHardy who wrote "The refinement of the low frequency power spectrum of NGC 5506, in particular filling in the gap between $\sim 10^{-5}$ and $\sim 10^{-7}$Hz, and its determination in other AGN to search for possible correlations of the knee frequency, for example with luminosity, remains an important observational task." We believe, that exactly this observational task is now the most important one for our diagnosis.

ACKNOWLEDGEMENTS. M.A.A. would like to thank J.P. Lasota, S. Kato, and Z. Loska for important suggestions which help to solve many problems. He thanks also Y. Osaki, H. Inoue, R. Matsumoto and J. Fukue for helpful discussions and J. Krolik for information about his unpublished results. F.W. acknowledges the crucial financial support from Lund University during his stay in Trieste.

REFERENCES.

Abramowicz M.A.: 1981 *Nature*, **294**, 235
Abramowicz M.A., Czerny B., Lasota J.P., and Szuszkiewicz E.: 1988, *Ap. J*, **332**, 646
Abramowicz M.A., and Kato S.: 1989 *Ap. J.*, **336**, 304
Abramowicz M.A., Lasota J.P., and Xu C.: 1986 in *Quasars*, eds. G. Swarup and V.K.Kaphai, (Reidel, Dordrecht)
Abramowicz M.A., Livio M., and Lu J.: 1986 in *Marcel Grossmann Meeting on General Relativity*, ed R.Ruffini (Elsevier, B.V.)
Abramowicz M.A., and Szuszkiewicz E.: 1989, in *Big Bang, Active Galactic Nuclei and Supernovae*, eds. Hayakawa S. and Sato K. (Universal Academic Press, Tokyo)
Abramowicz M.A. and Zurek W.H.: 1981, *Ap. J.*, **246**, 314
Bath G.T. and Pringle J.E.: 1982, *Month. Not. Roy. astr. Soc.*, **199**, 267
Begelman M.C. and Frank J.: 1989, this volume
Brown G.E.: 1988, *Nature*, **336**, 519
Condon J.J. and Mitchell K.J.: 1984, *Ap. J.*, **276**, 472
Cox A.N. and Steward J.N.: 1970, *Ap. J. Supp.*, **19**, 243
Edelson R.A. and Krolik J.H.:1988, *Ap. J.*, **333**, 646
Hasinger G.: 1988, in *Physics of Neutron Stars and Black Holes*, ed. Y. Tanaka (Universal Academic Press, Tokyo)
Lamb F.K.: 1988, *Adv. Space Res.*, **8**, 421
Lewin W.H., van Paradijs J., and van der Klis M.: 1988, *Space Sci. Rev.*, **46**, 273
Lightman A.P.: 1974, *Ap. J.*, **194**, 429
Lightman A.P.: 1974, *Ap. J. Letters*, **187**, L1
Kato S., Honma F., and Matsumoto R.: 1989, *Publ. Astr. Soc. Jap.*, **40**, 709
Malkan M.A.: 1989, this volume
Matsumoto R., Kato S., and Honma F.: 1989, this volume
Matsumoto R., Kato S., Fukue J., and Okazaki A.T.: 1984, *Publ. Astr. Soc. Jap.*, **36**, 71
McHardy I.M.: 1988, in *X-ray Astronomy with Exosat*, ed. R. Pallavicini,
Miyamoto S., Kitamoto S., Kazuhisa M., and Dotani T.: 1988, *Nature* **336**, 450
Narayan R.: 1989, this volume
Okazaki A.T., Kato S., and Fukue J.: 1987, *Publ. Astr. Soc. Jap.*, **39**, 447
Osaki Y.: 1989, this volume
Paczyński B.: 1987, *Nature*, **325**, 572
Paczyński B. and Bisnovatyi-Kogan G.: 1981, *Acta Astr.*, **31**, 283
Piran T.: 1978, *Ap. J.*, **221**, 652
Pounds K.A. and McHardy I.M.: 1988, in *Physics of Neutron Stars and Black Holes*, ed. Y. Tanaka (Universal Academic Press, Tokyo)
Pringle J.E., Rees M.J., and Pacholczyk A.G.: 1973, *Astr. Ap.*, **29**, 179
Rees M.J.: 1978, *Month. Not. Roy. astr. Soc*, **184**, 61P
Siemiginowska A. and Czerny B.: 1989, this volume

Tanaka Y.: 1989, talk at the *Spectrum-X Conference*, Moscow
White N.E.: 1989, this volume

NONLINEAR PULSATION IN THE TRANSONIC REGION
OF GEOMETRICALLY THIN ACCRETION DISKS.

Ryoji MATSUMOTO[1], Shoji KATO[2], and Fumio HONMA[2]

[1] College of Arts and Sciences, Chiba University, Yayoi-cho, Chiba 260, Japan

[2] Department of Astronomy, Kyoto University, Sakyo-ku, Kyoto 606, Japan

ABSTRACT. Nonlinear hydrodynamical simulations are performed to study the evolution of axisymmetric instabilities in viscous accretion disks around a black hole or a non-magnetized neutron star. When the viscosity parameter α, is sufficiently large ($\alpha >$ 0.05), quasi-periodic pulsations are excited in the innermost transonic region of the disk. Numerical results also show that after the onset of the thermal instability in radiation pressure dominated region, a geometrically thin accretion disk evolves into a slim accretion disk.

1. Introduction.

Observed time variabilities of active galactic nuclei (AGN) and X-ray binaries possibly are related to the instabilities in accretion disks (e.g., Abramowicz and Szuszkiewicz 1988). Since most of the gravitational energy is released within the innermost region of accretion disks, it is worth studying its stability. Here, we study the nonlinear evolution of axisymmetric instabilities, taking into account the dymamical and thermal effects of transonic accretion flow onto the central object.

The innermost region of accretion disks around a black hole or a non-magnetized neutron star has transonic nature; the disk has a critical radius where the radial flow velocity exceeds the sound speed (e.g., Liang and Thompson 1980). The critical radius is close to the radius of the last stable circular orbit ($\varpi = r_{ms}$). Transonic Shakura-Sunyaev type steady disk models have been constructed by several authors (Muchotrzeb 1983, Matsumoto et al. 1984, Abramowicz et al. 1988). Muchotrzeb-Czerny (1986) suggested that the transonic flow is unstable when the viscosity is so large that the sonic point is nodal. Later, Kato et al. (1988a,b) showed that the cause of the instability is the viscous pulsational instability discussed by Kato (1978).

The transonic region of accretion disks is a suitable place for the pulsational instability because local oscillations develop into a global one owing to the reflection of

F. Meyer et al. (eds.), Theory of Accretion Disks, 167–172.

low frequency outgoing acoustic waves. This reflection occurs because the epicyclic frequency has a maximum around $\varpi = 3.7r_g$, where r_g is the Schwarzschild radius (Kato et al. 1988a). This possibility is confirmed by Matsumoto et al. (1988) by performing one dimensional numerical hydrodynamic simulations for isothermal accretion disks. They also found that the amplitude of oscillation is modulated in the time scale of the wave crossing time across the trapped region ($3r_g < \varpi < 4r_g$). This modulation period is close to the observed period of QPO's in low mass X-ray binaries.

In this paper, we present the results of nonlinear hydrodynamical simulations of accretion disks including the viscous heating and radiative cooling.

2. Basic Equations and Numerical Methods.

We assume that the disk is axially symmetric and geometrically thin. Geneneral relativistic effects are simulated by using the pseudo Newtonian potential $\Psi = -GM/(r - r_g)$. For viscosity, we adopt the Shakura and Sunyaev type α-viscosity, $t_{\varpi\varphi} = \alpha P$, where P is the total pressure.

The basic time dependent equations written in the conservation form are as follows;

$$\frac{\partial}{\partial t}(\varpi\Sigma) + \frac{\partial}{\partial\varpi}(\varpi\Sigma v_\varpi) = 0$$

$$\frac{\partial}{\partial t}(\varpi\Sigma v_\varpi) + \frac{\partial}{\partial\varpi}(\varpi\Sigma v_\varpi^2 + \varpi W) = W(1 - \frac{dln\Omega_k}{d\varpi}) + \Sigma(2v_\varphi' v_k + v_\varphi'^2)$$

$$\frac{\partial}{\partial t}(\varpi\Sigma v_\varphi') + \frac{\partial}{\partial\varpi}(\varpi\Sigma v_\varpi v_\varphi' + \alpha\varpi W) = -(\Sigma v_\varpi v_\varphi' + \alpha W) - \Sigma v_\varpi\frac{dl_k}{d\varpi}$$

$$\frac{\partial}{\partial t}(\varpi\Sigma e) + \frac{\partial}{\partial\varpi}(\varpi\Sigma v_\varpi e + \varpi v_\varpi W + \alpha\varpi W v_\varphi') = -\varpi^2(\Sigma v_\varpi v_\varphi' + \alpha W)\frac{d\Omega_k}{d\varpi} - \varpi F$$

where, Σ and W are the surface density and the vertically integrated pressure, respectively. For the vertical integration, we assumed that the distribution of matter is given by a polytropic relation $P = K\rho^{4/3}$. The radiative cooling rate F is approximated by the diffusion flux,

$$F = \frac{16acT_0^4}{3\kappa\rho_0 h}$$

where T_0 and ρ_0 are the equatorial temperature and density, respectively, κ is the mean opacity and h is the half thickness of the disk. The energy density e is given by

$$e = [3(1 - \beta) + \frac{\beta}{\gamma - 1} + \frac{1}{2}]\frac{W}{\Sigma} + \frac{1}{2}(v_\varpi^2 + v_\varphi'^2)$$

where β is the ratio of the gas pressure to the total pressure. Because of numerical reason, $v'_\varphi = v_\varphi - v_k$ is used instead of v_φ, where v_k is the Keplerian rotation velocity. The notation Ω_k and l_k denote the Keplerian angular velocity and Keplerian angular momentum distribution, respectively.

Free parameters are the mass of the central object M, the accretion rate $\dot{M}$, and the viscosity parameter α. In the following, we use the accretion rate $\dot{m}$, normalized by the critical accretion rate $\dot{M}_{crit} = 32\pi c r_g/\kappa_{es}$, where κ_{es} is the electron scattering opacity.

The initial condition is the transonic steady solution (Matsumoto et al. 1984) with small velocity perturbation localized around the sonic point. The basic time dependent equations are integrated numerically by using classical Lax-Wendroff method with artificial viscosity. All the quantities at the inner boundary ($\varpi = 2.7r_g$) are calculated from the quantities at the mesh point next to the boundary because the inner boundary is in the supersonic region. The outer boundary at $\varpi = 100r_g$ is treated as a fixed boundary.

3. Numerical Results.

3.1. MODEL 1: TRAPPED RADIAL PULSATION IN LOW LUMINOSITY DISK.

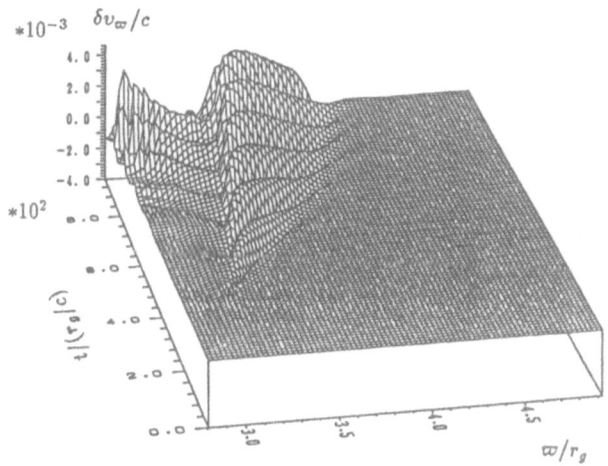

Figure 1: Time variation of velocity perturbation for model 1 ($M = 10M_\odot, \alpha = 0.2, \dot{m} = 0.1$).

Figure 1 shows the result of the numerical simulation for $M = 10M_\odot$, $\alpha = 0.2$ and $\dot{m} = 0.1$. The amplitude of the outgoing inertial acoustic wave generated by the perturbation imposed around the sonic point grows with propagation owing to the viscous pulsational instability (Kato et al. 1988a,b). Because the epicyclic frequency has a maximum at $\varpi = 3.7r_g$, low frequency components of the outgoing wave are reflected inward and they grow into shock waves. When $t \sim 1000r_g/c$, global oscillations confined in $3r_g < \varpi < 4r_g$ are established. Although the inner edge of the disk is not a reflecting wall, partial reflection of the ingoing wave maintaines the global oscillation (Kato et al. 1988a). The frequency of the oscillation ($\sim 80r_g/c$) is close to the maximum of the epicyclic frequency.

3.2. MODEL 2: PULSATIONAL INSTABILITY IN HIGH LUMINOSITY DISK.

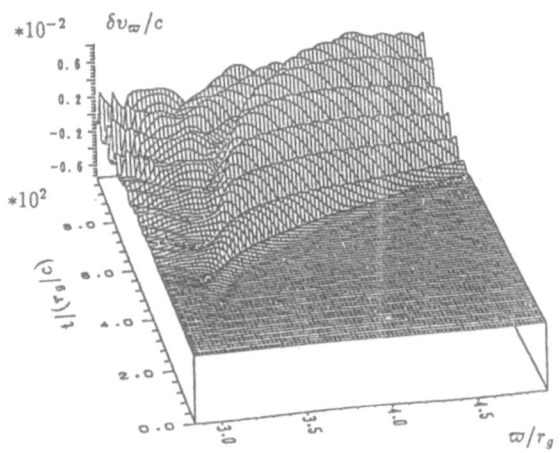

Figure 2: Time variation of velocity perturbation for model 2 ($M = 10M_\odot, \alpha = 0.2, \dot{m} = 0.3$).

Figure 2 shows the result for $M = 10M_\odot$, $\alpha = 0.2$ and $\dot{m} = 0.3$. In contrast to model 1, inertial acoustic waves propagate beyond the maximum of the epicyclic frequency. This is because the sound speed (or the temperature) is larger in this model than in model 1. Note that the dispersion relation for the inertial acoustic wave is $\omega^2 = \kappa^2 + c_s^2 k^2$, where c_s is the sound speed and κ is the epicyclic frequency (Kato et al. 1988a).

3.3. MODEL 3: THERMAL INSTABILITY IN PULSATIONALLY UNSTABLE DISK.

Figure 3 shows the time variation of the equatorial temperature for model 3 ($M = 10M_\odot, \alpha = 0.1, \dot{m} = 0.1$). In this model, numerical calculation is performed much longer time than in other models up to $t = 115000r_g/c$. Since the initial model is dominated by the radiation pressure in $5r_g < \varpi < 30r_g$, thermal instability takes place (e.g., Shakura and Sunyaev 1976). The onset of the thermal instability needs longer time than the thermal time scale [$\sim (\alpha\Omega)^{-1}r_g/c$]. After the random

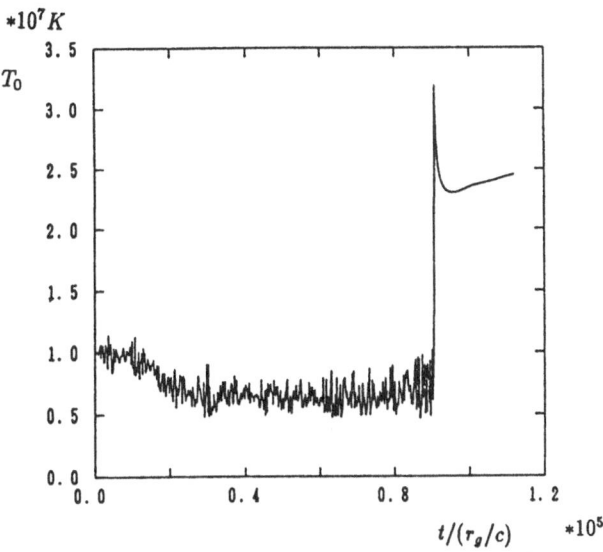

Figure 3: Time variation of the equatorial temperature at $\varpi = 3.2 r_g$ for model 3 ($M = 10 M_\odot, \alpha = 0.1, \dot{m} = 0.1$).

walk fluctuation due to the pulsational instability, the temperature decreases and the disk evolves into the gas pressure dominated state. The accretion rate in the thermally unstable region decreases and thus mass accumulates just outside this region. Later, the equatorial temerature slowly increases because the accumulated gas infalls due to viscosity. When the temperature increases up to the ignition point of the thermal instability ($t = 89000 r_g/c$), the equatorial temperature raises by factor 3 within the thermal time scale.

Dashed curves in figure 4 show the evolution of model 3 in $\Sigma - \dot{m}$ plane. Solid S-curves denote the steady disk models (Abramowicz et al. 1988) at $\varpi = 3 r_g$ and $\varpi = 5 r_g$. After the onset of the phase transition to the upper branch (slim disks), surface density decreases drastically and the disk evolves into the left turning point of the S-curves. The symbols in figure 4 denote the location and the types of sonic point of steady disk models. Since the criterion for the pulsational instability is identical with that the sonic point is nodal (Kato et al. 1988b), the transonic region of the slim disk is pulsationally stable in this case ($\alpha = 0.1$). This explains why the pulsation ceases after the onset of thermal instability.

4. Summary.

We have shown numerically that global radial pulsations are excited in the inner-most transonic region of accretion disks when the viscosity is sufficiently large. Although the oscillation period is the dynamical time scale and thus an order of magnitude smaller than the observed period of QPO's in low mass X-ray binaries, preliminary results suggest that the oscillation period modulates in longer time

172

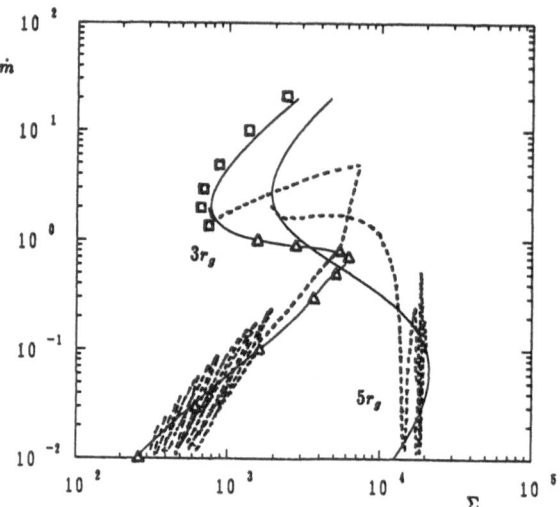

Figure 4: Evolution of model 3 on the Σ-$\dot{m}$ plane (dashed curves). Solid curves denote the steady disk models at $\varpi = 3r_g$ and $5r_g$. Symbols denote the saddle type sonic points ( ) and nodal type sonic points ( ).

scale. Further study is necessary to check whether the observational facts such as the luminosity-period relation can be explained.

ACKNOWLEDGEMENTS. The authors thank Dr. M.Abramowicz for discussions. Numerical computations were performed in FACOM M380 and VP200 at the Institute of Plasma Physics, Nagoya University.

REFERENCES.

Abramowicz,M.A., Czerny,B., Lasota,J.P., and Szuszkiewicz,E. 1988, *Astrophys. J.*, **332**, 646

Abramowicz,M.A. and Szuszkiewicz,E. 1988, in *Big Bang, Active Galactic Nuclei and Supernovae* ed. S.Hayakawa and K.Sato, (Universal Academy Press), p.265

Kato,S., 1978, *Mon. Not. R. astr. Soc.*, **185**, 629

Kato,S., Honma,F., and Matsumoto,R., 1988a, *Mon. Not. R. astr. Soc.*, **231**, 37

Kato,S., Honma,F., and Matsumoto,R., 1988b, *Publ. Astron. Soc. Japan*, **40**, 709

Liang,E.P.T. and Thompson,K.A., 1980, *Astrophys. J.*,**240**, 271

Matsumoto,R., Kato,S., Fukue,J., and Okazaki,A.T., 1984, *Publ. Astron. Soc. Japan*, **36**, 71

Matsumoto,R., Kato,S., and Honma,F., 1988, in *Physics of Neutron Stars and Black Holes*, ed. Y.Tanaka (Universal Academy Press), p.155

Muchotrzeb,B., 1983, *Acta Astron.*, **33**, 79

Muchotrzeb-Czerny,B., 1986, *Acta Astron.*, **36**, 1

Shakura,N.I., and Sunyaev,R.A., 1976, *Mon. Not. R. astr. Soc.*, **175**, 613

ONE-ARMED OSCILLATIONS OF DISKS AND THEIR APPLICATIONS.

Shoji KATO

Department of Astronomy, Kyoto University, Sakyo-ku, Kyoto 606, Japan

ABSTRACT. The presence of low frequency oscillation modes in disks is demonstrated. The frequency is much smaller than the dynamical one of the disks. Such low-frequency modes are one-armed oscillations. In relativistic disks, in particular, they are one-armed corrugation waves fluttering in the vertical direction. Possible applications of these waves to some astronomical objects are discussed.

1. Introduction.

There are many kinds of astronomical objects which have quasi-periodic time variations whose origin is attributed to surrounding disks. In some objects these time variations have time scales much longer than the dynamical time (the rotation time) of the disks. Then a natural question is whether disks can have oscillation modes with such low frequencies. The purpose of this paper is to point out that the answer is yes and such oscillations are one-armed ones.

First we shall show two observational examples of low frequency variations of disks. One is the long term spectrum variations in Be stars. Another is the low frequency noises in time variations of X-ray stars. Then we shall show how these variations can be interpreted in terms of one-armed oscillations.

2. Observational Evidence of Long-Term Variabilities of Disks.

2.1. LONG-TERM SPECTRUM VARIATIONS IN BE STARS.

Be stars are B-type stars with emission lines. The emission lines are attached to both red and blue sides of each absorption line. A characteristic point is that the relative intensity of emission lines of both sides changes semi-regularly with a long-term interval ($\sim$ several years) which is much longer than the rotation period of disks. This long term spectrum variation is called V/R variation. An example of time variation of the V/R intensity ratio is shown in figure 1.

F. Meyer et al. (eds.), Theory of Accretion Disks, 173–181.

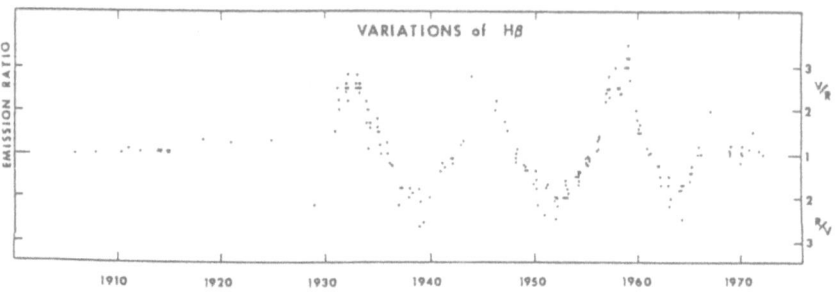

Figure 1: The time variation of the V/R intensity ratio of $H\beta$ in β^1 Mon (after Cowley and Gugula 1973).

A promising explanation of V/R variations is that the disk surrounding the central object is deformed into an eccentric form and the deformation rotates slowly around the central object (McLaughlin 1961). This is nothing but the model that one-armed waves rotate with a long period on rapidly rotating disks.

2.2. LOW FREQUENCY NOISES (LFN) IN X-RAY STARS.

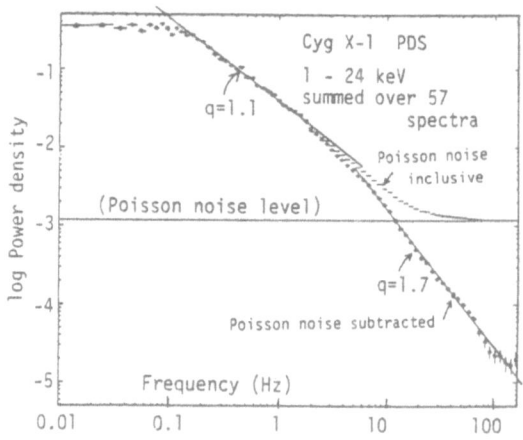

Figure 2: The power-density spectrum of low frequency noises in Cyg X-1 (after Makishima 1988).

The second example is taken from X-ray stars. One of the charactistic features of low mass X-ray binaries is chaotic, strong X-ray variations. A typical example of the result of power-density spectrum analyses of these time variations is shown in figure 2. A point to be emphasized is that the typical frequencies ($0.1Hz \sim 10Hz$) of these variations are much lower than the rotation frequency ($\sim 10^3 Hz$) of disks.

From the observational point of view, the origin of these time variations is attributed to the rapidly rotating innermost part of accretion disks (e.g., Inoue 1988).

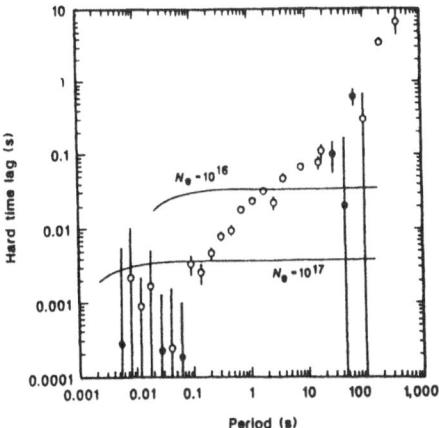

Figure 3: The frequency dependence of the time delay of the hard X-ray component to the soft component (after Miyamoto et al. 1988). Open circles represent the positive time delay, while filled circles the negative one.

Another important point to be explored is that in the case of Cyg X-1, in particular, hard X-ray variations delay in time in comparison with soft X-ray variations and the delay is longer as the period is longer (Miyamoto et al 1988). This is shown in figure 3. The relation between the time delay τ (in sec) and the period P (in sec) seems to be approximated as

$$\tau = AP^\alpha \qquad (2.1)$$

with $\alpha = 0.5 \sim 0.7$ and $A = 0.015 \sim 0.025$.

3. One-Armed Oscillations of Newtonian Nonselfgravitating Disks.

Geometrically thin, nonselfgravitating disks are considered. They rotate around central objects by angular velocity close to the Keplerian one. Possible global modes of oscillations in such disks are one-armed ones alone (Kato 1983). Other types of modes have necessarily short wavelengths in the radial direction and will be uninteresting observationally as well as theoretically. This comes from the following situation.

Roughly speaking, frequencies of radial oscillation of disks in the equatorial plane are of the order of the epicyclic frequency κ, because the restoring force upon perturbations comes from the disk rotation. The epicyclic frequency κ, however, varies strongly with radius. This means that when we consider a wave of a given frequency ω, the frequency difference between ω and κ becomes comparable with κ

as the wave propagates in the radial direction, even if they are close at a particular radius. This frequency differnce should be compensated by the pressure restoring force. This means that the wavelength of oscillations becomes very short. Such waves with very short wavelength, however, will be uninteresting theoretically, because excitation of such short wavelength waves will be difficult in general, and also uninteresting observationally because such short wavelength perturbations will not be observable.

The above argument, however, is not applicable to one-armed oscillations. One armed oscillations have a particular position in the problem of thin disk oscillations (Kato 1983). This is related to the fact that in thin disks the epicyclic frequency κ is close to the angular velocity Ω of disk rotation at any radius.

Let us consider disks with no temperature as an extreme case. They rotates exactly with the Keplerian angular velocity and $\kappa = \Omega$ holds at any radius. Let us further consider on such cold disks a bunch of particules with the same angular momentum. They are assumed to have been at the same position A at an initial epoch and have started to turn around the disk center at the same time in various directions slightly deviated from the direction of circular orbit. After one turn around the disk center, they gather again at the initial point A at the same epoch. This implies that the one-armed pattern is maintained without time change. Hence if the pressure restoring force due to presence of temperature is taken into account, the one-armed pattern can become global coherent oscillations with low frequency.

The frequency of such one-armed global oscillations can be estimated easily. As is well-known, the local dispersion relation for perturbations whose radial wavenumber k is

$$(\omega - m\Omega)^2 - \kappa^2 = c_s^2 k^2 \quad , \tag{3.1}$$

where c_s is the sound speed in the disk. Here eigen-modes have been taken in the form $exp[i(\omega t - m\varphi)]$, m being the number of arms. In the particular case of $m = 1$ and $\Omega \sim \kappa$, equation (3.1) gives (Kato 1983, Okazaki and Kato 1985)

$$|\omega| \sim \Omega \left(\frac{c_s}{\Omega \varpi}\right)^2 \quad , \tag{3.2}$$

if $c_s^2 k^2$ and $\Omega^2 - \kappa^2$ are comparable in magnitude and $k \sim 1/\varpi$ is taken.

As typical values of mass M and radius R of B stars we take $M = 10 M\odot$ and $R = 10 R\odot$. Furthermore $T = 10^4 K$ is adopted as the disk temperature. Then for $\Omega \sim (GM/\varpi^3)^{1/2}$ with $\varpi = 1.1 R$, the period of one-armed oscillations becomes

$$\tau = 2\pi |\omega|^{-1} \sim 5.5y \quad , \tag{3.3}$$

which is comparable with the period of V/R variations observed in Be stars.

Detailed forms of normal modes of one-armed oscillations in isothermal disks have been calculated by Okazaki (1989). His results seem to explain well the observed V/R variations. The result concerning the parameter dependences of variation period is shown in figure 4.

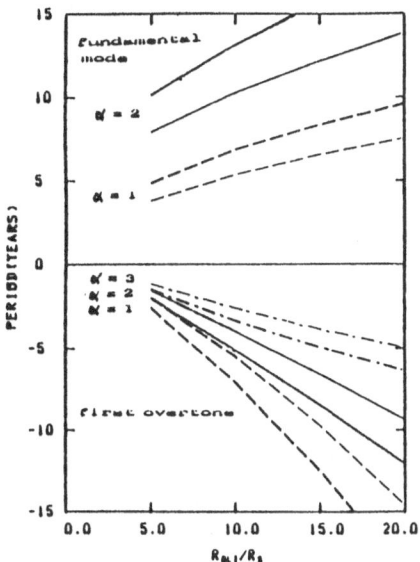

Figure 4: The disk size dependence of variation period. R_{out} is the outer radius of the disk and R_s the inner (stellar) radius. $\alpha = -d\ln \rho_{00}/d\ln \varpi$, ρ_{00} being the unperturbed density on the equatorial plane. Thick curves are for B0 stars, while thin curves for B5 stars.

4. One-Armed Corrugation Waves in Relativistic Disks.

In relativistic disks the Keplerian angular velocity $\Omega_K(\varpi)$ of rotation is not proportional to $\varpi^{-3/2}$. It increases faster than $\varpi^{-3/2}$ with decreasing ϖ. Because of this, the epicyclic frequency κ is no longer close to Ω in the innermost part of disks surrounding neutron stars and/or black holes. Hence the one-armed oscillations in the disk plane mentioned in the previous section are no longer low frequency modes in relativistic disks. However, we have still another type of low frequency one-armed oscillations. This is one-armed corrugation waves fluttering in the vertical direction. Corrugation waves are wave motions by which the disks make up-and-down motions in the vertical direction with some wavelength in the radial direction.

Let us consider displacements of a particle from the equatorial plane in the vertical direction. The particles feel the gravitational restoring force toward the equator. The force is proportional to the vertical displacement when the displacement is small. Hence the particle makes harmonic oscillations around the equator with a frequency. This frequency is equal to the (relativistic) Keplerian angular velocity Ω_K of disk rotation, if the deviation of the gravityational field from spherical symmetry due to rotation of the central object is neglected. In other words, the vertically fluttering period of a particle coincides with the Keplerian rotation period by which the particle rotates around the central object. This means that in the limit of no temperature one-armed corrugation pattern is a pattern maintained without time change. Hence if the pressure restoring force is present it makes the pattern a coherent mode of global oscillations with low frequency characterized by acoustic waves, as in the case of one-armed oscillations in the equatorial plane.

To confirm the above argument, we derive the dispersion relation for one-armed corrugation waves fluttering in the vertical direction. If the local approximation in the radial direction is adopted, we have (Kato 1989)

$$[(\omega - m\Omega)^2 - \kappa^2][(\omega - m\Omega)^2 - \Omega_K^2] = k^2 c_s^2 (\omega - m\Omega)^2 \quad , \tag{4.1}$$

where k is the wavenumber in the radial direction.

The above dispersion relation (4.1) can be understood as follows. The dispersion relation for local perturbations whose motions are restricted in the equatorial plane is given by

$$(\omega - m\Omega)^2 - \kappa^2 = 0 \quad , \tag{4.2}$$

if the gaseous pressure is neglected. On the other hand, the oscillations in the vertical direction due to gravitational restoring force are given by

$$(\omega - m\Omega)^2 - \Omega_K^2 = 0 \quad . \tag{4.3}$$

The coupling of these two oscillations through the pressure restoring force gives the dispersion relation (4.1).

Let us estimate the frequency of the waves. First we should know the amount of the difference between Ω^2 and Ω_K^2. The radial force balance in the equatorial plane is described by

$$-\frac{1}{\rho_0} \frac{\partial p_0}{\partial \varpi} + \Omega^2 \varpi - \Omega_K^2 \varpi = 0 \quad . \tag{4.4}$$

This gives

$$|\Omega^2 - \Omega_K^2| \sim \frac{c_s^2}{\varpi l} \quad , \tag{4.5}$$

where l is the characteristic radius by which the pressure changes appreciably. Hence when

$$|-2\omega\Omega| > \frac{c_s^2}{\varpi l} \tag{4.6}$$

holds, we have from equation (4.1)

$$\omega \sim -\frac{1}{2}\Omega \frac{\Omega^2}{\Omega^2 - \kappa^2} \frac{k^2 c_s^2}{\Omega^2} \quad . \tag{4.7}$$

This result means that in the innermost region ($\kappa \ll \Omega$) of relativistic disks there are very low frequency modes of oscillations whose phase rotates in the direction opposite to the disk rotation with frequencies smaller than Ω by the factor $(k c_s/\Omega)^2$.

Here we should know in what cases the condition (4.6) is satisfied. If equation (4.7) is substituted into inequality (4.6), we know that the condition (4.6) is reduced to

$$k^2 > \frac{1}{\varpi l} \quad . \tag{4.8}$$

This shows that equation (4.7) is valid as long as local perturbations ($kl > 1$) are considered (ϖ is usually comparable with or larger than l).

5. Application to Time Variability of Cyg X-1.

The frequency given by equation (4.7) is compared with the observed low frequency noises (LFN) of Cyg X-1. Observations show that the X-ray spectrum consists of two components (soft and hard components) and the component which variates with time is the hard one. Considering this situation, we take a relatively high temperature of $T = 10^9 K$. Then for $k = 1.0/r_g$ and $\varpi = 4r_g$ with $M = 10 M\odot$, we have

$$\omega \sim 16.0 s^{-1} \quad . \tag{5.1}$$

This is in good agreement with frequencies of LFN in Cyg X-1.

Next we should consider whether the hard time delay observed in Cyg X-1 can be explained. Let us take the image that the hard time delay is due to inward propagation of the waves from the region where the soft X-ray is emitted to the region where the hard X-ray is emitted. Then the frequency dependence of the time delay τ is a result of the frequency dependence of propagation speed of waves. In other words, the relation $\tau \propto P^\alpha$ implies that the group velocity $\partial \omega / \partial k$ of waves is proportional to ω^α. Actually from equation (4.7) we have

$$\frac{\partial \omega}{\partial k} \simeq \pm \frac{\Omega}{(\Omega^2 - \kappa^2)^{1/2}} \left(-\frac{2\omega}{\Omega} \right)^{1/2} c_s \quad . \tag{5.2}$$

This means $\alpha = 1/2$, which is consistent with the results obtained by Miyamoto et al. (1988).

The proportional coefficient A of $\tau = AP^\alpha$ is estimated. The time delay is given by $D/(\partial \omega / \partial k)$ when the propagation distance is D. Hence A is given by

$$A = \frac{(\Omega^2 - \kappa^2)^{1/2}}{\Omega} \left(\frac{\Omega}{4\pi} \right)^{1/2} \frac{D}{c_s} \quad ,$$

where $-\omega = 2\pi/P$ has been used. If the physical quantities adopted before are taken with $D = 0.5 r_g$, we have

$$A \sim 0.03 \quad ,$$

which is comparable with that derived from observational results (see section 2.2).

6. Discussion on Validity of Application to Cyg X-1.

The presence of low frequency corrugation waves discussed in the previous section is free from detailed structures of disks and perturbations. The necessary factor for the presence of the waves is the pressure restoring force alone. In other words, so long as the pressure restroring force is present, the difference of adiabatic or isothermal perturbations does not bring no essential difference in the order of frequencies. This will be understood from the physical argument concerning the origin of the waves.

It is noted, however, that the adiabatic approximation adopted will be relevant. This is because a given fluid element feels a time periodic perturbation of the time scale of $1/\Omega$, which is shorter than the thermal time scale. The latter is in general of the order of $1/\alpha\Omega$, where $\alpha(< 1)$ is the viscosity parameter of the conventional α-viscosity.

Next the effects of drift motion of the unperturbed medium on wave motions are examined. The main effect is to change the frequency ω to $\omega + kv_d$, where v_d is the drift speed of the medium toward the central object and of the order of $\alpha c_s^2/\Omega\varpi$ if α-viscosity is adopted. Since ω is given by equation (4.7), the ratio $|kv_d/\omega|$ is found to be of the order of $\alpha(k\varpi)^{-1}$. Hence the effect of drift motion is unimportant as long as local perturbations are considered.

If relatively low frequency perturbations (which have long wavelength) are considered, however, the local approximations adopted here is irrelevant. More careful discussion on frequency and on propagation is necessary.

So far the origin of the corrugation waves was not discussed. They might be generated in the geometrically thin disks by pulsational instability (Kato 1989) or might come from the outer region where they would be generated by potential perturbations due to the secondary star. The image that the waves propagate from the geometrically thin disk to the hot corona will be natural to explain the prevalence of the hard delay in comparison with the soft delay. Related to this problem, we should also remind the following situation. A wave propagating in a lower temperature region will be more difficult to be observed than a wave propagating in a higher temperature region. This is because in the former case the wavelength becomes smaller with propagation and the phase mixing among various modes will smear out perturbations.

Our basic image is that in the innermost regions of disks many kind of wave motions are superposed. The low frequency noises (LFN) observed in low mass X-ray binaries might be their results.

ACKNOWLEDGEMENTS. The author thanks Dr. A.T. Okazaki for permitting to present a figure before publication.

REFERENCES.

Aydin, C. and Faraggiana, R. 1978, *Astron. Astrophys. Suppl.*, **34**, 51

Cowley, A. and Gugula, E. 1973, *Astron. Astrophys.*, **22**, 203

Inoue, H. 1988, *COSPAR Symposium No.14 (Helsinki, Finland)*

Kato, S. 1983, *Publ. Astron. Soc. Japan*, **35**, 249

Kato, S. 1989, *Publ. Astron. Soc. Japan*, submitted

Makishima, K. 1988, *in Physics of Neutron Stars and Black Holes, ed. Y. Tanaka (Universal Academic Press)*, p.175

McLaughin, D.B. 1961, *Royal Astr. Soc. Canada*, **55**, 73

Miyamoto, S., Kitamoto, S., Mitsuda, K., and Dotani, T. 1988, *Nature*, **335**, 450

Okazaki, A.T., and Kato, S. 1985, *Publ. Astron. Soc. Japan*, **37**, 683

Okazaki, A.T. 1989, to be published

THERMAL AND TIDAL INSTABILITIES IN ACCRETION DISKS OF DWARF NOVAE.

Yoji OSAKI

Department of Astronomy, Faculty of Science, University of Tokyo,
Bunkyo-ku, Tokyo 113, Japan

ABSTRACT. Various outbursting phenomena in dwarf novae are now thought to be caused by variable rates of accretion onto white dwarf components through accretion disks in cataclysmic binary systems. In this review, I discuss the thermal instability (the ordinary disk instability) and the tidal instability in accretion disks as causes of outbursts in dwarf novae. In the first part, we discuss the disk instability model for normal outbursts of dwarf novae; in particular we consider following three problems: the narrowness of transition fronts, the delay of UV flux to the optical at the rise of outburst, and the variation of disk radius in an outburst-quiescence cycle.

We then consider the superoutburst and superhump phenomenon in SU UMa stars. We particularly discuss the tidal instability of accretion disks recently discovered by White-hurst (1988a), by which instability an accretion disk evolves into a non-axisymmetric form (an eccentric accretion disk). It is shown that the tidal instability in accretion disks is caused by a parametric resonance of particle orbits in accretion disks in binary systems. I finally propose a new working model of the superoutburst phenomenon based on the combined effects of the ordinary disk instability and the tidal instability. It is concluded that the disk instability model can explain the basic features of both normal- and super-outbursts in dwarf novae.

1. Introduction.

Dwarf novae are eruptive variables showing repetitive outbursts characterized by amplitude of $2 - 6$ magnitude, duration of a few days to 20 days, and recurrence time of $20 - 300$ days. There are three sub-classes in dwarf novae: the ordinary U Gem-(or SS Cyg-) type, the Z Cam-type showing occasional standstills, and the SU UMa-type showing superoutbursts. Dwarf novae are one of sub-groups of cataclysmic binary systems, in that a Roch-lobe filling cool dwarf star loses mass through the inner Lagrangian point and a (presumably non-magnetic) white dwarf accretes it through an accretion disk. Various outbursting phenomena occurring in these stars are believed to be caused by variable accretion onto white dwarfs. Dwarf novae are one of the best laboratories among various celestial objects to study

F. Meyer et al. (eds.), Theory of Accretion Disks, 183–205.

accretion disks, since the time dependent accretion plays an essential role in dwarf nova outbursts and details of accretion disks can only be revealed through time dependent phenomena. Besides that, observations in a wide range in wavelength extending from IR, optical, UV to X-ray are available for some of dwarf novae in various stages of outburst cycle, and some systems are eclipsing binaries that can offer a unique opportunity to map spatial brightness distributions within accretion disks by use of eclipse light curves.

It is now well established that the outbursts of dwarf novae are caused by a sudden brightening of accretion disks due to increased accretion. Two different models have been proposed to explain sudden increased accretion onto the white dwarf: one model is the mass transfer burst model advocated by Bath (1973) and his group and the other one is the disk instability model first proposed by myself (Osaki, 1974) and extensively pursued by several groups (too many to mention all, but Hoshi 1979; Meyer and Meyer-Hofmeister 1981, 1984; Smak 1982, 1984c; Cannizzo, Ghosh, and Wheeler 1982; Cannizzo and Wheeler 1984; Faulkner, Lin, and Papaloizou 1983; Papaloizou, Lin, and Faulkner 1983; Mineshige and Osaki 1983, 1985; Mineshige 1986).

In this talk, I am mainly concerned with instabilities in accretion disks as causes of outbursts in dwarf novae and little will be mentioned about the mass transfer burst model. Two kinds of instabilities are so far known in accretion disks of dwarf novae: the thermal instability (the ordinary disk instability) and the tidal instability recently discovered by Whitehurst (1988a). I discuss the ordinary disk instability in connection with the normal outburst of U Gem stars in section 2. I then discuss in section 3 the tidal instability in accretion disks in connection with the superoutbursts and superhumps in SU UMa-type dwarf novae. Finally I present a working model to explain the superoutburst phenomenon in SU UMa stars based on combined effects of disk instability and tidal instability.

2. Thermal Instability in Accretion Disks of Dwarf Novae.

It was found in late 1970s (Hoshi, 1979) and early 1980s (Meyer and Meyer-Hofmeister, 1981; Smak, 1982; Cannizzo, Ghosh and Wheeler 1982) that the standard accretion disk theory allows double solutions for a given surface density in outer parts of accretion disk; one solution is a low-viscosity cool disk and the other is a high-viscosity hot one. This is produced due to the effects of strong variations in opacity around 10^4 K where the ionization of hydrogen occurs. It was suggested by them that the resulting cyclic relaxation oscillation may be the cause of dwarf nova outburst. The thermal limit cycle mechanism for the dwarf nova outburst is best depicted by the "S-shaped" thermal equilibrium curve. Based on this thermal limit cycle instability, time evolutions of unstable disks have been calculated by various groups (e.g., Smak 1984c; Meyer and Meyer-Hofmeister 1984; Papaloizou et al 1983; Lin et al. 1985; Mineshige and Osaki, 1983, 1985), reproducing observed outburst light curves.

Since there are good reviews on the disk instability model for dwarf nova outbursts and on its comparison with the mass transfer burst model (see, e.g., Smak 1984a; Meyer 1986a; Verbunt 1986), I shall not repeat discussing all aspects of the disk instability model but rather I will concentrate on three remaining problems that I think important for the interpretation of outbursts in dwarf nova. They are; (1) narrowness of the transition fronts, (2) UV delay to optical in the rise phase of outburst, and (3) variation in the disk radius throughout an outburst-quiescence cycle.

2.1. TIME-DEPENDENT FORMULATION OF ACCRETION DISKS.

First I briefly review the basis of the disk instability model for the dwarf nova outburst. It is supposed in this model that the mass transfer to the disk from the secondary occurs steadily but the accretion onto the white dwarf occurs intermittently because of an intrinsic instability within the accretion disk. The basic instability mechanism in the accretion disk is the thermal instability mentioned above.

To calculate time evolution of unstable accretion disks and to obtain light curves and so on, we must solve the equations of time-dependent accretion disks. The standard procedure to do so is to use the Shakura and Sunyaev (1973) formulation of the so-called α disk. It is assumed in this prescription that the disk is geometrically thin and axi-symmetric. Then its vertical (local) structure and radial (global) structure are solved separately. The basic equations governing the global structure of a time-dependent disk are equations of mass and angular momentum conservation and energy equation. The first two equations are combined to give a single equation describing the time evolution of the surface density Σ within the disk;

$$\frac{\partial \Sigma}{\partial t} = \frac{3}{r} \frac{\partial}{\partial r} \left[r^{1/2} \frac{\partial}{\partial r} (r^{1/2} f) \right], \tag{1}$$

where $f = \nu \Sigma$ is the vertically integrated viscosity and ν is the kinematic viscosity. The more exact equation must have additional terms describing the effect of angular momentum removal from the disk by the tidal torque exerted by the secondary and the effects of mass supply due to inflowing stream from the secondary, and it is given by equation (26) of Smak's (1984c) paper [there is a misprint in his equation (26): a factor $r^{1/2}$ is missing in the square bracket of the third term of the right-hand side].

This equation is the diffusion-type equation for the surface density, and the kinematic viscosity ν plays a role as the diffusion coefficient. The kinematic viscosity ν is related to the surface density Σ through the local energy equation:

$$<c> \Sigma \frac{dT}{dt} = Q^+ - Q^-, \tag{2}$$

where $< c >$ is an appropriate specific heat, T is the disk (mid-plane) temperature, Q^+ and Q^- are viscous heat generation and the radiative loss from the surface of the accretion disk, respectively, and the time derivative in equation (2) is the Lagrangian time derivative. By equating the right-hand side to zero, we obtain the condition of thermal equilibrium. The thermal equilibrium curves have been calculated by various groups (Meyer and Meyer-Hofmeister 1983; Faulkner et al 1983; Mineshige and Osaki 1983; Cannizzo and Wheeler 1984; Smak 1984c). Since the S-shaped equilibrium curves are the basis of the disk instability mechanism, we discuss the energy equation in more detail.

The viscous heat generation is given in the α model by

$$Q^+ = \frac{9}{4}f\Omega^2 = \frac{3}{2}\alpha(R_g/\mu)T\Sigma\Omega, \tag{3}$$

where $\Omega = (GM/r^3)^{1/2}$ stands for the Keplerian angular velocity of the disk material, R_g and μ are the gas constant and the mean molecular weight. This means that it is proportional to the disk temperature for a given surface density Σ if α is a fixed constant. On the other hand the radiative loss from the surface of the accretion disk is obtained only by solving the equation of energy transport in the vertical direction and the numerical result is formally expressed as

$$Q^- = 2F = 2\sigma T_{eff}^4 = 2F(\Sigma, T, r), \tag{4}$$

where F is the radiative flux and σ is the Stefan-Boltzmann constant. Figure 1 illustrates an example of the effective temperature T_{eff} as functions of the disk temperature T with surface density as a parameter at a given radius, which were calculated by Mineshige and Osaki (1983). Since $Q^+ \propto T$, the thermal instability occurs when the slope in the $(log\,T, log\,T_{eff})$ plane is less than $1/4$. As seen in Figure 1, the radiative loss function has a plateau around the temperature $T \sim 10^4$ K where hydrogen undergoes change in the state of ionization and this is the place where the thermal instability occurs.

The local thermal instability propagates as transition fronts throughout the accretion disk by mass exchange with neighbouring zones via equation (1). It has been found that the viscosity parameter α on the hot branch of the thermal equilibrium curve has to be chosen at least a factor three higher than that of cool branch in order to have the spatially coherent outburst within the disk and to reproduce outbursts observed in dwarf novae by the disk instability model.

There are basically two types of outbursts in the disk instability model: Type A in Smak's (1984c) notation in which the outburst sets in from the outer parts and the heating front propagates inward, and Type B in which the outburst starts from the inner parts and the heating front propagates outward. Type A outburst shows generally an asymmetric light curve with the faster rise than the decline while the rise and fall are more or less symmetric in Type B outbursts.

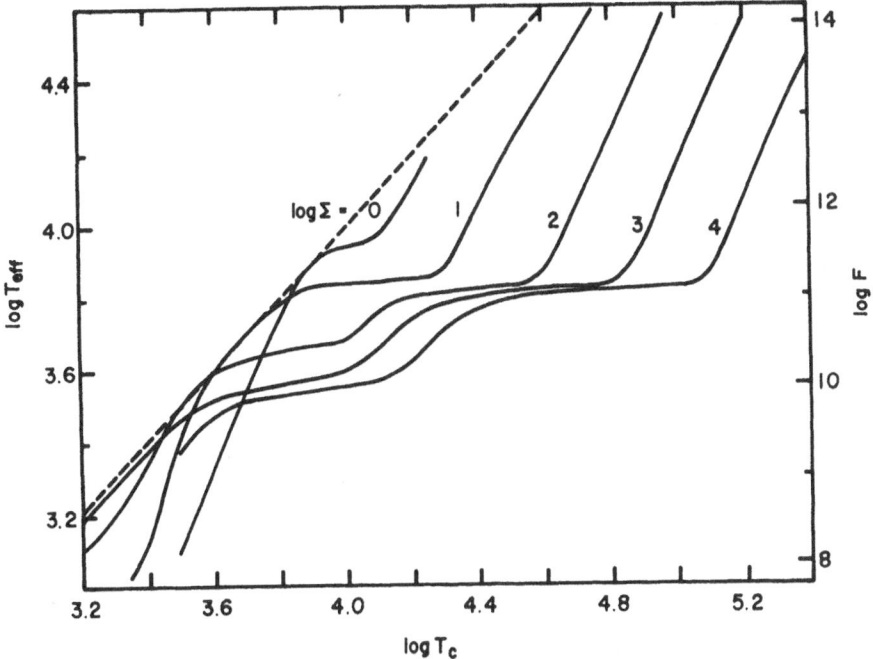

Figure 1: Radiative loss function or effective temperature T_{eff} of an accretion disk as a function of the disk (mid-plane) tempereture T with the surface density Σ as a parameter (from Mineshige and Osaki, 1983).

2.2. NARROWNESS OF TRANSITION FRONTS.

One of the unique features of the disk instability is the propagation of the transition front which transforms the disk from cool state to hot state or vice-versa. It is crucial to understand this phenomenon correctly for the disk instability model.

Meyer (1984) has argued that the width w of the transition front is very narrow, as narrow as the vertical thickness of the disk, and he has then formulated the propagation of transition fronts by using a local front approximation applying appropriate jump conditions. Papaloizou and Pringle (1985) criticized Meyer's formulation, by claiming that his formulation violated conservation of mass and angular momentum. Meyer (1986) replied to their criticism by arguing that his formulation did not violate the basic conservation law. The controversy has not yet been settled.

Here I present my view on this controversy. First I present my conclusion and then I describe the reason why I think so. My conclusion is as follows; the propagation of transition fronts is basically a global phenomenon and it can not be described locally. As for narrowness of the fronts, there exist two kinds of front widths: the viscosity front width w_v over which the viscosity changes greatly and the temperature front width w_T over which the disk temperature changes from hot state to cool state. These two widths are in general different and it is very important to make distinction between them. It is argued that either one of the two front widths can be as wide as the disk radius.

To show this, I first note that both parties (Papaloizou and Pringle 1985; Meyer 1986) agree on one point, that is, the propagation velocity V_F of the transition front is inversely proportional to the width w of the front [i.e., equation (9) of Papaloizou and Pringle 1985 and equation (10) of Meyer 1986]:

$$V_F \sim \frac{\nu}{w_v}, \tag{5}$$

where ν is the kinematic viscosity for the high-viscosity state. Here I add the subscript v to w, because the width appearing in equation (5) applies for the width over which the viscosity jumps greatly. On the other hand, if the transition front propagates with the speed V_F, we have from energy equation

$$V_F \sim \frac{w_T}{t_{th}}, \tag{6}$$

where t_{th} is the thermal time and w_T stands for the front width over which the disk temperature jumps from cool state to hot state. By combining equations (5) and (6), we obtain

$$w_v w_T \sim \nu t_{th}. \tag{7}$$

Equation (7) is the very equation which constrains the widths of the transition front.

All of previous studies implicitly assume $w_v \sim w_T$. However these two widths can in general be different. Let us now examine what happens if the two widths are assumed to be the same order of magnitude as commonly supposed. We then find (Meyer 1984)

$$w \sim (\nu t_{th})^{1/2} \sim \eta^{1/2} H = \varepsilon r \sim (1/30)r, \tag{8}$$

where H is the half thickness of the disk and the factor η typically about five takes into account the increase of the thermal time scale due to hydrogen ionization. If we write the viscous diffusion time scale and the front propagation time scale as t_d and t_F, respectively, and the viscous drift velocity of matter in the high-viscosity hot state as v_d, we find from equations (5) and (8)

$$\frac{t_F}{t_d} = \frac{v_d}{V_F} \sim \varepsilon. \tag{9}$$

Suppose that the disk is initially in hot viscous state and the surface density at the outer edge has just touched the critical density below which no hot state exists and the cooling transition has just set in there. Then the cooling front propagates inward and transforms the whole disk into the cool state with the timescale t_F given by equation (9). The mass accreted onto the white dwarf during this interval is estimated as $\Delta M/M_{disk} \sim t_F/t_d \sim \varepsilon$, that is, only about 3% or less of the disk mass.

However, the density distribution within the disk must change drastically during this interval from a nearly steady state disk with $\Sigma \propto r^{-3/4}$ to a quiescent disk with $\Sigma \propto r^1$. The total angular momentum of the disk must basically be conserved during this interval because the angular momentum carried away with matter accreted onto the white dwarf may be negligibly small. If we require the conservation of total angular momentum during this transition, we can show (Osaki 1989) that about five-sixth of matter in the hot region must flow out into the cool region through the front (the "snow-shovel" effect of Lin et al, 1985) while the remaining one-sixth of the hot disk mass must be accreted onto the white dwarf. In other words, we must get rid of matter from the hot disk as much as 1/6 of the original mass in order to have a transition completed from the quasi-stationary viscous disk to the cool disk. This evidently contradicts with that estimated above. This suggests the propagation time scale of the cooling front is $t_F \simeq (1/6)t_d$ and hence $V_F \simeq 6v_d$. We have therefore obtain the viscous width of the cooling front $w_v \simeq (1/6)r$ from equation (5). This means that $w_v \sim 25 \times w_T$ in this case from equation (7).

The difficulty of the local front approximation becomes more serious in the case for the heating front propagating outward because the heating front, by traversing all the way to the outer edge, transforms a cool disk into a hot disk in a very short time without leaving no surface density depression behind the front. This could be possible only if the front were the source of mass and angular momentum.

In Meyer's local front approximation, the propagation velocities of transition fronts are locally determined, and they are independent of the direction of propagation. However, those other people who solved explicitly the equations of time dependent accretion disks numerically (e.g., Smak 1984c; Mineshige and Osaki 1985; Cannizzo, Wheeler, and Polidan 1986) all observe that the heating fronts propagating outward are much slower than those propagating inward as evidenced by symmetric light curves of Type B outbursts and asymmetric ones of Type A outbursts. This suggests that $w_v \gg w_T$ for heating fronts propagating outward while $w_v \ll w_T$ for those inward. Figure 2 schematically illustrates this situation.

My main point is that the propagation of transition fronts is determined by the global evolution of the disk and that the front widths are determined from the propagation velocity V_F of the front in such a way to satisfy equations (5) and (6). Meyer's (1984) local front approximation may probably be applicable only to the onset stage of the instability. When the transition front traverses over an appreciable fraction of the disk, non-linear effects due to the propagation of the front may be dominated and the density distribution of the disk is appreciably modified.

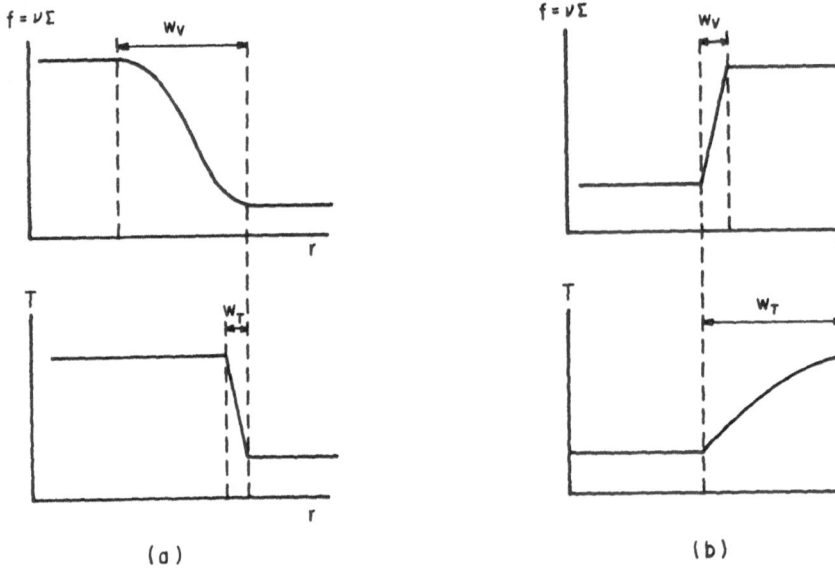

Figure 2: Schematic drawings for variations in viscosity $f = \nu\Sigma$ (upper figure) and in the disk temperature T (lower figure) within transition fronts; (a) for the case of a heating front propagating outward (or cooling front propagating inward), (b) for the case of a heating front propagating inward.

2.3. UV DELAY TO OPTICAL ON THE RISE PHASE.

The disk instability model is very successful in explaining the basic observational features of dwarf nova outbursts. A few problems however remain to be solved in the framework of disk instability. One of the most serious difficulties in the disk instability model concerns about observed UV delay to optical in the rise part of outburst (e.g., VW Hyi by Hassall et al. 1983; SS Cyg by Cannizzo, Wheeler, and Polidan, 1986; Verbunt 1987); the rise of UV flux is delayed about 0.5 ~ 1 day to that of optical flux. Pringle, Verbunt, and Wade (1986) have argued that this observation is against the disk instability model because the disk should jump to a hot state with an effective temperature higher than 10,000 K in a very short time (i.e., in the thermal time scale) in the disk instability model once instability sets in and the heating front propagates inward only after the outer region had already been in a hot state. In such a situation, UV flux should increase almost simultaneously with optical flux, which contradicts with observations.

Several proposals have been made to solve this difficulty (Meyer-Hofmeister, 1987; Mineshige 1988; Meyer and Meyer-Hofmeister 1988). Since I believe that Mineshige's (1988) explanation is the right path to solve this problem, I shall discuss his explanation. Mineshige (1988) has shown that the outer cool region of the disk does not jump directly to the hot state but it makes a halt in an intermediate warm state with $T_{eff} \sim 6000$K before it climbs to the final hot state above 10,000K. He called this the stagnation stage.

In order to have the stagnation stage, the outer disk should remain at an effective temperature around 6000 K for a period during which the heating front traverses an appreciable fraction of the disk. In the local front approximation this is not possible because the front propagation time t_F is always longer than the thermal timescale t_{th}. However, as discussed in the previous subsection, the temperature front width w_T could be as wide as an appreciable fraction of the disk radius. In such a case, the propagation time of the heating front could be of the same order of magnitude as the thermal time scale. Then an appreciable fraction of the disk undergoes in some sense a thermal transition from cool to hot state simultaneously. Now the question is how long the outer disk can stay in a transition stage with the warm effective temperature around $T_{eff} \sim 6000$ K.

Mineshige (1988) has shown that the stagnation stage is produced by following three effects combined: one effect is the increase of thermal timescale due to ionization energy of hydrogen around the disk temperature $T \sim 10,000$ K, and the second effect is the existence of a plateau in the radiative loss function at a level of $T_{eff} \sim 6000$ K (see Figure 1). The latter effect causes the disk to stay at an effective temperature near 6000 K for a while even if the disk mid-plane temperature increases considerably during the thermal transition stage. The third effect is the existence of an intermediate stable (or more exactly a metastable) state which prolongs the stay of the disk in the warm stage.

Mineshige (1988) has suggested that previous calculations (e.g., Pringle et al. 1986; Cannizzo and Kenyon 1987) have failed to get a sufficiently long delay of UV flux to optical because they have used an approximate form of energy equation. It is thus essential to solve the energy equation explicitly in order to explain the UV delay phenomenon.

A sufficiently long delay of UV to optical can be reproduced either (1) by a long thermal time scale, or (2) by the existence of an intermediate stable branch. Since the thermal timescale is given by

$$t_{th} \sim \frac{\eta}{\alpha\Omega}, \tag{10}$$

the first possibility is realized if the viscosity parameter α in the transition stage is small.

As for the second possibility, several people have found the intermediate stable branch in the thermal equilibrium curve (Mineshige and Osaki 1983; Cannizzo and Wheeler 1984; Meyer-Hofmeister 1987). Such an equilibrium curve shows a double S-shaped figure (or more like the greek letter "ξ") having an intermediate stable branch around $log_{10} T_{eff} = 3.8$. This curious feature in turn originates from the radiative energy loss function Q^- which has already been shown in Figure 1.

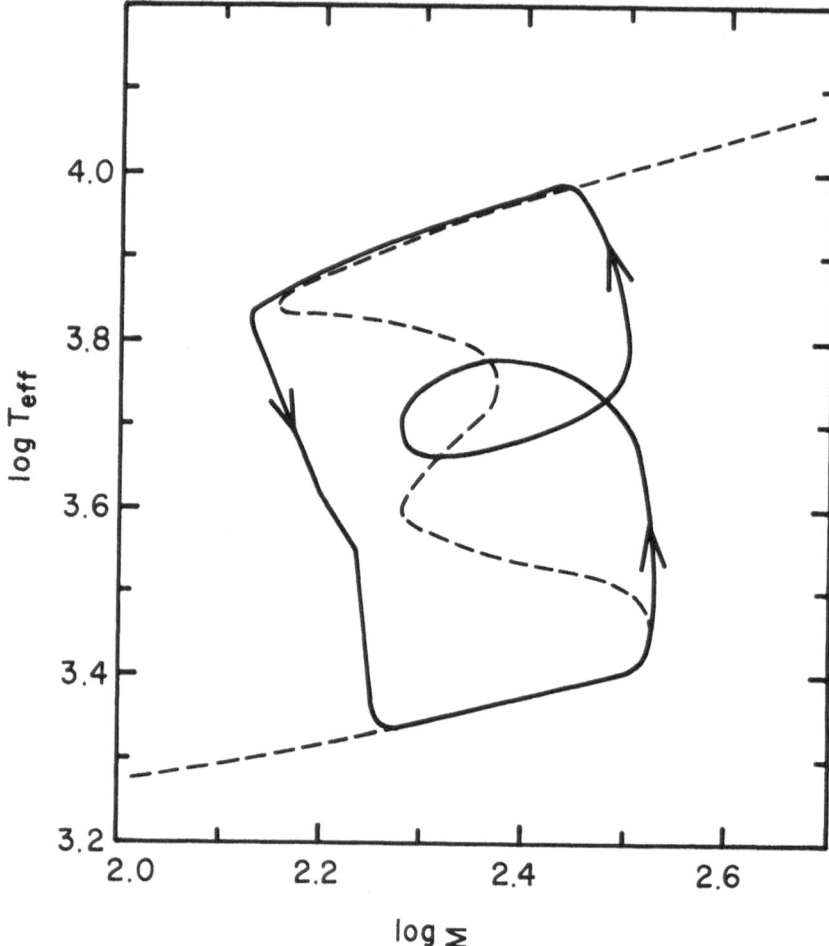

Figure 3: Evolutional path in the (Σ, T_{eff})-plane in the outer part of the disk for a type A outburst (from Mineshige and Osaki, 1985).

Based on this radiative loss function but by assuming $\alpha_{hot} = 0.1$ and $\alpha_{cool} = 0.03$, Mineshige and Osaki (1985) and Mineshige (1988) have calculated the time evolution of unstable disks. Figure 3 exhibits the time evolution of an outer disk region during a type A outburst in the surface density Σ and effective temperature T_{eff} plane. When the thermal instability sets in, the path in this diagram for the outer disk does not go straightly upward but it curves toward the metastable state and it stays for some time in this stage.

On the other hand, Meyer and Meyer-Hofmeister (1988) have recently suggested a new model to solve the difficulty of UV delay. One of interesting features of their model is the suggestion that the inner parts of the accretion disks may be continually depleted during quiescence. I think this may be quite conceivable because the irradiation heating by the white dwarf (and/or the boundary layer) dominates over the viscous heating at least in the inner parts when the disk is optically thin (Smak, 1984d). Consequently, the thermal equilibrium curves are modified in such a way that their double-valued forms are suppressed, enabling continuous depletion of matter during quiescence.

In any way, the result must depend on details of the thermal equilibrium curves. I believe that observed UV delay is not a difficulty for the disk instability model but it is rather a good tool to diagnose the viscosity parameter α and the thermal equilibrium curves in accretion disks.

2.4. VARIATION IN THE DISK RADIUS IN AN OUTBURST CYCLE.

The variation in the disk radius in the outburst cycle of a dwarf nova is another important observational constraint to the theory of outburst. Observations (Smak, 1984b ; O'Donoghue, 1986) indicate that the disk expands during outburst and it shrinks gradually during quiescence.

Smak (1984c) showed that the disk instability model can reproduce observed variations of the disk radius. This phenomenon is quite naturally understood in the disk instability model because the accumulation of matter with low angular momentum in the outer parts of the accretion disk during quiescence leads to a gradual contraction of the disk radius while the sudden increase of viscosity and resulting accretion of stored matter during outburst must be accompanied with an expansion of the disk radius in order to conserve the total angular momentum of the disk. By using a simpler model for the disk instability, Anderson (1988) has essentially confirmed Smak's conclusion. Figure 4 reproduces Anderson's calculation and its comparison with observations of U Gem.

On the other hand, Livio and Verbunt (1988) have considered the same problem based on the mass transfer burst model. They have argued that the mass transfer burst model can equally well reproduce the observed variation in the disk radius. However, there is a serious flaw in their discussion and it is very unlikely that the mass transfer burst model can explain the observed variation in the disk radius.

To show this, we examine Livio and Verbunt's (1988) discussion. They have argued in the first place that the equilibrium radius of the accretion disk is larger when the mass transfer rate is higher. They have then argued in the framework of mass transfer burst model that the disk expands to a larger radius after a transient contraction when the mass transfer rate is enhanced during outburst and the disk contracts gradually to a smaller radius corresponding to the lower rate of mass transfer in quiescence.

My objection to Livio and Verbunt's (1988) model concerns about their supposition that the equilibrium radius is the function of the mass transfer rate and that it is larger for a higher mass transfer rate. The general expectation for the steady-state disk is that its outer radius is basically given by the tidal radius which

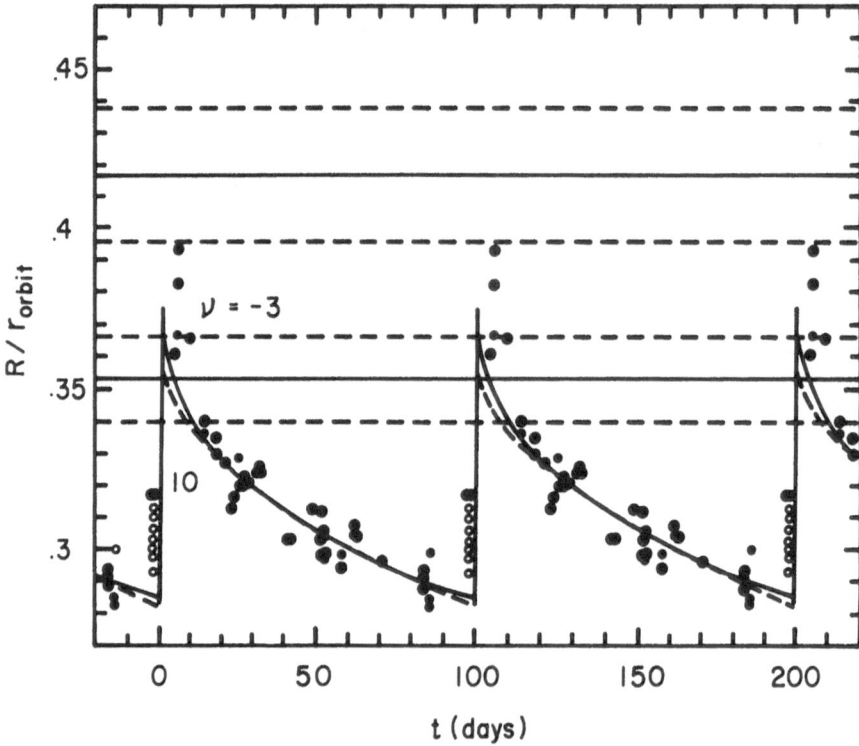

Figure 4: Variation in the disk radius during quiescence in U Gem and theoretical calculation based on the disk instability model (after Anderson, 1988).

is approximately given by the last non-intersecting periodic orbit of the restricted problem of the three body (Paczynski, 1977; Papaloizou and Pringle 1977) and it should be independent of mass transfer rate.

The equilibrium radius of the accretion disk is determined by the equation of angular momentum conservation which is given by

$$\frac{dJ_d}{dt} = j_0 \, \dot{M}_0 - j_{in}\dot{M}_{acc} - \dot{J}_{tidal}, \tag{11}$$

where J_d is the total angular momentum of the accretion disk, $\dot{M}_0$ and $\dot{M}_{acc}$ stand for the mass supply rate from the secondary to the disk and the mass accretion rate from the disk to the central white dwarf, respectively, and $\dot{J}_{tidal}$ is the total tidal torque exerted on the accretion disk by the secondary, while $j_0 = (GMr_0)^{1/2}$ and $j_{in} = (GMr_{in})^{1/2}$ are the specific angular momenta of the stream (the infalling matter from the secondary) and of matter at the inner edge, respectively. The first term and the second term in the right-hand side of equation (11) represent the angular momentum supply to the disk from the secondary and the angular momentum loss from the disk which is carried with accreting matter onto the white dwarf, respectively, while the third term gives the angular momentum removal by the tidal torque.

Let us now consider the tidal torque $\dot{J}_{tidal}$ in equation (11). Generally speaking, the tidal torque is determined by the strength of tidal dissipation and thus it is proportional to the viscosity coefficients. In fact, Papaloizou and Pringle (1977) have shown that the tidal torque is related to the tidal dissipation in the form;

$$\dot{J}_{tidal} = \int_{disk} \frac{\rho \varepsilon_\nu dV}{\Omega - \omega}, \tag{12}$$

where ε_ν is the viscous heat production per unit mass as a result of the tidal perturbations of accretion disk flow by the secondary's gravitational field and Ω and ω are the angular velocities of matter in the accretion disk and of the secondary, respectively. If the viscous dissipation is mainly due to the shear viscosity with the kinematic viscosity coefficient ν (or if the bulk viscosity coefficient ζ is proportional to the shear viscosity coefficient ν), equation (12) can be written as

$$\dot{J}_{tidal} = \int_{r_{in}}^{R_d} 2\pi r \nu \Sigma \omega g(r) \, dr, \tag{13}$$

where R_d and r_{in} are the outer disk radius and the inner disk radius, respectively, and $g(r)$ is a non-dimensional function of the radial coordinate r, which is determined by the flow pattern due to tidal perturbation (Papaloizou and Pringle 1977).

In the steady state, the right hand side of equation (11) must vanish and the mass accretion rate $\dot{M}_{acc}$ onto the white dwarf must equal the mass transfer rate $\dot{M}_0$ and the angular momentum supply by incoming stream must balance with the angular momentum removal by the tidal torque.

In the standard accretion disk theory, the vertically integrated viscosity $\nu \Sigma$ is related to the steady mass accretion rate by

$$\nu \Sigma = \frac{1}{3\pi} \dot{M}_{acc} \left[1 - (r_{in}/r)^{1/2} \right]. \tag{14}$$

If we substitute equation (14) into equation (13), we find

$$\dot{J}_{tidal} = \dot{M}_{acc} \left[\frac{2\omega}{3} \int_{r_{in}}^{R_d} r[1 - (r_{in}/r)^{1/2}]g(r) \, dr \right]. \tag{15}$$

Equation (15) means that the tidal torque consists of two factors; one factor given by the mass accretion rate $\dot{M}_{acc}$ and another factor which is a function of the outer disk radius R_d only for a given binary parameter. Then the mass transfer rate simply cancels in steady state in the right-hand side of equation (11) and the equilibrium disk radius is thus independent of mass transfer rate. The disk radius in units of the binary separation a is determined completely if the mass ratio of the binary is given. Two dimensional hydrodynamical simulations (e.g., Lin and Pringle, 1976; Whitehurst, 1988b; Hirose and Osaki, 1989a) also show that the accretion disk in steady state expands to the tidal radius which is approximately given by the last non-intersecting particle orbit in the restricted problem of three body (Paczynski 1977; Papaloizou and Pringle 1977) and that the tidal radius is independent of mass transfer rate.

The difference between our conclusion and that of Livio and Verbunt (1988) is traced to different formulas for the tidal torque. In fact, Livio and Verbunt's (1988) formula for the tidal torque (i.e., their equation [1]) did not include any viscosity coeffients (neither shear ν nor bulk ζ viscosity coeffients). It is evident that the tidal torque is proportional to the viscous dissipation and hence it must include the viscosity coefficients (e.g., consider the case of the tidal torque exerted on the earth by the moon). Thus, Livio and Verbunt's (1988) argument on the variation of the disk radius is incorrect. It is very unlikely that the mass transfer burst model can reproduce the observed variations in disk radii of dwarf novae if the tidal torque is treated properly.

3. SU UMa stars and Tidal Instability in Accretion Disks.

SU UMa stars constitute a sub-class of dwarf novae. They exhibit two distinct type of outbursts; a short outburst ("normal outburst") usually lasting two to three days and a long outburst ("superoutburst") lasting ten days or longer. In addition to this defining character of the subclass, they have following distinct characteristics; (1) they occupy the short-period side of the "period-gap" in the orbital period distribution of cataclysmic variables, and (2) they exhibit the periodic light humps called "superhumps" during supermaxima which repeat with a period very close to the binary orbital period, but always a few percent longer than that. Besides them, (3) the superoutburst periodicity is more regular than the periodicity of normal outbursts, (4) every superoutburst seems to be triggered by a normal outburst (Marino and Walker 1979), and (5) during superoutbursts of one of SU UMa star Z Cha, narrow absorption lines are observed and their γ-velocity is found to vary with the beat period between the orbital period and the superhump period (Vogt, 1982; Honey et al, 1988). Similar phenomena are also observed in WZ Sge and TU Men.

3.1. TIDAL INSTABILITY IN ACCRETION DISKS.

Various attempts were made to explain the superoutburst and superhump phenomenon of SU UMa stars but they were not very successful. However, a major breakthrough has recently occurred in this problem as Whitehurst (1988a) has discovered a new kind of instability in accretion disks which is caused by the tidal action of the secondary star for a cataclysmic binary system with a low mass secondary.

Whitehurst (1988a) has made numerical simulations of two-dimensional hydrodynamics of the accretion disk flow corresponding to SU UMa stars and he has shown that the accretion disk in a binary with a small mass ratio $q = M_2/M_1$ less than 0.25 is tidally unstable if the outer disk radius exceeds a critical value and that it evolves into a non-axisymmetric (eccentric) disk which slowly rotates in the inertial frame of reference. The superhump phenomenon is then explained by the periodic tidal dissipation in the non-axisymmetric eccentric disk with the synodic period between the secondary and the eccentric disk. The variation in the γ-velocity is naturally explained by the slowly precessing eccentric accretion disk and the phase relation between the superhump light variation and the γ-velocity observation is correctly explained by this model. Whitehurst's (1988a) results are basically confirmed by Hirose and Osaki (1989a) in a similar calculation and their results are presented separately in these proceedings.

The Whitehurst model is very encouraging because it can naturally explain why the SU UMa stars are restricted to those cataclysmic variable stars with short orbital periods below the period gap of cataclysmic variable stars. Based on the Whitehurst tidal instability, the present author (Osaki, 1989) has proposed a new model for the superoutburst phenomenon as a combined effect of the thermal instability (i.e., the ordinary disk instability) and the tidal instability. This model will be explained in the last sub-section.

3.2. CAUSE OF THE TIDAL INSTABILITY.

Let us now discuss the cause of the tidal instability in accretion disks. Whitehurst (1988a) has shown that the cause of tidal instability lies in the instability of periodic orbits in the restricted problem of three body, which instability had been known for some time (Paczynski 1977) but its significance with respect to the superhump phenomenon had not been appreciated.

It is known (Lubow and Shu 1975; Paczynski 1977) that the stream lines of the flow in accretion disks are well approximated by simple (non-intersecting) periodic orbits in the restricted problem of three body. In fact, Paczynski (1977) has calculated the non-intersecting periodic orbits for various mass ratios of binaries and examined their stability.

In order to find the cause of the tidal instability, we (Hirose and Osaki, 1989b) repeat the same calculation as that of Paczynski (1977). Figure 5 exhibits the stability results thus obtained for various mass ratios. The abscissa of the figure is the mean angular velocity Ω^* of periodic orbits in the corotating frame of binary in units of angular frequency of the binary and the ordinates shows the stability parameter a of Henon (1965). If $\mid a \mid < 1$, then the periodic orbit is stable while it is

198

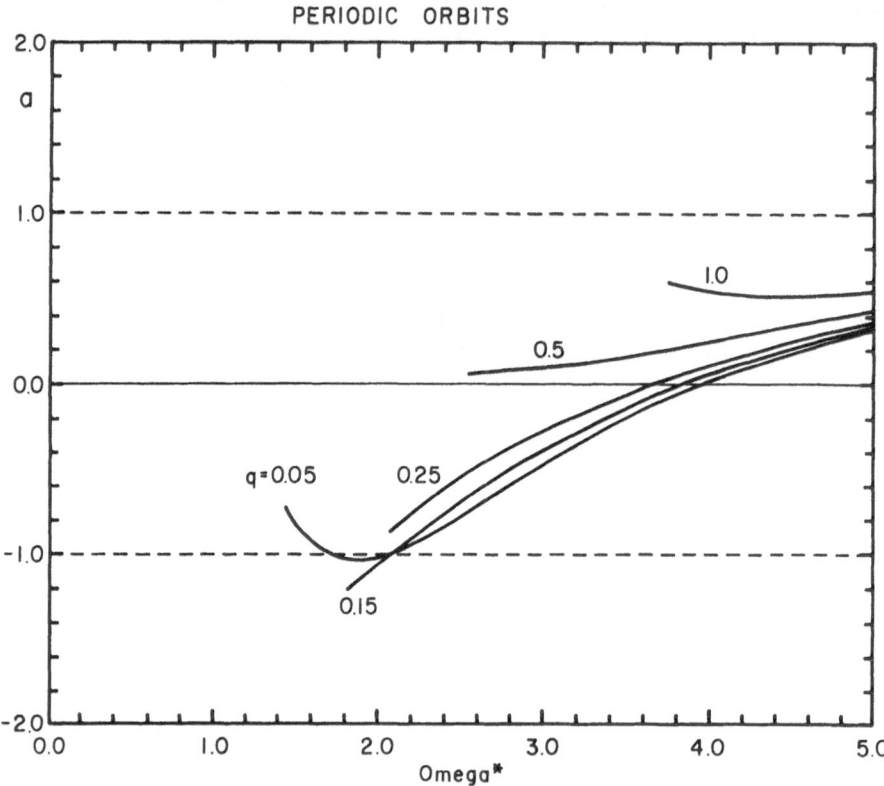

Figure 5: The stability parameter a of Henon is shown as a function of Ω^* for periodic orbits in accretion disks with various mass ratios of binaries. Here Ω^* is the angular velocity (or mean motion) of the periodic orbit of interest in the corotating frame of the binary in units of the angular velocity of revolution of the binary.

unstable if $|a| > 1$. As seen from figure 5, the instability appears always when Ω^* is near 2.0. In other words, the tidal instability occurs when the revolution period of a test particle in the accretion disk is just near the one-third of the revolution period of the secondary in the inertial frame of reference, that is, 3:1 resonance between the particle orbit and the binary motion. This suggests that some kind of resonance phenomenon is responsible for the tidal instability. In fact, we can show that the parametric resonance is indeed responsible for the tidal instability. It is also noted that the condition of $\Omega^* \sim 2$ corresponds to that of $r_{crit}/a \sim 0.46$ for the mean radius of the test particle in the periodic orbit. This condition is satisfied only in those binary systems with small mass ratios $q = M_2/M_1 < 0.25$ in which the radius of the critical Roche lobe around the primary star is large enough to accommodate the critical particle orbit.

To examine the stability of a periodic orbit, we take a test particle in a periodic orbit and superimpose on it a small perturbation. We write the position vector of a test particle in the corotating frame of binary as $r(t) = r_0(t) + \boldsymbol{\xi}$, where $r_0(t)$ is the position vector of the particle in the periodic orbit and $\boldsymbol{\xi}$ is the displacement vector of perturbation. By linearizing the equation of motion, we obtain perturbation equations for $\boldsymbol{\xi}$.

To solve the perturbation equations, we use a successive approximation. As the first step, we assume that the periodic orbit is very near to the Keplerian circular orbit and we neglect the non-axisymmetric part of the perturbing potential due to the secondary. We then obtain the well known equation of the epicyclic oscillation;

$$\frac{d^2\xi}{dt^2} + \kappa^2\,\xi = 0, \tag{16}$$

where ξ is the radial component of the perturbation vector $\boldsymbol{\xi}$, $\kappa_0 \simeq \Omega^* + 1$ is the epicyclic frequency, and the subscript 0 is added to κ to indicate the zeroth order approximation.

The next step is to take into account the neglected terms in the perturbation equations. By evaluating those terms with substitution of the solution of the zeroth order epicyclic motion, we obtain the same equation of epicyclic oscillation as equation (16), but this time the square of the epicyclic frequency is not a constant but a periodic function of time. In other words, the epicyclic frequency is not a constant for a given periodic orbit as in the ordinary formulation, but it is a periodic function of time. Thus, when a test particle moves around near the periodic orbit, it feels time-varying (periodic) restoring force.

Since the mean motion of the periodic orbit is denoted by Ω^*, we may write

$$\kappa^2 = \kappa_0^2\left(1 + \sum_{n=1}^{\infty} \epsilon_n\,cos\ n\Omega^*t\right), \tag{17}$$

where ϵ_n is the coefficient of Fourier expansion to the square of the epicyclic frequency for periodic orbits. Equation (16) is then

$$\frac{d^2\xi}{dt^2} + \kappa_0^2\left(1 + \sum_{n=1}^{\infty} \epsilon_n\,cos\ n\Omega^*t\right)\xi = 0. \tag{18}$$

Equation (18) is the very equation that is known as Hill's equation in the lunar theory. If we retain only the third harmonics of the Fourier expansion, we have

$$\frac{d^2\xi}{dt^2} + \kappa_0^2\left(1 + \epsilon_3\,cos\ 3\Omega^*t\right)\xi = 0. \tag{19}$$

Equation (19) is the Mathieu equation. It is well known as the parametric resonance that the Mathieu equation gives unstable solution if the disturbing frequency, $3\Omega^*$, of the restoring force is in the neighbourhood of twice the natural frequency κ_0 ;

$$3\Omega^* \approx 2\,\kappa_0. \tag{20}$$

Here κ_0 is the time independent part of the epicyclic frequency and its approximate expression is given by that of the Keplerian circular orbit: $\kappa_0 \approx \Omega^* + 1$. The condition of parametric resonance will therefore be $\Omega^* \approx 2$. This is the very condition we have sought for. It may be noted here that the ϵ_3 term arises from P_3 term (the third harmonic function) of the perturbing potential of the secondary. More details for its derivation will be given elsewhere (Hirose and Osaki 1989b).

3.3. A MODEL OF THE SUPEROUTBURST PHENOMENON.

Let us now come back to the problem of the superoutburst phenomenon of SU UMa stars. Although the tidal instability discovered by Whitehurst (1988) can successfully explain the superhump phenomenon, the basic mechanism of the superoutburst phenomenon remains to be clarified; that is, what causes the long duration of superoutbursts (enhanced mass transfer? or some other mechanisms) and what distinguishes the superoutburst from the normal outburst. The present author (Osaki, 1989) has recently proposed a working model of the superoutburst phenomenon of SU UMa stars based on a combined mechanism of the thermal instability (the ordinary disk instability) and Whitehurst's tidal instability in the accretion disk. The basic idea of this model is that a superoutburst cycle (a "supercycle") of SU UMa stars is the relaxation oscillation cycle for the angular momentum of the disk due to variable tidal torque in a cataclysmic variable star with a low mass secondary. In what follows we explain this model.

This model seeks the seat of both types of outbursts in instabilities of accretion disks and therefore the mass transfer rate from the secondary is assumed to be constant all the time. We suggest that both types of outbursts in SU UMa stars are basically caused by the thermal instability in the accretion disk (i.e., the ordinary disk instability mechanism) in that the alternation of quiescence and outburst is a thermal relaxation oscillation of an accretion disk between a low-viscosity cool state and a high-viscosity hot state. The superoutburst is understood in this model as an outburst accompanied with tidal instability while normal outbursts are those without it.

As for the Whitehurst tidal instability, we suppose that an accretion disk in a binary with a mass ratio $q = M_2/M_1$ less than about 0.25 becomes tidally unstable and it evolves into a precessing eccentric disk once the outer disk radius exceeds a critical radius (i.e., $R > R_{cr} \sim 0.46a$) where R denotes the outer disk radius and a is the binary separation. The elongated tail of the eccentric disk passes very near the secondary star once every synodic period between the precessing eccentric disk and the secondary (i.e., the superhump period), which gives rise to greatly enhanced tidal dissipation and thus greatly enhanced tidal torque on the accretion disk. We propose in our model that the tidal removal of angular momentum from the accretion disk occurs predominantly in such a phase (supermaximum) when a non-axisymmetric structure is formed in the accretion disk.

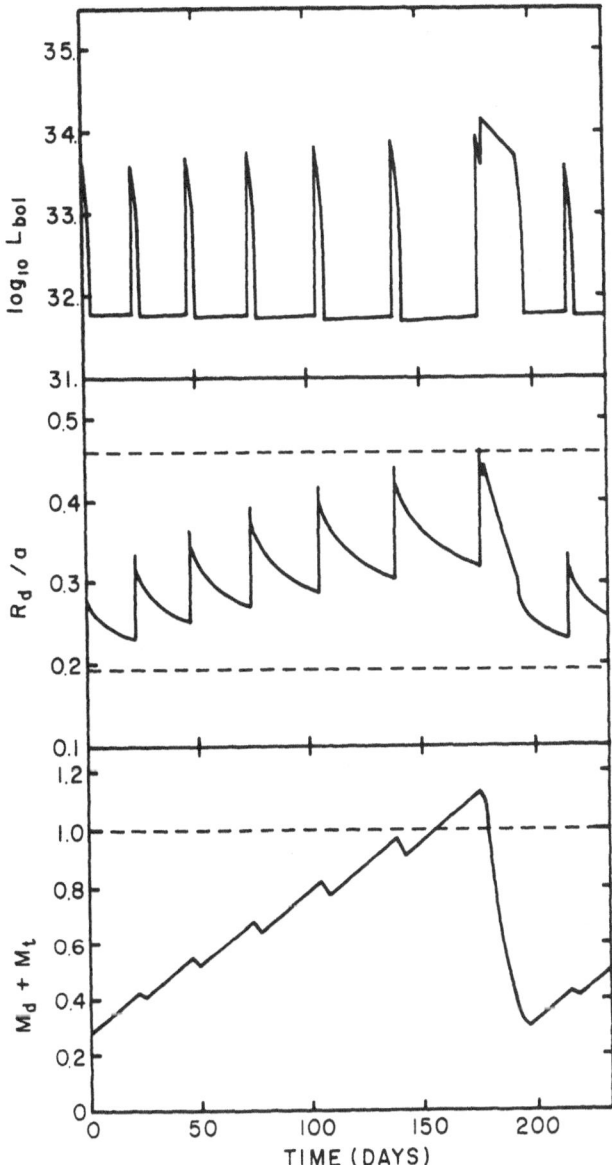

Figure 6: Time evolution of a thermally and tidally unstable accretion disk during a supercycle. The model parameters used are for VW Hyi, a proto-type SU UMa star. Three figures exhibit, respectively, (a) the bolometric light curve, log_{10} L_{bol}, (b) the disk's outer radius in units of the binary separation, and (c) the total mass of the disk (sum of the disk mass and the torus mass), which is normalized by the critical mass above which the disk is tidally unstable.

The essential difference between the superoutburst and the normal outburst in our model is whether the disk is tidally unstable or not, and this in turn is the difference in the outer disk radius.

If the disk is compact, then it is tidally stable and the angular momentum removal from the disk is small because of inefficient tidal torque. Rapid removal of mass with low angular momentum from the inner edge of the disk but without efficient tidal torque during an outburst leads to an expansion of the disk in such a case. It is shown that only a small fraction (about 10%) of the disk mass is accreted during such an outburst and therefore the outburst is relatively of small scale (i.e., a normal outburst).

On the other hand, if the disk radius exceeds the critical radius R_{cr}, the disk becomes tidally unstable and it develops into a non-axisymmetric (eccentric) disk (the superhump phenomenon). If the removal of angular momentum from the outerpart of eccentric disk due to greatly enhanced tidal torque is more efficient than the angular momentum transfer by shear viscosity from the inner parts to the outer parts in the disk, the disk will contract. The long duration of the superoutburst is explained in our model by the snow-plowing effect of material in the outer disk due to contraction of the disk edge.

The accretion disk is therefore rather compact at the end of the superoutburst (i.e., at the start of a new cycle). As a new superoutburst cycle begins, each normal outburst pushes the disk radius larger and larger and eventually triggers a tidal instability (i.e., a superoutburst). The "supercycle" of SU UMa stars is understood as a cyclic variation in the disk radius.

We can formulate this scenario by using a simple two-zone model consisting of a disk component and a torus component at its outer edge (see, Anderson 1988). The basic equations governing time evolution of an unstable accretion disk are equations of conservation of mass and angular momentum in the disk. If we write the total mass and total angular momentum of the disk component and of the torus by (M_d, J_d) and (M_t, J_t), these equations are written by

$$\frac{d(M_d + M_t)}{dt} = \dot{M}_0 - \dot{M}_{acc}, \tag{21}$$

and

$$\frac{d(J_d + J_t)}{dt} = j_0 \dot{M}_0 - j_{in}\dot{M}_{acc} - \dot{J}_{tidal}. \tag{22}$$

Equation (22) is the same equation as equation (11) except that we distinguish the torus component from the disk component.

To solve equations (21) and (22), we use a method first developed by Anderson (1988) in which method we assume power-law functions for the surface-density distribution within the disk. These equations are then reduced to ordinary differential equations and they can be easily integrated. In so doing, we have adopted following simplifying assumptions:

(1)During the quiescence, the accretion disk is completely inviscid and the matter transferred from the secondary is simply accumulated in the torus at the outer edge of the disk. The surface density distribution in the disk in quiescence is described by $\Sigma \propto r$.

(2) The thermal instability sets in when the density in the torus exceeds the critical value above which the cool state does not exist any more. The thermal instability, propagating through the disk as the heating front, transforms in a very short time the disk into the hot viscous state in which the surface density distribution is described by that of the steady state disk with $\Sigma \propto r^{-0.75}$. The decay of outburst is described by the propagation of the cooling front.

(3) During quiescence and normal outbursts in which the disk radius is well below the critical radius, the disk is essentially circular and the tidal removal of angular momentum is negligibly small.

(4) The tidal instability sets in when the outer disk edge exceeds the critical radius $R_{cr} \simeq 0.46a$. Then the disk is transformed into a non-axisymmetric eccentric disk and the tidal torque is greatly enhanced during this phase (supermaximum). The effects of angular momentum removal from the accretion disk due to the tidal torque during the superoutburst are taken into account as a contraction of the disk edge.

By using these assumptions, we have solved equations (21) and (22). It is first shown that approximately 3.5 power of the disk radius at the end of normal outbursts increases linearly with the elapsed time after the previous superoutburst, and the disk radius eventually exceeds the critical value above which the disk is tidally unstable, thus triggering a superoutburst. The results of calculations are reproduced in figure 6 for a particular set of model parameters, intended to reproduce VW Hyi, a proto-type SU UMa star.

Figure 6a exhibits the light curve, which reproduces very well the cyclic behavior of normal- and superoutbursts of VW Hyi. Figure 6b illustrates the time variation in the disk radius, showing how the disk radius varies during a supercycle and how a superoutburst is triggered by the last normal outburst. Our model thus predicts a definite pattern of variation in the disk's outer radius and this can be tested in future observations. Figure 6c shows the variation of the disk mass, and it shows that only a small fraction of the disk mass is accreted during normal outbursts while more than a half of the disk mass is accreted during the superoutburst.

Finally we note that our model does not require any enhanced mass transfer from the secondary in order to explain the supermaximum but the cyclic behavior of SU UMa stars follows as a natural consequence of the thermal instability and the tidal instability in accretion disks.

I conclude this talk by saying that the disk instability mechanism can explain the basic features of both normal- and super-outbursts of dwarf novae.

ACKNOWLEDGEMENTS. I would like to thank Drs. F. Meyer and E. Meyer-Hofmeister, and organizers of the workshop for kindly inviting me to participate in this workshop and

204

to give this talk. I owe to Mr. Hirose for producing Figure 5 of this article and for many helpful discussions.

REFERENCES.

Anderson, N. 1988, *Astrophys. J.*, **325**, 266
Bath,G.T. 1973, *Nature Phys. Sci.*, **246**, 84
Cannizzo, J.K., Ghosh, P., and Wheeler, J.C. 1982, *Astrophys. J. Letters*, **260**, L 83
Cannizzo, J.K., and Wheeler, J.C. 1984, *Astrophys. J. Suppl.*, **55**, 367
Cannizzo, J.K., Wheeler, J.C. and Polidan, R.S., 1986, *Astrophys. J.*, **301**, 634
Cannizzo, J.K., and Kenyon, S.J. 1987, *Astrophys. J.*, **320**, 319
Cannizzo, J.K., Shafter, A.W., and Wheeler, J.C. 1988, *Astrophys. J.*, **333**, 227
Faulkner, J., Lin, D.N.C., and Papaloizou, J., 1983, *Monthly Notices Roy. Astron. Soc.*,**205**, 359
Hassall, B.J.M., Pringle, J.E., Swarzenberg-Czerny, A., Wade, R.A., Whelan, J.A.J., and Hill, P.W., 1983 *Monthly Notices Roy. Astron. Soc.*, **203**, 865
Honey, W.B., Charles, P.A., Whitehurst, R., Barrett, D., and Smale, A. P., 1988, *Monthly Notices Roy. Astron. Soc.*, **231**, 1
Henon, M., 1965, *Ann., d'Ap.*, **28**, 992
Hirose, M. and Osaki, Y. 1989a, in *these proceedings*
Hirose, M. and Osaki, Y. 1989b, in preparation
Hoshi, R. 1979, *Prog. Theor. Phy.*,**61**, 1307
Lin, D.N.C. and Pringle, J.E., 1976, *IAU Symp. No. 73*, p237, eds Eggleton, P., Mitton, S., and Whelan, J., Reidel, Dordrecht, Holland
Lin, D.N.C. Papaloizou, J., and Faulkner, J., 1985, *Monthly Notices Roy. Astron. Soc.*,**212**, 105
Livio, M. and Verbunt, F. 1988, *Monthly Notices Roy. Astron. Soc.*,**232**, 1p
Lubow, S.H., and Shu, F.H. 1975, *Astrophys. J.*, **198**, 383
Marino, B.F., and Walker, W.S.G. 1979, *IAU Colloquium No.* 45, p29
Meyer, F. 1984, *Astron. Astrophys.*, **131**, 303
Meyer, F. 1986a, in *Radiation Hydrodynamics in Stars and Compact Objects*, eds. D. Mihalas and K.-H.A. Winkler , p249 (Springer-Verlag, Berlin)
Meyer, F. 1986b, *Monthly Notices Roy. Astron. Soc.*, **218**, 7p
Meyer-Hofmeister, E. 1987, *Astron. Astrophys.*, **175**, 113
Meyer, F., and Meyer-Hofmeister, E. 1981, *Astron. Astrophys.*, **104**, L10
Meyer, F., and Meyer-Hofmeister, E. 1983, *Astron. Astrophys.*, **128**, 420
Meyer, F., and Meyer-Hofmeister, E. 1984, *Astron. Astrophys.*, **132**, 143
Meyer, F., and Meyer-Hofmeister, E. 1988, submitted to *Astron. Astrophys.*
Mineshige, S. 1986, *Publ. Astron. Soc. Japan.*, **38**, 831
Mineshige, S. 1988, *Astron. Astrophys.*, **190**, 72
Mineshige, S., and Osaki, Y. 1983, *Publ. Astron. Soc. Japan.*, **35**, 377
Mineshige, S., and Osaki, Y. 1985, *Publ. Astron. Soc. Japan.*, **37**, 1
O'Donoghue, D. 1986. *Monthly Notices Roy. Astron. Soc.*, **220**, 23p
Osaki, Y. 1974, *Publ. Astron. Soc. Japan.*, **26**, 429
Osaki, Y. 1989, submitted to *Publ. Astron. Soc. Japan.*
Paczynski, B. 1977, *Astrophys. J.* , **216**, 822
Papaloizou, J. Faulkner, J., and Lin, D.N.C. 1983, *Monthly Notices Roy. Astron. Soc.*, **205**, 487
Papaloizou, J. and Pringle, J.E. 1977, *Monthly Notices Roy. Astron. Soc.*, **181**, 441
Papaloizou, J. and Pringle, J.E. 1985, *Monthly Notices Roy. Astron. Soc.*, **217**, 387

Pringle, J.E., Verbunt, F., and Wade, R.A. 1986, *Monthly Notices Roy. Astron. Soc.*, **221**, 169

Shakura, N.I., and Sunyaev, R.A. 1973, *Astron. Astrophys.*, **24**, 337

Smak, J. 1982, *Acta Astronomica*, **32**, 199

Smak, J. 1984a, *Publ. Astron. Soc. Pacific*, **96**, 5

Smak, J. 1984b, *Acta Astronomica*, **34**, 93

Smak, J. 1984c, *Acta Astronomica*, **34**, 161

Smak, J. 1984d, *Acta Astronomica*, **34**, 317

Verbunt, F. 1986, in *The Physics of Accretion onto Compact Objects* p.59, eds Mason, K.O., Watson, M.G., and White, N.E., Springer-Verlag, Berlin

Verbunt, F. 1987, *Astron. Astrophys. Suppl.*, **71**, 339

Vogt, N. 1982, *Astrophys. J.*, **252**, 653

Warner, B. 1985, In *Interacting Binaries,* p367, eds. Eggleton, P.P., and Pringle, J.E., Reidel, Dordrecht, Holland

Whitehurst, R. 1988a, *Monthly Notices Roy. Astron. Soc.*, **232**, 35

Whitehurst, R. 1988b, *Monthly Notices Roy. Astron. Soc.*, **233**, 529

HYDRODYNAMIC SIMULATION OF ACCRETION DISKS IN CATACLYSMIC VARIABLES.

M. HIROSE and Yoji OSAKI

Department of Astronomy, Faculty of Science, University of Tokyo,
Bunkyo-ku, Tokyo 113, Japan

ABSTRACT. The non-axisymmetric configurations of accretion disks in cataclysmic variables are studied by two-dimensional hydrodynamical calculations. The particle-hydrodynamic method first developed by Lin and Pringle (1976) is used, and time evolution of accretion disks under a constant mass supply rate from the secondary is followed until the system settles to a quasi-steady state. It is found that a binary system with comparable masses of component stars settles to a steady state of an elongated disk fixed in the rotating frame of the binary while in the case of low mass secondary it settles to a periodic oscillating state in which a non-axisymmetric (eccentric) structure is formed and it rotates slowly in the inertial frame of reference. The period of oscillation is a few percent longer than the orbital period of binary, and it offers a natural explanation to the "superhump" periodicity of SU UMa stars. Our results thus confirm basically those of Whitehurst (1988) who has discovered the tidal instability of an accretion disk in a low mass secondary.

1. Introduction.

One of the most important but unsolved problems in accretion disks is the problem of angular momentum removal from accretion disks by tidal torque. Lin and Pringle (1976) have shown by numerical simulations that the accretion disks have clear-cut outer boundaries and the tidal torque can remove angular momentum from the disk to ensure steady accretion of matter onto the primary star. Whitehurst (1988) has recently discovered that an accretion disk in a cataclysmic binary system with a small mass ratio is tidal unstable and it develops into a non-axisymmetric (eccentric) disk. Since we have also examined the non-axisymmetric configurations of accretion disks by hydrodynamic simulations, we report here our results.

2. Method of Calculations.

Our method of numerical calculations is that of the two-dimensional particle-

207

F. Meyer et al. (eds.), Theory of Accretion Disks, 207–212.
© 1989 by Kluwer Academic Publishers.

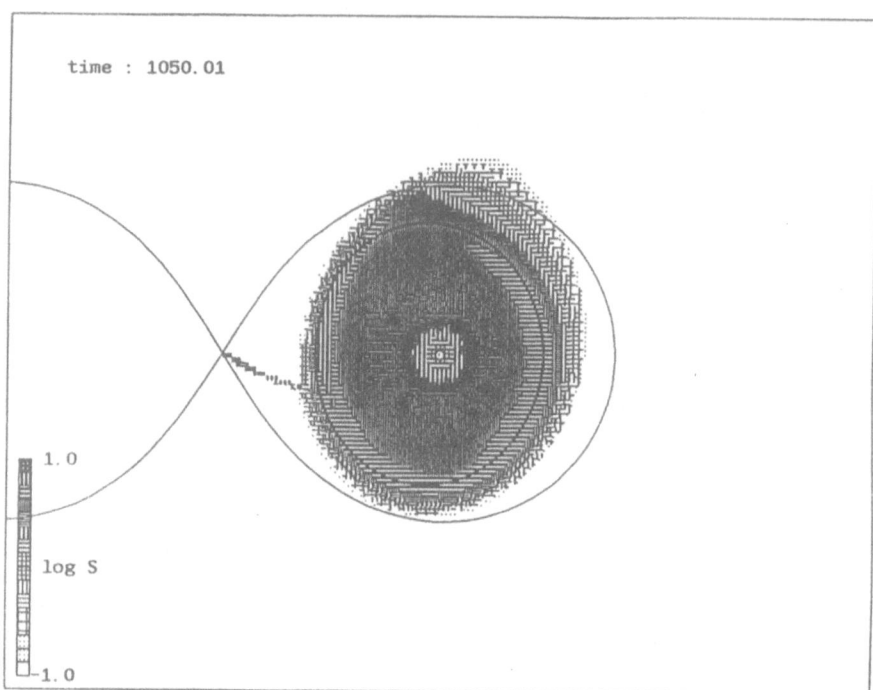

time : 1050.01

1.0

log S

-1.0

Figure 1: The density distribution of matter in steady state disk for a binary with mass ratio $q = 1.0$ (at $t = 1050$), which is shown by the number of particles in each cell (the scale is given at the left). The thin line shows the Roche lobes of the binary. The thick line in the disk indicates the last non-intersecting periodic orbit for the restricted problem of three body.

hydrodynamics first developed by Lin and Pringle (1976). In this formulation, it is assumed that the disk is geometrically thin so that only the motion in the disk plane is considered and that the pressure-gradient term in the equation of motion is neglected because the flow in accretion disks are highly supersonic. The motion of fluid is then followed by solving the ordinary differential equation for many particles representing fluid elements under the gravitational fields of the binary and the effects of viscosity are taken into account through the averaging of velocities of particles within a pre-specified cell mimicking the viscous interaction.

All calculations are performed in the corotating frame of the binary. The units of length, mass and time used in the calculation are the binary separation a, the total mass $M = M_1 + M_2$ and the reciprocal of the angular velocity of revolution of the binary $\Omega_b{}^{-1}$. The standard size of the spatial grid and the time step for the viscous interaction used are $l = 0.01$ and $\Delta t = 0.01$ in these units. The effective viscosity is given in this formulation by $\nu = C\,l^2/\Delta t$, where the numerical constant C is estimated later from the result of calculations. The mass is supplied continuously into the primary's Roche lobe from the inner Lagrangian point with one particle per unit time step Δt. The radius of the white dwarf is taken to be $R_{WD} = 0.01$.

3. Results.

3.1. THE CASE OF $Q = M_2/M_1 = 1$.

The system reaches the steady state at time around 500 in these units, that is about 80 orbital periods. Figure 1 illustrates the configuration of the disk in steady state. As seen in figure 1, it takes the form of a symmetric oval but its axis is slightly tilted because of the tidal dissipation which produces the tidal torque necessary for the disk to reach the steady state. Our result in this case is essentially the same as that of Lin and Pringle (1976). The only difference between the two calculations is the size of the spatial grid; our spatial grid is twice as fine as theirs. Consequently, the viscosity in our calculation is one-fourth of theirs. From comparison with the theory of standard accretion disk, we find $\nu = 2.02 \times 10^{-4}$ and we then estimate the constant $C \simeq 1/50$ in this case.

The stream lines of flow in accretion disks are thought to be approximated by periodic orbits in restricted problem of three body (Paczynski 1977). There exists an upper bound for such a periodic orbit above which periodic orbits intersect with other orbits. It is considered that the last non-intersecting orbit corresponds to the tidal radius above which the accretion disk is truncated due to greatly increased tidal dissipation. The last non-intersecting orbit is shown in figure 1 by the solid line. We see that the most dense parts of the disk correspond to it.

3.2. THE CASE OF $Q = 0.15$

We have made the same calculation as above for the case of $q = 0.15$. Figure 2a exhibits time evolution for the number of particles in the accretion disk while figure 2b shows six snap-shots of the disk during this evolution. As seen from the figures, the disk grows in its size as the number of particles in the disk increases steadily with time until $t \sim 400$ ($N \sim 23000$). However, the number of particles begins to decrease at this point and continues to do so until $t \sim 500$ and it reaches a steady state with $N \sim 17000$. To see what happens here, we show in the lower part of figure 2b the number of particles that hit the surface of the secondary star, thus being removed from the calculation.

It shows that the beginning of the decrease in total number of particles in the disk corresponds to the beginning of collision of disk matter on the surface of the secondary and that the system settles to a periodically oscillating state with a period nearly equal to, but slightly longer than the orbital period. Figure 2c shows the energy liberated by the viscous dissipation in the outer parts of the disk. It also oscillates with the same period and this corresponds to the superhump light curve of SU UMa stars. In producing figure 2c, we have artificially stopped mass transfer from the secondary in order to see the tidal dissipation more clearly.

We illustrate the accretion disk at ten phases of this period (the superhump

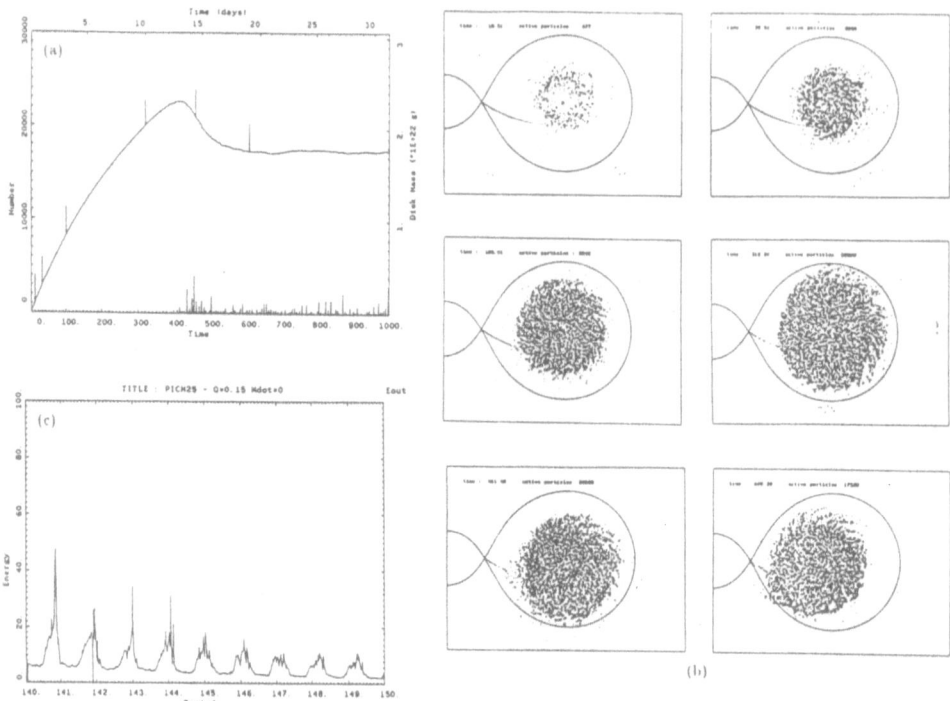

Figure 2: (a): Time evolution of total number of particles in the disk ($Q = 0.15$). Scales at the upper and right axes indicate time in days and mass of the disk in grams for a binary corresponding to Z Cha ($P_{orb} = 0.074499$ day, and $\dot{M} = 1.0 \times 10^{17}$ gs^{-1}). The arrows indicate times when the snap shots in fig. 2(b) are taken. The number of particles absorbed by the secondary in 10 time steps is also plotted with ten times magnified scale. It shows a spike-like periodic variation. The period corresponds to the superhump period. (b): The snap shots of density distribution in the disk ($Q = 0.15$). It is seen that a ring is first formed at the radius corresponding to the Keplerian circular orbit with the specific angular momentum of the incoming stream ($t = 10.01$). The number of particles in the disk increases with time and the disk expands accordingly. The tidal instability sets in around $t = 400$, and the disk then becomes non-axisymmetric and begins to rotate. (c): Time variation of the tidal dissipation in the outer parts ($r > 0.2a$) of the disk. It is seen that the light maxima ("superhumps") march against the orbital phases. This light curve is calculated in a model in which the mass transfer from the secondary is stopped. In the case of constant mass transfer rate, the light curve is slightly complicated because there exists another light maximum corresponding to Vogt's(1982) super-spot.

period) in figure 3a. In this quasi-steady state, the disk is transformed to a non-axisymmetric eccentric disk and it rotates in the opposite direction to the orbital motion of the binary in the corotating frame and that the elongated tail of the disk hits the surface of the secondary with this period. If we see this phenomenon from the observer's frame, then the eccentric disk slowly rotates and the synodic

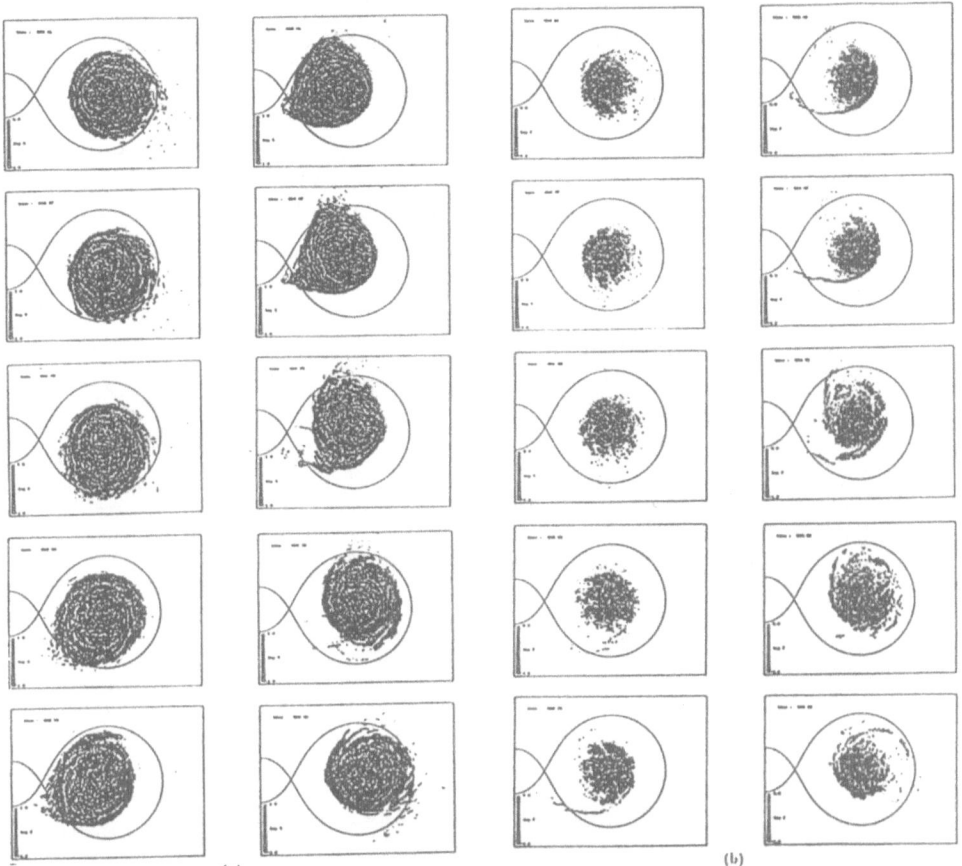

Figure 3: (a): The density contours in the disk at ten phases in the superhump period are shown by the number of the particles in each cell. The eccentric accretion disk rotates in the opposite direction to the orbital motion of the binary in the co-rotating frame of the binary. 3(b): The spatial distribution of radiative flux from the disk. The light maximum occurs at the ninth frame. The shock-like features are seen from fifth to eighth frames.

period between the orbital motion of the secondary and the elongated disk gives the oscillation period.

Figure 3b illustrates the spatial distributions of radiative flux from the disk. The maximum light (that is, the superhump maximum) occurs on the ninth frame which corresponds to the phase when the eccentric disk lies on the direction nearly perpendicular to the line joining two stars after the passage of the elongated tail of the eccentric disk near the secondary (the phase of the apse angle of the eccentric disk to the secondary = 0.81). This frame and a few frames ahead of it show the shock-like structure. It is formed because the elongated tail of the eccentric accretion disk is greatly distorted by the gravitational field of the secondary when the former passes near the latter, and the greatly enhanced dissipation occurs when perturbed matters fall near the periastronon of their orbits.

Our result thus confirms that of Whitehurst (1988) in that the accretion disk in cataclysmic binary with a low mass secondary is tidally unstable and it evolves into a non-axisymmetric eccentric disk and that the periodically varying tidal dissipation in the outer parts of non-axisymmetric disk offers a natural explanation to the superhump phenomenon. The superhump period in our simulation is longer by 5.8% than that of the orbital period while Whitehurst has got about 3.8% for it. The difference may probably originates from the difference in the distribution of matter within the disk as our disk has more mass in outer parts than his because of the form of the viscosity.

More details will be given elsewhere.

REFERENCES.

Lin, D.N.C. and Pringle, J.E., 1976, *IAU Symp. No. 73*, p237, eds Eggleton, P., Mitton, S., and Whelan, J., Reidel, Dordrecht, Holland
Paczynski, B. 1977, *Astrophys. J.*, **216**, 822
Whitehurst, R. 1988, *Monthly Notices Roy. Astron. Soc.*, **232**, 35

SIMULATIONS OF ACCRETION FLOW IN CLOSE BINARY STARS.

Robert WHITEHURST
Astronomy Department, The University,
Leicester LE1 7RH, United Kingdom.

ABSTRACT. The dynamics of accretion flow in close binary stars is investigated via a numerical simulation of the accretion flow. The importance of the secondary's tidal torque is discussed with respect to the radius of the accretion disc. It is found that the steady-state disc has a fixed radius, which is within the primary's Roche lobe, a result which differs from that of Livio & Verbunt (1988).

1. Introduction.

The role of the tides raised by the secondary in the accretion discs of close binary stars is important in determining several properties of the disc. Notably the 'superhumps' of SU UMa stars, (Whitehurst 1988a) and the disc's equilibrium radius (e.g. Papaloizou & Pringle 1977, Livio & Verbunt 1988). To understand this role fully the gas dynamics of the accretion flow must be considered. This is because for a free streamline in an symmetric orbit around the star's line-of-centres *no* net transfer of angular momentum occurs. In a disc the tidal dissipation arises from the interaction between neighbouring streamlines in the disc which are pinched together by the secondary's gravitational field, giving rise to an asymmetric mass distribution around the primary. This in turn allows transfer of angular momentum and a net torque.

2. Angular Momentum.

Two types of angular momentum need to be considered within the star-disc-stream system. One is the orbital angular momentum of the whole system around its centre-of-mass in the non-rotating frame of reference; (an inertial frame). This (ignoring external agencies) is always conserved and in steady-state indeed has a constant distribution between the two stars, the disc and the stream. The orbital angular momentum's distribution can only vary on time-scales comparable to the orbital period during periods of non-steady accretion.

F. Meyer et al. (eds.), Theory of Accretion Disks, 213–220.

A more convenient quantity, (used henceforth), is the spin angular momentum of the disc-primary system. This is measured in a non-inertial frame of reference centred on the primary, typically the co-rotating frame. This spin angular momentum will be fixed in steady-state, balance being achieved between that carried in to this system by the stream and that carried away via tides to the secondary.

The redistribution of angular momentum within the disc is the mechanism by which material accretes on to the primary, and so clearly the tides can have a role in driving the accretion rate. Osaki (1989) has postulated that high tidal torques may be responsible for the duration of superoutbursts in SU UMa stars. What is certain is that the size of the accretion disc is determined by the balance between the tidal torque and the outflow of angular momentum. Livio & Verbunt (1988) have argued that this size is determined by the global integral of angular momentum loss over the disc; which they balance by the angular momentum input by the stream. Unfortunately this approach is invalid because it neglects the disc's 'viscosity', (i.e. angular momentum transport), which is in fact *fundamental* to the correct treatment of this problem. This is because the net tidal torque is due to the asymmetries introduced in to the disc by viscosity. A better approach is to use a two-dimensional numerical simulation of the accretion disc, which allows the viscosity's influence to be calculated.

In the following section the results of such a simulation of an accretion disc is described. This is done using the numerical simulation due to Whitehurst (1988b) which describes the flow as a set of interacting test particles.

3. The Simulation.

Initial parameters for the simulation were chosen to be equivalent to those selected by Livio & Verbunt (1988) as appropriate for the U Gem system, i.e. mass ratio $q = M_2/M_1 = 0.5$, with primary mass $1M_\odot$. The model differs in that a Gaussian profile is used for the initial mass burst rather than a step function.

In Figure 1 the equilibrium disc is shown for the simulation, in the form of a velocity plot. Each line represents both the position of a particle at that instant and its velocity. (Particles moving anti-clockwise.) Here some 4,200 particles remain from a peak population of 20,000. The disc is fixed in the co-rotating frame and is tidally extended perpendicularly to the line-of-centres. Two weak density-waves are visible at the disc's edge. Note furthermore that the disc is also twisted slightly askew to the true perpendicular and it is this that allows a net transfer of angular momentum.

Figure 2 shows the corresponding plot of tidal torques, (represented as the angular momentum change), Figure 2a is a contour plot, Figure 2b a surface plot. These show a torque distribution very strongly concentrated at the edge of the disc, (the tidal radius). The pattern of the torques is interesting, revealing a four-fold symmetry with regions where angular momentum is alternately given and removed

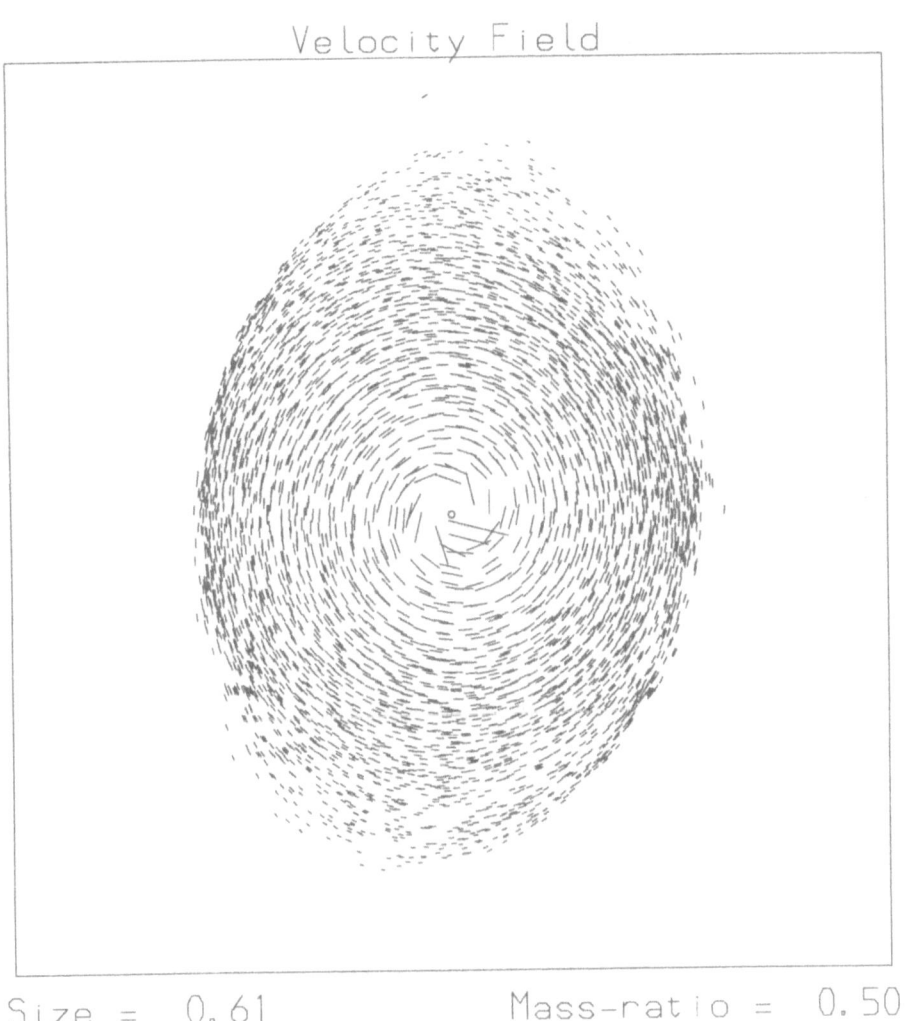

Velocity Field

Size = 0.61   Mass-ratio = 0.50

Figure 1: Velocity field snapshot of disc.

from the disc.

Figure 3 presents the azimuthal average of the loss in angular momentum around the disc versus radius. Note that at small radii particles in nearly circular orbits lose angular momentum, (albeit in small amounts), to the secondary. Furthermore at the disc's outer edges material will lose angular momentum. However in between these regions is a strong resonance where the disc is *gaining* angular momentum from the secondary. The radius where the change in angular momentum crosses from a gain to a loss marks the (radially averaged) edge of the disc, (the tidal

216

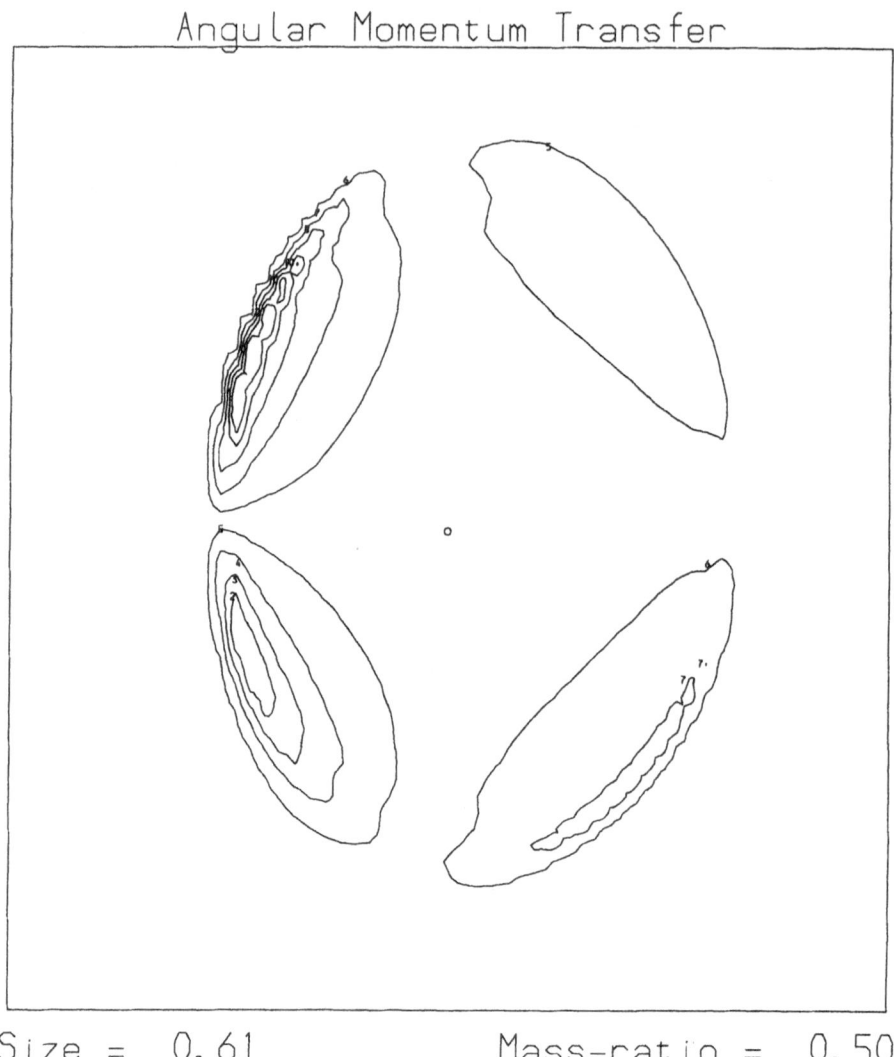

Angular Momentum Transfer

Size = 0.61   Mass-ratio = 0.50

Figure 2a: Tidal torque strength, contour plot.

radius). Within this radius the other cross-over point marks an area which the torque is attempting to void.

Figure 4 gives the variation of the disc's radius through the course of the simulation. The quantity plotted is the radius within which 99 % of the disc's mass lies, and so initially it is just a measure of the stream's domination. The first few values, (at intervals of one period), are therefore not plotted, but the decrease in the early part is due to the relative decline of the stream's mass. When the stream ceases to be significant the disc has a radius 0.3a, (a is the binary separation). It

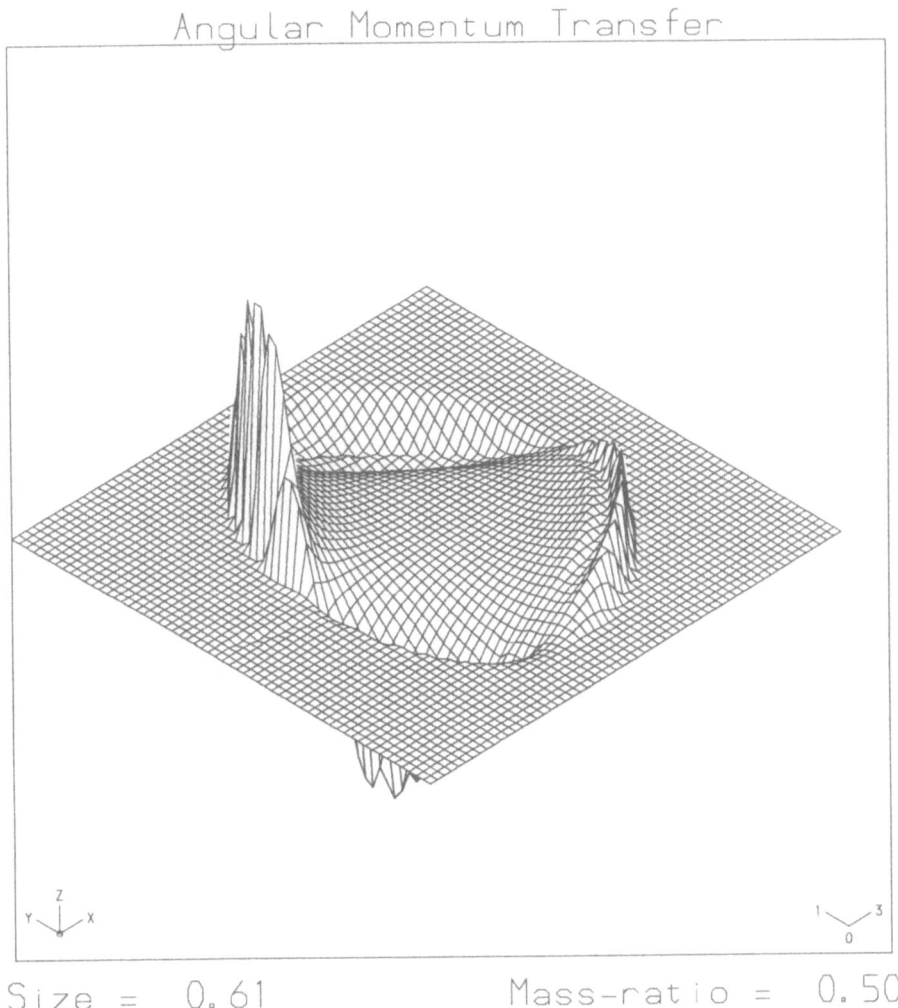

Figure 2b: Tidal torque strength, surface plot.

then grows under the influence of the viscosity until steady-state is achieved at a radius of about 0.46a. This is not the tidal radius but rather the disc's maximum radius. Note that this is the final value of the radius.

4. Conclusion.

Several authors, (e.g. O'Donoghue 1987, Whitehurst 1988a, Livio & Verbunt 1988), have argued that the change of the disc's size as measured by bright-spot eclipses

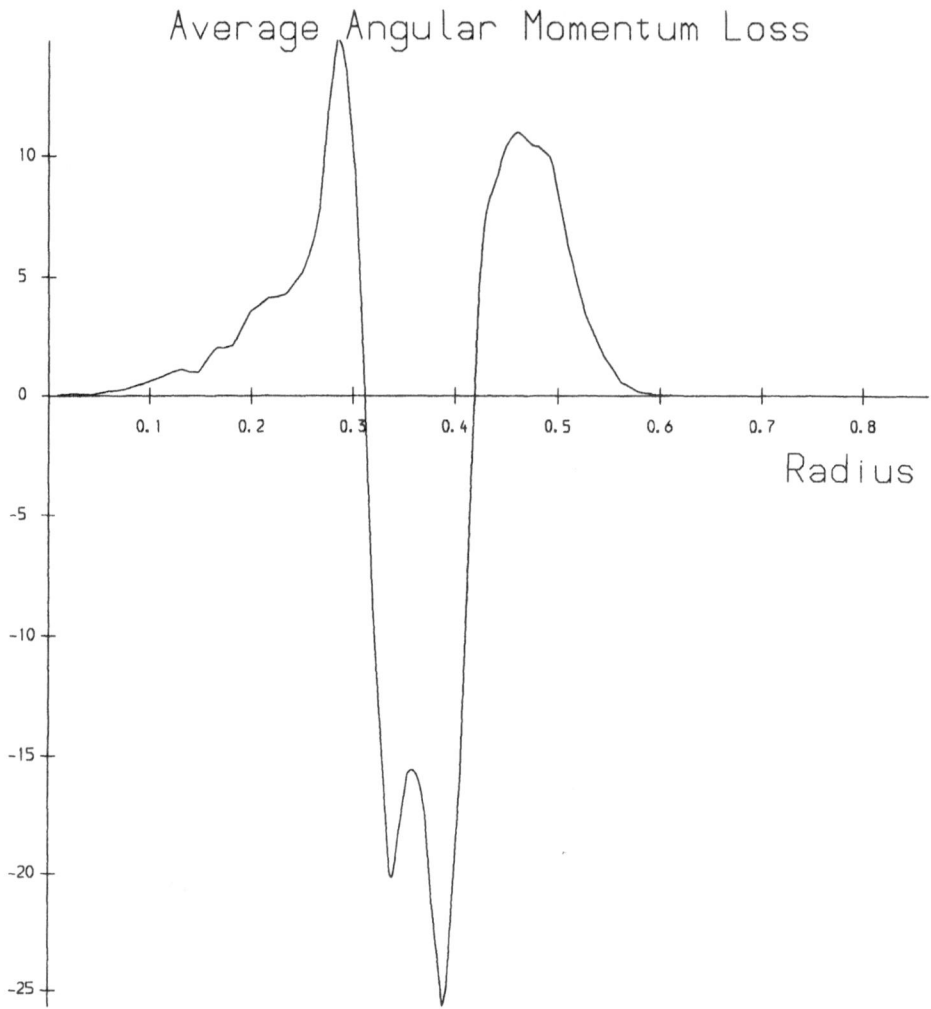

Figure 3: Average azimuthal torque.

can be used as a test of models of dwarf nova outbursts. Using the results above the behaviour of the disc's radius in a mass-transfer driven outburst-quiescence cycle can be predicted. Quite simply put the disc will stay at a relatively large, fixed radius during quiescence. Its only variation will be a rapid decrease in size during the burst phase of the outburst, with a slower recovery after the end of the mass-burst to that value determined by the tidal 'pinch' of the secondary.

Note that this differs from the description of Livio & Verbunt (1988). There are two reasons for this; firstly their numerical model does not model a steady-state

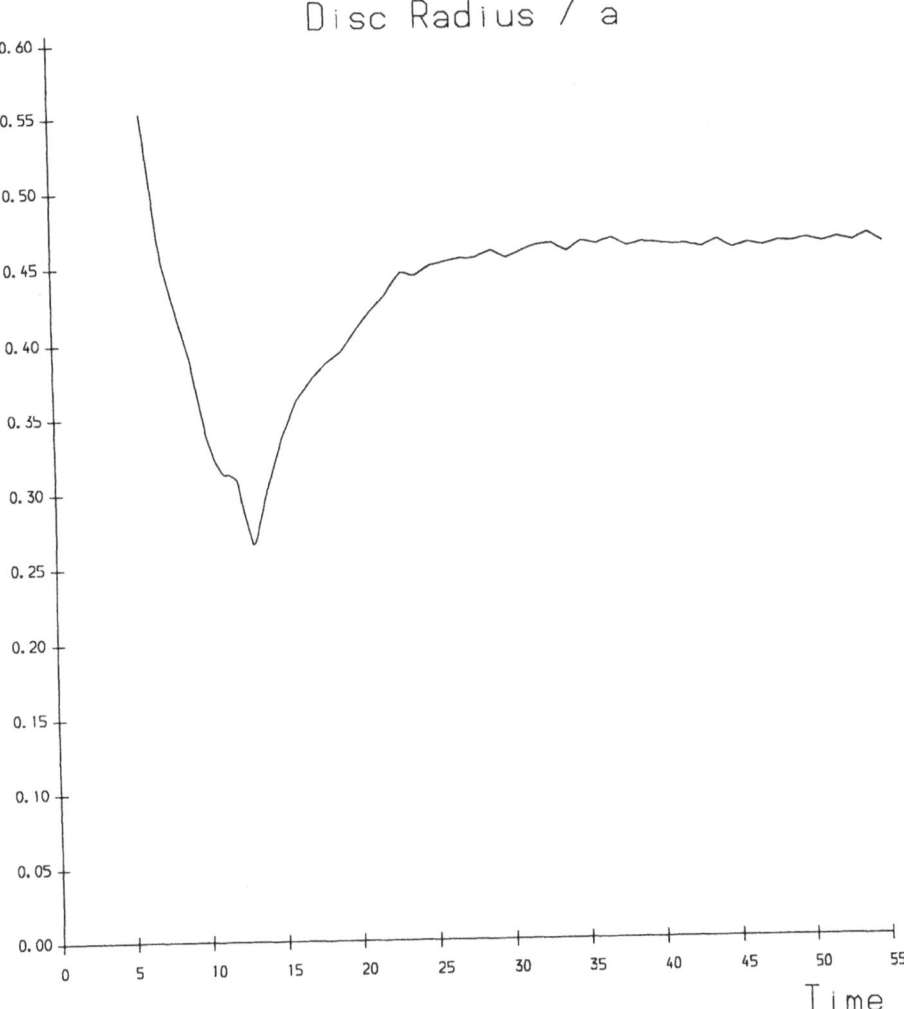

Figure 4: Maximum disc radius against time.

system but rather one in decline from outburst. (This is a well known problem of numerical models of mass-transfer bursts which always produce long declines.) In addition their assumption that the tidal torque can be calculated without knowledge of the effects of the disc's own internal angular momentum transport is an over-simplification as it ignores the vital properties which balance the disc against the secondary's torque.

REFERENCES.

Livio, Mario & Verbunt, Frank, 1988. Mon. Not. R. astr. Soc., **232**, 1p.

O'Donoghue, D., 1986. Mon. Not. R. astr. Soc., **220**, 23p.
Osaki, Y. 1989. Preprint.
Papaloizou, J. & Pringle, J.E., 1977. Mon. Not. R. astr. Soc., **181**, 441.
Whitehurst, Robert, 1988a. Mon. Not. R. astr. Soc., **232**, 55.
Whitehurst, Robert, 1988b. Mon. Not. R. astr. Soc., **233**, 539.

VISCOUS EVOLUTION OF ACCRETION DISCS IN THE QUIESCENCE OF DWARF NOVAE.

Shin MINESHIGE[1] and Janet H. WOOD[1,2]
[1] Astronomy Department, University of Texas at Austin, USA
[2] Institute of Astronomy, Cambridge, England

ABSTRACT. The viscous evolution of accretion discs in the quiescence of dwarf novae is investigated semi-analytically. There are two key factors: the inward flow of the initial mass in the disc and the diffusion of the material added into the disc later. The results are compared with the brightness temperature distributions obtained by the eclipse mapping of OY Car and Z Cha.

1. Introduction.

Since the standard accretion disc model (so-called α disc model) was proposed by Shakura and Sunyaev (1973), the theory of accretion discs has been making great progress (see the review by Pringle 1981). The unknown physics of the viscosity is placed into one parameter α, and the $r\varphi$ component of the shear stress tensor is defined to be $w_{r\varphi} = -\alpha p$, where p denotes the pressure. Many groups have calculated the time evolution of the accretion discs based on the α model and compared with observations of dwarf novae though not with eclipse (see the review by Verbunt 1987).

Horne (1985) has developed the technique of eclipse mapping. Now the direct comparison of the theory to the observations, in terms of the brightness temperature profiles of the disc, is possible. In this paper we show the results of modelling dwarf novae in quiescence with a time dependent α-disc model and discuss its implications for the viscosity in the disc.

2. Basic Behaviour of the Disc.

There are two key factors for the viscous evolution of the disc in the quiescence of dwarf novae. One is the inflow of the initial mass. We assume that the integrated viscous stress changes as $W \propto \Sigma(r/r_0)^\gamma$, where r_0 is the radius where the specific angular momentum is equal to that of the stream from the secondary star, and γ

F. Meyer et al. (eds.), Theory of Accretion Disks, 221–225.

is a numerical constant. If the initial temperature is flat (this is justified in the disc-instability model; Osaki 1974), then the surface density distribution and the temperature profile at later times are

$$\Sigma/\Sigma_0 \sim \xi^{3-2\gamma}\left[1 + 42\ \tau\ \xi^{2\gamma-1}\right], \qquad (1)$$

$$log(T_e/T_0) \sim 4.6\ \tau\ \xi^{2\gamma-1}, \qquad (2)$$

where $\tau = t/t_0$ and $\xi = \sqrt{r/r_0}$ are dimensionless time and radius and T_0 and t_0 are constants specifying the initial effective temperature and the viscous time scale. Thus the temperature increases faster at the outer radius for $\gamma > 0.5$, and faster at the inner radius for $\gamma < 0.5$.

The other factor which affects the viscous evolution is the mass added in the quiescent stage. If we input mass with a δ-function type density distribution at $t = 0$, then the evolution at later times is

$$\Sigma/\Sigma_0 \sim \beta\frac{\xi^{-5}}{\sqrt{4\pi\tau}}\ exp\left[-\frac{(ln\ \xi - \tau)^2}{4\tau}\right]. \qquad (3)$$

for $\gamma = 0.5$, where β is a parameter representing the mass-transfer rate from the secondary star. The peak in the surface density moves inward on a time scale of $t_0/9$ and also diffuses on a time scale of t_0. The basic behaviour of the solutions of $\gamma \neq 0.5$ are similar to that of equation (3).

Since the basic equation is linear in terms of the surface density, the general solutions are superpositions of these solutions.

3. Modelling of Z Cha and OY Car.

We show the results of modelling Z Cha in Fig. 1 and OY Car in Fig. 2. The observed brightness temperature profiles obtained by eclipse mapping are also depicted (from Wood et al. 1986; 1989). To reproduce the flat temperature profile, γ should be close to 0.5, otherwise the temperature will rapidly go up in the inner regions ($\gamma \ll 0.5$) or in the outer regions ($\gamma \gg 0.5$). The obtained mass-transfer rate is $\dot{M}_0 \sim 10^{15.5} g\ s^{-1} (= 5\times10^{-11} M_\odot yr^{-1})$, and the viscosity parameter is $\alpha \sim 0.01-0.03 \times (r/r_0)^{0.3}$ for both stars. In both figures the fitting inside the radius r_0 are excellent, while outside this radius appreciable discrepancies are evident probably due to the additional heating of the disc by the inflow material (we omit this effect) and due to the effect of the outer boundary condition (analytical solutions assume that the disc extends to infinity). For details of fitting, see Mineshige and Wood (1989).

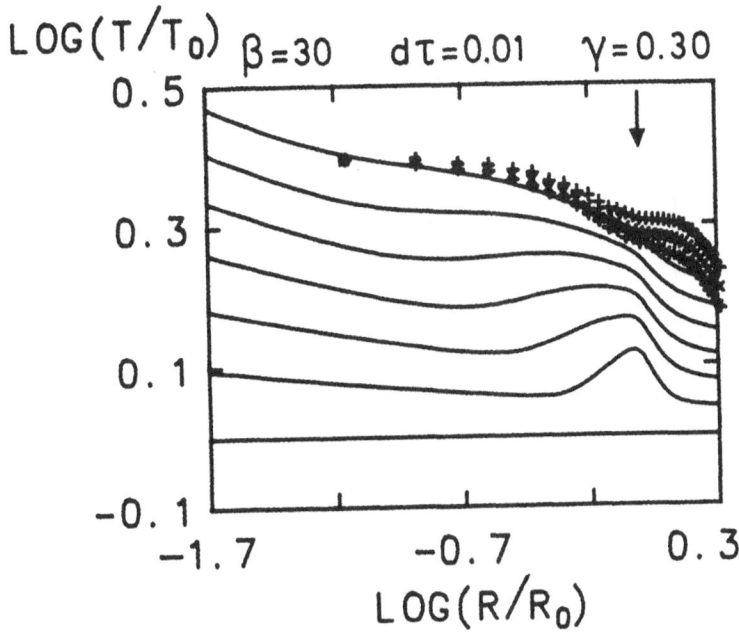

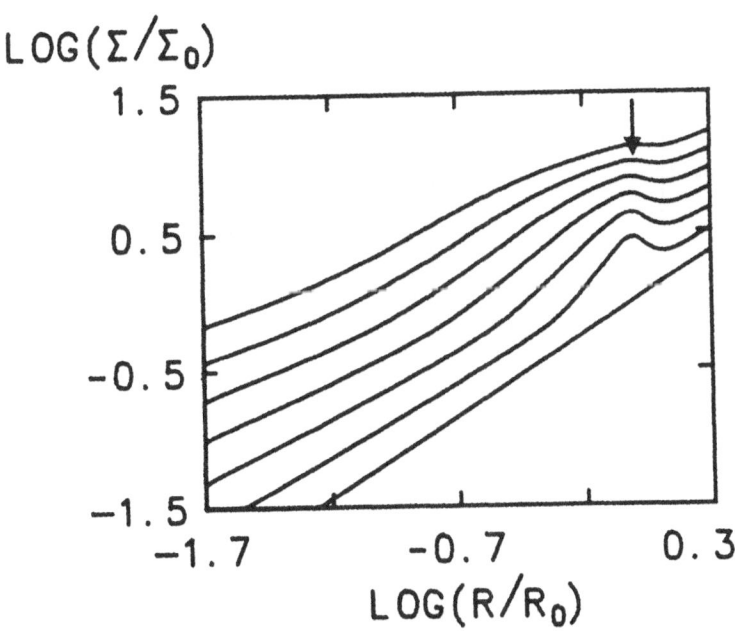

Figure 1: Model for Z Cha. The time interval between each line is $\sim 30(\alpha_0/0.01)^{-1}d$. The observed brightness temperatures of Z Cha (Wood *et al.* 1986) are also depicted by symbols + (maximum temperatures and minimum temperatures) and × (azimuthally averaged values).

224

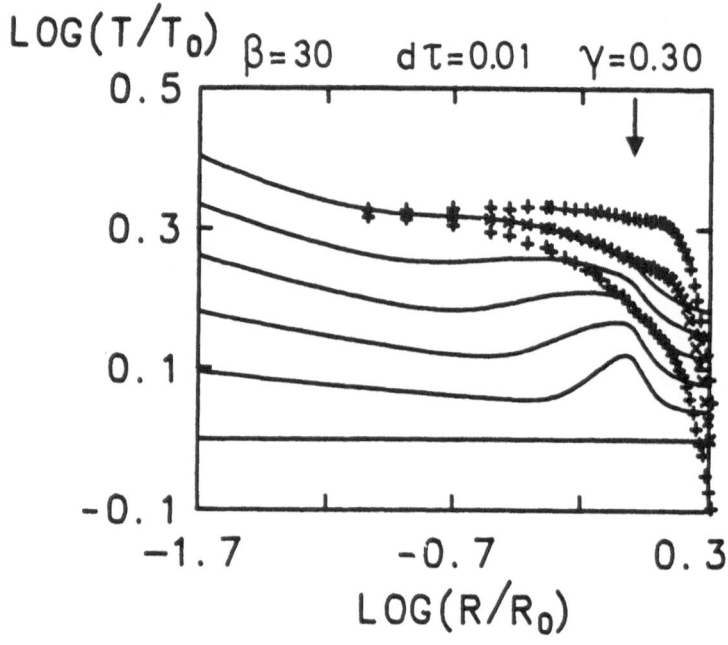

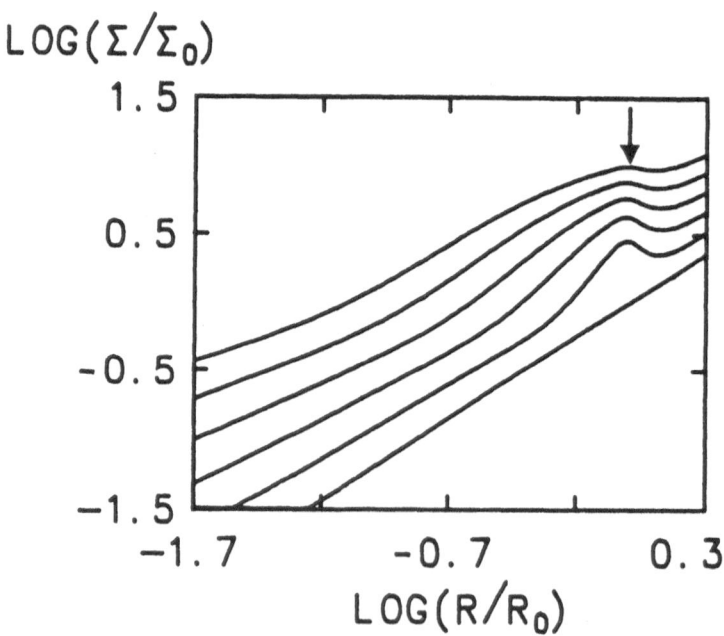

Figure 2: Model for OY Car. The time interval between each line is $\sim 35(\alpha_0/0.01)^{-1}d$. The observed brightness temperatures of OY Car (Wood *et al.* 1989) are also displayed.

4. Conclusions.

1. We can fit the observed brightness temperature gradients in Z Cha and OY Car taking into account two components: the radial inflow of the initial mass and the diffusion of the added mass from the secondary star.

2. If we scale the viscosity parameter as $\alpha = \alpha_0 (r/r_0)^\gamma$, $\alpha_0 \sim 0.01 - 0.03$ and $\gamma \sim 0.3 - 0.4$.

REFERENCES.

Horne, K. 1985, *Mon. Not. R. astr. Soc.*, **213**, 129

Mineshige, S. and Wood, J.H. 1989, submitted to *Mon. Not. R. astr. Soc.*

Osaki, Y. 1974, *Publ. Astron. Soc. Japan*, **26**, 429

Pringle, J.E. 1981, *Ann. Rev. Astr. Astrophys.*, **19**, 137

Shakura, N.I. and Sunyaev, R.A. 1973, *Astron. Astrophys.*, **24**, 337

Verbunt, F. 1987, in *The Physics of Accretion onto Compact objects*, eds. K.O. Mason, M.G. Watson, N.E. White, Springer, Berlin, p.59

Wood, J., Horne, K., Berriman, G., Wade, R., O'Donoghue, D., and Warner, B. 1986, *Mon. Not. R. astr. Soc.*, **219**, 629

Wood, J.H., Horne, K., Berriman, G., and Wade, R.A. 1989, *Astrophys. J.*, **341**, in press

BLACK HOLE ACCRETION DISC INSTABILITY AND SOFT X-RAY TRANSIENTS.

J. Craig WHEELER and Shin MINESHIGE
Astronomy Department, University of Texas at Austin, USA

ABSTRACT. Time dependent behaviour of the accretion discs in low-mass X-ray binaries are studied as a model for soft X-ray transients. For relevant accretion rates, the disc suffers a thermal instability leading to intermittent accretion onto the central compact object. It is shown that the effect of X-ray irradiation of the outer disc by the central disc is small in the case of black hole binaries, and so the propagation of the cooling waves is not prevented.

1. Introduction.

Low-mass X-ray binaries (LMXB) are semi-detached binary systems consisting of mass-losing late-type stars and compact objects (neutron star or black hole), which are surrounded by accretion discs fed by mass loss from the late-type companions. Soft X-ray transients are unique in this group by showing outbursts with recurrence time of 0.5 – 50 years, rise time scale 2 – 10 days, and decline time scale of order of a month (see the recent review by Priedhorsky and Holt 1987). The connection between soft X-ray transients and dwarf novae, eruptive semi-detached binaries containing white dwarfs, was discussed by van Paradijs and Verbunt (1984). Two models are proposed for outbursts of soft X-ray transients in analogy with dwarf novae: the disc-instability model (Cannizzo et al. 1985), and the mass-transfer burst model (Hameury et al. 1986).

2. Disc-Instability Model.

The disc-instability model was originally proposed for outbursts of dwarf novae (Osaki 1974; Hōshi 1979; Meyer and Meyer-Hofmeister 1981), and is now the most widely accepted model. It is known that the discs suffer the thermal instability due to the ionization of the hydrogen and the helium. The application of this model to soft X-ray transients was put forward by Huang and Wheeler (1989) and Mineshige and Wheeler (1989). Assuming that the effect of X-ray illumination of the disc is

F. Meyer et al. (eds.), Theory of Accretion Disks, 227–230.

negligible, they calculated the spatial propagation of the thermal instability over the disc, which leads to intermittent accretion of the gas in the disc onto the central compact object.

The results are summarized as follows:

(1) The calculated light curve based on the disc-instability model gives a good fit to the observations of the black hole candidate A0620-00 (Whelan *et al.* 1977).

(2) The central mass of $\sim 10\ M_\odot$ is favoured over $\sim 1\ M_\odot$ to reproduce the observed recurrence time of A0620-00 ($\sim 60\ yr$). The compact object in A0620-00 is more likely to be a black hole.

(3) If we scale the viscosity parameter as $\alpha \propto (h/r)^n$, where n is a numerical constant and h represents the semi-thickness of the disc, larger values of n (> 1.0) are preferable to fit the observations of A0620-00.

3. Effect of X-ray Irradiation.

The heating rate of the disc due to the X-ray irradiation originating from the central part of the disc in black hole binaries is given by

$$Q_{irr} = \frac{1}{2\pi r}\frac{dL_*}{dr}, \tag{1}$$

and $L_* \sim A\ L_0(h/r)^2$, where L_* denotes the part of X-ray luminosity which is caught by the disc, $A(\sim 1/2)$ the fraction of energy absorbed by the disc, and L_0 the X-ray luminosity of the inner regions (Shakura and Sunyaev 1973). Note that this is by a factor of $\sim (h/r)$ smaller than in neutron star binaries.

Lyutyi and Sunyaev (1976) showed that for $Q_{irr} < Q_{vis}\ \tau$ (Q_{vis} denotes the visous heating rate, and τ the optical depth), the temperature of the disc in the central plane will be practically independent of external heat sources. In this paper we thus consider the following two limiting cases: the case that the X-ray irradiation penetrates deeply into the disc and heats the entire vertical structrue of the disc (case A), and the case that the X-ray irradiates only the surface of the disc and the internal structure is not modified at all (case B). The actual situations will be between these two extreme cases.

The results are displayed in Fig. 1 by the solid line (the case without X-ray irradiation), the short dashed line (case A), and the long dashed line (case B). The mass-transfer rate into the disc is taken to be $\dot{M}_0 = 10^{15.0}g\ s^{-1}$, and $\alpha = 10^2(h/r)^{1.5}$. Calculations show that even at peak Q_{irr} is a few times smaller than Q_{vis}. Since the change in the internal structure is small even in case B, the propagation of the cooling wave is not prevented by X-ray irradiation. The theoretical light curves fall more slowly than the observations for time $\geq 100\ d$ even with the effect of X-ray irradiation included. The reason for the prolonged slow decline remains to be explained (Mineshige and Wheeler 1989). Osaki (1989) showed that in the systems which have a mass ratio of $M_1/M_2 > 4$ tidal instability causes the

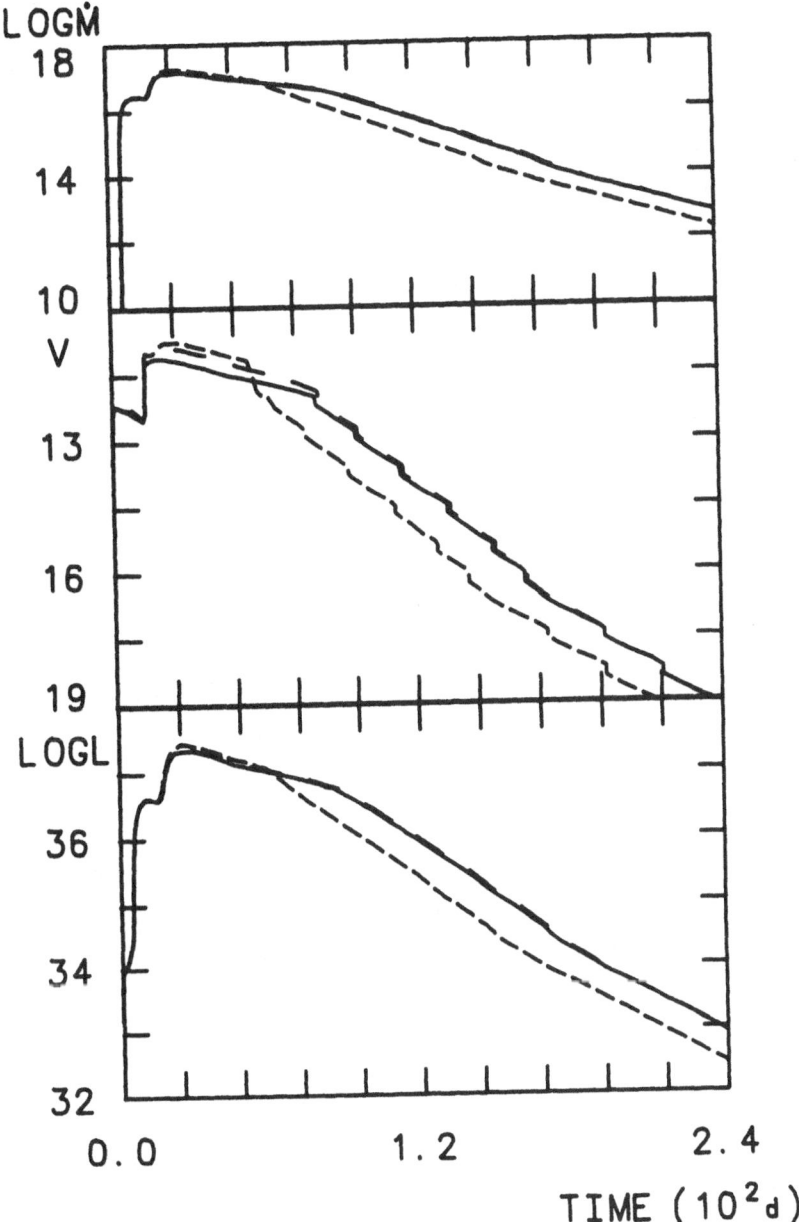

Figure 1: Effect of X-ray irradiation of the outer disc by the inner disc in black hole binaries. From the top, the mass-flow rate at the inner radius, the V-magnitude, and the bolometric luminosity are illusrated for the case without X-ray irradiation (solid lines), and the case with the X-ray irradiation (dashed lines). For details see text.

superoutbursts. If the outburst observed in A0620-00 was a superoutburst, then we may be able to naturally explain the long outburst duration.

4. Conclusions.

(1) The disc-instability model is more promising for outbursts of soft X-ray transients, particularly for those containing black holes. It can reproduce the basic features of the light curve of A0620-00.

(2) Heating of the disc by the X-ray irradiation originating from the inner part of the disc is small compared with the viscous heating. As a result, the internal structure of the disc will not change significantly, and so the propagation of the cooling wave is not prevented.

REFERENCES.

Cannizzo, J.K., Wheeler, J.C., and Ghosh, P. 1985, in *Proc. Cambridge Workshop on Cataclysmic Variables and Low-Mass X-Ray Binaries*, ed. D.Q. Lamb and J. Patterson (Dordrecht: Reidel), p.307

Hameury, J.M., King, A.R., and Lasota, J.P. 1986, *Astron. Astrophys*, **162**, 71

Hōshi, R. 1979, *Prog. Theor. Phys.*, **61**, 1307

Huang, M. and Wheeler, J.C. 1989, *Astrophys. J.*, **343**, in press

Lyutyi, V.M. and Sunyaev, R.A. 1976, *Sov. Astron.*, **20**, 290

Meyer, F. and Meyer-Hofmeister, E. 1981, *Astron. Astrophys*, **104**, L10

Mineshige, S. and Wheeler, J.C. 1989, *Astrophys. J.*, **343**, in press

Osaki, Y. 1974, *Publ. Astron. Soc. Japan*, **26**, 429

Osaki, Y. 1989, in this volume

Priedhorsky, W.C. and Holt, S.S. 1987, *Space Sci. Rev.*, **34**, 291

Shakura, N.I. and Sunyaev, R.A. 1973, *Astron. Astrophys*, **24**, 337

van Paradijs, J. and Verbunt, F. 1984, in *High Energy Transients in Astrophysics*, ed. S.E. Woosley, AIP Conf. Proc. No. 115, p.49

Whelan, J.A.J., Ward, M.J., Allen, D.A., Danziger, I.J., Fosbury, R.A.E., Murdin, P.G., Penston, M.V., Peterson, B.A., Wampler, E.J., and Webster, B.L. 1977, *Mon. Not. R. astr. Soc.*, **180**, 657

NON-AXISYMMETRIC SHEAR INSTABILITIES IN THICK ACCRETION DISKS.

Ramesh NARAYAN[1] and Jeremy GOODMAN[2]

[1]Steward Observatory, University of Arizona, Tucson, Arizona 85721, USA

[2]Department of Astrophysical Sciences, Princeton University,
Princeton, NJ 08544, USA

ABSTRACT. Recent results on the Papaloizou-Pringle non-axisymmetric instability in thick accretion disks is reviewed. Considerable work has been done on "slender" tori and annuli; these are systems whose radial widths are small compared to their mean radii. The linear instabilities in slender configurations fall into two classes, (i) the principal branch, and (ii) higher-order modes. The physical mechanism by which these modes become unstable has been identified. The idealized slender models have been extended in several ways; some work has been reported on (i) modes in wide tori/annuli, (ii) non-linear development of unstable modes, (iii) role of accretion, (iv) effect of self-gravity, and (v) effect of entropy gradient. These studies have brought us closer to the point where we can tell whether or not the PP instability is important for thick disks in nature. This issue is expected to be resolved by three-dimensional hydrodynamic simulations.

1. Introduction.

The subject of this review began with a seminal paper by Papaloizou and Pringle (1984) entitled "The dynamical stability of differentially rotating discs with constant specific angular momentum." These authors analyzed the global normal modes of differentially rotating fluid tori of constant entropy, uniform specific angular momentum and a polytropic equation of state, and concluded: "Our overall result is that the tori are unstable to low order non-axisymmetric modes and that the modes grow on a dynamical time-scale. We argue that because of the strength of the instability, similar unstable modes must exist in tori of non-uniform entropy or of non-uniform specific angular momentum." Many workers have since explored the physical mechanism of the instability and have sought to identify the range of models that display the instability. This body of work is reviewed here.

The importance of this subject is due to the extreme violence of the instability—for instance, the fastest-growing mode in a particular model system, the slender torus (section 3.1), grows in amplitude by a factor of 6 in just one orbit. Because of

F. Meyer et al. (eds.), Theory of Accretion Disks, 231–247.

this, it is expected that fluid configurations that display the most violent modes of the Papaloizou-Pringle instability (hereafter PP instability) cannot form at all in nature. The precise range of thick disk models that is excluded by this argument is still not known very well. As of today, it appears that extremely thick tori with a near-constant angular momentum profile may be ruled out. On-going numerical simulations should provide a better understanding of this important question.

2. Linear Stability Analysis.

Except for non-linear computer simulations, most of the theoretical work in this subject has been based on linear analysis. Here, one assumes that the perturbations are sufficiently small that terms in the equations that involve products of perturbations may be neglected. Linear stability analysis can be done in a local approximation or may be used to calculate the full global modes of a system. The former is much easier and leads to such famous criteria as the Rayleigh criterion, Schwarzschild criterion and Toomre criterion. Unfortunately, the local analysis can miss instabilities that involve global motions of the fluid and therefore must be supplemented by the global analysis before a definite conclusion can be reached regarding the stability of a fluid configuration. The PP instability in thick disks is a case in point. This instability is revealed only by a global analysis—there is no hint at all of its existence in the local analysis. We discuss the general principles of both kinds of stability analysis.

2.1. LOCAL STABILITY.

In the local analysis, one considers a small region of the fluid system and derives a dispersion relation between the wavevector k of a disturbance and its frequency ω. For instance, an isentropic non-gravitating fluid in differential rotation has a dispersion relation of the form

$$\omega^2 = c^2 k^2 + \kappa^2, \tag{1}$$

where c is the sound speed, and the epicyclic frequency κ is defined in terms of the angular velocity profile $\Omega(R)$ by

$$\kappa^2 = \frac{2\Omega}{R} \frac{d}{dR} (R^2 \Omega) = \frac{2\Omega}{R} \frac{d\ell}{dR}, \tag{2}$$

where $\ell(R)$ is the specific angular momentum. When $\kappa^2 > 0$, ω is real for all k and the fluid is locally stable. However, when $\kappa^2 < 0$, then for sufficiently low k, ω^2 becomes negative and the fluid is unstable. We thus have the criterion due to Rayleigh that, for local stability, the specific angular momentum must either be constant or increase outward (see Tassoul 1978 for references to the original literature on various local stability criteria).

For a non-rotating fluid with an entropy gradient, the dispersion relation is

$$\omega^2 = c^2 k^2 + N^2, \tag{3}$$

where the Brunt-Vaissala frequency N is defined by

$$N^2 = \frac{g}{\Gamma} \frac{d\ell n(P/\rho^\Gamma)}{dR} \propto \frac{ds}{dR}. \tag{4}$$

Here g is the acceleration due to gravity and s is the specific entropy. From this we derive Schwarzschild's criterion for convective stability, viz. that the entropy should increase outward (for inward-pointing gravity). When both an entropy gradient and differential rotation are simultaneously present, local stability requires $N^2 + \kappa^2 > 0$. However, this rule is valid only for axisymmetric disturbances. Cowling (1951) showed that if non-axisymmetric perturbations are allowed in a uniformly rotating fluid (where $\kappa = 2\Omega$), local convective instability sets in for some set of modes whenever $N^2 < 0$, regardless of the sign of $N^2 + \kappa^2$.

In the presence of self-gravity, a non-rotating isentropic fluid of density ρ has the dispersion relation

$$\omega^2 = c^2 k^2 - 4\pi G\rho. \tag{5a}$$

The corresponding relation for a sheet of surface density Σ is

$$\omega^2 = c^2 k^2 - 2\pi G\Sigma 3k3. \tag{5b}$$

In both cases the fluid becomes unstable at sufficiently low k. This is the Jeans instability. If there is rotation, the fluid is locally stable if the Toomre criterion is satisfied, viz. $Q \equiv c\kappa/\pi G\Sigma > 1$.

All the fluid systems we consider in this article (except those in section 6.2) are stable according to the above local criteria. Nevertheless they all have one or more globally unstable modes.

2.2. GLOBAL STABILITY.

In analyzing the global modes of a fluid system, the perturbations in the fluid are described by the three components of the velocity fluctuation, u (radial), v (tangential), w (vertical), the density fluctuation, ρ', and the pressure fluctuation, P'. Each of these is a function of three spatial coordinates and time. In the systems of interest here, the unperturbed fluid is axisymmetric and independent of time; therefore, the general mode can be assumed to have a simple dependence on ϕ and t, viz.

$$u(R, \phi, z, t) = u(R, z) \, exp[i(m\phi - \omega t)], \; etc. \tag{6}$$

The wavenumber m is an integer and the frequency ω is in general complex. Solutions of the form (6) correspond to a pattern of fluid motion that, in the inertial frame, rotates with a pattern angular speed, Ω_p, and grows with time as $exp(st)$, where

$$\Omega_p = Re(\omega)/m, \qquad s = Im(\omega). \tag{7}$$

The corotation radius R_c is defined to be that radius where the unperturbed fluid has the same angular speed as the pattern:

$$\Omega(R_c) = \Omega_p. \tag{8}$$

To solve for the five functions, u, v, w, ρ', P', each of which is a function of R and z, there are five equations, viz. the 3 components of the Euler equation of motion, the continuity equation, and the equation of state of the fluid. Under the assumption that the perturbations are weak, these equations may be linearized by neglecting all terms that involve products of perturbations. The resulting equations then lead to a partial differential equation in R, z with eigenvalue ω. This eigenvalue problem needs to be solved with the appropriate boundary condition to obtain the frequencies ω of the global modes. A growing mode corresponds to a complex ω with positive imaginary part. If ω is real, then the mode is neutral.

In some cases, it may be permissible to neglect variations in the fluid variables in z and to consider motions only in the $R\phi$ plane. The problem then reduces to an ordinary differential equation with specified boundary conditions and eigenvalue ω. This is a considerably easier problem to solve.

The key point about global stability analysis is that the modes involve motions of the whole system and not just of some localized region. The physics of global instabilities is often strongly tied to the boundary condition and so the edges of the system tend to be important. This is particularly true of the PP instability, as we discuss in section 4.

In general, global modes make a complete set only if singular modes are included (Case 1960). Therefore, the absence of an unstable nonsingular global mode does not necessarily imply complete stability—there could still be algebraically growing disturbances. We ignore this complication.

3. Slender Accretion Disks.

Much of the effort to understand the PP instability has been devoted to idealized "slender" systems, which have radial widths much smaller than their mean radii. The discussion below refers to isentropic systems.

3.1. SLENDER TORUS.

Consider a torus that orbits a point mass and extends from radius $R_o - a$ to $R_o + a$, with $a \ll R_o$ (the slender approximation). Let the angular velocity profile be parametrized in the form

$$\Omega(R) = \Omega \left(\frac{R}{R_o} \right)^{-q}.$$ (9)

The parameter q can range from $3/2$, which corresponds to a Keplerian disk, to 2, which corresponds to a thick constant-angular-momentum disk. Models with $q > 2$ are locally unstable by the Rayleigh criterion and are not of interest. The vertical half-thickness of the torus is given by $a\sqrt{2q-3}$ and ranges from 0 (thin disk) at $q = 3/2$ to a (circular cross-section) at $q = 2$. The equation of state of the fluid may be conveniently written in the polytropic form

$$P = K\rho^{1+\frac{1}{n}}.$$ (10)

The equilibrium slender torus is thus described by the two parameters, q and n. The pressure maximum is at $R = R_o$, $z = 0$.

The modes of the slender torus depend on a scaled azimuthal wavenumber,

$$\beta = ma/R_o.$$ (11)

Blaes (1985) made a partial analysis of the modes of the $q = 2$ system in the limit $\beta \to 0$. Goldreich, Goodman and Narayan (1986) carried out a more complete analysis of the entire spectrum of three-dimensional modes of the slender torus with $q = 2$, $n = 0$ (incompressible). Figure 1 shows the dispersion relation of this system, with solid and dashed lines indicating the neutral and unstable modes, respectively.

The unstable modes fall into two classes:

The Principal Branch: This class of unstable modes is present for all $\beta < 0.59$ and has its corotation radius at the pressure maximum of the torus, i.e. $R_c - R_o$. The horizontal velocity components, u, v, in these modes are almost independent of z, which shows that the modes are essentially two-dimensional in nature. The fastest-growing mode in this branch occurs at $\beta = 0.38$ and grows at a rate $s = 0.29\Omega$.

Higher-Order Modes. These modes occur at higher values of β. They involve distinctly three-dimensional fluid motions and have lower growth rates than the principal branch. Corotation is not necessarily at the pressure maximum, though it is always between the inner and outer edges of the torus, i.e. $R_o - a < R_c < R_o + a$.

Goldreich, Goodman and Narayan (1986) showed that the principal branch has two-dimensional motions because there is nearly perfect vertical hydrostatic equilibrium in each column of fluid. Furthermore, this condition is true even for tori with general q and n so long as $\beta \lesssim 0.5$. Therefore, the principal branch in more

236

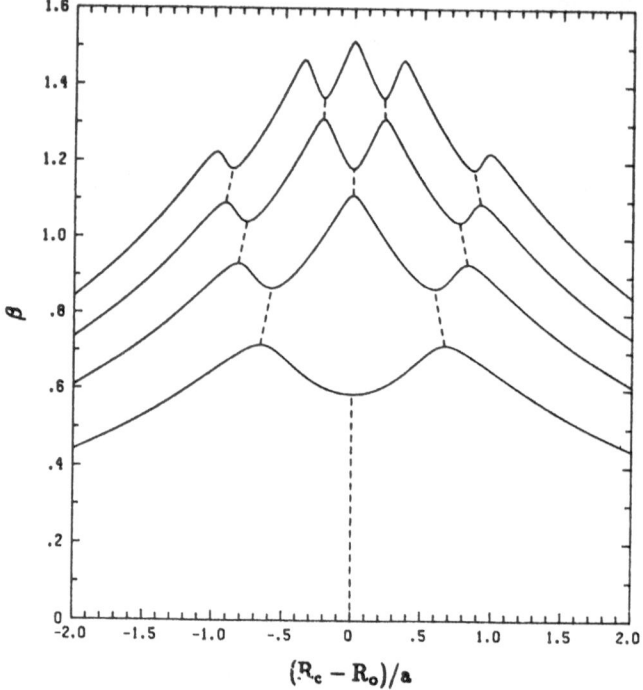

Figure 1: Dispersion relation of a slender isentropic, incompressible, constant- angular-momentum torus (Goldreich, Goodman and Narayan 1986). The vertical axis gives the scaled azimuthal wavenumber, β. The horizontal axis shows the position of the corotation radius in units such that ± 1 refer to the outer/inner edge of the torus, respectively. The solid lines correspond to neutral modes and the dashed lines to growing-decaying pairs of unstable modes. The vertical dashed line at $(R_c - R_o)/a = 0$, extending from $\beta = 0$ to $\beta = 0.59$, is the principal branch. The other dashed lines are higher order modes. Modes occur more and more densely at higher values of β. The diagram has been truncated at large β to avoid overcrowding.

general tori may be investigated by simplifying the geometry to that of a slender annulus and neglecting motions in z. The correspondence becomes nearly exact if the fluid in the annulus is given an effective polytropic index,

$$n_2 = n + 1/2. \tag{12}$$

Although this mapping from the torus to the annulus is valid only for the principal branch, it has nevertheless prompted a vigorous study of the complete mode structure of the two-dimensional slender annulus.

3.2. SLENDER ANNULUS.

The slender annulus is a fluid system that extends from cylindrical radius $R_o - a$ to $R_o + a$, with $a \ll R_o$, and is infinitely elongated along z. The angular velocity profile is as in eq. (9) and the equation of state is as in eq. (10) with polytropic index n_2 (equation 12). Only motions in the $R\phi$ plane are considered. As discussed above, the principal branch of the slender torus will be closely similar to that of the slender annulus; the study of this branch of instability may therefore be simplified by investigating the two-dimensional annulus. The higher-order modes of the two systems will probably have qualitative similarities but will certainly differ greatly in details.

A number of workers have investigated the modes of slender annuli, with some differences in the assumed boundary conditions and in the form of the radial gravitational force due to the central mass. A brief list of the systems that have been analyzed follows:

Blaes and Glatzel (1986): $q = 2$, $n_2 = 0$.

Glatzel (1987a,b; 1988): general q, n_2.

Goldreich, Goodman and Narayan (1986): principal branch for general q, n_2.

Goldreich and Narayan (1985), Narayan, Goldreich and Goodman (1987): higher-order modes for general q, n_2.

Kato (1987): corotation resonance in higher-order modes.

Papaloizou and Pringle (1984): $q = 2$, $n_2 > 1/2$.

Papaloizou and Pringle (1985): $q = 2 - \epsilon$, $q = \sqrt{3} + \epsilon$, $n_2 > 1/2$.

Papaloizou and Pringle (1987): general q, n_2.

The results may be summarized as follows. Slender annuli/tori have unstable principal modes whenever $q > \sqrt{3}$, regardless of n_2 (Papaloizou and Pringle 1985). For the particular case of an incompressible annulus ($n_2 = 0$), the growth rate can be expressed analytically as a function of q and β (Goldreich, Goodman and Narayan 1986). For the general n_2, the growth rate must be obtained by solving an eigenvalue problem based on an ordinary differential equation. However, the principal branch in these cases bears a very strong similarity to the incompressible mode. Indeed, the growth rates of the modes are approximated quite well by the incompressible annulus results provided β in the incompressible dispersion relation is replaced by $m(\Delta R_{1/4})/R_o$, where $\Delta R_{1/4}$ is the radial half-width of the annulus where the fluid density falls to one-quarter its central value.

The slender annulus has a whole sequence of higher-order unstable modes, all of which have corotation lying in the range $R_o - a < R_c < R_o + a$. In contrast to the principal mode, whose eignfunction has no nodes (in the real part of the enthalpy perturbation), the eigenfunctions of these higher-order modes have one or more nodes; indeed, the sequence of these modes may be labeled by two integers, ℓ_1 and ℓ_2, describing the number of nodes between $R_o - a$ and R_c, and R_c and $R_o + a$, respectively. In addition, there are other modes that are made unstable by a resonance at R_c. In contrast to the principal branch, the higher-order modes continue to be unstable even for $q < \sqrt{3}$ (Goldreich and Narayan 1985).

4. Physical Mechanism of the Instability.

Any flow with shear has excess free energy compared to an equivalent non-shearing flow. Ultimately, it is this free energy that drives shear instabilities. Based largely on a study of the slender annulus, some understanding has been gained of the manner in which this free energy is tapped by an unstable mode. The following discussion is based on ideas developed in detail by Narayan, Goldreich and Goodman (1987), but known for some time in the fluid mechanics community (Pierce 1974, Cairns 1979).

Role of the Corotation Radius: It may be shown that a mode has negative action (e.g. energy or angular momentum) for radii $R < R_c$ and positive action for $R > R_c$. This means that, compared to the undisturbed flow, the disturbed fluid has less energy/angular momentum for $R < R_c$ and more for $R > R_c$. Since the total action has to be conserved, a growing mode must have equal amounts of positive and negative action. This requires that the corotation radius must lie within the boundaries of the system. This is seen to be explicitly true for the slender torus in Fig. 1 and has been verified in all the slender annuli analyzed. (Rather surprisingly, the corotation radius of unstable modes may lie outside the fluid when the fluid is self-gravitating, see section 6.2.) When there is a growing mode, negative angular momentum increases with time in the fluid at radii $R < R_c$ and positive angular momentum increases for $R > R_c$. Consequently, all unstable modes transfer angular momentum from smaller to larger radii, thus acting in some sense like normal viscosity.

Corotation Amplifier: For each mode, there is a "forbidden" range of R around R_c where the radial part of the eigenfunction is evanescent, surrounded by two "permitted" regions where the eigenfunction is wave-like. Consider now a wave-packet that is incident on the corotation barrier from the permitted region at $R < R_c$. This packet carries negative action. When the packet is reflected at the corotation barrier, some amplitude tunnels through to $R > R_c$ by the normal wave-tunneling process. The transmitted packet has positive action and, therefore, the reflected packet will have more negative action than in the original packet. A similar super-reflection must clearly occur also for wave packets incident from $R > R_c$. The corotation region therefore behaves like an amplifier. The existence of this amplifier was recognized by Drury (1980) (see also Mark 1976, who considered the amplifier in the context of spiral density waves in self-gravitating disks).

Feedback and Instability: If the fluid system has an edge, then the output of the corotation amplifier will be reflected back as input, thus providing a feedback loop. Provided the loop satisfies a particular phase condition (integral number of waves), the amplifier will become a runaway oscillator. This is the simplest form of shear instability. Note that there can be an instability only if there is a sufficiently good reflecting boundary. This fact implies that the mode is a global one that will not be seen in a local analysis.

Single Boundary Mode: When the fluid system is semi-infinite in extent with a single edge, the situation is particularly simple. The phase condition in the feedback loop requires an integral number of wavelengths in the eigenfunction, going from corotation to the edge and back, and this is easily achieved by adjusting the position of R_c relative to the edge. There is thus an infinite sequence of discrete modes, all of which will be growing modes. The slender annuli do not have these modes, but some of the very wide systems may, particularly if there is viscosity to damp the wave amplitude at large R.

Two-Boundary Modes: When the system is finite and there are two edges, then there is a feedback loop on each side of corotation. If only one of these loops satisfies the phase condition, then a neutral mode is obtained. The mode action is small on the side of corotation for which the phase condition is violated. If both phase conditions are satisfied, then a pair of unstable modes is obtained, one growing and one decaying. To achieve this, it is necessary to tune both the position of R_c and the wavenumber β of the mode. Therefore, unstable modes occur in narrow bands of β. This is seen in Fig. 1 and is true in the case of the annuli as well.

Corotation Resonance: If the equilibrium flow has a gradient of specific vorticity, then there will be a resonance at corotation that can absorb positive or negative action, depending on the sign of the gradient. If positive/negative action is absorbed at corotation, then it is possible to have growing modes with predominantly negative/positive action and decaying modes with positive/negative action. (Such unbalanced modes with non-zero action would be neutral in the absence of the resonance.) When the growth rates are small, the modes are expected to saturate at very small amplitudes, because the mass of absorbing fluid near corotation is proportional to the growth rate. Consequently, after a rather small amount of action has been absorbed the flow near corotation is modified so as to prevent further absorption; the process is similar to Landau damping in plasmas.

Understanding the Principal Branch: The above ideas describe the physics of the higher-order modes in tori and annuli. In principle, they are relevant even for the principal branch. However, the modes in the principal branch are different because they are nodeless and another picture, due to Blaes and Glatzel (1986) and Goldreich, Goodman and Narayan (1986), has turned out to be more useful. According to these authors, the principal branch may be viewed as a superposition of traveling surface waves residing on the inner and outer edge of the slender torus or annulus. The shear (due to differential rotation) advects the wave on the inner edge forward with a velocity

$$V_s = q\Omega a \tag{13}$$

with respect to the fluid at R_o. The wave at the outer edge is advected backward with a similar velocity. Counteracting this shear are the phase velocities of the two waves with respect to the fluid at their respective edges, which for $\beta \ll 1$ and $q \neq 2$ are given by

$$3V_p3 = \frac{(2q-3)}{(2-q)}\Omega a. \tag{14}$$

The fact that V_p depends on $(2q-3)$ shows that the phase speed is a function of the effective inward (i.e. toward the pressure maximum or mean radius of the annulus) radial "gravity" at the edge, which is $(2q-3)\Omega^2 a$. The result also depends on the epicyclic frequency, $\kappa^2 = 2(2-q)\Omega^2$. One now makes the reasonable assumption that there can be an unstable mode only if the backward-moving wave on the inner edge and the forward-moving wave on the outer edge can couple across corotation. In order to communicate, i.e. to form parts of the same linear mode, the two waves must have a common total phase velocity. This requires $3V_p3 > V_s$, which explains why the principal branch is unstable only for $q > \sqrt{3}$. The rule given here, viz. that there must be coupling between the two edge waves, replaces the feedback phase condition discussed above for the higher-order modes. The reason for having a different prescription in this case is because the radial eignfunction of the principal mode is evanescent everywhere; the mode owes its very existence—not only its growth—to the dynamics of the boundary.

5. Numerical Simulations of the Instabilities.

5.1. NON-LINEAR EVOLUTION OF THE SLENDER ANNULUS.

Hawley (1987) used a finite-difference two-dimensional numerical hydrodynamic code to evolve the principal modes of slender annuli into the non-linear regime. He obtained the remarkable result that the annulus breaks up into a number of almost detached independent blobs, or "planets". (An unstable mode of wavenumber m initially makes m planets.) Hawley (1989, this volume) has computed the evolution of a slender torus with a three-dimensional code and has obtained similar results.

The dramatic change in the fluid configuration in the non-linear development of the principal mode suggests that the fluid may be heading toward another equilibrium. Goodman, Narayan and Goldreich (1987) found an analytical ellipsoidal constant-vorticity equilibrium configuration that strongly resembles Hawley's planets. The motion consists of z-independent solid-body-like retrograde rotation. Each fluid element goes round the center of the planet along elliptical streamlines with the same angular frequency.

The axis ratios of the analytical planets depend on the specific vorticity, and the sequence of configurations has a maximum vorticity, viz. $(2 - \sqrt{3})\Omega$. The planet with the maximum vorticity is infinitely elongated in the tangential direction and closely resembles a torus. Since the principal branch is two-dimensional in character, and since specific vorticity is conserved in such motions, it is interesting to compare the vorticity of planets with that of tori, given by $(2 - q)\Omega$. We then find the interesting result that only tori with $q > \sqrt{3}$ have equivalent planet-like configurations available, while those with $q < \sqrt{3}$ do not. Since $q > \sqrt{3}$ is also the condition for the principal mode of instability in the slender torus, Goodman, Narayan and Goldreich (1987) suggested that the instability must be driven by the bifurcation in equilibria. This is yet another way of understanding the physics of the principal branch.

An interesting question to ask is why the fluid prefers the planet rather than the torus configuration. It is suggestive that for the same vorticity, the planet has less shear than the torus. Since shear contributes to excess free energy, this probably implies that the planet has less free energy than the torus and therefore is the preferred state.

The planet configuration is itself not stable, but has numerous instabilities. These are however not as violent as the principal branch in the slender torus.

5.2. WIDE ANNULI/TORI.

Blaes and Glatzel (1986) obtained analytical results for non-slender annuli with $q = 2$, $n_2 = 0$. They found that, beyond a certain critical ratio of outer to inner radius, the principal branch becomes stable. This has been confirmed numerically for general q and n_2. The effect is explained by the interacting edge-wave interpretation of the principal mode described in section 4. When the system is wide, Ω is small at the outer edge, and so is the effective radial gravity. Consequently the phase velocity of the outer edge wave is low. The coupling of the two edge waves then depends on whether the inner wave can move backward with sufficient phase speed to counteract almost the entire shear. As the annulus gets wider this requirement becomes progressively more difficult to achieve and therefore the minimum q for the principal branch instability increases above $q = \sqrt{3}$. In fact, for a wide enough annulus, even $q = 2$ becomes stable. In addition, at a given q, the range of β over which the principal branch is unstable becomes less in wide annuli relative to a slender annulus.

Most of the higher-order modes do survive, however. The most unstable of these have corotation near the inner edge, with no nodes for $R < R_c$ and have one or more nodes for $R > R_c$. For the really wide systems, where the outer R becomes very large, the modes take on the character of the single-edge modes discussed in section 4.

Blaes and Hawley (1988) investigated the non-linear evolution of instabilities in wide annuli through numerical simulations. Their results may be summarized as follows:

1. When the principal branch is linearly unstable, the annulus evolves into planets, just as in the slender case. If the most unstable linear mode has an azimuthal wavenumber, m, then initially m planets form, but at late times these tend to coalesce into a single planet.

2. When the dominant instability is a higher-order mode of the type described above, with no node between the inner edge and R_c, the system evolves non-linearly into spiral waves of finite amplitude. Near the inner edge the spiral pattern terminates on a planet-like "head" with a relatively small density perturbation (e.g. 25% in one case).

Vertical hydrostatic equilibrium is probably not a good approximation in the case of wide systems and so these results on wide annuli, while suggestive, may deviate significantly from those for the equivalent tori even in the case of the principal branch. A few numerical investigations of wide tori have been reported. Some workers (Papaloizou and Pringle 1984, 1987, Kojima 1986a,b, Frank and Robertson 1988) investigated the fastest growing modes in wide tori by rewriting the linearized equations as an initial value problem and solving numerically. The results are qualitatively similar to those for wide annuli. In brief, for sufficiently small tori, the principal branch is the most unstable mode, but as the width increases, successive higher-order modes become dominant. These modes usually have corotation near the inner edge. At large R the wave action in tori tends to be concentrated near the surface, away from the equatorial plane. This is a clear indication that these modes cannot be described with a two-dimensional annulus approximation.

Zurek and Benz (1986) investigated the non-linear evolution of a moderately wide torus using a smooth particle hydrodynamics code.

6. Additional Effects.

The discussion so far was restricted to idealized inviscid, non-accreting, non-self-gravitating, isentropic tori and annuli. These conditions are unlikely to be realized in real systems. Some work has been done to investigate how the instabilities are modified when some of the restrictions are relaxed.

6.1. EFFECT OF ACCRETION.

Blaes (1987) studied annuli orbiting in a particular pseudo-Newtonian potential developed by Paczynski and Wiita (1980). This potential simulates the effects of the Schwarzschild metric near a point mass and, in particular, has a critical "cusp" radius such that when the inner edge of the annulus is inside this radius, there is accretion on to the central mass. Blaes found that the growth rates of all modes declined very rapidly as the accretion rate increased. The last mode to stabilize did so when the radial velocity at the pressure maximum was only $\sim 10^{-5}$ of the orbital velocity. There are at least three possible reasons for this effect:

1. The fluid that accretes at the inner edge carries with it some negative action. If the rate of growth of action due to the instability is less than the loss rate due to advection, the mode will decay rather than grow.

2. The radial velocity has a sonic point at the inner edge of the torus. Since no fluid disturbances can be communicated outward across the sonic point, any flux of mode action incident on the sonic point is totally absorbed. (Globally, partial reflection can still occur outside the sonic point because of variations in the sound speed, and other background quantities, on scales smaller than the WBK wavelength of the mode.) Modes trapped inside corotation grow only if they radiate through corotation faster than through the sonic point. Modes trapped outside corotation do not exist in Blaes's accreting models, since the outer radius of the torus is infinite.

3. When the inner edge approaches the cusp radius the effective radial gravity acting on this edge becomes less than the Newtonian effective gravity. Because of this, the phase velocity of the inner edge wave is reduced and this decreases the effectiveness of the principal branch.

The third explanation is probably not correct, since Blaes's marginally accreting torus, for which the effective gravity just vanishes at the inner edge, has a growth rate only slightly smaller than the maximum rate found in any of his tori.

For the marginally stable torus, the flow time from corotation to the sonic point is $\sim 10^5$ times larger than the growth time in the marginally accreting torus. (The ratio becomes $\sim 10^3$ if we use, instead of corotation, the radius at which the action density peaks.) Thus explanation (1) is probably also wrong: the modes are not stabilized by advection alone. Explanation (2) is clearly related to (1) but is distinct, in our minds at least, and is the one we favor.

6.2. EFFECT OF SELF-GRAVITY.

Self-gravitating thick disks may occur during star formation (e.g. Bodenheimer, 1989, this volume), when two white dwarfs merge (e.g. Livio, 1989, this volume), or possibly in accretion disks in AGN (e.g. Begelman, 1989, this volume). Goodman and Narayan (1988) studied the effect of self-gravity in the fluid and studied two model problems: (1) A slender torus with $n = 0$, $q = 2$, (2) a slender annulus with $n_2 = 0$, general q. They found the initially surprising result that the principal branch becomes less unstable in the presence of self-gravity. Indeed, for sufficiently strong self-gravity, the principal branch is completely stabilized for all $q \leq 2$. However, the higher-order modes of the slender torus do survive. In addition, for sufficiently strong self-gravity, two new unstable modes appear. One of these modes is closely related to the Jeans instability. The other mode is unusual in that its corotation radius does not lie inside the boundaries of the system, thus violating a theorem discussed in section 4. The reason for this is not understood; it may be related to the long-range nature of the gravitational interaction.

The stabilization of the principal branch by self-gravity may be understood as follows. Self-gravity has a tendency to encourage inhomogeneities and this reduces the effective radial restoring force on the edge waves. Consequently, the phase velocity of these waves is reduced, making it harder to achieve the necessary coupling of the two waves to make an unstable principal mode. It is worth noting that the effect of self-gravity on the principal branch is exactly the opposite of that expected from local stability arguments (section 2.1). For instance, if a fluid is locally dynamically unstable because $\kappa^2 < 0$, then including self-gravity would only make the local instability more violent. This is yet another example of the deep difference that exists between local and global stability.

6.3. EFFECT OF ENTROPY GRADIENT.

Frank and Robertson (1988) numerically investigated the effect of including entropy gradients in wide tori. Kojima, Miyama and Kubotani (1989) and Kubotani, Miyama and Sekiya (1989) analytically studied the effects in slender annuli. Glatzel (1989, this volume) has made a comprehensive analysis of g-modes in slender annuli with entropy gradients.

In order to ensure local convective stability, the entropy variation in these models was chosen to have a minimum at the pressure maximum. It is found that in the slender systems the principal branch becomes more unstable in the presence of a "stable" entropy gradient. The results are less clear in wide systems. The effect on the slender systems is again easy to understand in terms of the coupled edge-wave picture. When the entropy increases towards the edges, there is an additional radial restoring force at the boundaries due to buoyancy. This leads to an increase in the phase velocity of the edge waves, leading to enhanced instability. Once again, the effect is exactly opposite to what one would expect from local stability analysis.

7. Questions for the Future.

A number of issues still remain unresolved in this field, of which we discuss three. Until these questions are settled we cannot tell how important the PP instability is for real accretion disks in nature. The first and second issues can be investigated by means of non-linear numerical simulations in three dimensions. It is heartening that the techniques and the technology to do this are already available now (Hawley 1989, this volume). It is not clear what the best technique would be to approach the third issue.

7.1. EFFECTS OF ACCRETION IN THREE-DIMENSIONAL TORI.

We discussed earlier that in a particular model of two-dimensional accretion studied by Blaes (1987) all the instabilities disappeared. A crucial question now is:

what will be the effect of viscous accretion in realistic three-dimensional tori?

Several effects were discussed in section 7.1 to explain Blaes's results. Some of these are likely to be less effective in three dimensions than in two. For instance, in three dimensions, the radial component of the flow becomes supersonic only near the cusp and most of the torus is subsonic. Therefore, the feedback loop is less likely to be disrupted. Also, the cusp, where the effective radial gravity becomes very small, is restricted to the equatorial plane and (in the case of a thick disk) most of the "inner" edge lies along steep funnels where the effective gravity is quite high. Thus, the phase velocity of the inner edge wave is not likely to be modified substantially in three dimensions, making the instability more robust than in two dimensions. On the other hand, the wave action of the most unstable modes in wide tori tends to be concentrated near the cusp so that the funnel may be less important than it might at first appear.

On balance, it appears that one will need higher values of $\dot{M}$ in three dimensions (than in two) to control the instabilities. Exactly what the critical $\dot{M}$ may be is not known. Only three-dimensional numerical simulations can answer this.

7.2. WHAT KINDS OF THICK DISKS CAN WE RULE OUT?

The thickest and astrophysically most interesting tori have q close to 2, i.e. nearly constant angular momentum. All the studies indicate that these are also the most violently unstable systems. Therefore, one would like to know how close a torus can get to constant-angular-momentum and still survive in nature.

From the studies carried out so far it appears that tori with $q \gtrsim 1.8$–1.9 may be too unstable to form. (This conclusion could be drastically revised when the effect of accretion is better understood.) Such tori can be reasonably thick, but may not provide the extreme super-Eddington luminosities envisaged in the original models of thick tori (Paczynski and Wiita 1981), nor will they have very narrow funnels. This is unfortunate since these particular features are the most interesting aspects of thick disks.

One possibility is that a torus that starts off with $\Omega \propto R^{-q}$ with $q \sim 2$ may evolve through the instabilities to a non-power-law angular velocity profile. Thus, the effective q at small radii may be reduced significantly, possibly to below 1.8, while the outer regions of the torus may be unaffected. This will have a large effect on the shape of the funnel without significantly changing the thickness of the disk.

These questions may be answered relatively soon through numerical simulations.

7.3 CAN THE INSTABILITIES PROVIDE TURBULENT VISCOSITY?

In the previous section we talked about the effect of the more violent instabilities, which could either destroy the torus altogether or modify the torus in such a way that the modes are stabilized. This still leaves open the possibility that weaker higher-order instabilities may persist. These modes could saturate at modest levels, possibly through shocks, and may provide a kind of viscosity for angular momentum transport. There has been some analytical work on the steepening of linear waves in simple two-dimensional systems (Hanawa 1988). Finite-difference numerical simulations may not be very helpful to study these effects because the higher-order modes have short radial wavelengths, requiring high spatial resolution, and have low growth rates, requiring integration over many orbits. Nevertheless, some work has been done with a two-dimensional code (e.g. Kaisig 1989).

Another possibility is that after one of the more violent global modes has saturated, for instance as planets in the case of the principal branch, there may be short-wave three-dimensional disturbances that could evolve directly to fully developed turbulence. This phenomenon, called shear-flow transition (e.g. Bayly, Orszag and Herbert 1988), is known to occur in simple shear flows such as plane Poiseulle flow and boundary layer flow. It is not clear that the same effect will occur in rotating flows in accretion disks. If it does, then we will ultimately make some progress toward understanding turbulent viscosity in accretion disks.

ACKNOWLEDGEMENTS. None of our work in this field would have been possible without the deep insight and inspiration of our collaborator, Peter Goldreich, and we wish to express our gratitude to him. This work was supported in part by NASA astrophysical theory grant NAGW-763 (to RN) and by Sloan Foundation and David and Lucille Packard Foundation fellowships (to JG).

REFERENCES.

Bayly, B.J., Orszag, S.A., Herbert, T., 1988. *Ann. Rev. Fluid Mech.*, **20**, 359
Blaes, O.M., 1985. *MNRAS*, **216**, 553
Blaes, O.M., 1987. *MNRAS*, **227**, 975
Blaes, O.M., Glatzel, W., 1986. *MNRAS*, **220**, 253
Blaes, O.M., Hawley, J.F., 1988. *Ap. J.*, **326**, 277
Cairns, R.A., 1979, *J. Fluid Mech.*, **92**, 1
Case, K., 1960. *Phys. Fluids*, **3**, 143
Cowling, T.-G., 1951. *Ap. J.*, **114**, 272
Drury, L. O'C., 1980. *MNRAS*, **193**, 337
Frank, J., Robertson, J.A., 1988. *MNRAS*, **232**, 1
Glatzel, W., 1987a. *MNRAS*, **225**, 227
Glatzel, W., 1987b. *MNRAS*, **228**, 77
Glatzel, W., 1988. *MNRAS*, **231**, 795
Goldreich, P., Goodman, J., Narayan, R., 1986. *MNRAS*, **221**, 339
Goldreich, R., Narayan, R., 1985. *MNRAS*, **213**, 7P
Goodman, J., Narayan R., 1988. *MNRAS*, **231**, 97

Goodman, J., Narayan, R., Goldreich, P., 1987. *MNRAS*, **225**, 695
Hanawa, T., 1988. *A, A*, **206**, 79
Hawley, J.F., 1987. *MNRAS*, **225**, 677
Kaisig, M, 1989. *A, A*, in press
Kato, S., 1987. *Proc. Astron. Soc. Japan*, **39**, 627
Kojima, Y., 1986a. *Progr. Theor. Phys. (Osaka)*, **75**, 251
Kojima, Y., 1986b. *Progr. Theor. Phys. (Osaka)*, **75**, 1464
Kojima, Y., Miyama, S.M., Kubotani, H., 1989. preprint
Kubotani, H., Miyama, S.M., Sekiya, M., 1989. preprint
Mark, J.W.-K., 1976. *Ap. J.*, **205**, 363
Narayan, R., Goldreich, P., Goodman, J., 1987. *MNRAS*, **228**, 1
Paczynski, B., Wiita, P.J., 1980. *A, A*, **88**, 23
Papaloizou, J.C.B., Pringle, J.E., 1984. *MNRAS*, **208**, 721
Papaloizou, J.C.B., Pringle, J.E., 1985. *MNRAS*, **213**, 799
Papaloizou, J.C.B., Pringle, J.E., 1987. *MNRAS*, **225**, 267
Pierce, J.R., 1974. *Almost All About Waves*, MIT
Tassoul, J.-L., 1978. *Theory of Rotating Stars*, Princeton University Press, Princeton
Zurek, W.H., Benz, W., 1986. *Ap. J.*, **308**, 123

INSTABILITIES OF ACCRETION FLOWS CAUSED BY THE INTERACTION OF INTERNAL GRAVITY MODES.

Wolfgang GLATZEL

Max-Planck-Institute for Astrophysics, Karl-Schwarzschild-Strasse 1,
D-8046 Garching bei München, FRG.

ABSTRACT. The modal structure of internal gravity modes in shearing systems is investigated using analytical methods. If the buoyancy frequency profile satisfies certain symmetry conditions, two types of instabilities are found, one of which can be explained by over - reflection, the second by mode resonances.

1. Introduction.

Since the discovery by Papaloizou & Pringle (1984) that accretion tori are violently unstable with respect to nonaxisymmetric perturbations, the role of sonic and surface wave type modes in generating this instability has been studied extensively (for a review and references see Narayan, this volume). Although entropy gradients have already been taken into account by Frank & Robertson (1988), the spectrum of gravity modes associated with them remains to be investigated. We shall give here a brief description of their modal structure based on a simplified model and using an analytical treatment. Further details can be found in Glatzel (1989).

Sonic instabilities in accretion tori can be explained by mode resonances which are caused by a distortion of the pattern speed of the standing sound waves of the system by the shear motion. Due to the supersonic pattern speed of the sonic modes these resonances occur for supersonic flow velocities. It has been speculated that a similar distortion of the pattern speed will also occur for gravity modes thus producing resonance instabilities in the gravity mode spectrum. Since the pattern speed of gravity modes is low, they are expected in particular for low shear rates. One of the main motivations for this study is, to find out under which conditions this conjecture concerning the gravity modes is true.

2. The Model.

2.1. STATIONARY CONFIGURATION.

In order to allow for an analytical treatment of the problem we have to simplify

F. Meyer et al. (eds.), Theory of Accretion Disks, 249–258.

our model considerably. The three dimensional structure of an accretion torus is replaced by cylindrical geometry, in which the perturbation problem is described by ordinary differential equations rather than by partial differential equations. Moreover, we shall restrict ourselves to slender tori or - in our simplified geometry - to thin cylindrical shells. A characteristic property of accretion tori is that the external force balances the centrifugal force at some point within the torus, i.e. the effective gravity vanishes at some radius r_0. Denoting the distance from r_0 by x the effective gravity may then within our approximation be taken as

$$\frac{1}{\bar{\rho}}\frac{\partial \bar{p}}{\partial r} = -gx \qquad (2.1)$$

where $g > 0$ is a constant and $\bar{\rho}$ and $\bar{p}$ are the density and the pressure of the stationary configuration respectively. According to equation (2.1) the pressure and, - for convective stability - also the density has a maximum at r_0. The position of the inner and outer boundary is determined by zero pressure or density. A mathematically simple density stratification describing this situation is a parabolic profile, which can always be achieved by a suitable choice of the entropy distribution. Denoting half of the thickness of the shell by x_0 and normalizing $\bar{\rho}$ to its maximum value the density may then be written as:

$$\bar{\rho} = 1 - (x/x_0)^2 \qquad (2.2)$$

Choosing a frame corotating with the flow at r_0 as frame of reference the tangential component of the velocity $\bar{V}$ is in the limit of a thin cylindrical shell given by

$$\bar{V} = \bar{V}_0 x = \left(\frac{\partial \bar{V}}{\partial r} - \frac{\bar{V}}{r}\right)\Bigg|_{r_0} x \qquad (2.3)$$

2.2. PERTURBATION EQUATION.

We consider two dimensional perturbations which due to the symmetries of the problem may be taken to be proportional to $exp(im(\omega t + \varphi))$. (Three dimensional perturbations tend to be less unstable than two dimensional ones.) We neglect self gravity (Cowling's approximation) and assume incompressibility (Boussinesq approximation) for the perturbations. The Boussinesq approximation is not essential for the spectrum of gravity modes we are interested in but will eliminate the sonic modes, which have been studied in detail in previous investigations. Measuring lengths in units of x_0 and defining the variable y as

$$y = r^{3/2}\bar{\rho}^{1/2}v_r \qquad (2.4)$$

where v_r is the radial component of the velocity perturbation the perturbation equation may within the approximation described in Section 2.1 be written as:

$$\frac{d^2y}{dx^2} + y\{\frac{1}{\bar{\sigma}^2}\frac{2x^2\,Ri}{1-x^2} + \frac{1}{\bar{\sigma}}[-\frac{\partial^2\bar{V}_p}{\partial x^2} + \frac{2x}{1-x^2}\frac{\partial\bar{V}_p}{\partial x}] - k_p^2 + \frac{1}{(1-x^2)^2}\} = 0 \quad (2.5)$$

with

$$\frac{\partial\bar{V}_p}{\partial x} = (\frac{\partial\bar{V}}{\partial r} + \frac{\bar{V}}{r})\bigg|_{r_0} /\bar{V}_0 \quad (2.6)$$

$$\frac{\partial^2\bar{V}_p}{\partial x^2} = (\frac{\partial^2\bar{V}}{\partial r^2} + \frac{1}{r}\frac{\partial\bar{V}}{\partial r} - \frac{\bar{V}}{r^2})\bigg|_{r_0} /\bar{V}_0 \quad (2.7)$$

$$k_p^2 = \frac{m^2 - 1/4}{r_0^2} \quad (2.8)$$

$$\bar{\sigma} = \tilde{\omega}_p + x \quad (2.9)$$

$$\tilde{\omega}_p = r_0\omega/\bar{V}_0 \quad (2.10)$$

The Richardson number Ri is defined as

$$Ri = g/\bar{V}_0^2 \quad (2.11)$$

For $\frac{\partial\bar{V}_p}{\partial x} = 1$ and $\frac{\partial^2\bar{V}_p}{\partial x^2} = 0$ equation (2.5) becomes identical with the perturbation equation for the corresponding linear shear layer. Since there is no essential difference between the modal structure of a plane and a rotating configuration we shall restrict ourselves to the discussion of the plane case in the following.

$\tilde{\omega}_p$ as defined by equation (2.10) is the pattern speed of the mode in units of the flow velocity at the edges of the shear layer. For small flow velocity, where normalization by gravity is more appropriate we shall discuss the modal structure in terms of

$$\omega_p = \frac{r_0\omega}{\sqrt{g}} = \tilde{\omega}_p Ri^{-1/2} \quad (2.12)$$

instead of $\tilde{\omega}_p$.

In order to allow for an analytical solution we adopt the approximation $1-x^2 \approx 1$ in equation (2.5), which means that the density is taken to be constant (see equation 2.2) but without neglecting its derivative. Boundary conditions of the eigenvalue problem for ω_p are then no longer provided by the regularity requirement of the solutions of equation (2.5). Instead we assume rigid boundary conditions ($y = 0$) at $x = \pm 1$. It has been checked a posteriori by a comparison with the numerical solution of the exact equations that this approximation yields qualitatively correct results of the right order of magnitude. By a transformation of the independent variable equation (2.5) can then be reduced to Whittaker's equation and the dispersion relation may be expressed in terms of confluent hypergeometric functions:

$$(\frac{\tilde{\omega}_p+1}{\tilde{\omega}_p-1})^{2\mu}e^{\zeta_- -\zeta_+}{}_1F_1(1/2-\mu+\kappa,1-2\mu,-\zeta_-){}_1F_1(1/2+\mu-\kappa,1+2\mu,\zeta_+)$$

$$-{}_1F_1(1/2-\mu+\kappa,1-2\mu,-\zeta_+){}_1F_1(1/2+\mu-\kappa,1+2\mu,\zeta_-)=0 \tag{2.13}$$

with

$$\zeta_- = (\tilde{\omega}_p-1)2(k_p^2-2Ri-2\frac{\partial \bar{V}_p}{\partial x}-1)^{1/2} \tag{2.14}$$

$$\zeta_+ = (\tilde{\omega}_p+1)2(k_p^2-2Ri-2\frac{\partial \bar{V}_p}{\partial x}-1)^{1/2} \tag{2.15}$$

$$\mu = (1/4-2\tilde{\omega}_p^2 Ri)^{1/2} \tag{2.16}$$

$$\kappa = \frac{-\frac{\partial^2 \bar{V}_p}{\partial x^2}-2\tilde{\omega}_p\frac{\partial \bar{V}_p}{\partial x}-4\tilde{\omega}_p Ri}{2(k_p^2-2Ri-2\frac{\partial \bar{V}_p}{\partial x}-1)^{1/2}} \tag{2.17}$$

We note that the dispersion relation (2.13) is multi - valued, which requires a careful determination of the correct branch.

3. The Richardson Criterion.

A sufficient condition for stability or, equivalently, a necessary condition for instability of the configurations studied is provided by the Richardson criterion:

$$\frac{1}{\bar{\rho}}\nabla\bar{p}\frac{1}{c_p}\nabla\bar{s}-\frac{1}{4}r^2(\nabla\Omega)^2 < 0 \tag{3.1}$$

somewhere in the flow is necessary for instability. c_p and $\bar{s}$ are the specific heat at constant pressure and the entropy respectively. Ω denotes the angular velocity of the stationary configuration. The Richardson criterion (3.1) is still valid when compressibility, which has a stabilizing effect, is taken into account (Hanawa 1987). Three dimensional perturbations tend to be less unstable than two dimensional ones and we have therefore restricted ourselves to two dimensional perturbations in Section 2.

Equation (3.1) shows that entropy gradients in a convectively stable configuration can in principle overcome the destabilizing effect of shear and thus might possibly allow for stable accretion tori, which was the main motivation to study the influence of entropy gradients. However, as already described in Section 2.1, in any accretion torus the gravity vanishes at some point and for convective stability the entropy gradient has to vanish there too. Accordingly the sufficient condition for stability can for topological reasons never be satisfied in a toroidal configuration.

This might cause the weak influence of entropy gradients on the growth rates of sonic modes as found by Frank & Robertson (1988).

A decrease of the shear rate or, alternatively, an increase of the sound speed has been found to decrease the growth rate of the sonic instabilities in accretion tori. Accordingly thick tori have been suspected and are likely to become stable against sonic and surface wave instabilities. If this is still true when entropy gradients are taken into account, is an open question. In this respect the behaviour of the gravity modes generated by entropy gradients is of interest in particular for low shear rates and will be studied here.

4. The Modal Structure of Gravity Modes.

4.1. MEDIUM AT REST.

For a medium at rest a simple analytical approximation can be derived for the pattern speed ω_p of the neutrally stable gravity modes. We find two sets of modes corresponding to symmetric and antisymmetric eigenfunctions respectively:

$$\omega_p \approx \pm \frac{1}{\sqrt{2}}(n\pi + \frac{5\pi}{8})^{-1} \quad ; \quad n = 0, 1, 2, ... \tag{4.1}$$

$$\omega_p \approx \pm \frac{1}{\sqrt{2}}(n\pi + \frac{7\pi}{8})^{-1} \quad ; \quad n = 0, 1, 2, ... \tag{4.2}$$

Eigenfunctions, i.e. y or v_r, of modes described by equation (4.1) are symmetric with respect to x, while those corresponding to equation (4.2) are antisymmetric. The symmetric modes can be eliminated by imposing the boundary condition at $x = 0$ and $x = +1$ or $x = -1$ instead of at $x = +1$ and $x = -1$. The pattern speed of the gravity modes has a maximum value and decreases with increasing number of nodes in the eigenfunction.

4.2. NEUTRALLY STABLE MODES IN THE SHEAR FLOW.

The pattern speed ω_p of some neutrally stable gravity modes in a plane linear shear flow ($\frac{\partial^2 \bar{V}_p}{\partial x^2} = 0$; $\frac{\partial \bar{V}_p}{\partial x} = 1$) is shown in Fig. 1 as a function of the flow velocity ($Ri^{-1/2}$) for fixed wavenumber $k_p = 0$. According to equations (4.1) and (4.2) we have labelled the modes with the corresponding values of n and indicated their symmetry properties and direction of propagation (parallel to the boundaries) with subscripts A or S and + or - respectively.

For zero flow velocity ($Ri = \infty$) we verify the occurrence of pairs of symmetric and antisymmetric modes as given by equations (4.1) and (4.2). In contrast to the behaviour of the sonic modes the pattern speed of the gravity modes increases as the flow velocity is increased. (Dashed lines in Fig. 1 denote the flow velocity at the

254

boundaries.) They remain neutrally stable and do not reach a critical layer within the flow. However, apart from the general trend we find with increasing flow speed for each mode several minima and maxima in ω_p. The occurrence of these extrema suggests an interpretation as the result of the crossing of two sets of neutral modes, where the crossings have unfolded into avoided crossings. The pattern speed of one of these sets follows the general trend and increases with the flow velocity, the pattern speed of the second set decreases with increasing shear similar to the sonic modes. If this interpretation is correct, we would expect to find modes having a critical layer within the flow whose pattern speed coincides with the extrapolation of the sequences of avoided crossings towards higher $Ri^{-1/2}$. In fact, these modes do exist and are described in Section 4.3.

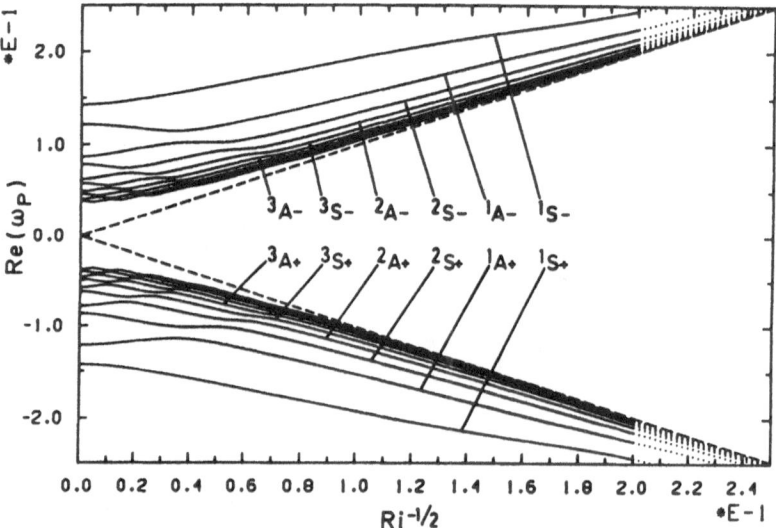

Figure 1: The pattern speed $Re(\omega_p)$ of neutrally stable gravity modes in the plane linear shear layer as a function of the shear rate $Ri^{-1/2}$ for zero wavenumber $k_p = 0$. Dashed lines denote the flow velocity at the boundaries. The modes are labelled with the order parameter n, where subscripts indicate the symmetry property and propagation direction. Note the sequences of avoided crossings.

4.3. MODES HAVING A CRITICAL LAYER WITHIN THE FLOW.

Real and imaginary parts of the pattern speed ω_p of gravity modes in a plane linear shear flow having a critical layer within the flow are shown in Figs. 2 and 3. Notation and parameters are as in Fig. 1. Values of ω_p shown in Figs. 2 and 3 which do not correspond to a critical layer within the flow have no physical meaning. They were obtained by a wrong connection of branches in equation (2.13) and have been plotted in order to demonstrate, that the extrapolation of the real parts of the pattern speed of modes having a critical layer within the flow exactly follows the sequences of avoided crossings in Fig. 1.

Modes having a critical layer within the flow are no longer neutrally stable. If the critical layer is close to the boundary, the modes are damped. With increasing shear damping rate and pattern speed decreases. Just before zero pattern speed is reached, i.e. if the local Richardson number at the critical layer is $\approx 1/4$, the damping rate turns into a growth rate, which undergoes a maximum and drops to zero when the pattern speed vanishes. For symmetry reasons the two modes corresponding to the same value of n and travelling in opposite direction parallel to the boundaries in the medium at rest cross at this point. From Fig. 3 we observe that this crossing has unfolded into an instability band. In contrast to the corresponding phenomenon of the sonic modes the g - modes do not uncouple again for higher shear rates. Instead the pattern speed remains zero and the modes become damped (except for $n = 0$) for $Ri \to 0$. As a consequence, there is only one resonance per mode. No multiple resonances as for the sonic modes do occur.

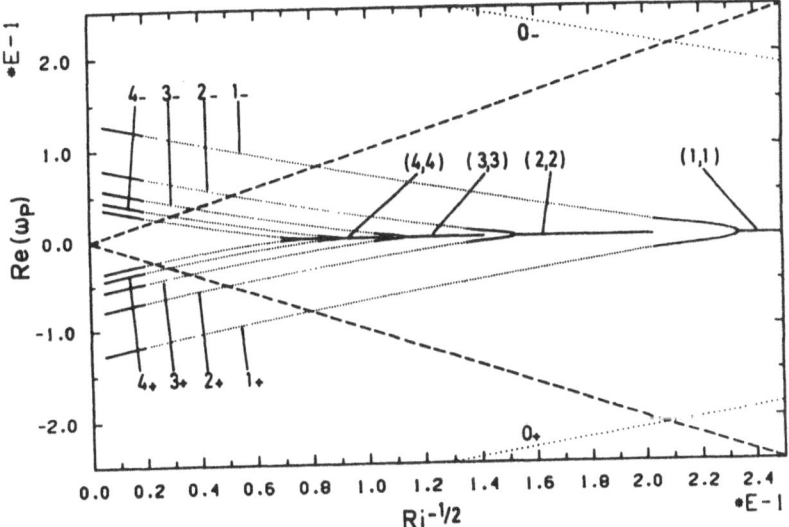

Figure 2: Same as Fig. 1 for modes with non vanishing imaginary part of ω_p and having a critical layer in the flow. Points outside the range confined by the dashed lines are shown for illustration and have no physical meaning. Note the mode resonances labelled (n, n), where n is the order parameter of the crossing modes.

The instability bands in Figs. 2 and 3 are confined by two Richardson numbers for which ω_p vanishes. For $\frac{\partial^2 \bar{V}_p}{\partial x^2} = 0$ this fact can be used to determine unstable and stable domains in a k_p - Ri - diagram (Fig. 4). A resonance instability band (n, n) is confined from above by the curve

$$k_p^2 = 2Ri + 2\frac{\partial \bar{V}_p}{\partial x} + 1 - \frac{\pi^2}{4}(2n + 1)^2 \quad ; \quad n = 0, 1, 2, \ldots \qquad (4.3)$$

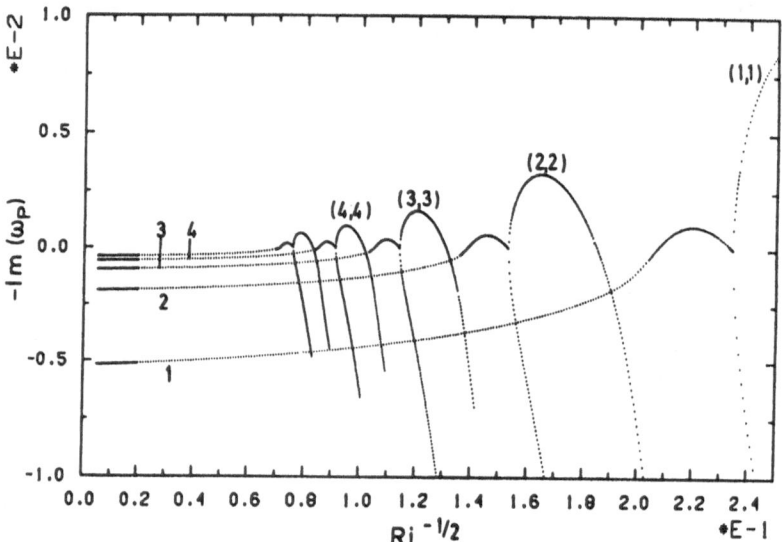

Figure 3: The imaginary part of the pattern speed $Im(\omega_p)$ for the modes shown in Fig. 2.

and from below by the curve

$$k_p^2 = 2Ri + 2\frac{\partial \bar{V}_p}{\partial x} + 1 - \frac{\pi^2}{4}(2n+2)^2 \quad ; \quad n = 0, 1, 2, ... \qquad (4.4)$$

The domains of resonance instability are indicated in Fig. 4 by hatched areas. We emphasize that regions between the different domains of resonance instability do not correspond to stability. A resonance instability (n, n) is followed by a domain of non resonant instability $(Re(\omega_p) \neq 0)$ of the two modes giving rise to the resonance (see Fig. 3), whose upper boundary is given by equation (4.4). For values of k_p above the curve given by equation (4.3) with $n = 0$ neither resonant nor non resonant instabilities exist. This domain corresponds to stability as indicated in Fig. 4.

For a plane shear layer the $(0, 0)$ resonance exists down to $Ri = 0$ and we find instabilities for any value of the Richardson number, whereas for rotating configurations the $(0, 0)$ resonance extends down to a finite value of Ri only, i.e. rotating flows with high shear rates can be stable with respect to gravity mode instabilities. With decreasing shear rate the number of unstable modes increases, while the maximum growth rate σ of the instabilities in physical units is proportional to the shear rate, indicating that the cause of the instabilities is the shear motion:

$$\sigma \approx 10^{-1}\bar{V}_0 \qquad (4.5)$$

5. Mechanisms for Instability.

The behaviour of gravity modes in a shear flow described in Section 4 indicates two different types of instability: Resonance instability for $Re(\omega_p) = 0$ which is caused

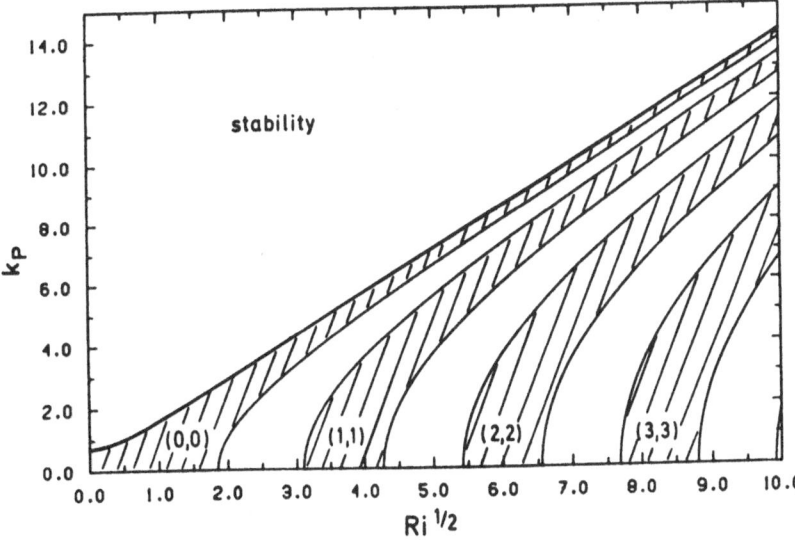

Figure 4: The relation between wavenumber k_p and Richardson number Ri for modes having zero pattern speed $\omega_p = 0$ in the plane linear shear layer. Domains of the (n, n) resonance instabilities are indicated by hatched areas which are followed by non resonant instabilities for higher Ri. No instability occurs in the region above the heavy line. The number of unstable modes increases with Ri and at least one unstable mode is found for any value of Ri.

by the interaction of two modes and non resonant instability for $Re(\omega_p) \neq 0$ caused by the interaction of a mode with the mean flow.

The non resonant instability is consistent with the over - reflection mechanism described by Lindzen & Barker (1985). Necessary for this mechanism to work are wave regions on either side of the critical layer and that the local Richardson number at the critical layer is smaller than 1/4. We note that the damping rate of a mode turned into a growth rate, if the local Richardson number at the critical layer was $\approx$ 1/4 (see Figs. 2 and 3). By considering a monotonic profile of the Brunt - Väisälä frequency, e.g. by imposing the boundary conditions at $x = 0$ and $x = +1$ or $x = -1$ instead of at $x = +1$ and $x = -1$, a wave region can exist only on one side of the critical layer. In this case no instability is found.

The instabilities for $Re(\omega_p) = 0$ can be explained by the resonant interaction of two modes whose energy has opposite sign. By exchange of energy the amplitude of both of them can grow, even if the total energy is to be conserved. Necessary for this kind of instability are mode crossings, which are caused by the shear. However, only the "symmetric" set of gravity modes is distorted in the appropriate way and provides mode crossings having a critical layer within the flow. By a monotonic profile of the Brunt - Väisälä frequency the "symmetric" modes are eliminated (see Section 4.1) and resonance instabilities do not exist in this case either.

REFERENCES.

Frank, J. & Robertson, J.A., 1988. *Mon. Not. R. astr. Soc.*, **232**, 1

Glatzel, W., 1989. in preparation

Hanawa, T., 1987. *Astron. Astrophys.*, **179**, 383

Lindzen, R.S. & Barker, J.W., 1985. *J. Fluid Mech.*, **151**, 189

Papaloizou, J.C.B. & Pringle, J.E., 1984. *Mon. Not. R. astr. Soc.*, **208**, 721

SIMULATIONS OF THREE-DIMENSIONAL SLENDER TORI.

John F. HAWLEY
Dept. of Astronomy, University of Virginia, Charlottesville, VA 22903, USA

ABSTRACT. Some considerations and conclusions that have been obtained from numerical simulations of slender, nonself-gravitating, fluid 'tori' orbiting in a fixed central potential. The numerical work on the nonaxisymmetric 'Papaloizou-Pringle' instability is summarized. The difficulties and pitfalls inherent in these time-dependent finite differencing calculations are reviewed. Preliminary three-dimensional simulations of slender torus modes are presented.

1. Introduction.

One of the long term goals in numerical research has been to develop and use codes capable of calculating three-dimensional (3D) simulations in sufficient resolution. The desirability of obtaining highly resolved, accurate three dimensional simulations for the investigation of black hole physics seems clear. However, to date only preliminary work has been done in this area. The main reason is that the computing resources were not adequate for detailed 3D simulations. Attempts were limited to short evolution times on grids employing crude resolution. Further, even if adequate hardware had been available, the results of a 3D simulation would be exceedingly complex; detailed 2D simulations are a necessary prerequisite.

The aim is to extract hard, usable information from a numerical calculation, and generally this means being quantitative. We must, however, give some sort of error bar with any numbers quoted, that is we must account for the uncertainty in the results due to the fact that it is a discrete numerical calculation. In addition to any dependence on the actual *physics* of the problem, a given result will depend upon (1) the fineness of the discretization (resolution in a finite-difference calculation, number of particles in an N-body calculation) (2) the algorithm used to represent the operator, i.e., the numerical scheme, (3) the initial conditions, (4) the techniques employed to sample and diagnose the results.

What we are really after is how the results depend upon the physics, i.e., the fundamental physical parameters of the problem. This is what researchers concentrate on in their simulations: how does the answer change with different adiabatic

F. Meyer et al. (eds.), Theory of Accretion Disks, 259–268.

index, initial energy, angular momentum, size, etc. The answers to such questions are the interesting ones. However, in the absence of information regarding solely numerical effects, the information on the physics is nearly worthless. In fact it can be worse than useless if unbeknownst to the researchers, or to the reader of their papers, numerical effects contribute more to the results than the physics does.

In this paper I describe numerical investigations of the nonaxisymmetric disk instability, commonly referred to as the Papaloizou-Pringle instability. The discovery of this instability has necessitated an evolution in descriptions of thick accretion disks. At the very least it serves as a reminder of the dangers of neglecting a spatial dimension in a problem! This instability is ideally suited for the development of a three dimensional (3D) hydrodynamics computer code. For slender tori, the instability can be described with two dimensional (2D) eigenmodes, yet the rate of growth will depend upon the full 3D hydrodynamics. Along the way to the full 3D simulation, I will use several 2D examples to show how numerical effects can be isolated and accounted for.

2. The Nonaxisymmetric Disk Instability.

The nonaxisymmetric disk instability is reviewed in these proceedings by Narayan. There are literally dozens of papers in the literature that deal with this instability; most use linear perturbation theory. Several attempts have been made to study these instabilities numerically. I have previously carried out simulations in the two dimensional limit, including slender tori orbiting black holes (Hawley 1987a), idealized slender tori (Hawley 1987b), and general relativistic 2D 'tori' orbiting black holes (Blaes and Hawley 1987). Zurek and Benz (1986) were the first to attempt a 3D simulation, using the technique of smooth particle hydrodynamics to consider a constant angular momentum torus orbiting a Newtonian point mass.

While several papers stand out for their insights and descriptions of the physics of the instabilities, I will refer to just one of them, the work of Goldreich, Goodman, and Narayan (1986, hereafter GGN), for this paper has the most direct bearing on the numerical work discussed below. In their study, GGN consider a slender torus (width of the torus a much less than the orbital radius r_o) in a 2D limit by assuming vertical hydrostatic equilibrium. The angular velocity profile through the torus is characterized by a simple power law $\Omega \sim r^{-q}$. A mode is characterized by the parameter $\beta = ma/r_o$. Slender 2D tori with a polytropic index n_{2D} are shown to be equivalent to 3D tori with polytropic index $n_{3D} = n_{2D} - 1/2$. The tori are described using quasi-Cartesian coordinates with $x = (r - r_o)$ and $y = r_o\phi$. This 2D system was studied numerically in Hawley (1987b). One of the results of these simulations is that the nonlinear modes transport significant angular momentum (as characterized by the evolution of the parameter q), and at mode saturation the slender tori form elliptical structures that have been named "planets". In this present paper, I will continue with simulations of the 2D slender torus and show

how that work can be used to 'bootstrap' 3D finite difference simulations of slender torus modes.

2.1. EIGENMODE INITIALIZATION.

An advantage of working with the idealized slender torus is that the linear perturbation equations have been worked out by GGN. Their solutions provide both mode growth rates and the perturbation eigenfunctions. The first series of simulations will examine the evolution of slender tori that are initialized with eigenmodes.

The questions to be examined are: (1) What mode growth rate does the code obtain? (2) At what amplitude does a mode saturate? (3) How effectively is angular momentum transported through the torus by the mode? (4) How do modes of different wavenumber couple, and how does mode energy cascade, to higher or lower wavelengths?

We begin with a constant angular momentum ($q = 2$) slender torus with a 2D adiabatic index of 1.5. The initial eigenmode corresponds to $\beta = 0.4$. This torus has been simulated on grids of 64^2, 96×64, 96^2, 128^2, and 192^2 zones in (x, y). The timestep is set by the courant condition with a courant factor of one half. In practice a vacuum zone outside the torus determines the timestep. The initial amplitude of the density perturbation varies throughout the torus, but is typically less than a percent. The density perturbation $\delta\rho$ is displayed in Figure 1, where one can see that the maximum perturbation is near the edge of the slender torus. Physically, this mode consists of a pair of waves on the inner and outer surface of the torus.

It is useful to monitor several quantities as a function of time. A measure of the power in a particular eigenmode is obtained by integrating over the angle

$$k(x) = \int_0^{2\pi} \rho(x, \phi) e^{im\phi} d\phi$$

and then averaging over the torus

$$\frac{1}{(x_{out} - x_{in})} \int_{x_{out}}^{x_{in}} ln(k^* k) dx$$

with the limits x_{out} and x_{in} chosen to correspond to the spatial location of the edge of the torus at the beginning of the calculation. In this set of calculations, the relative strengths in the $m = 1$, 2, and 3 modes are monitored. Mode growth rates are calculated from this information.

The angular momentum parameter q is obtained by averaging the total mass and angular momentum over angle for each x to produce an angle-averaged angular velocity as a function of x. A least-squares fit is then used to find q such that $\Omega = x^{-q}$.

The code used for this work has been briefly described elsewhere (Hawley 1986a). A more detailed description of the numerical techniques is given in Hawley Smarr, and Wilson (1984). In this paper I will be comparing results using two different advection techniques, one the formally second order "monotonic" advection scheme of van Leer (1977), and the other the simple first-order "donor cell" scheme. The later results are intended to demonstrate the effects of numerical diffusion.

TABLE 1

Eigenmode Initialization with "Mono" Scheme

Grid	$m = 1$	$m = 2$	$m = 3$	t_{sat}	q_{sat}	q_{min}
64^2	0.2601	0.4923	0.5638	3.00	1.759	1.743
96×64	0.2621	0.5100	0.6820	2.98	1.750	1.726
96^2	0.2627	0.5128	0.6873	2.98	1.750	1.727
128^2	0.2630	0.5298	0.7683	2.97	1.752	1.719
192^2	0.2617	0.5269	0.7868	2.97	1.749	1.716

The the slender torus simulation results are in Table 1. The columns labeled by mode number give the observed mode growth rate in units of the angular velocity Ω^{-1}. The $m = 1$ mode growth is averaged over the first orbit, while the $m = 2$ and $m = 3$ rates are taken between 1.6 and 2.2 orbits. These times were chosen as intervals over which the modes were undergoing relatively steady exponential growth. The growth rates are obtained by a least squares fit to time-sampled mode amplitudes. Between 50 and 100 points are used. While the formal error of the least squares fit is quite low, growth rates calculated at specific times by means of centered differencing of the data show variations on the order of one to six percent even for the highest resolutions. The linear perturbation growth rate of 0.2627 is within the estimated error from the numerical calculation. The mode amplitude curves for these runs are shown in Figure 1. The lowest resolution simulation shows larger variations in mode amplitude during the exponential growth phase, and saturates at a reduced amplitude. However, the curves show systematic convergence.

The larger variation with respect to resolution for the growth rates of the $m = 2$ and $m = 3$ mode is not surprising. Shorter wavelengths will naturally require greater grid resolution. In the results of Blaes and Hawley (1987) it was found that unstable acoustic modes (as opposed to the unstable surface wave modes) can have several wavelengths within the outer part of the disk, requiring considerable radial resolution.

The error bars for the growth rates can be estimated from the results in Table One, and from the formal errors in the least square fit. These errors are certainly nonzero, but they appear to be independent of the grid resolution at the highest resolutions used. Obviously this need not have been the case, and the resolution required for any given problem will vary. As an instructive example, consider a

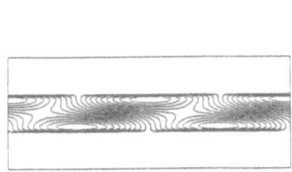
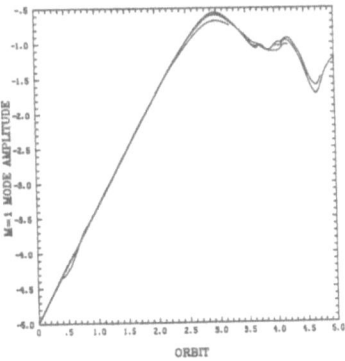

Figure 1: Density perturbation $\delta\rho$ in the slender torus problem, and time history of mode growth for four grid resolutions, 64^2, 96^2, 128^2, and 196^2.

case that is *not* well resolved. For this we replace the piecewise-linear monotonic transport scheme used in the code with the simple first-order upwind "donor cell" scheme. No other change is made. The results from a series of simulations on this same slender torus problem are given in Table 2.

TABLE 2

Eigenmode Initialization with "Donor Cell" Scheme

Grid	$m = 1$	t_{sat}	q_{sat}
64^2	0.1113	1.21	1.952
96×64	0.1962	3.57	1.824
128^2	0.2230	3.50	1.865
192^2	0.2410	3.39	1.866

In many numerical studies the emphasis is upon the bulk behavior of a fluid. In an instability simulation we are concerned with the behavior of (initially) low amplitude waves in a given system. This makes an instability calculation particularly susceptible to diffusive errors. There are several hallmarks of numerical diffusion, as illustrated by the donor cell simulations. Diffusion tends to decrease or damp out the growth of unstable modes (diffusion has a stablizing effect). The amplitude at which a mode saturates can be reduced substantially from its "true" value. Figure 2 is a plot of the evolution of the q parameter for the various resolution donor cell runs. For comparison, the 64^2 and 196^2 mono scheme results are shown as dashed lines. It is obvious from this plot that diffusion alone can transport angular momentum quite effectively. This is purely a numerical effect and dependent on resolution. There is evidence of slow convergence (at a rate *less* than first order) towards the q curve obtained with the highest resolution mono scheme simulation. It would be difficult, if not impossible, to separate out this physical momentum transport from the numerical in one given simulation. However, numerical effects depend on numerical details while the physics does not. Multiple resolution simulations can therefore distinguish between what is numerical and what is physical. While it may be that at high enough resolution the donor cell scheme will converge to a more nearly grid independent result, Figure 2 clearly demonstrates that 192^2 zones are not sufficient.

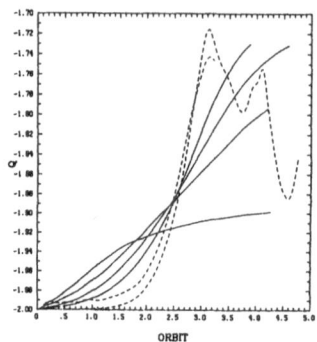

Figure 2: Evolution of the q parameter for first-order "donor cell" simulations of resolutions 64^2, 96×64, 128^2, and 192^2. The dashed lines, included for comparison, show the q evolution for the 64^2 and 192^2 "mono scheme" simulations.

Obviously numerical schemes have limitations; the donor cell scheme has especially significant limitations. The point is we are not at the mercy of the numerical scheme we employ. One simulation tells us nothing; it is impossible to distinguish between physics and numerics. But, as shown with these examples, a set of convergence tests quickly reveals when the answers one obtains are grossly dependent upon numerical resolution. Numerical effects depend on numerical details while the physics does not. In cases like that one must either increase the resolution or improve the scheme. Resolution is not the only numerical variable that should be tested. For example, if there is a parameterized function, e.g., a parameterized viscosity, then that too should be varied, and its numerical dependencies ferreted out.

2.2 RANDOM INITIALIZATION.

In general we do not have available a linear perturbation analysis to provide eigenmodes as initial conditions. Rather, we must chart the rather uncertain waters of stability analysis by means of direct simulation. For the slender tori we will ignore the fact that we have a linear analysis, and initialize with noise.

The procedure is to add enthalpy perturbations of the same magnitude as in the eigenmode initialization, except that in each grid zone there is an additional random scale factor in the range $(+1/2, -1/2)$. There will be power in all wavelengths down to $2\Delta x$ regardless of the resolution of the simulation. Thus the initial conditions will be dependent upon resolution; what is meant by "convergence" in such a situation becomes more problematic. However, this type of study should provide clues as to the nonlinear coupling between modes.

Some data for these models is summarized in Table 3. The $m = 1$ growth rates are obtained by considering the mode growth for one orbit after the exponential growth phase is well established. This occurs at different times for the different resolutions. For the 64^2 and 96^2 grids the growth rate is measured between orbits 3 and 4. For the higher resolutions mode growth sets in about half an orbit earlier. As can be seen from the table, the $m = 1$ growth rates are lower than those obtained in the eigenmode initializations: two percent lower for the highest resolution case. The fractional variation in slope is larger in the random initialization than in the eigenmode case, varying by as much as twenty percent. The systematic trend, is to a value below that of the linear perturbation theory.

The time history of the $m = 2$ mode is interesting. The mode grows at a rather low rate until orbit 4.5 at which point it grows at a more rapid value until saturation. The $m = 2$ growth rate listed in the table correspond to this second, larger rate. Presumably this enhanced growth results from nonlinear mode-mode coupling. Further analysis will be required to understand this behavior.

TABLE 3

Random Initialization with "Mono" Scheme

Grid	$m = 1$	$m = 2$	t_{sat}	q_{sat}	q_{min}
64^2	0.2610	0.4406	5.58	1.744	1.735
96^2	0.2502	0.4612	5.49	1.742	1.722
128^2	0.2562	0.3788	5.14	1.737	1.712
192^2	0.2552	0.3061	5.23	1.749	1.719

3. Three-Dimensional Simulations.

The ultimate goal for these numerical simulations is to carry out the disk instability studies in the full three dimensions. Preliminary 3D computations have now been done for the slender torus. As before, consider an idealized "torus" in Cartesian geometry (x, y, z). Assume periodic boundary conditions on y and equatorial reflection boundary conditions for z. The 3D domain is the upper half of the slender torus along one wavelength.

The problem considered is a $\gamma = 5/3$, constant angular momentum torus, with $\beta = 0.4$. As for the 2D case, the torus is initialized with an eigenmode obtained from the linear perturbation theory of GGN. There is no z dependence in the perturbation. For this mode, the linear growth rate is $0.2490\Omega^{-1}$. The simple relationship between 2D and 3D polytropic indices means that the equivalent 2D problem is a $\gamma = 1.5$ torus.

First we consider the 2D case. A simulation with a grid of 96×64 zones produces a growth rate of 0.2482 over one orbit. An $m = 2$ mode becomes organized by the end of one orbit, and begins an exponential growth phase with growth rate of 0.4694. Mode growth, angular momentum transport, and planet formation all follow the now familiar pattern.

As an additional test, consider a 2D torus initialized with this same mode, but run with the 3D equation of state, $\gamma = 5/3$. The result of this experiment is that within one orbit the mode initially imposed upon the grid dramatically changes its structure, and the growth rate over the first orbit is $0.2609\Omega^{-1}$. This supports the claim that the code is indeed calculating the proper eigenmodes; despite an incorrect $m = 1$ mode initialization, growth proceeds at the rate appropriate to the equation of state!

This last experiment is important as a test for the 3D simulations, for it helps to insure that three dimensional effects are actually playing a role in the results, and that the code is not simply calculating a stack of 2D planes. A correct simulation will distinguish between the two- and three-dimensional equation of state.

For this preliminary set of experiments, the simulations are limited to two grid resolutions: $96 \times 64 \times 32$ in (x, y, z), with $dx = dz$, and $96 \times 64 \times 64$ with $dx = 2dz$. Only the higher resolution simulation was run to and beyond mode saturation. Execution time on the Cray 2 is approximately 10^5 grid zones per cpu second, and about 1000 timesteps are required per orbit of evolutionary time. Some of the results are summarized in Table 4.

TABLE 4

3D Torus Simulations

Grid	$m = 1$	$m = 2$	t_{sat}	q_{sat}	q_{min}
96×64	0.2482	0.4694	3.02	1.761	1.730
$96 \times 64 \times 32$	0.2574	0.2901	—	—	—
$96 \times 64 \times 64$	0.2535	0.4646	3.01	1.769	1.718

Figure 3 is a series of isodensity plots for the torus at points in its evolution. The results are similar to those obtained in 2D. As the mode grows the torus begins to assume an 'S' shape. At mode saturation, the bulk of the gas is concentrated in a 'planet', the name given to the ellipsoidal structures that were observed in the 2D simulations (Hawley 1987a,b). A pair of shocks form that begin to move through the planets in a manner that is qualitatively like the 2D result. A density contour slice through the planet (Fig. 4) shows the location of the shock, as well as the degree to which the torus has been flattened by angular momentum redistribution.

There are an inadequate number of simulations and experiments at this time to claim convergence for the 3D results, except by analogy to the 2D cases. However, some general preliminary conclusions can be reached. First, the 2D simulations provide a quite adequate description of the full 3D torus evolution, at least as

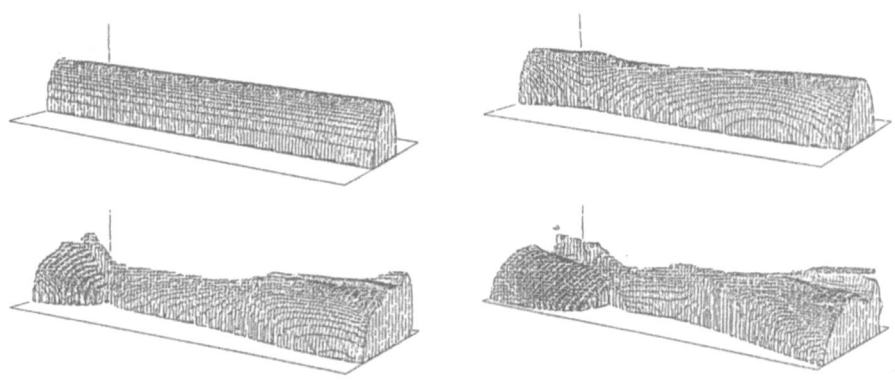

Figure 3: Three-dimensional isodensity plots for the slender torus evolution. The times are $t = 0$, 2.53, 3.26, and 3.72 orbits.

far as shock formation. While it was always likely that planet formation was the correct endpoint of the mode growth phase, there was a chance that planets were an artifact of the restriction to two dimensions. The 3D simulations remove any lingering doubt.

As a test bed for 3D code development, this problem is outstanding. It is a fully nonlinear, fully three-dimensional system for which nearly analytic solutions exist, specifically, the growth rates and eigenmodes of the nonaxisymmetric instability. In these first simulations, the growth rates obtained for the 3D torus are consistent with the GGN theory, and the polytropic relationship between two- and three-dimensional evolutions has been reproduced. Angular momentum evolution, as measured by the evolution of the q parameter, is quite similar for both the 2D and 3D cases.

For the unstable slender torus, further 3D simulations will provide more details on the evolution of the planets. As has been clear from the 2D simulations, they are formed far out of equilibrium. Since the shocks that form will generate vertical entropy gradients, planet evolution is truly a 3D problem. Perhaps the most important result from the 3D simulation is the confirmation that the much simpler 2D system provides an adequate description of the physics of the unstable modes in slender tori.

To conclude, the availability of large memory supercomputers such as the Cray 2 makes possible reasonably well resolved three dimensional simulations. This means that many questions not previously amenable to numerical investigation, can now be investigated. However, even the simplest 3D simulations will require a rather substantial investment of computer resources. This will be especially true if it is recognized that a single simulation on its own cannot provide useful information except about the coarsest features. Any detailed numbers extracted from the simulation will require some form of convergence tests.

268

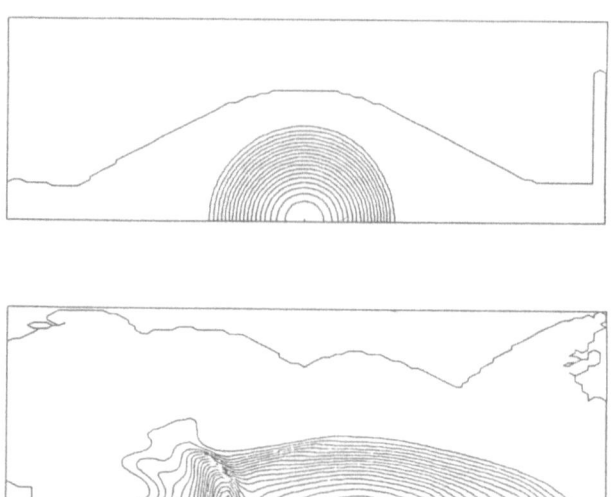

Figure 4: Density contours in cross-section of the torus at times $t = 0$ and 3.72 orbits. The slice is taken through one of the 'planets'. At left is a shock wave that has begun to move radially through the torus.

ACKNOWLEDGEMENTS. This work was supported in part by the NSF, grant PHY88-02747, and utilized the Cray 2 and Cray XMP at the National Center for Supercomputing Applications at the University of Illinois. A portion of this work was carried out while the author was a participant in the program on Computational Fluid Dynamics at the Institute for Theoretical Physics, Santa Barbara.

REFERENCES.

Blaes, O. M. and J. F. Hawley, *Astrophys. J.*, **326,** 277 (1987)

Goldreich, P., J. Goodman, and R. Narayan, *Monthly Notices Royal Astron. Soc.* **221,**339 (1986)

Hawley, J. F., *Monthly Notices Royal Astron. Soc.* **225,**677 (1987)

Hawley, J. F., in *Proceedings of the George Mason Conference on Supermassive Black Holes,* ed. M. Kafatos (Cambridge: Cambridge Press) 325 (1988)

Hawley, J. F., L. L. Smarr, and J. R. Wilson, *Astrophys. J. Supp.*, **55,**211 (1984)

van Leer, B., *J. Comp. Phys.*, **23,** 276 (1977)

Zurek, W. H., and W. Benz, *Astrophys. J.*, **308,**123 (1986)

ACCRETION DISKS IN LOW MASS X-RAY BINARIES.

Nicholas E. WHITE
EXOSAT Observatory, Astrophysics Division,
Space Science Department of ESA, ESTEC,
2200 AG Noordwijk, The Netherlands.

ABSTRACT. The observed properties of accretion disks in LMXRBs (low mass X-ray binaries) are reviewed, at the outside where the disk interacts with the gas stream from the companion, and in the inner region close to the compact object. X-ray orbital modulations from the low mass X-ray binaries, LMXRBs, include eclipses by the companion and/or periodic dipping behaviour from structure at the edge of the disk. The dipping behaviour gives a direct probe into the gas stream/disk interaction. In some high inclination systems the central X-ray emitting compact object is shadowed by the rim of the disk, but X-rays are scattered to the observer via an ADC (accretion disk corona). A sinusoidal-like orbital modulation can be modelled by structure at the disk rim and places tight constraints on how the gas stream merges into the disk. Moving into the central emission region, the observations of the spectra of the LMXRB can be tested against models for the spectra from accretion disks and the interaction with a neutron star. Blackbody disk models do not give a good fit to the spectra. Comptonisation models seem to better represent the data, but currently are rather idealized.

1. Introduction.

Low mass X-ray binaries are neutron stars (or in a few case stellar mass black holes) accreting material from a late type binary companion star. The energy source is the conversion of the gravitational potential of the matter into radiative emission, in the inner regions of the accretion disk and, when a neutron star is present, in a boundary layer between the disk and the solid object. The X-ray heated accretion disk and companion star dominate the optical light and most LMXRB are identified with faint blue stars. For a long time the study of the LMXRB was hampered by the fact that few showed X-ray pulsations or orbital modulations. X-ray bursts, believed to be the result of thermonuclear flashes in the accreted material, were the only direct indication of the presence of neutron stars. But these give little insight into the binary system itself. In the last 5 years this situation has changed radically.

The search for orbital modulations benefitted from the continuous coverage of up to 76 h provided by the deep orbit of EXOSAT. At the same time optical

F. Meyer et al. (eds.), Theory of Accretion Disks, 269–282.

CCD photometry of the faint blue optical counterparts has also revealed orbital modulations in a number of objects. A total of 25 LMXRB now have well established orbital periods (Parmar and White 1988). These range from 11 min up to 9 days. The vast majority have periods between 2.9 and 9 h and contain main sequence stars. There is a small group with periods between 11 min and 50 min which contain degenerate, or close to degenerate companions. A few have been found with periods of between 1 and 9 days, these contain late type giants.

The discovery of quasi-periodic oscillations, QPO, from many of the bright galactic center LMXRB by EXOSAT (beginning with van der Klis *et.al.* 1985) has provided the first temporal diagnostic into the inner emission regions in these objects. This subject has blossomed into a major industry and it is not possible to include it here. The reader is referred e.g. to the review by Lewin, van Paradijs and van der Klis (1988) for further information. Instead this review deals with the equally important problem of understanding the nature of the X-ray spectra from LMXRB.

2. X-ray orbital modulations.

Total X-ray eclipses are rare from LMXRB. This became apparent in the late 1970s (Joss and Rappaport 1979) and prompted Milgrom (1978) to suggest that the companion star in each system is in the shadow of a thick accretion disk surrounding the X-ray source. This picture has proved to be basically correct, but with two unexpected features.

First, the interaction between the disk and the gas stream causes a turbulent thickening at that point. In the higher inclination systems material thrown well above the disk plane causes absorption dips, that predominantly occur when the confluence of the gas stream and disk passes through the line of sight to the X-ray source. These periodic absorption dips were first discovered by White and Swank (1982) and Walter *et.al.* (1982) from XB1916-05 using various instruments on the *Einstein* observatory. Since then many more examples of this phenomenon have been found, mostly with the EXOSAT observatory.

Second, the highest inclination systems where the disk completely blocks our view of the central source at all orbital phases can still be observed because an accretion disk corona, ADC, evaporated by X-ray heating, allows X-rays to be scattered into the line of sight. For higher luminosity sources, close to the Eddington limit, the ADC may also be optically thick when viewed at higher inclinations and also occult the central X-ray source before the disk rim gets in the way. The X-ray source appears extended with dimensions comparable to the size of the accretion disk and a partial X-ray eclipse is seen. The first of these sources to be found was X1822-37 by White *et.al.* (1981).

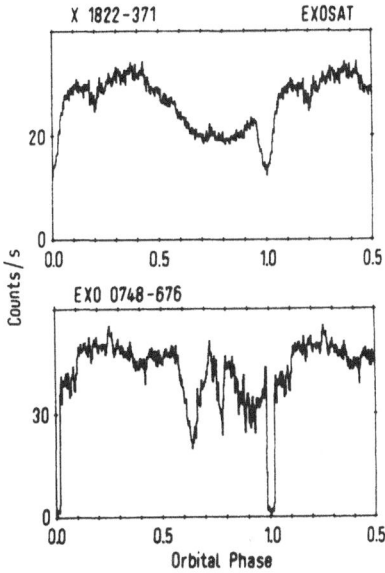

Figure 1: The orbital lightcurves of the partially eclipsing LMXRB X1822-37 and the totally eclipsing LMXRB XBT0748-676 (EXO0748-676). Both are folded at the orbital period from observations that lasted several orbital cycles. Taken from Parmar *et.al.* (1986).

2.1. PARTIAL AND TOTAL ECLIPSES.

The lightcurve of X1822-37 (Fig. 1, taken from data presented in Hellier and Mason 1989) displays a partial eclipse (at phase 0) and a sinusoidal modulation with a minimum that precedes the eclipse (White *et.al.* 1981). The partial eclipse arises because X-rays from the point source are scattered in an ADC. The sinusoidal modulation in the X-ray lightcurve results from obscuration by a thickened region at the accretion disk edge where the gas stream impacts. Detailed modelling of this by White and Holt (1982) gave the size of the X-ray scattering cloud and the profile of the structure at the disk edge. An important outcome of this modelling is that the radius of the central extended X-ray source is about half that of the outer radius of the disk where most of the structure is located. White and Holt (1982) conclude that the ADC is optically thick and this, not the accretion disk rim, hides the central X-ray source. The outer disk radius is similar to that obtained from modelling the optical eclipse (Mason and Cordova 1982; see below). Because the observed X-ray emission is scattered, the observed luminosity is only a few percent of the central, unobscured X-ray source.

There are two other systems where partial eclipses have been discovered: XB2129+47 (White and Holt 1982; McClintock *et.al.* 1982) and X0921-63 (Mason *et.al.* 1987). Another object where the compact X-ray source is probably hidden from direct view by an ADC or the disk rim is Cyg X-3 (White and Holt 1982). The energy dependence and shape of the sinusoidal-like modulation can be explained as the occultation of an ADC by structure at the edge of a disk, giving a sinusoidal-like modulation similar to that of X1822-37 (White and Holt 1982; Molnar and Mauche 1986).

Not all eclipses from LMXRB are partial. Cominsky and Wood (1984, 1989) find a 15 min total eclipse every 7.1 h from the transient X-ray burst source XBT 1658-298, as well as considerable dipping activity for one quarter of an orbital cycle preceeding the eclipse. The quality of this discovery suffered somewhat from poor coverage. A much cleaner example is the transient source XBT0748-676 discovered by EXOSAT (Parmar *et.al.* 1986). The lightcurve of this source folded on the 3.82 h orbital period is compared with that of X1822-37 in Fig. 1. An eight minute eclipse is seen along with considerable dipping activity at the phases preceeding it. The comparison of this with X1822-37 is quite striking (Fig. 1). The dipping activity in XBT0748-676 corresponds nicely to the minimum in the sinusoidal modulation in X1822-37. The inclination of XBT0748-676 must be a few degrees lower, such that the line of sight to the X-ray source just grazes the top of the companion and also the rim of the disk.

2.2. THE DIPPERS.

EXOSAT discovered many examples of LMXRB that show periodic dipping activity, but no eclipses (see Parmar and White 1988 for a recent overview). An example is shown in Fig. 2. This is the *big dipper* X1624-490 (Watson *et.al.* 1985) with a period of 21 h. This is the longest of the dip periods and each dip interval can last up to 8 h. These sources are similar to XBT0748-676, but viewed at lower inclinations ($\sim$ 70–80°) so that the companion does not eclipse the X-ray source. The morphology of the dips is quite varied. Sometimes they become barely detectable and then reappear strongly within one cycle (see e.g. Smale *et.al.* 1988). Also on occasion anomalous dips can occur 180° out of phase with the main events.

In all but one case the dips are energy dependent and are associated with an increase in absorption. However, the spectra do not fit a simple cold absorber model. In some cases there is an excess at low energies (e.g. Parmar *et.al.* 1986), in others the dips seem to show an energy independent reduction, in addition to the decrease at low energies caused by photoelectric absorption. The most extreme example of the latter is X1755-33 where the dips are completely energy independent (White *et.al.* 1984), although their duty cycle and structure (but not their depth) appear similar to those of other dippers. These effects have been variously attributed to lower abundances of the medium z-elements (White and Swank 1982), scattered emission in an ADC (e.g. Parmar *et.al.* 1986), partial ionization of the absorbing medium (Frank, King and Lasota 1987) and rapid (relative to the accumulation time of the spectral data) variations in absorption (Swank and Inoue 1989). The

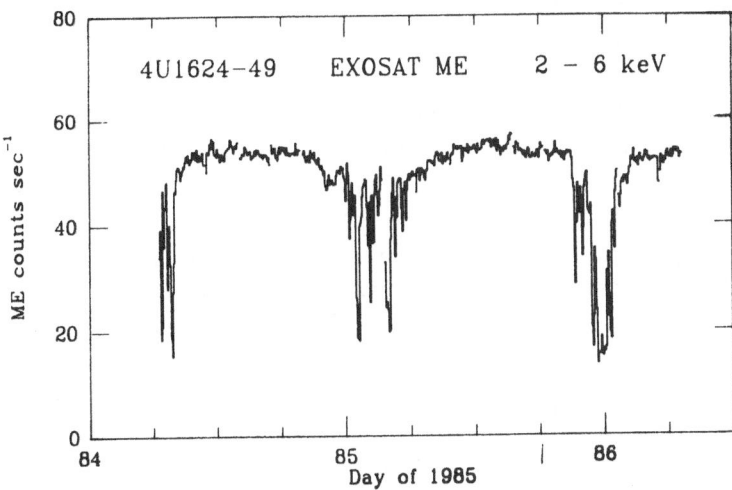

Figure 2: The lightcurve of the big dipper, X1624-490. The 21 h dip period is evident. Taken from Watson *et.al.* (1985).

relative importance of these effects seems to vary from source to source. While it seems likely that all will be present at some level, we must await the better spectral resolution and sensitivity given by the upcoming missions (e.g. Astro D), where the ionization state and abundance can be directly measured from the edge energies and depth.

2.3. OPTICAL MODULATIONS.

All of the optical counterparts to the dipping/eclipsing LMXRB show a corresponding orbital modulation. Two modulation components have been identified. In the X-ray eclipsing systems a sharp optical minimum is seen at the time of eclipse and a second smooth lower amplitude modulation is also evident, with a minimum that leads the eclipse (e.g. Schmidtke and Cowley 1987). These two features were first noted by Mason *et.al.* (1980), from observations of the lightcurve of X1822-37, and modelled in detail by Mason and Cordova (1982), using the profile obtained for the outer disk rim given by the X-ray modelling. This shows that the optical light comes from (i) the inner X-ray illuminated rim, (ii) the X-ray illuminated inner disk (iii) the outer disk rim and (iv) the X-ray heated companion star. The varying aspect of all four gives the overall modulation whereas the eclipse of (ii) and (iii) by the companion gives the sharp eclipse. A more recent paper by Hellier and Mason (1989) reconsiders the earlier lightcurve modelling using more recent EXOSAT data which gives better orbital phase coverage plus contemporaneous optical data. They simultaneously fit the optical and X-ray lightcurves. Essentially they find good agreement with the earlier work. One difference is that they do not formally require any contribution from the companion star to the optical modulation, although it may contribute up to 30% of the optical light.

In the X-ray dipping systems the optical lightcurves do not show a sharp eclipse (e.g. Motch *et.al.* 1987). Two of the dipping sources XB1254-69 and X1755-33 were monitored by EXOSAT simultaneously with optical observations to establish the phase of the dips with respect to the optical modulation (Motch *et.al.* 1987; Mason, Parmar and White 1985). The results for both were very similar and show that the dips occur 0.1-0.2 in phase before the optical minimum (Fig. 3). Motch *et.al.* (1987) consider that the optical modulation can arise either from the varying aspect of the X-ray heated poles of the companion (the equator is shadowed by the disk) and/or by the varying aspect of the X-ray illuminated bulge. A partial eclipse of the accretion disk is not likely to be a strong contributor since it would result in much narrower minimum than observed. The modulation from the X-ray heated secondary predicts the observed phase lag of the optical minimum with respect to the dips, whereas the heated disk bulge does not. Motch *et.al.* (1987) did not consider the possible contribution from the outer disk rim found to be important by Mason and Cordova (1982) and Hellier and Mason (1989). There is a need to more carefully reconcile the work on the eclipsing and dipping systems.

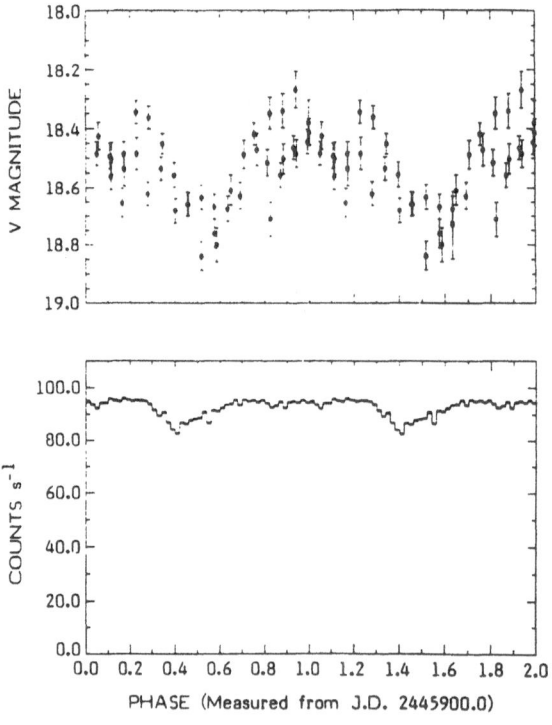

Figure 3: The X-ray and optical lightcurves of the dipping source X1755-33 taken from Mason, Parmar and White (1985). The data has been folded at the 4.4-hr orbital period. Note that the optical minimum leads the X-ray by ~0.15 in phase.

There are many LMXRB that only show optical modulations with no dips or eclipses in the X-ray lightcurve (e.g. XB1636-536; Pedersen, van Paradijs and Lewin 1981; Smale and Mukai 1988). These are low inclination systems where the turbulence at the disk edge does not pass through the line of sight.

2.4. THE THEORY OF THE OUTER DISK STRUCTURE.

A turbulent region in the disk at the confluence of the gas stream was first discussed in detail by Lubow and Shu (1975, 1976). They pointed out that the thickness of the gas stream will exceed the hydrostatic disk thickness at the confluence by a factor 3-4. Part of the stream may sweep over the disk, leaving behind a turbulent wake, which may be responsible for the periodic dips. The material in the stream would prefer to orbit the compact object at an *inner ring* corresponding to the specific angular momentum at the inner Lagrangian point. Accretion will take place only if angular momentum can be lost, or transferred back to the binary orbit. This will cause the disk radius to increase close to its tidal limit (where the orbits start to intersect). Lubow and Shu (1975,1976) used the *inner ring*, not the tidal radius. Frank, King and Lasota (1987) considered this point and found for a disk radius close to the tidal limit the stream may still be thicker than the disk, by up to a factor of 3. It is worth noting that the calculations of the relative height of the gas stream and disk did not include the effects of X-ray heating or turbulence on the disk, both of which could considerably swell its thickness. Nonetheless they suggest that there may be two impact points, one at the outer edge and a second closer in where the stream interacts with the inner ring. Because the second impact occurs much closer to the X-ray source the material may form a two phase medium with cold clouds in a hot photo-ionized plasma (similar to that invoked in AGN's), so providing an explanation for the clumpiness of the medium inferred from the erratic behaviour of the dips and the spectral absorption anomalies.

While parts of the Frank-King-Lasota model are attractive, it is not consistent with the modelling of the X1822-37 lightcurve. This requires that the majority of the disk structure be confined to the outer rim, not the inner ring; putting it much closer in would result in a very different lightcurve (Hellier and Mason 1989; White and Holt 1982). But the radius found for the ADC in X1822-37 is a factor of two smaller than the disk radius and comparable to the expected inner ring radius. It may be that the X-ray source is hidden behind this ring, rather than an optically thick ADC as proposed by White and Holt (1982). Nonetheless the X1822-37 lightcurve requires the bulk of the material responsible for the dipping activity to be located near the disk rim, where a two phase medium will not form.

2.5. THE UNUSUAL CASE OF XB1916-05.

The optical identification of the 50 min periodic dip source XB1916-05 object had a troubled history. The original *Einstein* HRI uncertainty region contained a relatively bright G star, that showed no evidence for the presence of an X-ray source companion, and a faint 22 mag star severely blended with the G star that was suggested as the counterpart (Walter *et.al.* 1982). Recently Grindlay and Cohn (1987) showed that the HRI position used by Walter *et.al.* (1982) is incorrect because of coma in the X-ray image (the source was close to the edge of the field of view) and that within the new uncertainty region they had identified the optical counterpart, a 21st magnitude star. The identification is beyond doubt since Grindlay and Cohn (1987) show that the star is modulated with the binary period.

A surprize is that the optical period is 1% longer than the X-ray dip period (Grindlay *et.al.* 1988). This initially did not seem to be a major concern because cycle to cycle variations in the dip morphology make it difficult to obtain an accurate X-ray period, with various values found in the range 49.70 to 50.06 mins (Smale *et.al.* 1989). However, the current optical period of 50.4567±0.007 is well outside the range of observed jitter in the X-ray period and the difference is hard to reconcile as an observational effect (see also Schmidtke 1988).

Grindlay *et.al.* (1988) chose the optical period as the orbital period and go on to suggest the 1% shorter X-ray dip period arises because XB1916-05 is a hierarchical triple. In a triple system the separation of a binary will undergo a small modulation which in turn may modulate the mass transfer rate (Bailyn 1987). To explain the 1% shorter X-ray period requires a retrograde orbital period for the third star of 2.5 day. To avoid seeing the orbital period in the X-ray dip period, Grindlay *et.al.* (1988) require that the mass transfer rate be very sharply pulsed at the shorter period. However, the X-ray lightcurve shows dipping activity for a considerable fraction of the orbit suggesting that the mass transfer is a continuous process.

The recurrence period of the dips has been clearly demonstrated to be equal to the underlying orbital period in XBT0748-676 (Parmar *et.al.* 1986) and XBT 1658-298 (Cominsky and Wood 1984) where the eclipse by the companion star provides an unambiguous fiducial marker. This is not surprising since the dip period reflects the orbital motion of the stream and its impact point on the disk, both of which are fixed to the binary reference frame. In addition in both XB1254-69 (Motch *et.al.* 1987) and X1755-33 (Mason, Parmar and White 1985) it has been equally well demonstrated that the optical modulation and X-ray dips are phase locked.

It seems more likely that the 50.0 min X-ray dip period represents the orbital period and that the 50.4 min optical modulation arises in some other manner. An obvious starting point is the superhump phenomenon found in the related SU UMa dwarf novae cataclysmic variables (CVs). The superhump period is always a few percent longer than the orbital period measured from eclipses. Amongst the CVs only SU UMa stars, which all have orbital periods of between 80 and 120 min, show superhumps. This is notable because XB1916-05 is the only dipping source with an orbital period less than 120 min. The most promising model for superhumps is that discussed by Whitehurst (1988). He finds that for the shorter orbital period systems where $M_c/M_x < 0.15$ (which includes XB1916-05), the accretion disk is tidally unstable and that this causes it to become asymmetrical in shape, with the axis of the asymmetry rotating slightly faster than the orbital period. Each superhump (maximum in the lightcurve) occurs at the time of maximum tidal stress. This is unlikely to work in XB1916-05 where the optical light is dominated by reprocessed X-ray emission. Another possibility is that the optical modulation reflects variations in the projected disk surface area around one disk rotation. This is different from the explanation for the modulation from the dippers XB1254-69 and X1755-33, where it has been attributed to the varying aspect of the X-ray heated polar regions of the companion (Motch *et.al.* 1987), but it might be consistent with the disk model used by Mason and Cordova (1982) and Hellier and Mason (1989). One further advantage of the tidally distorted disk model is that the cyclic variation in the position of the gas stream/disk impact point, as the eccentric disk rotates, would provide a natural explanation for both the small jitter in the measured X-ray period and the extreme variations in the dip morphology.

3. The Spectra of LMXRB.

While considerable effort was made early on towards understanding the spectral properties of X-ray pulsars and the black hole candidate Cyg X-1, attempts to understand the spectra of the non-pulsing LMXRB came relatively late. Only in the past five years has there been any serious attempt to model their spectra. Early observations of Sco X-1, the prototype LMXRB, had suggested that its entire spectrum from the infra-red to the hard X-ray could be modelled as simple thermal bremsstrahlung from a plasma cloud with an optical depth of a few (e.g. Chodil et al 1969). However, higher quality spectra obtained in the late 1970s showed the spectrum of Sco X-1 and several other LMXRB to require at least two components to adequately model the spectrum.

The late 1970s saw the acceptance that the LMXRB contain weakly magnetised neutron stars surrounded by an accretion disk. It seemed likely that the softer spectrum of the LMXRB compared to X-ray pulsars somehow reflected the different geometry of the accretion flow. The first serious attempt to model the spectra of the luminous (10_{38} erg s$_{-1}$) LMXRB in terms of accretion disk spectra was by Mitsuda et al (1984) who fitted the sum of blackbodies from an accretion disk plus another blackbody from a boundary layer between the disk and the neutron star. The fit to the *Tenma* GSPC spectra was reasonably good. Around the same time White, Peacock and Taylor (1985) and White et al (1986) found a similar "two component" model, but disagreed on the details; they found a better fit to the EXOSAT GSPC spectra if the "disk" emission comes from the Comptonisation of cool photons on hot electrons (similar to the Cyg X-1 Comptonisation model).

The spectra of the less luminous (10_{37} erg s$_{-1}$) LMXRB, mostly the ones that show X-ray bursts, were found by White et al (1986) to be much simpler. These have "harder" spectra than found from the higher luminosity systems and are well modelled as simple powerlaw spectra with exponential high energy cutoffs. This spectral shape could also be understood in terms of either a Comptonised spectrum (White et al 1986), or a sum of blackbodies from the disk and neutron star (Mitsuda et al 1989).

These studies clearly established the presence of two spectral components from the most luminous LMXRB and that one component is a highly variable blackbody associated with the neutron star. The form of the other "disk" component is still not agreed and has been the subject of much debate. The most recent attempt to resolve this disagreement was made by White, Stella and Parmar (1988) who fitted a variety of different accretion disk models to the spectra of a sample of LMXRB and compare the results with similar fits to the Comptonised model. The results from these papers are discussed in the following sections.

3.1. ACCRETION DISK X-RAY SPECTRA.

The main uncertainty in applying accretion disk models to the spectra of LMXRB is the nature of the viscosity in the inner radiation pressure dominated disk, where the X-rays come from. In the standard Shakura-Sunyaev prescription the viscosity uncertainties are hidden by scaling the viscous stress with the disk pressure. The problem is that it is not known if the viscosity in the radiation pressure dominated region scales with the radiation pressure, or only the gas pressure. If it is the former then the disk is unstable to thermal and viscous instabilities, which may drive the disk into a puffed-up state quite different from the standard Shakura-Sunyaev solution. If it is the latter then the disk is stable, but its not clear that this is a valid disk model.

An additional complication is the opacity of the disk in its inner regions will be dominated by electron scattering, rather than free-free. This will cause the spectrum from any radius to be a modified blackbody, rather than a blackbody. Since the emissivity is reduced this will make the disk hotter at a given radius. White, Stella and Parmar (1988) point out that the use by Mitsuda et al (1984) of a simple blackbody disk spectrum is incorrect in the context of the classical accretion disk models. However, Mitsuda et al (1989) continue to use this form arguing that it is justified given the large uncertainties in the accretion disk theory in the radiation pressure dominated regime.

Along the same lines the problems in the early modelling of the spectrum of Cyg X-1 need to be considered. The low state spectrum cannot be well described by any of the above optically thick accretion disk models. In this case it seems likely that the emission is dominated by the Compton cooling of electrons by an unspecified source of uv, or longer wavelength photons (Shapiro, Lightman and Eardley 1976). The LMXRB are different because of the presence of the neutron star, instead of the black hole in Cyg X-1, and it has been suggested by Czerny,

Czerny and Grindlay (1986) that the photons from the neutron star may stabilise a disk where the viscous stress is proportional to the radiation pressure. Nonetheless, Cyg X-1 demonstrates that Compton losses can in certain circumstances dominate in the inner disk and, given the overall uncertainties in the inner disk structure, must be considered a possibility in the case of LMXRB.

The spectra calculated following the prescriptions given in Shakura and Sunyaev (1973) are specified primarily by the viscosity parameter α, the mass accretion rate, and the distance to the source (including a cos i term to account for the unknown inclination). In addition a blackbody component from the boundary layer was included if an acceptable fit was not obtained. Three disk models were fit. The pure blackbody model, a modified blackbody for a Shakura-Sunyaev disk where viscous stress scales as radiation pressure and a modified blackbody where viscous stress scales as the gas pressure (from Stella and Rosner 1984). The α parameter was fixed at 10_{-3} for the Shakura-Sunyaev disk model, for larger values the inner disk would have an effective optical depth less than unity and Comptonisation would dominate. For the other modified blackbody disk α was fixed at unity; reducing it would tend to require higher mass accretion rates to obtain the same spectral shape. The X-ray source mass was fixed at one solar value.

In figure 4 the disk models for a blackbody disk and modified blackbody disk are compared. An additional blackbody to represent emission from the boundary layer is also required to obtain a good fit. The contribution of the blackbody is quite different in the two cases. For a pure blackbody disk the boundary layer dominates at high energies, whereas for a modified blackbody disk model the boundary layer contribute most at intermediate energies. This highlights a fundamental difference in the approach taken by Mitsuda et al (1984). The other modified blackbody disk model for a gas pressure viscosity scaling, gives a similar spectral shape and fit to the Shakura-Sunyaev disk model.

While all the disk models can be made to give good fits, they all require mass accretion rates a factor of 2 or more in excess of the Eddington limit. This arises because the temperatures required are high. Decreasing α or increasing the X ray source mass only makes the situation worse (unless the X-ray source mass is decreased to an implausibly low value). This seems to be a fundamental problem with the simple disk calculations.

The Comptonisation model gives a fit with a spectral shape that is very similar to the modified blackbody spectrum shown in Figure 4. The optical depth and temperature of the scattering plasma is typically 13 and 4 keV, with a Comptonisation y parameter of order 3. While the optical depth and temperature of the scattering medium are quite different to that of Cyg X-1 (where they are 5 and 27 keV respectively) the y parameter is very similar. The Comptonisation model places no contraint on the overall luminosity of the source so the problem of Super-Eddington luminosity does not arise in this case. The problem with this model is that it conveys no information on the geometry of the emission region, nor on the

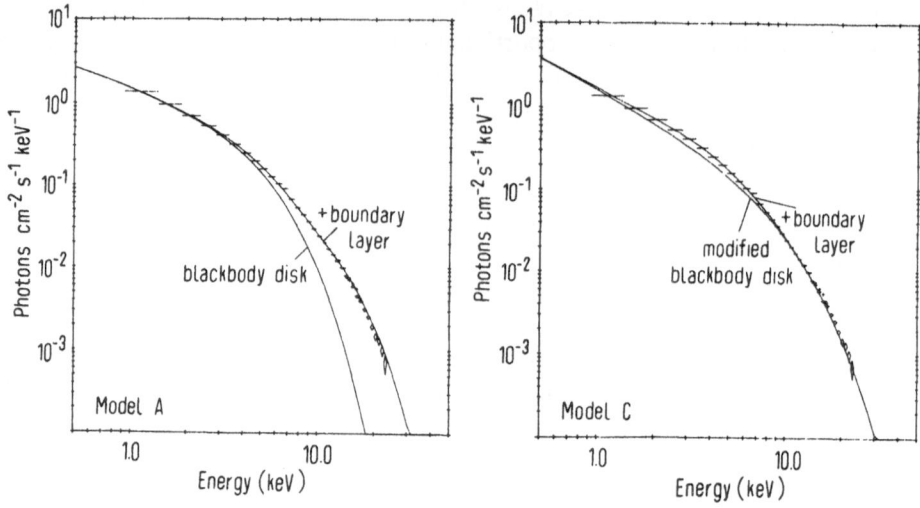

Figure 4: The fit to the spectrum of Cyg X-2 of the blackbody disk (model A) and modified blackbody disk (model C) models. A blackbody component from a boundary layer with the neutron star are also included.

source of the hot electrons or cool photons.

A blackbody component with a luminosity of 16% to 34% the total was also required to obtain an acceptable fit for the higher luminosity systems. In the lower luminosity X-ray burst sources the blackbody is absent with in some cases limits less than 5% the total (except of course during X-ray bursts). The expected emission from the boundary layer for a neutron star rotating far away from its break-up period should be 70% the total, so the detected blackbody is under-luminous by a factor of 2 or more. Why this should be is not clear, but it may suggest that the boundary layer is not a pure blackbody, or the neutron star rotates close to break-up.

White, Stella and Parmar (1988) note that the spectra of the black hole candidate LMC X-3 can also be well fit to either the modified blackbody disk model, or the Comptonisation model. In this case no additional blackbody from a boundary layer is required. However, the spectrum of LMC X-3 is much softer than that of Cyg X-1 in the low state (but like the Cyg X-1 high state spectrum). It is also softer than the LMXRB that contain neutron stars. This much softer spectrum may be a consequence of the fact that the temperature of an accretion disk at a given radius and mass accretion rate decreases as the mass of the central source increases.

4. For the Future.

The study of accretion disks in LMXRB has made good progress over the past 5 years, but there are still many outstanding problems. The observations of the dipping sources by EXOSAT has provided a means of testing and developing models for the gas stream/disk interaction, but this has yet to be fully exploited. The high quality X-ray spectra of the LMXRB obtained by EXOSAT, *Tenma* and most recently by *Ginga* can be used to constrain the nature of the emission from the inner region of an accretion disk. More detailed models for the inner accretion disk and its interaction with the neutron star now need developing to simultaneously account for the time variability properties (e.g. QPO) and the spectral behaviour described here.

REFERENCES.

Bailyn, C.: 1987, *Astrophys. J.* **317**, 737.
Chodil, G., et al. 1968, *Astrophys. J.* **197**, 457.
Cominsky, L.R., Wood, K.S.: 1984, *Astrophys. J.* **283**, 765.
Cominsky, L.R., Wood, K.S.: 1989, *Astrophys. J.* **337**, 485.
Czerny, B., Czerny, M., Grindlay, J.E., 1986, *Astrophys. J.* **311**, 241.
Frank, J., King, A.R., Lasota, J.-P.: 1987, *Astron. Astrophys.* **178**, 137.
Grindlay, J.E., Cohn, H.: 1987, *IAU Circ.* 4393.
Grindlay, J.E., Bailyn, C.D., Cohn, H., Lugger, P.M., Thorstensen, J.R., Wegner, G.: 1988, *Astrophys. J. Letters* **334**, L25.
Hellier, C., Mason, K.O.: 1989, *Monthly Notices Roy. Astron. Soc.* (in press).
Joss, P. C., Rappaport, S. A.: 1979, *Astron. Astrophys.* **71**, 217.
Lewin, W.H.G., Van Paradijs, van der Klis, M.: 1988, *Space Sci. Rev.* **46**, 273.
Lubow, S.H., Shu, F.H.: 1975, *Astrophys. J.* **198**, 383.
Lubow, S.H., Shu, F.H.: 1976, *Astrophys. J. Letters* **207**, L53.
Mason, K.O., Middleditch, J., Nelson, J.E., White, N.E., Seitzer, P., Tuohy, I.R., Hunt, L.K.: 1980, *Astrophys. J. Letters* **242**, L109.
Mason, K.O., Cordova, F.: 1982, *Astrophys. J.* **262**, 253.
Mason, K.O., Parmar, A.N., White, N.E.: 1985, *Monthly Notices Roy. Astron. Soc.* **216**, 1033.
Mason, K.O., Branduardi-Raymont, G., Cordova, F.A., Corbet, R.H.D.: 1987, *Monthly Notices Roy. Astron. Soc.* **226**, 423.
McClintock, J.E., London, R.A., Bond, H.E., Grauer, A.D.: 1982, *Astrophys. J.* **258**, 245.
Milgrom, M.: 1978, *Astron. Astrophys.* **208**, 191.
Mitsuda, K., et al., *Pub. Astr. Soc. Japan*, **36**, 741.
Mitsuda, K., Inoue, H., Nakamura, N, and Tanaka, Y., 1989. *Pub. Astr. Soc. Japan*, in press.
Molnar, L.A., Mauche, C.W.: 1986, *Astrophys. J.* **310**, 343.
Motch, C., Pedersen, H., Beuermann, K., Pakull, M.W.: 1987, *Astrophys. J.* **313**, 792.
Parmar, A.N., White, N.E., Giommi, P., Gottwald, M.: 1986, *Astrophys. J.* **308**, 199.
Parmar, A.N., White, N.E.: 1988, in *X-ray Astronomy with EXOSAT*, N.E. White, R. Pallivicini, eds., Memoria S.A.It, **59**, p. 147.

Pedersen, H., van Paradijs, J., Lewin, W.H.G.: 1981, *Nature* **294**, 725.

Schmidtke, P.C.: 1988, *Astron. J.* **95**, 1528.

Schmidtke, P.C. Cowley, A.P.: 1987 *Astron. J.* **92**, 374.

Shakura, N.I. Sunyaev, R.A., 1973, *Astrophys. J.* **24**, 337.

Shapiro, S.L., Lightman, A.P., Eardley, D.M., 1976, *Astrophys. J.* **204**, 187.

Smale, A.P., Mukai, K.: 1988, *Monthly Notices Roy. Astron. Soc.* **231**, 663.

Smale, A. P., Mason, K. O., White, N. E., Gottwald, M.: 1988, *Monthly Notices Roy. Astron. Soc.* **232**, 647.

Smale, A.P., Mason, K.O., Williams, O.R., Watson, M.G.: 1989, *Pub. Astron. Soc. Japan* (in press).

Stella, L., Rosner, R., 1984, *Astrophys. J.* , **277**, 312.

Swank, J.H., Inoue, H.: 1989 *Bulletin of the AAS* **20**, 1106.

van der Klis, M., Jansen, F., Van paradijs, J., Lewin, W.H.G., Van den Heuvel, E.P.J., Truemper, J., Sztajno, M.: 1985, *Nature* **316**, 225.

Walter, F. M., Bowyer, S., Mason, K. O., Clarke, J. T., Henry, J. P., Halpern, J., Grindlay, J. E.: 1982, *Astrophys. J. Letters* **253**, L67.

Watson, M.G., Willingale, R., King, A.R., Grindlay, J.E., Halpern, J.: 1985, *IAU Circ.* 4051.

White, N.E., Becker, R.H., Boldt, E.A., Holt, S.S., Serlemitsos, P.J., Swank, J.H.: 1981, *Astrophys. J.* **247**, 994.

White, N. E., Holt, S. S.: 1982, *Astrophys. J.* **257**, 318.

White, N. E., Parmar, A. N., Sztajno, M., Zimmermann, H. U., Mason, K. O., Kahn, S. M.: 1984, *Astrophys. J. Letters* **283**, L9.

White, N.E., Peacock, A.P., Taylor, B.G., 1984, *Astrophys. J.* **296**, 457.

White, N.E., Peacock, A.P., Hasinger, G., Mason, K.O., Manzo, G., Taylor, B.G., Branduardi-Raymont, G., 1986, *Astrophys. J.* **218**, 129.

White, N.E., Stella, L., Parmar, A.N., 1988, *Astrophys. J.* **324**, 363.

White, N. E., Swank, J. H.: 1982, *Astrophys. J. Letters* **253**, L61.

Whitehurst, R.: 1988, *Monthly Notices Roy. Astron. Soc.* **232**, 35.

REFLECTED ACCRETION DISK EMISSION LINES IN CATACLYSMIC VARIABLES.

Frederic V. HESSMAN,

Max-Planck-Institut für Astronomie, D-6900 Heidelberg / Königstuhl, FRG

ABSTRACT. The broad emission lines in cataclysmic variables are potentially powerful probes of accretion disk structure. However, the line profile shapes are still not fully understood and the radial velocity variations of these lines often exhibit strange phase shifts with respect to the white dwarfs. I show how one might use the accretion disk's spectrum - scattered out of the orbital plane by the secondary star - to explain some of this behavior.

1. Introduction.

Cataclysmic variables are close binary systems with orbital periods of typically a few hours. A late-type star (the *secondary*) fills its Roche lobe and transfers matter through the inner Lagrangian point onto a white dwarf (the *primary*) via an accretion disk (see Figure 1).

The broad emission lines in cataclysmic variables are an important source of information about the binary system parameters and the structure of the accretion disks. Their radial velocity variations often provide the only information about the orbital motion of the primary. The Doppler-broadened profiles contain information about the geometric and thermal structure of the disk which is unavailable from even photometric eclipses. However, problems in the interpretation of the line profile shapes (eg. the fact that the profiles rarely exhibit the double-peaks expected for a rotating disk) and the unexpected phase shifts between the radial velocity curves and the orbital motion of the primary seen in many systems (Watts, 1985) hampers our use of these probes.

2. The Reflected Spectrum.

The late-type secondary star intercepts a few percent of the total light from the accretion disk and roughly half of that is scattered back out into space (Figure 1).

F. Meyer et al. (eds.), Theory of Accretion Disks, 283–287.
© *1989 by Kluwer Academic Publishers.*

284

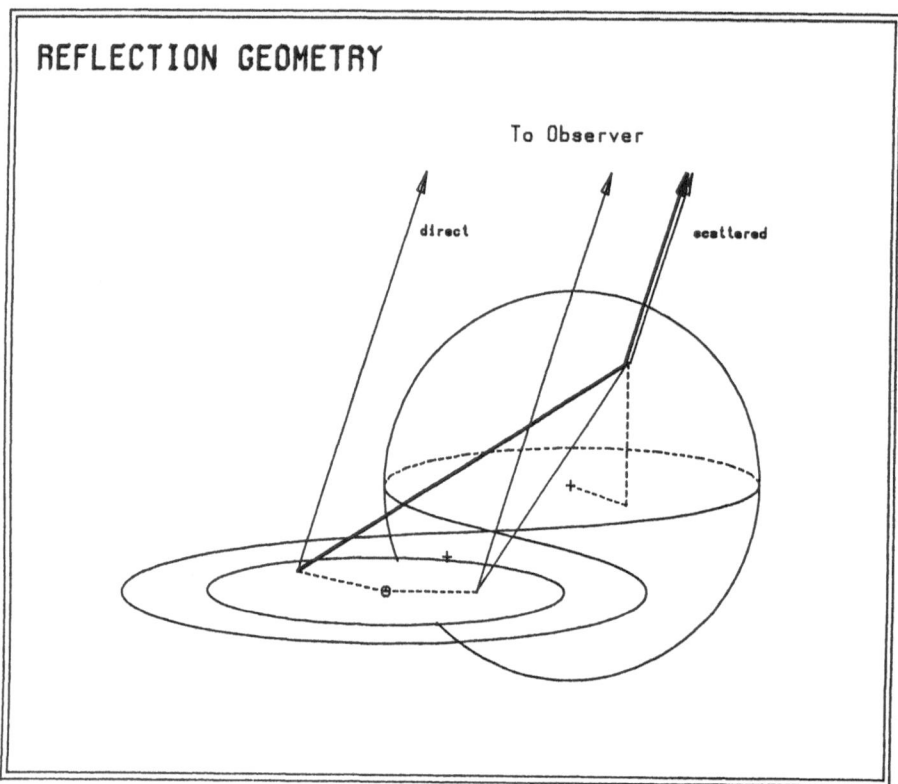

Figure 1

The observable reflection effect in some integrated flux is roughly:

$$\frac{F_{scat}}{F_{direct}} \approx \frac{\frac{1}{2}\frac{1}{2}\frac{1}{2}\frac{1}{4} \, \pi R_2^2 \, F_{disk} \left(\frac{R_{disk}}{a}\right)^2}{cosi \, \pi R_{disk}^2 \, F_{disk}} \approx \frac{10^{-3}}{cosi}$$

where the factors of $\frac{1}{2}$ and $\frac{1}{4}$ symbolically represent the albedo, the fraction scattered into the line of sight, the visible fraction of the secondary star, and a mean cos i for the secondary surface. This effect is thus small in the continuum, particularly if we consider the *monochromatic* albedo of a typical M or K star atmosphere. However, the atmospheres (or at least the *upper* atmospheres) of the secondary stars in cataclysmic variables should differ - perhaps markedly - from those of normal stars due to the very high-intensity and -temperature fluxes coming from the white dwarf and accretion disk. Disks around white dwarfs are very efficient at producing UV radiation, and the incident flux on an M-star can be an order-of-magnitude larger than the star's intrinsic flux. This external flux must ionize a substantial part of what would have been the normal atmosphere, creating an electron "mirror" (albeit a rather poor one) for the optical radiation. Signs that this mechanism indeed works has shown up in observations of the secondary stars in SS Cyg (Hessman *et al.* 1984; Robinson, Zhang, and Stover, 1986), Z Cha (Wade and Horne, 1988), and, in a somewhat different manner, in BE UMa (Ferguson *et al.* 1987).

For the purposes of the present contribution, however, an additional aspect of the reflection effect from such a scattering layer is important : *the Doppler broadened emission line profile seen by the secondary star in the plane of the orbit is much broader and potentially different from that seen by an observer at low inclinations.* Thus, a large fraction of any observed emission at high velocities in such systems could be scattered light.

Using a standard model for the Roche geometry and a simplified accretion disk spectrum, one can approximately compute the reflection effect in the lines. First, the surfaces of the secondary and the disk are divided into a large number of elements. Next, one finds all of the *visible* secondary elements which, at a particular orbital phase, can scatter some disk flux. Finally, the total amount of scattered flux as a function of observed velocity is computed. In order to simplify and speed up the numerical integration of the scattered light, I assumed that the distribution of line emission is given simply by a power-law $I(R) \propto R^{-1.5}$ (typical of many observed profiles). The flux from any segment of the disk bound by the radii R_1 and R_2 and the azimuths ϕ_1 and ϕ_2 can then be expressed as an analytic function (here for the 1st quadrant):

$$F(v) = \frac{cosi}{D^2} \int_{\phi_1}^{\phi_2} d\phi \int_{R_1}^{R_2} dR \; I(R_D) \left(\frac{R}{R_D}\right)^{-\frac{1}{2}} \delta(v - v_{Kepler}(R) \; sin\phi \; sini)$$

$$= \frac{2 \; cosi \; R_D^2}{D^2 \; v_D \; sini} \; u^{-2} \; \sqrt{1 - w^2} \; \left|_{u^2 min(max(r_1,(sin\phi_1/u)^2))}^{u^2 max(min(r_2,(sin\phi_2/u)^2))}\right.$$

where R_D is the outer disk radius, $r \equiv R/R_D$, and $u \equiv v/v_D \sin i$. The secondary and disk was divided up into about 100 and 500 sub-areas, respectively.

Spectra computed using this method are shown in Figure 2. While the scattered profile is negligible in the centers of the wings, it can become dominant in the wings for systems with low inclinations. Whether or not this effect can be *seen* depends upon the amount of continuum radiation underneath (whose Poisson noise could make it impossible to see). The most uncertain aspect of this problem is the exact structure of the secondary's atmosphere: a low optical albedo - despite the large amount of incident UV radiation - could push this mechanism from being marginally observable to utterly negligable. Further theoretical work on the structure of UV illuminated atmospheres needs to be done.

3. Conclusions.

I have shown that it is possible that scattered line emission contaminates the wings of the observed accretion disk profiles in cataclysmic variables. Since the high

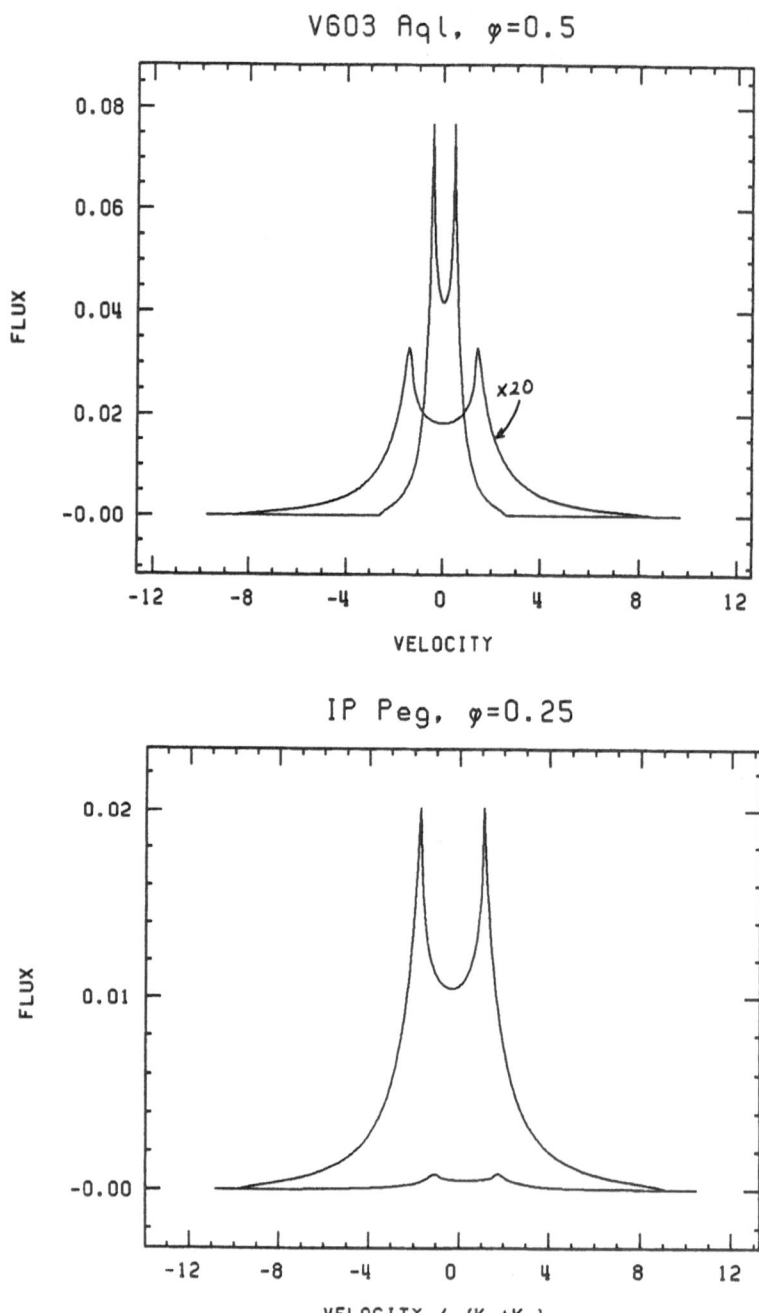

Figure 2

velocity emission is classically used to determine the orbital motion of the primaries, it is obvious that this effect could - in principle - introduce spurious changes in the amplitudes and phases. Unfortunately, the expected amount of polarization is too small to be measureable with present instrumentation, so it is currently impossible to isolate the reflected profile.

REFERENCES.

Ferguson, *et al.*: 1987, *Astrophys. J.* **316**, 399.

Hessman, F. V.: 1985, Ph.D. dissertation, Univ. Texas at Austin.

Hessman, F. V., Robinson, E. L., Nather, R. E., and Zhang, E.-H.: 1984, *Astrophys. J.* **286**, 747.

Robinson, E. L., Zhang, E.-H., and Stover, R. J.: 1986, *Astrophys. J.* **300**, 794.

Wade, R. A., and Horne, K.: 1988, *Astrophys. J.* **324**, 411.

Watts, D. J.: 1985, *Recent Results on Cataclysmic Variables*, Bamberg (ESA SP-236).

THE STRUCTURE OF THE BOUNDARY LAYER OF ACCRETION DISKS IN CATACLYSMIC BINARIES.

Wilhelm KLEY
Universitäts-Sternwarte München, Scheinerstr.1, D-8000 München 80, FRG

1. Introduction.

The boundary layer (BL) is the innermost part of the accretion disk, where the disk grazes the central star, a white dwarf (WD) in Cataclysmic Variables (CV). There frictional interaction decelerates the disk material to the stellar rotational velocity. Knowing the detailed structure of this BL is of great importance for accretion disk physics because, depending on the rotation of the central star, up to one half of the total available accretion energy can be released here. The emitting area is much smaller than the surface of the accretion disk so that the dissipated energy will be radiated away predominantly in the high energy part of the spectrum. Simple models of the BL (Pringle, 1977; Pringle and Savonije, 1979; Tylenda, 1981) lead to the expectation of an optically thin BL emitting hard X-rays in the case of low mass flow through the disk (the quiescent state) and an optically thick BL emitting soft X-rays in the case of high mass flow through the disk (outburst state).

In fact, most of the CVs show hard X-ray emission during the quiescenct (and outburst) state. Some systems (with high $\dot{M}$, high M_{WD}, low distance) were found to emitt soft X-rays during outburst. The spectra of those systems (U Gem and SS Cyg) confirm the view of an optically thick soft X-ray emitting BL. The UV spectra of these objects resemble closely the expected slope of standard stationary accretion disks $F_\lambda \sim \lambda^\alpha$, $\alpha = 2.33$, while in the Voyager band ($\lambda = 500\text{-}1100\text{Å}$) the slope changes roughly to 0.5. Furthermore the observed PCyg line profiles of many CVs during outburst point to a fast mass loss from these systems. The question of heating the WD and mixing of disk matter with the upper stellar layers is very important for the theory of nova eruptions.

All of these topics, the soft X-ray and FUV spectral distribution, the origin of the wind, the flow structure, and the mixing/heating processes with the WD are intimately connected to the structure of the optically thick BL. The investigations so far have basic shortcomings, such as an one-dimensional treatment or, in the

F. Meyer et al. (eds.), Theory of Accretion Disks, 289–295.

case of two-dimensional calculations, an inadequate treatment of energy transport through radiation. For further details and the references see the review article by Shaviv (1987).

Accordingly we have developed a numerical algorithm which can treat time dependent hydrodynamic flows simultaneously together with radiation processes. It is a mixed explicit/implicit method with the radiation transport treated in the flux limited diffusion approximation. The details of the method and the test calculations are described by Kley (1989).

Here we present some calculations performed with this program to study the structure of the interaction region of the inner accretion disk with the stellar surface layers in the case of a high mass flow rate through the disk (*i.e.* the optically thick outburst case).

The model parameters used in the calculations are $M_{WD} = 1.0 \ M_\odot/yr$, $T_{eff}=$ $1.5 \cdot 10^4$ K, $\dot{M}= 3 \cdot 10^{-8} \ M_\odot/yr$, $\nu = 10^{15} cm^2/s$, which are typical parameters for CVs (in outburst). To investigate the influence of rotation of the WD we have studied two cases, first no rotation (model LDTR3) and second $\Omega^* = 0.33\Omega_k^*$ (model LDTR9), where Ω_k^* is the angular Keplerian velocity at the stellar radius, *i.e.* the WD rotates with one third of its break up velocity.

As initial conditions in the radial direction we have taken a stellar atmosphere in hydrostatic equilibrium around the WD. Near to the equatorial plane we imposed a vertical equlilibrium disk model. The computational domain has the extent R_{WD} $\leq r \leq 1.5R_{WD}$, and $0 \leq \alpha \leq \pi/2$, and the boundary conditons are similar to those described in Kley and Hensler (1987). References to lengths are given from now on in units of the stellar radius R_{WD} .

The intial model is then evolved in time under the imposed boundary conditions until a quasistationary state is reached.

2. Results.

Figure 1 shows in the central region the final density contours with velocity arrows superimposed. Larger values of the numbers labelling the contour lines indicate higher densities. The actual values in the the legend to the figures are given in cgs units. The unit of time is given by the Keplerian rotational period at the surface of the star divided by 2π. As can been seen from Fig.1, the matter flows in the disk towards the central star with the largest velocity at $r =1.05$. Nearly all the matter accumulates in an equatorial belt around the star. This behaviour was not seen in the pure hydrodynamic calculations (without any cooling mechanism for the gas), where the matter formed a dense shell completely enveloping the central WD. Only a very small amount of gas with low density is flowing towards higher stellar latitudes, where it partly leaves the computational domain. The mass loss through the outer boundary is generally less than 0.1% of the mass inflow rate. While the radial extent is quite small, < 0.004 (see also Fig.2), the BL extends

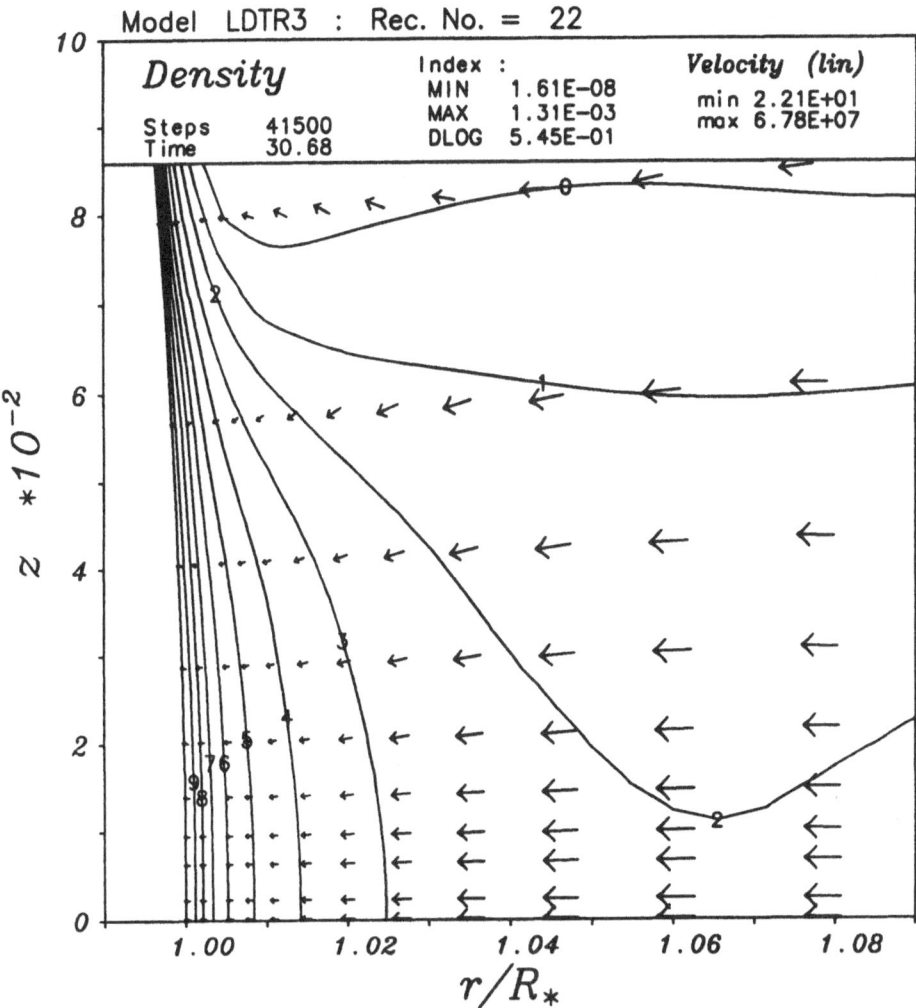

Figure 1: The final density contours of the inner region for the non-rotating model. Superimposed are the velocity arrows (in cm/s).

in the vertical direction up to $z \approx 0.08$, which is essentially a consequence of the different pressure scale heights in the vertical and radial directions. The plot of the radiation energy (Fig.2) shows that in the BL the energy increases strongly, and that the whole region is a source of intense radiation. As expected by the (flux limited) diffusion approximation, the radiative flux is always directed perpendicular to the contours of equal radiation energy. One should remark here that although the arrows correspond to the comoving flux, the difference from the inertial flux is entirely negleglible (at least in this case of accretion onto a WD).

Because the amount of energy liberated in the BL depends on the rotation of the WD, it is important to investigate the behaviour of the BL structure for different

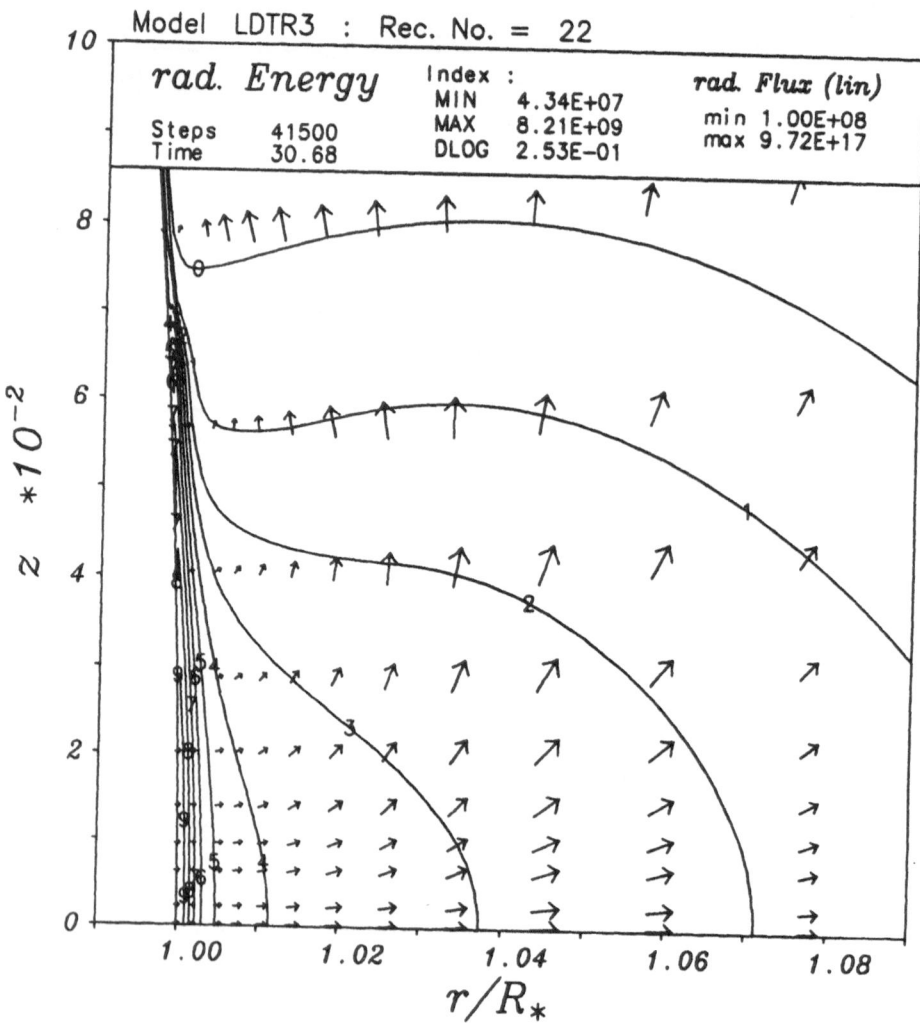

Figure 2: Radiation energy contour lines with the radiation flux vectors superimposed.

values of the stellar rotation. As mentioned above, we took two different values Ω^* = 0 and $\Omega^* = \Omega_k^*$. Fig.3 shows the specific dissipation rate (dissipated energy per second and gram) in the disks midplane (the equatorial plane) for the two models. For $r < 1.13$ the dissipation is much higher in the non-rotating case because of the larger rate of shear near the surface of the WD. The largest difference occurs at the position of the strongest gradient of $\Omega(r)$ at $r \approx 1.05$. There the dissipation in the non-rotating case is higher by roughly a factor of 3. This increased release of energy results into a gas temperature which is about $50000\,K$ higher with absolute values around $(3-4)\cdot 10^5 K$. At the point of maximum dissipation (at $r \approx 1.05$) the gas flows towards the WD supersonically (with the radial Mach number ≈ 6). This is a consequence of the large viscous information speed as implied by the viscosity

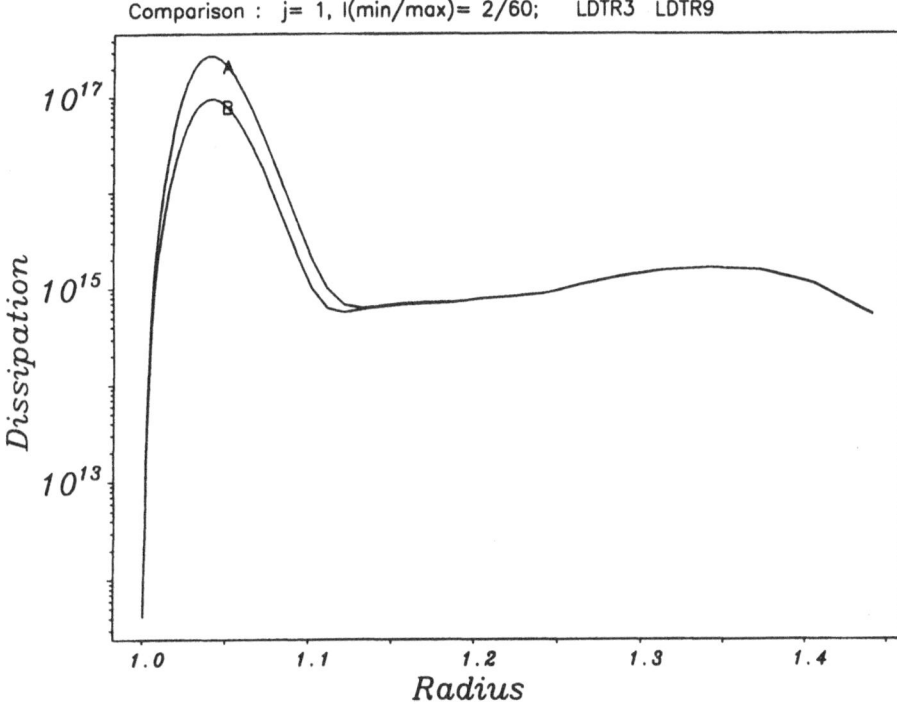

Figure 3: Radial dependence in the equatorial plane of the specific dissipation rate Ω/ρ. The curve A corresponds to the non-rotating star (model LDTR3), and the curve B to the rotating case (model LDTR9).

coefficient ν.

3. Observational consequences and conclusion.

To estimate the observational properties of the BL we have applied a crude procedure to obtain a spectrum of the BL. To do so, first the surface of the disk-star interaction region was determined by integrating over the inverse mean free path until the value $\tau = 1$ was reached. The height of the disk obtained in this way is plotted in Fig.4. It has a minimum value of $\approx 3 \cdot 10^{-2} R_{WD}$ at $r \approx 1.05$, because there the matter flows fastest towards the WD. Outwards the height increases linearly with an opening angle of about 6.5°. Inwards there is a continuous transition into the stellar surface. The maximum temperature in the surface reaches $3 \cdot 10^5$ K in the in the rotating case in and $3.8 \cdot 10^5$ K in the non-rotating case at minimum disk height.

To calculate the spectra, at each surface point the disk/BL is assumed to radiate a black-body (BB) spectrum at the corresponding local radiation emperature $T(\tau=1)$. Then, taking into account limb darkening, we integrated over the surface

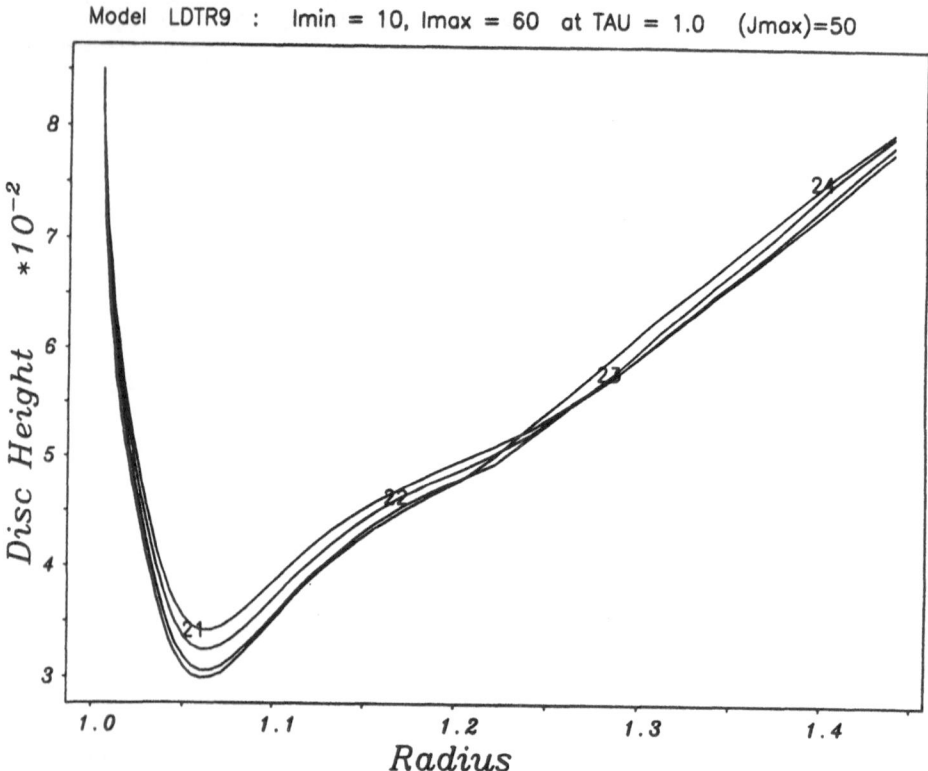

Figure 4: The height of the disk (see text).

of the disk/BL-star under different inclinations to obtain a BB spectrum. One thing to note is that the inclusion of the BL radiation leads to a steeper slope in the UV band, even larger than $\alpha = -2.33$ especially at higher inclinations.

To get an impression how our results compare with observations, we have superimposed the soft X-ray data of SS Cyg taken with HEAO-1 (Cordova et al., 1982) with a model spectrum calculated for the appropiate inclination of $\approx 40°$. The spectrum shown in Fig.5 was corrected for different values of the interstellar absorption until the best agreement with the observational data points was acchieved. The best value is $N_H = 1.5 \cdot 10^{20} cm^{-2}$, which agrees with the values obtained so far. It should be noted that the model with stellar rotation did not agree well with the observations.

To conclude, it seems thus possible (if the stellar parameters of the WD are known) to determine the rotation of the WD and last not least the mass inflow rate by comparing model calculations of this kind with observed spectra. Especially suited for this study are systems with a continuously high accretion rate like the UX Uma systems, which are thought to be permanently in outburst.

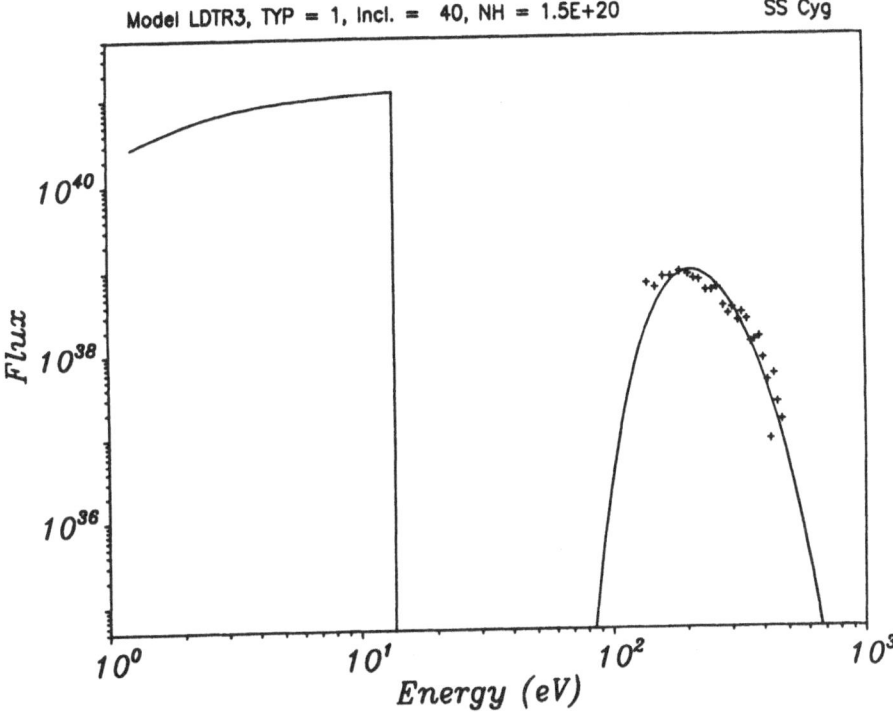

Figure 5: Fit of a calculated BL spectrum to the soft X-ray observations of SS Cyg. The spectrum is corrected for interstellar absorption and the best fit value is $N_H = 1.5.10^{20} cm^{-2}$. The flux is given in arbitrary units.

REFERENCES.

Cordova, F.A., Chester, T.J., Tuohy, I.R., Garmire, G.P.: 1982, Astrophys. J. **235**, 163.
Kley, W., Hensler, G.: 1987, Astron. Astrophys. **172**, 124.
Kley, W.: 1989, Astron. Astrophys. **208**, 98.
Pringle, J.E.: 1977, Monthly Notices Roy. Astron. Soc. **178**, 195.
Pringle, J.E., Savonije, G.J.: 1979, Monthly Notices Roy. Astron. Soc. **197**, 777.
Shaviv, G.: 1987, Astrophys. Space Sci. **130**, 303.
Tylenda, R.: 1981, Acta Astron. **31**, 267.

THE DYNAMICS OF TWISTED ACCRETION DISCS.

Sanjiv KUMAR

Max-Planck-Institute for Astrophysics, Karl-Schwarzschild-Strasse 1
D-8046 Garching bei München, FRG.

ABSTRACT. The theory of twisted discs and its application is briefly reviewed. Such twisted discs are probably quite common, though their current observational relevance is limited to explaining long term periodicities in High Mass X-Ray binaries, the eclipse dip in ϵ Aur, and perhaps the S-shapes of jets of radio galaxies.

1. Introduction.

In accretion disc theory, it is usual to consider thin planar discs. The implicit assumption is that the accretion process is predominantly in the equatorial plane of the accreting compact object, that can be a black hole at the centre of AGNs, or a collapsed star in a binary, or in starforming regions. This enables the formulation of disc theory which is axisymmetric, and often stationary, implying considerable simplification of the disc equations. The rationale is that the complexity of microphysics, e.g. equation of state, opacity, energy transport, can then be handled to explain the spectra observed from such systems. However, it is more likely than not, that the accretion process is slightly out of the equatorial plane, with substantial dynamical consequences. For instance, (1) how large is the region of axisymmetry in the compact object-accretion disc-jet structure; or (2) how do the long term periodicities of High Mass X-Ray Binaries (HMXBs) light curves arise.

The dynamics of twisted discs were first used (Bardeen and Petterson 1975) to understand disc alignment in the presence of Lense-Thirring torques, in the centres of AGNs. This approach was extended to all thin discs with small tilt angles in the presence of arbitrary external torques (Petterson 1977a,b; 1978) with application to Her X-1. These equations were for the Euler angles β and γ, where β (R, t) is the tilt angle and γ(R, t) is the azimuthal angle of the line of nodes. They characterise the tilt of a rigid ring (locally axisymmetric) at each radius. The equations in (β, γ) are non-linear. Considerable simplification is effected if variable $W = -i\beta e^{i\gamma}$ is used - the equations are then linear; however there is a problem with global angular momentum conservation (Hatchett, Begelman and Sarazin 1981). This led

297

F. Meyer et al. (eds.), Theory of Accretion Disks, 297–306.

to Eulerian perturbations of the equations of motion, abandoning the assumption of rigid rings at each radius, thereby conserving the global disc angular momentum (Papaloizou and Pringle 1983), with applications to discs in AGNs (Kumar and Pringle 1985), and in HMXBs (Kumar 1986). From observations, the geometry of tilted discs has been used to model the 35-day periodicity in Her X-1/HZ Her (see Priedhorsky and Holt 1987 for a review of such observations in HMXBs - e.g. Her X1, SMC X1, LMC X4). A twisted circumbinary disc in ϵ Aur, a possible triple, may explain the flat dip with a hump in the middle (Eggleton and Pringle 1985; Lissauer and Backman 1984; Kumar 1987).

2. Equations.

A simplified derivation of the twisted disc equations in cylindrical coordinate (R, ϕ, z) may be given if we assume 2 kinematic viscosities - ν for the usual angular momentum transport in planar discs, and σ for the angular momentum transport due to the twists. However σ is related to ν in a way given below; to find this relationship requires a proper treatment of the equations of motion.

The disc is thin - H/R $\ll$ 1, where H is the local vertical scale height, and angular momentum transport is slow. The relevant time scales are (1) Keplerian $t_K = \Omega^{-1}$, (2) thermal $t_{th} = (\alpha\Omega)^{-1}$, (3) radial drift $t_R = R^2/\nu = (\alpha(H/R)^2\Omega)^{-1}$ and (4) diffusion $t_D = R^2/|\sigma|$. It is usual to consider the α-disc model (Shakura and Sunyaev 1973), with variables Σ (R, t), the surface density ; v_R (R, t), the radial velocity, and the angular momentum per unit surface area $\vec{j} = \Sigma R^2\Omega(\tilde{j}_1,\tilde{j}_2,1)$ in cartesian coordinate (x, y, z). Here $\tilde{j}_1,\tilde{j}_2 \ll 1$. Now define $W = \tilde{j}_1 + i\tilde{j}_2$. The ordering of time scales $t_K \ll t_{th} \ll t_R, t_D$ enables us to neglect the energy equation.

The continuity equation is

$$\frac{D\Sigma}{Dt} = \frac{\partial\Sigma}{\partial t} + \frac{1}{R}\frac{\partial}{\partial R}(R\Sigma v_R) = 0 \tag{1}$$

and the angular momentum transport equation is

$$\frac{D\vec{j}}{Dt} = \frac{\partial\vec{j}}{\partial t} + \frac{1}{R}\frac{\partial}{\partial R}(Rv_R\vec{j}) = \frac{1}{2\pi R}\frac{\partial\vec{G}}{\partial R} + \vec{T} \tag{2}$$

where $\vec{G}(R,t)$ is the viscous torque, and $\vec{T}$ is the external torque per unit surface area which is responsible for disc precession, and therefore twist. For a disc composed of rings with small tilts, the viscous stresses $t_{R\phi}$ and t_{RZ} contribute, the latter with a $cos^2 \phi$ dependence which is averaged over ϕ and gives the factor 1/2. Then writing $1 = \Sigma R^2\Omega, \vec{j} = (\tilde{j}_1,\tilde{j}_2,0)$

$$\vec{G} = -3\pi\nu\vec{j} + \pi\sigma lR\frac{\partial\vec{j}}{\partial R} \tag{3}$$

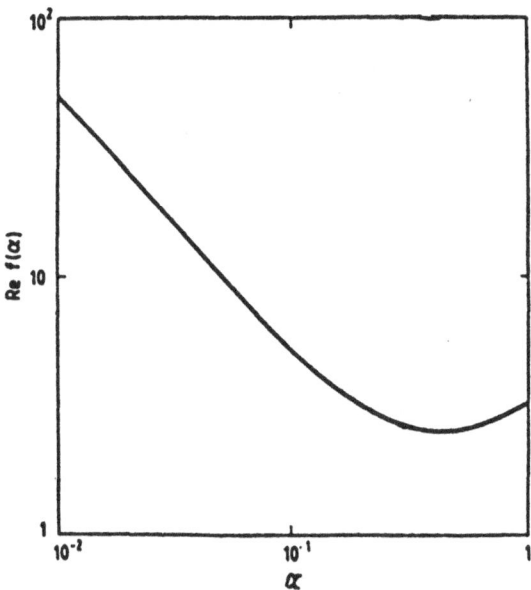

Figure 1: The real part of the diffusion coefficient $\tilde{\sigma}(\alpha)$ is shown as a function of α. Note that for $10^{-1} \lesssim \alpha \lesssim 1$, Re $(\tilde{\sigma})$ is approximately constant, and for $\alpha < 10^{-1}$, Re $(\tilde{\sigma}) \sim 1/\alpha$.

The zeroth order terms in (2) give

$$\frac{Dl}{Dt} = \frac{\partial l}{\partial t} + \frac{1}{R}\frac{\partial}{\partial R}(Rv_R l) = -\frac{3}{2R}\frac{\partial}{\partial R}(\nu l) \tag{4}$$

(1) and (4) can be used to eliminate v_R, which then gives the equation for surface density evolution

$$\frac{\partial \Sigma}{\partial t} = \frac{3}{R}\frac{\partial}{\partial R}\left[R^{1/2}\frac{\partial}{\partial R}(\nu \Sigma R^{1/2})\right] \tag{5}$$

Using (1), (3) and (4) in first order terms of (2) gives the required equation for the evolution of the twist,

$$\frac{DW}{Dt} = \frac{\partial W}{\partial t} + v_A\frac{\partial W}{\partial R} = \frac{1}{2lR}\frac{\partial}{\partial r}\left(\sigma lR\frac{\partial W}{\partial R}\right) + i\Omega_p W \tag{6}$$

where $v_A = v_R + \frac{3\nu}{2R}$ is the radial advection velocity of the tilt W and Ω_p is the precession frequency induced by $\vec{T}$. In general, the boundary condition at the outer edge is $W(R_{out}) = 1$; at the inner edge, the vanishing of viscous torques gives $\frac{\partial \ln W}{\partial \ln R} = \frac{3\nu}{\sigma}$.

Like $\nu = \alpha H^2\Omega$ we can write $\sigma = \tilde{\sigma}H^2\Omega$ where

$$\tilde{\sigma} = \tilde{\sigma}(\alpha) = (6i\alpha - 2 + i/\alpha)/(2i - \alpha) \tag{7}$$

The form of $Re(\tilde{\sigma})$ is shown in figure 1. Note that for $10^{-1} \leq \alpha \leq 1, Re(\tilde{\sigma}) \sim$ constant; for $\alpha < 10^{-1}, Re(\tilde{\sigma}) \sim \alpha^{-1}$. In $(\tilde{\sigma})$ is small, and while it gives rise to viscosity induced internal precession of fluid orbits, it does not have significant bearing on the disc alignment radius R_a, that is of central interest.

For applications, it is convenient to consider the scaled version of (6). If the disc is steady, $\frac{\partial \Sigma}{\partial t} = 0$, we use the scaling

$$\frac{H}{H_*} = (R/R_*)^g \left[1 - (R_*/R)^{1/2}\right]^{1/2} \tag{8}$$

where H_* and R_* are values at the inner edge for H and R to avoid unphysical singularities at the inner edge.

Using $\chi = (R/R_*)^{-1/2}, \tau = \Omega_*(H_*/R_*)^2 t$, (6) becomes

$$\frac{\partial W}{\partial \tau} + \frac{3\alpha}{4}\chi^{9-4g}\frac{\partial W}{\partial \chi} - i\omega_p\chi^{2p}W = \frac{\tilde{\sigma}}{8}\chi^{9-4g}\frac{\partial}{\partial \chi}\left[(1-\chi)\frac{\partial W}{\partial \chi}\right] \tag{9}$$

where ω_p is a constant, and the exponent p indicates the radial dependence of the precession frequency. The boundary conditions now become W $(\chi_{out}) = 1$, and at $\chi = 1 \frac{\partial \ln W}{\partial \ln \chi} = \frac{-6\alpha}{\tilde{\sigma}}$.

Note that if $\nu = 0$ at R_*, then the boundary condition ist automatically satisfied.

3. Some Applications.

We now briefly consider how equations (6) or (9) together with (5) can be used to find the alignment radius R_a, within which the disc effectively lies in the equatorial plane of the spinning compact object. Two roughly equivalent measures exist (1) where the tilt angle $\beta = |W| = 1/2$, gives $|W(R_{out})| = 1$ and (2) $\gamma = \arg W = 1$ given arg W $(\chi_{out}) = 0$. The same estimate can be made from time-scale arguments for a steady state density distribution $\Sigma = \Sigma(R)$. From (6), assuming $\frac{\partial W}{\partial t} = 0$, and using $v_A \approx 0$ which is certainly true for $R/R_* \gg 1, \frac{DW}{Dt} \approx 0$. The right hand terms balance to give $\Omega_p W \sim \sigma W/R^2$, or equivalently

$$\Omega_p^{-1} \approx t_p \sim t_D \approx R^2/|\sigma|. \tag{10}$$

This time-scale argument implictly determines R_a. This is used below for thin discs around black holes, in HMXBs and in $\in$ Aur.

3.1. DISCS IN AGNS.

If the spin axis of a supermassive black hole in the centre of an AGN is misaligned, with respect to the symmetry plane of the gas, in the innermost parts of the galaxy, which eventually accretes then the tilted fluid orbits will be precessed prograde by

the rotation of space-time, which in the far field limit is called the Lense-Thirring precession. This relativistic effect can be calculated in the Boyer-Lindquist coordinates for the Kerr metric. It has the same value as the angular velocity of the zero angular momentum observer,

$$\Omega_{ZAMO} = \frac{2amr}{(r^2 + a^2)^2 - a^2 \sin^2 \theta (r^2 - 2mr + a^2)}$$

(see Bardeen 1973 for this standard notation), in the limit $r \to \infty, \theta \approx \pi/2$ (near the equatorial plane), and is called the Lense Thirring frequency $\Omega_{LT} = 2GJ/c^2 R^3$, where $J = am$ is the spin angular momentum of the hole.

The accreting matter will then torque the hole spin axis into alignment with the symmetry plane of the gas-matter distribution surrounding the hole-disc-jet configuration on a time scale $t_a \sim t_{gas}(J/J_M)(R_*/R_a)^{1/2} \sim 10^8$ yrs, the same as the overall lifetime of radio galaxies (Rees 1978). To keep the hole misaligned requires the presence of another hole, as might be expected in merger remnants like cD galaxies (see Blandford 1988 for an up to date review).

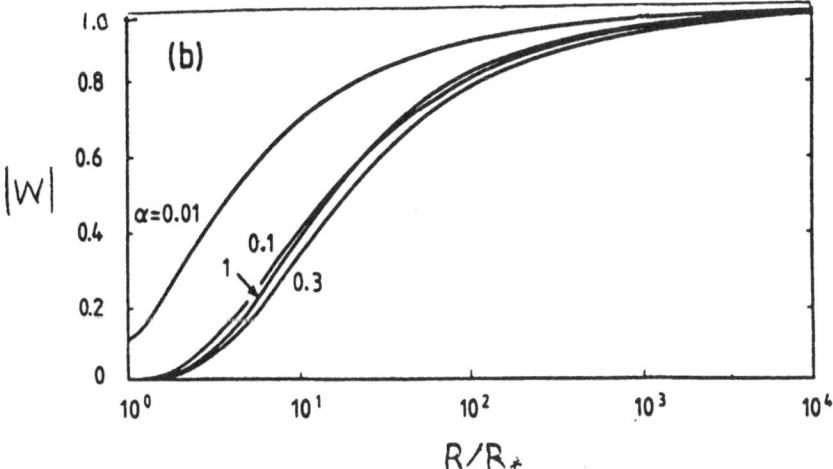

Figure 2: The variation of the tilt angle $\beta = |W|$ with radius R/R_* is shown for different values of α. $|W| = 1$ at the outer radius. The parameter values used are g $= 9/8, \omega_{LT} = 75, H_*/R_* = 10^{-1}$.

For a thin disc within this framework, the old argument for the estimate of alignment radius is that the radial inflow time balances the precession time, $\Omega_{LT}^{-1} = t_p \sim t_R = R/v_R$. Here v_R is calculated from the standard disc model. For outer region, $v_R \sim \alpha^{4/5} R^{-1/4}$; for the middle region $v_R \sim \alpha^{4/5} R^{-2/5}$, and implies that $R_a \sim \alpha^{-16/35}$ in the outer region, $R_a \sim \alpha^{-1/2}$ in the middle region. This has the tantalizing consequence that for small viscosity R_a can be large. This means

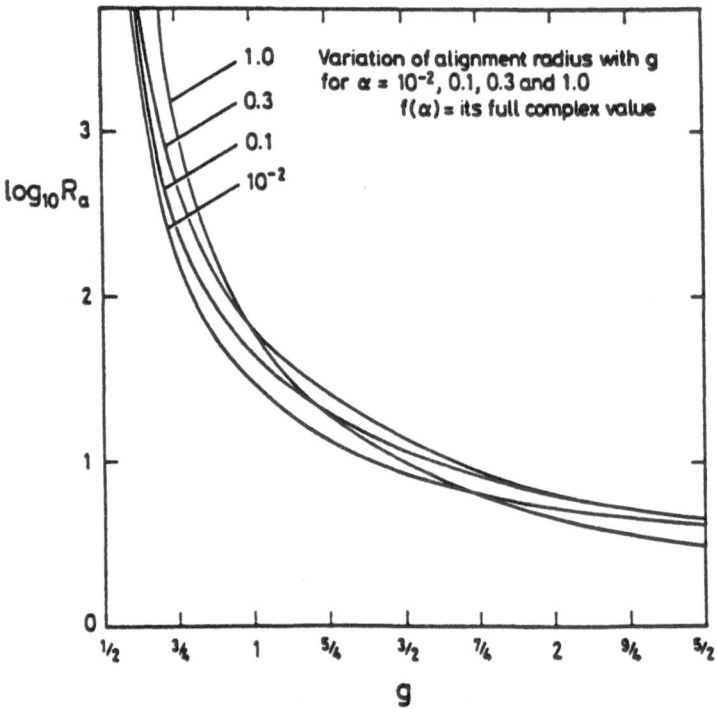

Figure 3: The alignment radius R_a is plotted against g, the scaling for the scale height $- H/H_* \approx (R/R_*)^9$. The surface density $\Sigma(R) \sim R^{3/2-2g}$.

that a general relativistic effect can be manifest at large distance. However, if $\alpha \to 0$, the tilted orbits should not smoothen out and remain so thereby indicating to the contrary a small R_a. That this is indeed so can be seen from a balance of the diffusion and precession times in equation (10). From this we find that for $\alpha \to 0, \tilde{\sigma} \sim \alpha^{-1}$, so that $R_a \sim \alpha^{3/2}$ for middle region, and $R_a \sim \alpha^{54/35}$ for the outer region of the disc. A precise value of R_a may be found from (9). For example, if the hole is maximal Kerr, then a $= 1$, $R_* = GM/c^2$ (the innermost stable orbit), $\omega_p = \omega_{LT} = 2(R_*/H)^2$ also p $= 3$. For $H_*/R_* = 10^{-1}, 10^{-1} < \alpha < 1, R_a/R_* = 15$, see Fig 2: for $H_*/R_* = 3.10^{-2}, R_a/R_* = 3 \cdot 10^2$. This then is scale of the axisymmetric central configuration. Fig. 3 shows how R_a varies with g. Note that if $\eta \overset{\circ}{M}/\overset{\circ}{M}_E \gtrsim 1$, we expect a radiation torus and if $\eta \overset{\circ}{M}/\overset{\circ}{M}_E \lesssim 10^{-2}$, we expect an ion torus (η is accretion efficiency). For these objects, $H/R \sim 1$ so that the thin disc approximation does not apply. Consequently the theory here does not apply. Alignment in tori is an important open problem.

3.2. DISCS IN HMXBS.

Perhaps the most persuasive argument for a twisted, precessing disc in Her X1 is the feature of the optical light curve, which advances in phase over the 35 day cycle. The only way to explain this is eclipse of the primary by a disc in retrograde precession around the secondary (van Paradijs 1983) This motivates us to apply the twist evolution equation (9) to answer the question of alignment. We find that the 35 day variability is well accounted for by this theory. While (9) needs to be solved numerically for a precise determination of R_a, the timescale argument gives the same result and is sufficient for our purpose. We use the steady state version of (9). Note that the Shakura-Sunyaev disc is equivalent to assuming $g = 9/8$. The input into (9) required are the values of α, p and ω_p. This brings us to the precession mechanisms in binary systems.

The two relevant mechanisms are forced precession (FP) and node recession (RN), both of which require a misaligned mass-losing primary, though RN can also arise if there is an instability in mass flow from the primary due to asymmetric illumination. Both are retrograde. For $\beta \ll 1, \Omega_{RN} \approx -\frac{3}{4}\frac{GMp}{a^3}\frac{R^{3/2}}{(GM_x)^{1/2}}s^{-1}$, where M_p is the primary mass, a is the stellar separation, M_x is the compact object mass. This gives $p_{RN} = -3/2$ and $\omega_{RN} = \frac{3}{4}10_q^8(R_*/a)^3$ for $H_*/R_* = 10^{-1}$, where $q = M_p/M_x$. Forced precession arises from the coupling of the compact object to the quadrupole moment of the primary which is counter precessing. $\Omega_{FP} \approx -\epsilon_p\Omega_{orb}$,where ϵ_p is the ellipticity of the primary and Ω_{orb} is the binary orbital frequency $\epsilon_c \sim \Omega^2 R_p^3/GM_p \sim 10^{-1}$ for $\Omega \approx \Omega_{orb}$ (corotation). From this, $p_{FP} = 0$ and $\Omega_{FP} = -10^4(R_*/a)^{3/2}\sqrt{1+q}$.

From observed orbital parameters for Her X1, use q = 1.7, a = $6.4.10^{12}$ cm, R_{out}/R_* (disc) = 2.4 $\times 10^3$, which give $\omega_{FP} = 3.2.10^{-5}$ and $\omega_{RN} = 4.9 \times 10^{-10}$. We assume that the neutron star surface $|\bar{B}| \sim 10^{12}$ gauss, so that $R_* \sim 10^2 R_{ns}$ is the magnetosphere radius. For $g = 9/8, 10^{-1} < \alpha < 1, t_{FP} \approx t_{RN} \approx t_D(R_{out})$, so that we have alignment with the neutron star. Since t_{RN} decreases with R, t_{FP} remains constant while t_D increases with R, it is important that the time scales are comparable near the disc outer radius if the observational features are to be explained. For $\alpha = 10^{-2}$, this balance requires that $g \lesssim 1$. Thus if the Shakura-Sunyaev disc is a reasonable model (and also this model), then $10^{-1} \lesssim \alpha < 1$ is necessary to explain the observations.

3.3. DISC IN ϵ AUR.

Two triple models have been suggested for ϵ Aur, with a close binary, a circumbinary disc, and a FOI supergiant primary in contraction phase. The close binary is inclined $\simeq 20°$ to the larger orbital plane. The latter plane is seen edge on. The eclipse every 27 years has a flat minimum with a central hump and has been modelled before as an opaque rectangular object along the line of sight with aspect ratio of $\approx$ 10 : 1. The primary mass $M_p = 1.3M_\odot$ or 16 $M_\odot$, while $M_c = 2.5M_\odot$ or 8 $M_\odot$ for the 2 different models. Precise values of these parameters are however not necessary to determine whether the circumbinary disc is aligned or not. This is because the close binary transfers angular momentum to the inner edge of the disc (Lin and Papaloizon 1979), and forces alignment due to non-vanishing of the viscous stresses there. The matter has a larger accretion time-scale than in the usual case of a single accreting object. The disc can then cool to an opaque cloud, if the primary is contracting and mass transfer due to Roche lobe overflow has stopped. This steady state disc has properties quite different from the usual disc which we now consider.

From (1) and (4),

$$v_R = -\frac{3}{\Sigma R^{1/2}} \frac{\partial}{\partial R}(\nu \Sigma R^{1/2}). \tag{11}$$

Since matter cannot accrete, at the inner boundary $v_R = 0$. With $\partial \Sigma / \partial t = 0$ and no accretion, $v_R = 0$ at the outer edge. Then $\nu \Sigma R^{1/2} = $ constant for the whole disc. To write (6) in a suitable form, assume $H/R = $ constant; then (6) becomes

$$\frac{\partial W}{\partial \tau} - \frac{3\alpha}{4} \chi^4 \frac{\partial W}{\partial \chi} - i\omega p \chi^{2p} W = \tilde{\sigma}(\alpha) \chi^4 \frac{\partial}{\partial \chi}(\partial \frac{\partial W}{\partial \chi}) \tag{12}$$

Note that although $v_R = 0$ throughout the disc, $v_A > 0$ i.e. the tilt is advected outwards.

All 3 relevant precession frequencies are retrograde: (1) node recession, $\Omega_{RN} \approx -\frac{3}{4a^3} \frac{GMpR^{3/2}}{(2GM_c)^{1/2}}$; (2) forced precession $\Omega_{FP} \lesssim \Omega_{orb}$ and (3) quadrupole moment of the close binary, $\Omega_{QM} \approx -\frac{3}{4}\Omega(\frac{a}{R})^2$, where Ω is the local Kepler frequency, and a is the close binary separation. The (p, ω_p) values are respectively, $RN(-\frac{3}{2}, -10^{-5})$, $FP(0, -2.5 \cdot 10^{-2})$ and $QM(7/2, -10^2)$ using $H/R = 10^{-1}, a \approx 5.10^{12}$ cm, a_2 (the larger binary separation) $\lesssim 10^{15}$ cm, $M_c \sim 2M_\odot, M_p \sim 4M_\odot$. It turns out from numerical solutions of (12) with $\partial W / \partial t = 0$ that the single most important parameter is α, and that there be the presence of the close binary quadrupole moment (which in the scaling used here always has the same (p, ω_p) values). $10^{-1} \lesssim \alpha \lesssim 1$ is required for substantial alignment. Unlike the twisted disc around compact objects, here the timescale argument for alignment is not relevant because of the different nature of disc alignment governed by the viscous dynamics near the inner edge.

4. Discussion.

In this treatment of twisted disc dynamics, the relevant radial infall and diffusion timescales allowed us to neglect the energy equation. It was then possible to use isothermal or polytropic local structure to simplify (6). The isothermal disc led to (10). If a polytropic disc is used, we find an additional advective term governed by $\tilde{\sigma}(\alpha)$. The modification of R_a is substantial only for small $\alpha \lesssim 10^{-2}$ for discs in AGN, larger by a factor of 10. For $10^{-1} \lesssim \alpha \lesssim 1$, the range of greatest interest, the difference is negligible.

The full time dependent equations (5) and (6) can also be solved. It turns out that the diffusive nature of the disc is strong enough to wipe out, say mass modulation of inflow at the outer edge, leaving R_a intact. If $\partial\Sigma/\partial t = 0$ but $\partial W/\partial t \neq 0$, then for a step profile of $W(t = 0)$, the disc settles into a steady state which does not depend on the initial profile. If there is a local thermal instability which leads to an outburst (Bath and Pringle 1982; Meyer and Meyer-Hofmeister 1981), and follows the usual S-curve in the (μ, Σ) plane, we speculate that the disc dynamics is unaffected by the switching actions. With the increase in viscosity in the matter infalling during the high state we expect that t_D decreases, and R_a will move inward for matter infalling during the high state. Note that $t_D < t_F$, where t_F is the front propagation time, if $\alpha > H/R$. Note that for $\alpha \sim H/R$, the dynamics would be affected. Similarly it moves outward during low state. However for a circumbinary disc, we expect the opposite.

The dynamics of the disc for small α is of interest as it sheds light on viscosity controlled disc response. The perturbed velocities (v'_R, v'_ϕ) satisfy $v'_\phi = -zf\left[W(R)\right]/8\alpha^2$, $v'_R = izf\left[W(R)\right]/4\alpha^2$. Compare this response with that of a particle orbiting in the symmetry plane of an axisymmetric potential $\psi = \psi(R, Z)$, which motion is then perturbed. Then $v'_\phi = \frac{3\mathcal{K}}{2\Omega}e^{i\pi}\delta R$ and $v'_R = \frac{3\mathcal{K}^3}{2\Omega^2}e^{i\pi/2}\delta R$ where $\delta R = e^{i\mathcal{K}t}\delta R_o$, δR_o is the initial displacement and $\mathcal{K}$ is the epicyclic frequency. We see that v'_R has the same phase lag in both cases. Since $\Omega = \mathcal{K}$ for Kepler disc, v'_ϕ is relatively smaller by a factor of 2 in a twisted disc. The intrinsic difference between the 2 cases however is that for the disc, (v'_ϕ, v'_R) depend on v'_z through W(R). For a particle, there is no relation between δR and δz; the z-motion is completely decoupled from that in (R, ϕ) directions. Bounds may be put on the energy contained in these perturbed shear-flows which are proportional to $1/\alpha^4$.

Observational relevance of the twisted accretion discs has been emphasized above, though at present this is restricted to explaining the long-term periodicities in binaries an radio-jets. Future X-ray polarisation measurements maybe expected to clarify the alignment of discs away from he outer-edge of the accretion disc (Rees 1975). In general, the tilt of the polarisation vector should increase with wavelength, or with radius.

ACKNOWLEDGEMENTS. I would like to thank J.E. Pringle for introducing me to this subject, and E. Meyer-Hofmeister, F. Meyer and R. Kippenhahn for their kind hospitability in Munich.

REFERENCES.

Bardeen, J.M. 1973. In: Black Holes, Les Houches, eds. DeWitt B. & DeWitt, C., Gordon & Breach, New York

Bardeen, J. & Petterson, J.A., 1975. Ap. J., 195, L 65

Bath, G. & Pringle, J., 1982, M.N.R.A.S., 199, 267

Blandford, R. 1988. In: 300 Years of Gravitation, eds. Hawking, S. & Israel, W., Cambridge University Press.

Eggleton, P. & Pringle, J., 1985, Ap. J., 288, 275

Hatchett, S., Begelman, M., & Sarazin, C., 1981. Ap. J., 247, 677

Kumar, S., 1986, M.N.R.A.S., 223, 225
Kumar, S., 1987, M.N.R.A.S., 225, 823
Kumar, S., 1988, M.N.R.A.S., 233, 33
Kumar, S. & Pringle, J., 1985, M.N.R.A.S., 213, 435
Lin, D.M.C. & Papaloizon, 1979, M.N.R.A.S., 188, 191
Lissauer, J. & Backman D., 1984, Ap. J., 286, L39
Meyer, F. & Meyer-Hofmeister, E., 1981, A. & A. 104, L10
Papaloizon, J. & Pringle, J., 1983, M.N.R.A.S., 202, 1181
Petterson, J.A., 1977a, Ap. J., 214, 550
Petterson, J.A., 1977b, Ap. J., 216, 827
Petterson, J.A., 1978, Ap. J., 226, 253
Priedhorsky, J. & Holt, S. 1987, Space Sci. Rev. 45,291.
Rees, M.J., 1975, Nature 171, 457
Rees, M.J., 1978, Nature 275, 516
Shakura, N. & Sunyaev, R., 1973 A & A, 24, 337
van Paradijs, J., 1983, in Accretion Driven Stellar X-Ray Sources, ed. Lewin, W. & van
    den Heuvel, E. C.U.P., Cambridge

A FREE ACCRETION DISK IN SN 1987A?

Friedrich MEYER and Emmi MEYER-HOFMEISTER
Max-Planck-Institut für Astrophysik, D–8046 Garching bei München, FRG

ABSTRACT. The observations of Middleditch et al. (1989) of a submillisecond pulsar in SN 1987A might indicate a remnant neutron star rotating near breakup. This suggests that sizeable angular momentum was present in the core of Sanduleak -69^0 202 before collapse (For a model see Hillebrandt and Meyer, 1989 (submitted to A&A Letters)). We speculate here that not all angular momentum could be accreted and some matter was left after the collapse around the neutron star in critical rotation. Further evolution would lead to the formation of an accretion disk in which the surplus angular momentum is removed to infinity while the mass gradually accretes inward.

For an assumed α viscosity such a disk develops a standard distribution in space and time described by a solution that is invariant under transformations of length and time which leave the governing differential equation and boundary conditions invariant.

If the present accretion rate from such a disk shows as a contribution to the X-ray luminosity of SN 1987A (Dotani et al., 1987, and Sunyaev et al., 1987) it is of order 10^{18} g/s. We discuss here the development of such an accretion disk.

1. The invariant solutions.

The partial differential equation for the surface density Σ is

$$\frac{\partial \Sigma}{\partial t} = \frac{3}{r} \frac{\partial}{\partial r} \left[r^{1/2} \frac{\partial}{\partial r} \left(r^{1/2} \nu \Sigma \right) \right]$$

(r distance from rotation axis, ν kinematic viscosity). In a α parametrization one has

$$\nu = \alpha (H/r)^2 \sqrt{GMr} = C r^{1/2}$$

(H scale height in the disk $\approx$ disk thickness, M central mass). H/r is usually a slowly varying function of r. We take for simplicity $C = const$ but note that the same procedure can be performed with any power law dependence on r. For a numerical example we choose characteristic values $\alpha \approx 1/3$ and $H/r \approx 1/30$.

Invariant solutions are of the form

$$r\Sigma(r,t) = \frac{A_n}{t^{n+1}} \varphi_n(\xi) \qquad A_n = const \quad n = 0,1,2,\ldots$$

F. Meyer et al. (eds.), Theory of Accretion Disks, 307–311.
© 1989 by Kluwer Academic Publishers.

with

$$\xi = \frac{4}{3C} r^{3/2}/t$$

as the invariant non-dimensional coordinate. The functions $\varphi_n(\xi)$ are solutions of a special type of linear differential equation and can be obtained in closed form,

$$\varphi_n(\xi) = \left[\prod_{s=1}^{n} \left(\xi \frac{d}{d\xi} + s \right) \right] \varphi_0(\xi) = P_n(\xi) \varphi_0(\xi) ,$$

$$\varphi_0(\xi) = \xi^{1/3} e^{-\xi} .$$

These solutions have zero angular momentum inflow at infinity and at the center and contain finite amounts of mass and angular momentum. Since time does not explicitly occur in the linear differential equation any time derivative of a solution is again a solution. The generating operators above are essentially these time derivations.

The P_n are polynomials of degree n. Thus the set of functions $\{\varphi_n\}$ is complete. The functions are also orthogonal,

$$\int_0^\infty P_n P_m \varphi_0 d\xi = C_n \delta_{nm} \quad C_n = const. ,$$

Therefore the general solution is

$$r\Sigma(r,t) = \sum_{n=0}^{\infty} \frac{A_n}{t^n} P_n(\xi) \xi^{1/3} e^{-\xi} .$$

This shows the exponential cut-off at large ξ, i.e. r. (By choice of the zero point of t the coefficient A_1 can be made to vanish.) For large time the higher order components become negligeable,

$$\lim_{t \to \infty} r\Sigma(r,t) = \frac{A_0}{t} \xi^{1/3} e^{-\xi} .$$

The (conserved) angular momentum is wholly contained in this component, which determines A_0 and thus the complete asymptotic evolution of a free disk (after the higher order transients have decayed). One thus obtains the following features
1) The asymptotic function is stable
2) The asymptotic evolution is completely determined by one number, the conserved angular momentum
3) The mass accretion rate at the center becomes a unique function of time, only dependent on the initial angular momentum.
    The fundamental solution was obtained by Lynden-Bell and Pringle (1974) as a special superposition of solutions of Bessel function type that are obtained when one

separates into products of space and time dependent functions. They showed that any δ-function type initial distribution asymptotically evolves into this solution. Here, this follows directly from the completeness of our set of functions.

Integrating over the disk one obtains the disk mass,

$$M_{disk} = const\ t^{-1/3}\ .$$

Time derivation yields a relation between present mass and accretion rate $\dot{M}$ at the center

$$M_{disk} = 3\dot{M}\ t\ .$$

For a time of two years and $\dot{M} \approx 10^{18}$ g/s this gives $10^{-8}\,M_\odot$. The figures show the relative distribution of mass and angular momentum in space for different times.

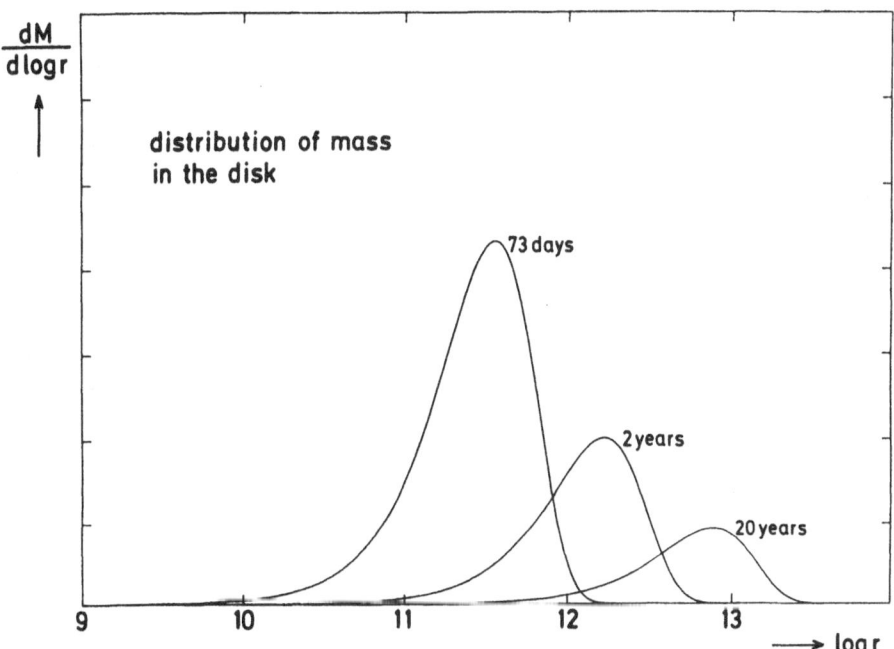

Figure 1

2. Shepherded disk.

The observation of a 8 hour period in the pulsar frequency variation $\Delta\nu/\nu \approx 10^{-6}$ (Middleditch et al., 1989) can be interpreted as due to a companion of mass $M_2 \approx 10^{-3}\,M_\odot$ in an orbit of $r_0 \approx 10^{11.2}$ cm. It's Roche radius R_2,

$$\frac{R_2}{r_0} \approx \left(\frac{1}{3}\frac{M_2}{M}\right)^{1/3} = 10^{-1.2}$$

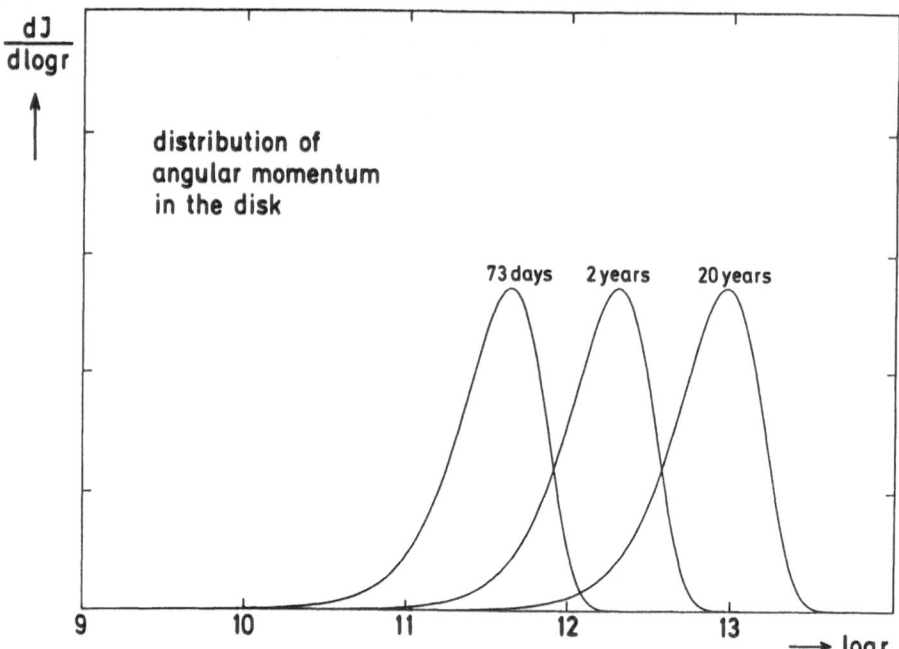

Figure 2

is then marginally larger than the disk thickness for a present accretion rate of 10^{18} g/s,

$$\frac{H}{r_0} = \left(\frac{2\Re T r_0}{\mu GM}\right)^{1/2} = 10^{-1.6}(T/T_{eff})^{1/2}$$

($\Re$ gas constant, μ molecular weight, $M = 1.4M_\odot$, T temperature in the disk, T_{eff} effective temperature at disk surface). Thus the inside disk is possibly bound (shepherded) by the companion that prevents mass to flow outwards across the orbit by picking up the angular momentum. (An originally outside part of the disk would complementary accrete its mass to the companion while removing its angular momentum to infinity, asymptotically like a free disk). If the companion does not feed in mass, i.e. $\dot{M}(r_0) = 0$, the invariance properties with this boundary condition requires

$$r\Sigma(r,t) = f(t)\,g(r)$$

with the usual solutions of the differential equation for the surface density in form of a superposition of Bessel functions,

$$r\Sigma(r,t) = B\,exp\{-3C\lambda t\}z J_{1/3}(z)$$

with

$$z = \frac{4}{3}\sqrt{\lambda}r^{3/4}\ .$$

The condition of vanishing mass flow at r_0, equivalent to $r^{1/2} \frac{\partial}{\partial r}(r\Sigma) = 0$ requires $z(r_0)$ to be a solution of

$$\frac{2}{3} J_{1/3}(z) - z J_{4/3}(z) = 0 .$$

The sequence z_n of zeros of this equation defines a sequence of decay constants $3C\lambda_n$ with

$$\lambda_n = \frac{9}{16} \frac{z_n^2}{r_0^{3/2}} .$$

The lowest one, $z_1 = 1.24$, yields the asymptotically dominant solutions and gives the exponential decay time

$$t_0 = \frac{16}{27} \frac{r_0^{3/2}}{C z_1^2} \approx 10^{6.7} s \approx 2 months$$

depending on the assumption about C, i.e. $\alpha(H/r)^2$. On this time scale the mass accretion rate also drops off.

3. Conclusions.

A free disk around a compact object develops with time into a standard distribution only determined by its angular momentum. It moves outward and broadens with time while diminishing in mass. A companion in orbit around the central mass will introduce an exponential decay on the time scale

$$t_0 = \frac{16}{27\alpha} \left(\frac{r}{H}\right)^2 \frac{P}{2\pi}$$

where P is the orbital period. The two time scales for SN 1987A are of the order of the age of the remnant (2 years now) and 2 months respectively, the latter depends on the assumed disk thickness. Temperatures in the disk would eventually fall in both cases sufficiently to form grains and possibly meteoric bodies, ending the evolution and leaving a fossil disk behind, provided an inner part of the disk can shield the grain forming regions from irradiation by the still hot neutron star.

REFERENCES.

Dotani, T., et al., 1987: Nature 330, 230.
Lynden-Bell, D., and Pringle, J.E., 1974: Mon. Not. R. Astr. Soc. 168, 603.
Middleditch, J., and al., 1987: IAU circular No. 4735.
Sunyaev, R., et al., 1987: Nature 330, 227.

THE SPECTRA OF RELATIVISTIC ACCRETION DISKS.

Ronald E. TAAM and Albert FU
Department of Physics and Astronomy, Northwestern University,
Evanston, Illinois 60208, U.S.A.

ABSTRACT. The influence of general relativistic effects on the continuum spectra of geometrically thin accretion disks surrounding weakly magnetized neutron stars or black holes in compact x-ray sources is considered. Specific attention is focused on the effect of the gravitational redshift and relativistic Doppler shift on the emission spectrum and on the enhancement of the apparent accretion disk area due to gravitational light bending for a range of system inclination angles. It is shown that the effects of general relativity can be significant as compared to the Newtonian model especially for those cases where the inner edge of the accretion disk extends to the innermost stable orbit. As an application of these results we investigate the constraints on the mass transfer rate imposed by the lack of soft x-rays emitted during the quiescent state of the black hole candidate A0620-00.

1. Introduction.

Accretion disks have become standard in the phenomological interpretation of interacting close binary star systems. Although the concept of an accretion disk was introduced twenty years ago (Lynden-Bell 1969) the viscosity describing its structure and evolution remains unknown. As a consequence, much of the work in the field has been based on the steady state approximation for which the emission properties of the disk are insensitive to this quantity (Shakura and Sunyeav 1973). Despite the general consensus regarding the utility of accretion disks, it is to be noted that general relativistic effects have not been incorporated in the standard model used in the interpretation of spectra from compact x-ray sources. From a fundamental physics point of view such relativistic effects would be especially important for the accretion disks surrounding weakly magnetized neutron stars or black holes. As an application of the early work of Page and Thorne (1974), Luminet (1979), and Fukue and Yokoyama (1988), Hanawa (1989) has shown that relativistic disks produce significantly softer spectra than their Newtonian counterparts in flat space. In particular, the inferred temperature characterizing the general relativistic disk, as measured by a distant observer, is about a factor of 3 lower than that inferred based upon a Newtonian model. As stressed by Hanawa

F. Meyer et al. (eds.), Theory of Accretion Disks, 313–319.
© *1989 by Kluwer Academic Publishers.*

(1989) this could have a significant impact on the understanding of the soft x-ray component in low mass x-ray binary systems. However, the application of such a model to this class of sources is limited since the inner regions of the disk, where relativistic effects are most important, may be disrupted for those neutron stars with magnetic field strengths $\gtrsim 10^{10}$ Gauss.

On the other hand, for accretion disks surrounding very weakly magnetized neutron stars ($\lesssim 10^9$ Gauss) or black holes relativistic effects may be directly applicable since in this case one expects the inner edge of the disk to extend to the innermost stable orbit at $3R_g$, where R_g is the Schwarzschild radius. Of particular interest in the present study is the application of steady relativistic accretion disk models to the black hole candidate source A0620-00 (McClintock and Remillard 1986). The recent analysis of de Kool (1988) suggests that the lack of x-ray emission in the 0.2 - 20 keV region found by Long, Helfand, and Grabelsky (1981) is consistent with the system transferring mass at rates less than $10^{-11} M_\odot yr^{-1}$. Relativistic effects would increase this upper limit, and combined with the optical data would constrain the mass transfer rate during the quiescent state more tightly. This is particularly relevant since a class of models invoked to describe the nature of the soft x-ray transient phenomena depends sensitively on the mass transfer rate during this state. Specifically, Hameury, King, and Lasota (1986, 1987) hypothesize that the transient activity is associated with a mass loss instability in the x-ray irradiated companion. For mass transfer rates in the range $\sim 10^{-14} - 10^{-10} M_\odot yr^{-1}$ the stellar atmosphere is unstable. Clearly, any constraints imposed upon the mass transfer rate would be important for the viability of this model. In this paper we calculate the emission expected from a relativistic disk in the steady state approximation for a range of mass transfer rates and binary inclination angles with the aim of providing constraints on the mass transfer rate in the quiescent state of A0620-00.

2. Results.

The method of analysis is similar to that presented in the study of the spectra from a rotating boundary layer situated on the surface of a weakly magnetized neutron star in Fu and Taam (1989a) and will not be presented here. We assume, in this preliminary study, that the steady accretion disk is optically thick and that its surface radiates like a blackbody for mass accretion rates in the range $10^{-12} \lesssim \dot{M}(M_\odot yr^{-1}) \lesssim 10^{-10}$. With the flux distribution of an accretion disk surrounding a nonrotating central object given as a function of radius by Page and Thorne (1974), the local effective temperature, T_{eff}, can be expressed as

$$T^4_{eff} = \frac{3GM\dot{M}}{8\pi\sigma R_g^3 r^{5/2}(r-3/2)} \left[\sqrt{r} - \sqrt{3} + \sqrt{3/8} ln \left(\frac{\sqrt{r}+\sqrt{3/2}}{\sqrt{r}-\sqrt{3/2}} \cdot \frac{2-\sqrt{2}}{2+\sqrt{2}} \right) \right]. \quad (1)$$

Here M is the mass of the compact object, r is the Schwarzschild radial coordinate in terms of R_g, and G and σ are the gravitational and Stefan-Boltzmann constants respectively. We note that for general relativistic disks the maximum surface temperature is lower by ~ 1.23 and occurs slightly further out in the disk at $\sim 4.78R_g$ as compared to $4.08R_g$

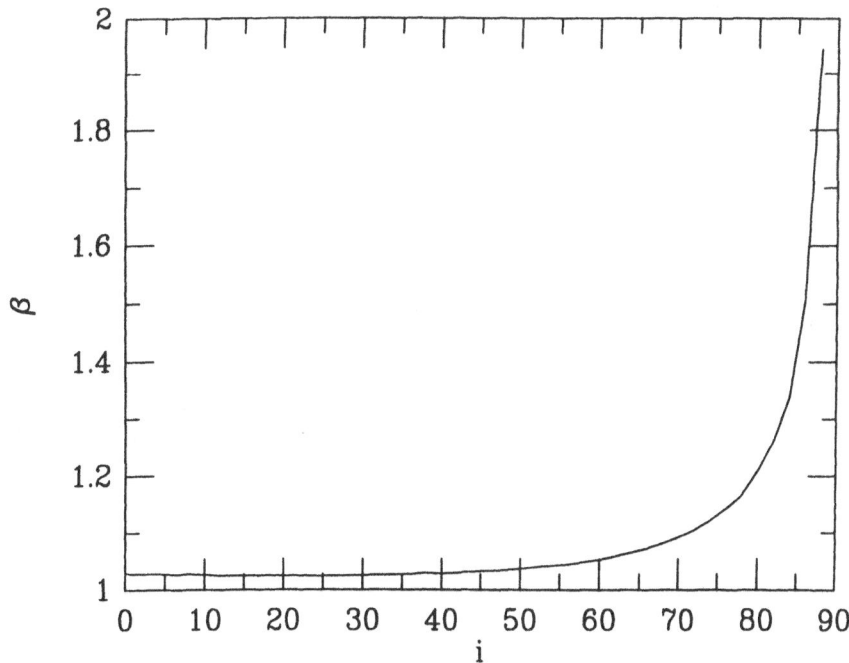

Figure 1: The ratio of the apparent area of the disk, β, with respect to the case in flat space for the x-ray emitting region ($r_x = 30$) as a function of inclination angle, i.

for the Newtonian case with a torque free boundary condition at $3R_g$. The radiation emitted from the disk is shifted in frequency due to the effects of longitudinal and transverse Doppler shifts as well as by gravitational redshift by an amount given by

$$\frac{\nu_\infty}{\nu_o} = \frac{L}{\gamma(1 - \vec{v} \cdot \vec{v}_\gamma)}, \tag{2}$$

where the lapse function $L = \sqrt{1 - r^{-1}}$, $1/\gamma$ corresponds to the transverse Doppler shift with $\gamma = (1 - v^2)^{-1/2}$, and $1/(1 - \vec{v} \cdot \vec{v}_\gamma)$ to the longitudinal Doppler shift. In the above expressions the rotational velocity of the gas, v, and the photon velocity v_γ are normalized to the speed of light. The flux is obtained by integration over the surface of the apparent image as outlined in Fu and Taam (1989a,b).

The ratio of the apparent x-ray emitting area of an accretion disk, β, with respect to the case in flat space is illustrated over a range of inclination angles in Figure 1. It is evident that the gravitational focusing of light rays increases the image area above that expected in Euclidean space. From this result it is clear that the dependence of the apparent area on inclination angle, i, is less steep than

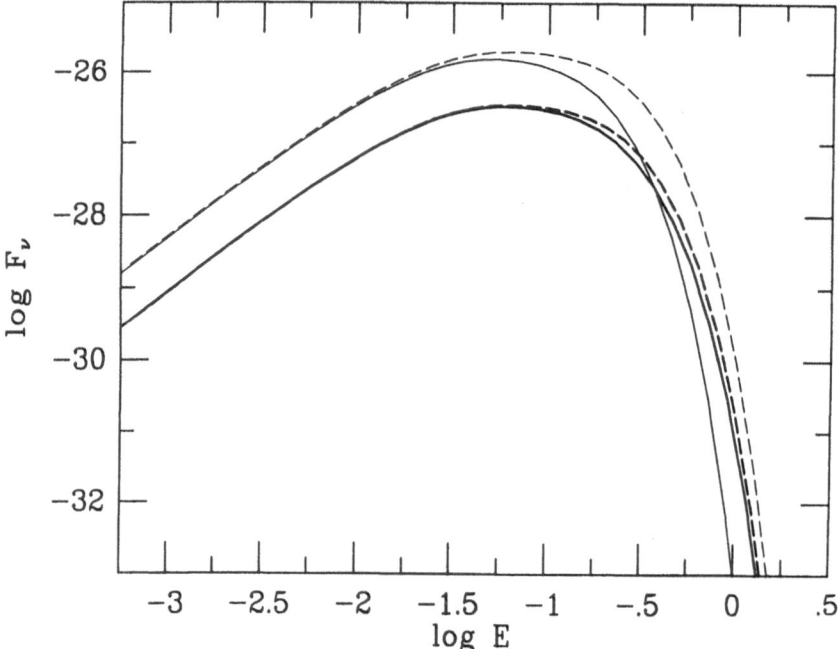

Figure 2: Spectral flux observed at infinity as a function energy in keV for an inclination angle of 0° (thin solid line) and 80° (thick solid line). The spectra for a Newtonian disk in flat space are displayed as a dashed lines for comparison.

the cos i factor expected without gravitational light bending. The differences are especially notable near the orbital plane ($i \gtrsim 80°$) where $\beta \gtrsim 1.2$.

The spectral energy distributions from an accretion disk surrounding a black hole of $7M_\odot$ accreting at $10^{-11}M_\odot yr^{-1}$ as viewed from inclination angles of 0° and 80° are shown in Figure 2. For comparison, the spectra for a Newtonian disk in flat space are also displayed. Note that the spectrum of a relativistic disk is softer than the corresponding one in flat space in both cases with the greater differences in the spectral shape corresponding to smaller inclination angles. This latter result reflects the lower temperatures in the relativistic disk and the effect of the gravitational and transverse Doppler shifts. The smaller differences with large inclination angles is attributable, in part, to the broadening of the spectrum at high energies by the longitudinal Doppler shifts (see Fu and Taam 1989a) and the larger apparent area of the x-ray emitting region. The differences are not as large as expected based upon the analysis of Hanawa (1989) since the zero torque condition at the inner edge was not incorporated in his Newtonian model disk. Our results imply that, when relativistic effects are included, the upper limits for the mass accretion rate in the system A0620-00 in its quiescent state are greater than that inferred by de Kool (1988). In particular, the flux in the 0.2–20 keV band for a system inclination angle of 0° and 80° are presented in Figure 3 with account taken of the attenuation in the interstellar medium by photoelectric absorption (with cross sections taken from the work of Morrison and McCammon 1983). For a column density of hydrogen equal to $3.9 \times 10^{21} cm^2$ and a distance of 870 pc we find that the upper limit to the mass flow rate within the inner regions of the disk, imposed by the lack of soft x-rays seen in the quiescent state (F $\lesssim 10^{-12} ergscm^{-2}s^1$), is about a factor of 3 larger [i.e., $\dot{M} \lesssim 3(8) \times 10^{-11} M_{\odot} yr^{-1}$ for $M = 7(13) M_{\odot}$] than that found previously. In contrast to the results based upon the Newtonian approximation, the upper limits are insensitive to the inclination angle of the system. The optical V magnitudes, corrected for 1.2 magnitudes of extinction, at an inclination angle of 0° and 80° for a range of mass accretion rates are illustrated in Figure 4. Upon comparison with the observed optical brightness of the system, with the star and disk contributing nearly equally, during quiescence ($V \sim 18.3$; see Oke 1977, Whelan et al.

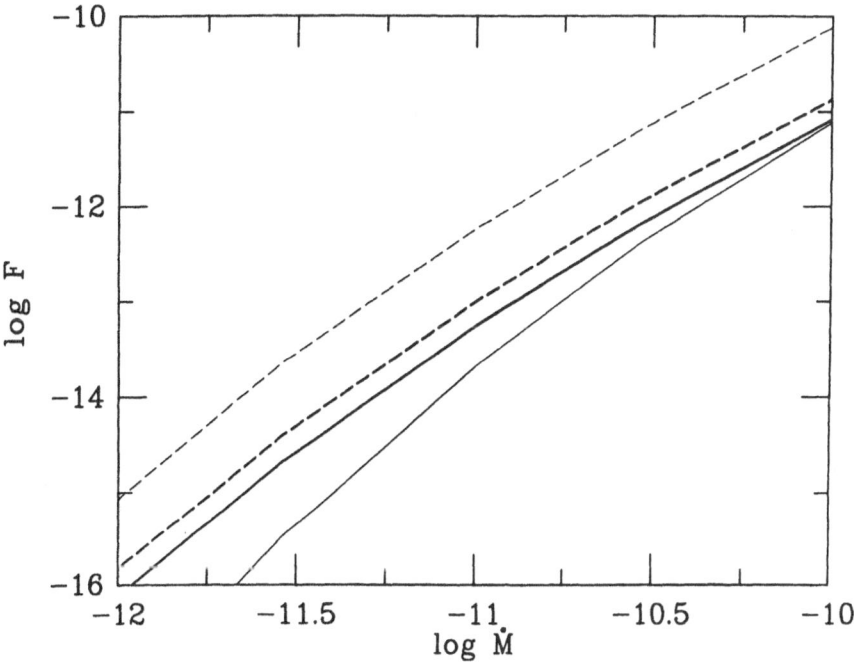

Figure 3: The flux in the 0.2 - 20 keV band from an accretion disk for a binary system inclination angle of 0° (thin solid) and 80° (thick solid) is illustrated as a function of mass transfer rate. The flux for the Newtonian disk in flat space is shown as a dashed line for comparison.

1977, Murdin et al. 1980) it is immediately seen that the optical data provides a constraint on the mass transfer rate of $\lesssim 10^{-11} M_{\odot} yr^{-1}$ for $i = 0$ and $\lesssim 10^{-10} M_{\odot} yr^{-1}$ for $i = 80$. We remark that this mass transfer rate is in the mass transfer range for which an illuminated stellar companion is unstable (Hameury et al. 1986). However, the lack of hard x-rays ($\gtrsim 2$ keV) in the quiescent state suggests that x-ray heating of the photosphere is ineffective and that models based

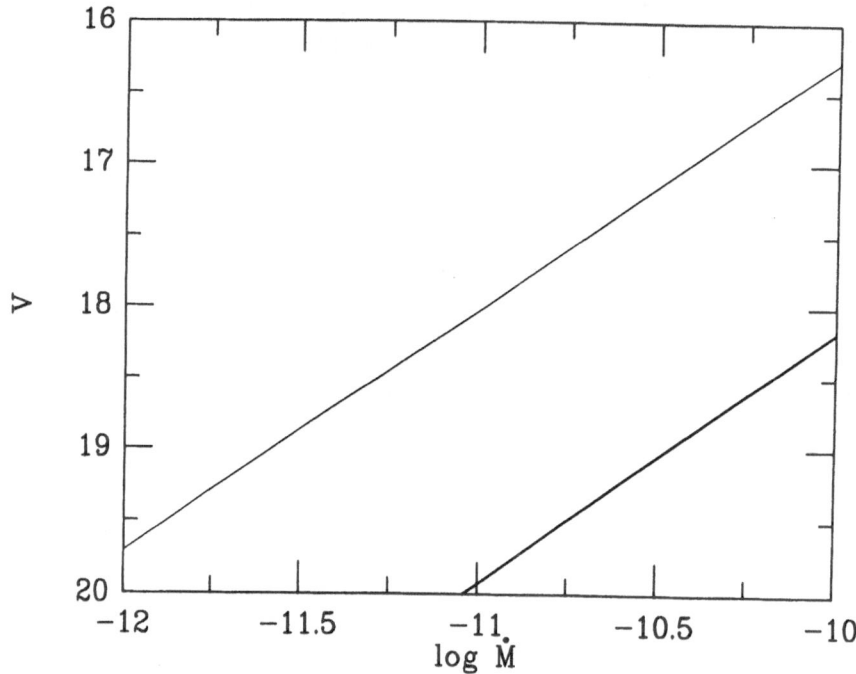

Figure 4: The visual magnitude, V, from an accretion disk for a system inclination angle of 0° (thin solid) and 80° (thick solid) is shown as a function of the mass transfer rate. The magnitude is corrected for 1.2 magnitudes of extinction.

upon a mass transfer instability situated in the stellar envelope may not be applicable to a system like A0620-00.

For black hole masses in the range of $7 - 13M_\odot$ and inclination angles between $0° - 80°$, the results of our calculations imply that, if the mass flow rate through the accretion disk is constant and lies between $\sim 10^{-11} - 10^{-10}\, M_\odot yr^{-1}$, the x-ray flux in the 0.2 − 20 keV band is predicted to lie between 10^{-15} and $10^{-12}\, ergs\, cm^{-2} s^{-1}$. If x-rays are not seen in future missions at the level expected our results would imply either that the contribution of the disk to the optical light is smaller than that based upon steady disk models or that the viscosity of the matter in the disk is too low to have allowed it to flow toward the inner regions within the time of the last outburst nearly 15 years ago. Because of the potential importance of general relativistic effects, analyses based upon this framework should be incorporated in the interpretation of the future x-ray measurements of A0620-00.

ACKNOWLEDGEMENTS. This research has been supported in part by NASA under grant NAGW-768.

REFERENCES.

de Kool, M. 1988, Ap. J. <u>334</u>, 336.

Fu, A., and Taam, R. E. 1989a, Ap. J., *submitted.*

Fu, A., and Taam, R. E. 1989b, Ap. J., *submitted.*

Fukue, J., and Yokoyama, T. 1988, Publ. Astr. Soc. Japan, 40, 15.

Hanawa, T. 1989, Ap. J., *submitted.*

Hameury, J. M., King, A. R., and Lasota, J. P. 1986, Astr. Ap., 162, 71.

Hameury, J. M., King, A. R., and Lasota, J. P. 1987, Astr. Ap., 171, 140.

Long, K. N., Helfand, D. L., and Grabelsky, D. A. 1981, Ap. J., 248, 925.

Luminet, J.-P. 1979, Astr. Ap., 75, 228.

Lynden-Bell, D. 1969, Nature, 223, 690.

McClintock, J. E., and Remillard, R. A. 1986, Ap. J., 308, 110.

Morrison, R., and McCammon, D. 1983, Ap. J., 270, 119.

Murdin, P., Allen, D. A., Morton, D. C., Whelan, J. A. J., and Thomas, R. M., 1980, M.N.R.A.S., 199, 709.

Oke, J. B. 1977, Ap. J., 217, 181.

Page, D. N., and Thorne, K. S. 1974, Ap. J., 191, 499.

Shakura, N. I., and Sunyaev, R. A. 1973, Astr. Ap., 24, 337.

Whelan, J. A. J., et al. 1977, M.N.R.A.S., 180, 657.

THE COMPACT OBJECT IN SS433: NEUTRON STAR OR BLACK HOLE?

Tomaž ZWITTER and Massimo CALVANI
International School for Advanced Studies, Strada Costiera 11, Miramare,
34014 Trieste, Italy

A large fraction of the X–ray emission from SS433 originates in the relativistic jets (Watson *et al.* 1986). Recent observations of the partial X–ray eclipse by the Ginga satellite can be used to constrain the size of the emission region; this in turn limits the size of the underlying accretion disk. Moreover, use of the Roche lobe geometry allows an estimation of the mass of the compact object.

Ginga observations were reported by Kawai *et al.* (1988, 1989); they observed a partial eclipse in the energy range 2 to 28 keV. The duration of the eclipse was 2.4 days. The flux reached the low state 0.5 – 0.6 days after ingress.

We explain the observations with the following geometrical model: the eclipsed part of the X–ray emission is assumed to be originating in two equal cylinder and their symmetry axis points in the direction predicted by the kinematic model (Margon 1984 and references therein). The accretion disk obscures the basement of the blue jet. The visible part of the jet starts at the distance L_d from the compact object where the jet emerges from the disk's funnel. The inner part of the red jet is additionally obscured by the rim of the accretion disk.

The cylinders should satisfy two constraints following from the observations: (i) They are partially eclipsed by the normal star for 2.4 days. (ii) The star completely covers them 0.5 – 0.6 days after the time of ingress. This explains the long-lasting low state in the X–ray intensity.

As the results may strongly depend on the actual shape of the accretion disk, we consider two extreme cases:

(i) The accretion disk has a maximal size, i. e. it fills the critical Roche lobe of the compact object. In this case the red jet is completely obscured (model A), or (if it is long enough) a part of it is seen (model B).

(ii) The accretion disk has a minimal size, i. e. it is a sphere with radius L_d (model C). Technical details of the calculations will be published elsewhere (Zwitter & Calvani 1989c).

The results are presented in Fig. 1. Permitted regions in the (disk thickness, mass of the compact object) plane for the three models are marked. R_* and R_x are the radii of the critical Roche lobes of the normal star and the compact object. A

F. Meyer et al. (eds.), Theory of Accretion Disks, 321–323.
© *1989 by Kluwer Academic Publishers.*

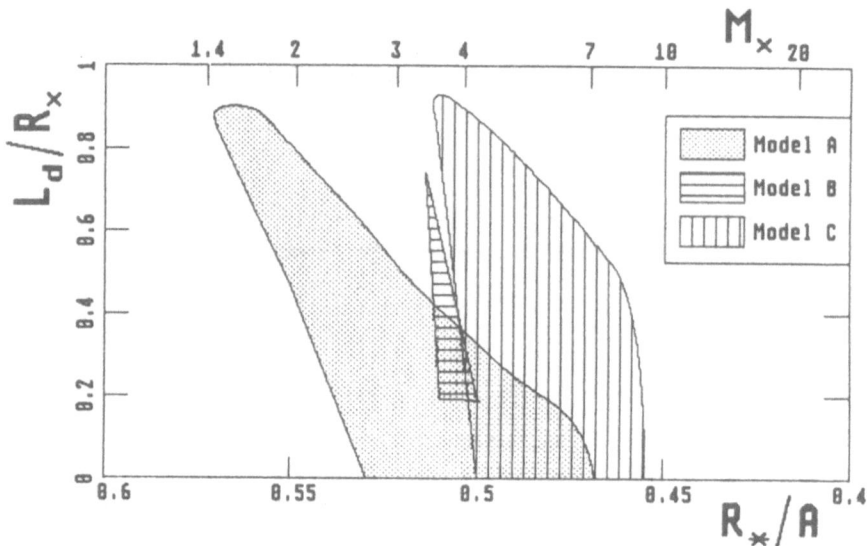

Figure : Permitted regions in the (disk thickness, mass of the compact object) plane. See text.

denotes the separation of the binary system. M_x is the mass of the compact object in solar units.

The Ginga observations show that the mass of the compact object is between 1.4 and $10 M_\odot$. This is consistent with the analysis of the EXOSAT data ($M_x > 2 M_\odot$, Zwitter & Calvani 1989a) and with the results from the optical data ($M_x > 9.1 M_\odot$, Leibowitz et al. 1984, Zwitter & Calvani 1989b). The weighted average of the three results suggests that the mass of the compact object is $\approx 5 M_\odot$, and therefore that it is probably a black hole.

The geometrical models presented above are relatively complicated because the shape of the accretion disk is not known. Unfortunately a small variation in the disk's shape can produce a large difference in the degree of obscuration of the red cylindrical emission region. The problem can be avoided by new measurements similar to the ones we discuss here, but taken at the time of the crossing of the moving lines. The particularly symmetrical geometrical situation at this precessional phase would make the models much simpler and, we believe, more accurate.

REFERENCES.

Kawai, N., Matsuoka, M., Pan, H. C., and Stewart, G. C. (1988). In *Physics of Neutron Stars and Black Holes*, Y. Tanaka (Ed.). (Universal Academy Press, Tokio), p. 231

Kawai, N., Matsuoka, M., Pan, H. C., and Stewart, G. C. (1989). *Pub. Astron. Soc. Japan*, **41**(2), in press

Leibowitz, E. M., Mazeh, T., and Mendelson, H. (1984). *Nature*, **307**, 341

Margon, B. (1984). *Ann. Rev. Astr. Ap.*, **22**, 507.

Watson, M. G., Stewart, G. C., Brinkmann, W., and King, A. R. (1986). *Mon. Not. R. astr. Soc.*, **222**, 261

Zwitter, T., and Calvani, M. (1989a). *Mon. Not. R. astr. Soc.*, **236**, 581

Zwitter, T., Calvani, M., Bodo, G., and Massaglia, S. (1989b). *Fund. Cosmic Phys.*, submitted

Zwitter, T., and Calvani, M. (1989c). ISAS preprint

PHYSICS OF ACCRETION BY SPIRAL SHOCK WAVES

H.C. SPRUIT
Max Planck Institut für Physik und Astrophysik
Karl-Schwarzschildstr. 1, 8046 Garching, West Germany,

ABSTRACT. The properties of disks accreting by global spiral shocks are reviewed. The discussion includes the formation of these waves, the way they transport angular momentum, a selfsimilar model for spiral shocked accretion and recent numerical simulations. These simulations suggest that at least in disks produced by mass overflow in close binaries the efficiency of accretion (in terms of an effective α-value) can be substantial.

1. Introduction

Spiral waves in disks have been studied extensively in the context of galactic dynamics (*e.g.* Binney and Tremaine, 1987), and to some extent that of protostellar disks (spiral waves generated by proto-Jupiters: Goldreich and Tremaine, 1980, Lin and Papaloizou, 1979, 1986; Larson, 1988). It has been known for a long time that such waves can carry a negative angular momentum, so that their dissipation would lead to accretion of the fluid supporting the waves onto the central object (Donner, 1979, Michel, 1984). Yet the idea that such waves provide a natural means by which an accretion disk may solve its 'angular momentum problem' without resorting to an enhanced viscosity has developed only recently. In the calculations by Donner (1979) and Lin and Papaloizou (1986) for example waves with negative angular momentum are present, but a viscosity is also included explicitly in these formulations, and is treated as the main angular momentum transporting process. The (nearly) inviscid numerical simulations by Sawada *et al.* (1985) and Matsuda *et al.* (1987) were the first to show accretion through spiral shaped shock waves. The existence of consistent accretion solutions of this kind was shown by more analytical means by Spruit (1987), and their properties were compared with those of viscous disks in Spruit *et al.* (1987). Applications to the protoplanetary case were made by Larson (1988).

In this paper I review the results obtained so far on accretion by spiral shocks. They give a reasonably optimistic picture about the ability of spiral shock waves to produce accretion speeds comparable to those inferred from observations. It is not at all clear at present however, if the mechanism can compete in practice with other mechanisms for producing an effective viscosity (like magnetic turbulence), in part because estimates are for these alternative processes are not yet very reliable.

F. Meyer et al. (eds.), Theory of Accretion Disks, 325–340.

2. Angular momentum transport

Consider a thin disk of large extent, in the sense that the outer radius r_o is very much larger than the inner radius r_i. Near both boundaries plausible mechanisms exist for producing perturbations that transport angular momentum. At the inner boundary for example an asymmetric magnetosphere may excite shock waves (Michel, 1984). A sound wave or surface wave instability of the kind considered by Drury (1977, 1980, 1985) and subsequently by, among others, Papaloizou and Pringle (1984, 1985), Kojima, (1986), Sekiya and Miyama (1988), Glatzel(1987, 1989), Narayan et al. (1987), Hawley and Blaes, (1988), Hawley (this volume) can develop at either boundary. Such mechanisms may easily produce angular momentum transport within some distance of these boundaries. It is not clear however that they also suffice to produce a significant transport at distances $r \ll r_o$ but $\gg r_i$, and it is in this range that the 'angular momentum problem' in cool (thin) disks is most acute.

Let us consider the effect of nonaxisymmetries at the outer disk boundary. For example, the mass distribution in the outer parts of a star forming disk might be irregular, depending on the history of the disk formation process. The orbital frequency at this distance is small and for simplicity consider these irregularities just as stationary in an inertial frame. Gas orbiting at a somewhat smaller distance however sees the nonaxisymmetries as time dependent, hence waves are excited in the disk. Being excited at the outer edge, they are inward propagating waves, wound up by the differential rotation into a *trailing* spiral pattern. As the wave propagates inward, its frequency as measured in a comoving frame increases. Such waves have the interesting property of possessing a 'negative angular momentum'. This just means that a disk with such a wave in it has a lower total angular momentum than the same disk without the wave. Let m be the azimuthal order of the wave, Ω the Kepler frequency, so that the wave frequency in a comoving frame, σ, is

$$\sigma = \omega - m\Omega = -m\Omega, \tag{1}$$

since we have assumed that the pattern frequency ω is negligible. For waves propagating in a differentially moving fluid a conserved quantity, the wave action s_w can be defined; in the short wavelength (WKB) approximation it is (*e.g.* Lighthill, 1978)

$$s_w = \int_V \frac{\frac{1}{2}\rho v_w^2}{\sigma} dV, \tag{2}$$

where ρ is the gas density, v_w the velocity amplitude of the wave, and the integral is over the volume V of the wave packet. This quantity is conserved along the path taken by the wave packet, as long as there are no (linear or nonlinear) dissipative processes affecting the wave. Note that the *energy* of the packet is not conserved even in the absence of dissipative processes; this expresses the fact that in a shear flow a wave exchanges energy with the mean flow. For a wave of azimuthal order m in a disk, the quantity

$$j = m s_w \tag{3}$$

is equally conserved, and can be identified with the angular momentum of the wave (Narayan et al. 1987). With (1)

$$j = -\int_V \frac{\frac{1}{2}\rho v_w^2}{\Omega} \quad < 0. \tag{4}$$

The minus sign has actual physical significance. It means that the disk with the wave in it has a lower angular angular momentum than without the wave (the total angular momentum being conserved in the absence of external torques and dissipation). When and where the wave dissipates, it reduces the angular momentum of the gas, causing the gas to sink towards the central object. When and where will the wave dissipate? Suppose the wave starts at the outer edge with a small amplitude. As it propagates in, Ω increases inward, and to conserve j, the wave amplitude v_w in general has to increase. To see this consider the volume occupied by the wave packet, which is

$$V = 2\pi r H \Delta r, \tag{5}$$

where H is the disk thickness and Δr is the radial extent of the packet. Since the front and back of the packet move in r with the radial component of the local group speed, which is proportional to the sound speed c_s, the distance between them increases at a rate $d\Delta r/dt \propto \Delta r dc_s/dr$, which can be integrated to yield

$$\Delta r \propto c_s. \tag{6}$$

With 4 this yields

$$v_w \propto [\Omega/(rHc_s\rho)]^{1/2}. \tag{7}$$

For the selfsimilar disk solution discussed below, $\Omega \propto r^{-3/2}$, $H \propto r$, $c_s \propto r^{-1/2}$, $\rho \propto r^{-3/2}$, so that

$$v_w \propto r^{-3/4}, \tag{8}$$

though this will be different for disks with a different radial structure. The amplitude of the wave *increases* inward, and eventually the wave turns into a shock. From then on the wave's angular momentum is no longer conserved, but slowly decreases (in absolute value) as the wave dissipates energy in the shock. The negative angular momentum transferred in the process to the fluid makes it accrete. As the wave's angular momentum decreases, the specific angular momentum of the gas it encounters also decreases (as $r^{1/2}$). In the consistent selfsimilar solutions, the two vary exactly in parallel, so that accretion goes on throughout the entire disk.

Where the wave meets the inner edge, a part of it will reflect. The reflected wave propagates outward, and again by angular momentum conservation *decreases* in amplitude as it propagates outward. At a distance $r \gg r_i$ the reflected wave therefore has no influence. It follows that there is a range in radius $r_i \ll r \ll r_o$ where the wave depends *neither on the outer nor on the inner boundary condition*. See figure 1. What does depend on these conditions is the extent of this range (being small or nonexistent if the initial amplitude of the wave at large radius is very small). Since the time scale for variations in the conditions at r_o ($\approx$ the orbital time scale there) is very long, the wave is nearly *stationary* in the range $r_i \ll r \ll r_o$ (compared with the orbital time scale). It follows that it must be a special kind of stationary solution of the hydrodynamic equations, namely one that contains no reference to inner or outer boundaries. For a simplified equation of state and energy loss terms, such solutions are known (Donner, 1979; Spruit, 1987) in the form of selfsimilar (scale invariant), spiral shaped shock waves. This kind of solution looks the same when seen at different magnifications, except for a rotation in azimuth over some angle; the spiral is logarithmic. Examples are shown in figure 2.

328

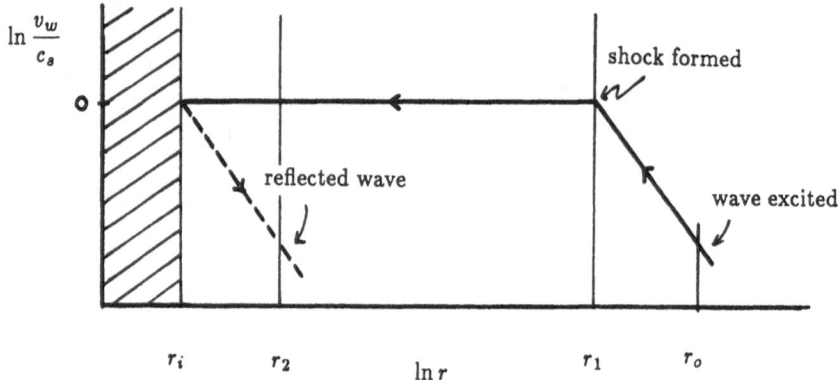

Figure 1: Formation of a spiral shocked accretion flow from a wave excited at the outer edge of a disk (schematic). The figure shows the amplitude (in units of the local sound speed) of a trailing wave generated at the outer boundary r_o by a stationary nonaxisymmetry there. As it propagates in, its amplitude increases due to wave angular momentum conservation, and at r_1 it turns into a shock of moderate strength. The reflected wave produced at the inner edge of the disk decreases outward (dotted line). Between r_2 and r_1 the accretion flow set up by the wave does not depend on the conditions at the inner or outer edge, and for a simple equation of state and energy loss term is selfsimilar.

The picture summarized by figure 1 is substantiated by numerical simulations. This can be seen in figures 4, 5. Figure 4 shows how the spiral shock is set up by the inward propagation of the external perturbation at the sound speed. It also shows how the reflected wave at the inner edge decreases outward in amplitude. Figure 5 shows how a weak disturbance increases in amplitude toward the center until it has the amplitude and form predicted from the selfsimilar solution.

The heuristic arguments above demonstrate that flows like the selfsimilar spiral shocks should be expected generically in cases where the inner radius of the disk is very much smaller than the outer edge, and where a finite amount of nonaxisymmetry exists at the outer edge. That is, the condition for their occurence is that the disk must be 'live': there must be something that adds mass nonaxisymmetrically at the the outer edge, or there must be another source of perturbation. Whereas other accretion mechanisms work plausibly near the disk edges, for example Michel's (1984) spiral shocks generated by an asymmetric magnetosphere of the central object, the present process is especially interesting because it works in its purest form at places that are neither close to the inner nor to the outer boundary.

The picture drawn here suggests that spiral shaped shock waves would be a generic feature in disks, contributing to (or even dominating) the accretion process. This does not exclude the possibility of 'dead', nonaccreting disks, however. Conceivably there might be 'fossil' disks which receive no mass, and are far from external perturbers (e.g. Kafka and Meyer, 1984). The level of perturbation in such disks could be too low to grow into a shock before reaching the inner edge of the disk.

The properties of selfsimilar accretion flows are discussed further in the next section. It must be stressed however that they are rather special solutions. In realistic situations, the presence of inner and outer boundaries at finite distances, a strong perturbing force in the form of a companion, and more realistic treatment of the equation of state and the energy loss modify the flow. In particular, though accretion by spiral shocks may still occur, the opening angle of the spiral will in general not be constant, and hence the flow not selfsimilar any more. For example, numerical simulations show that accretion in binaries with mass ratios of order unity is more effective in the outer parts of the disk than expected from a selfsimilar flow; this is due to the high amplitude of the wave generated by the companion (see further discussion in section 6).

3. Selfsimilar solutions

Numerical simulations of spiral shocked accretion has the advantage of easy adaptation to realistic conditions but is less suited for a discussion of the general properties and physics. In this section I show how (semi)analytical results from a simpler model problem provides insight that nicely complements the numerical results.

The model makes the following assumptions and approximations:
1. Selfgravity is neglected, the gravitational field is that of a (Newtonian) point mass.
2. Viscosity is neglected.
3. The equation of state is that of an ideal gas of constant ratio of specific heats γ.
4. The dimension perpendicular to the disk is treated in a two-dimensional form of 'slim disk' approximation. That is, the physical quantities are represented by their values at the midplane, the disk is assumed vertically in hydrostatic equilibrium in a gravitational field that is approximated by expansion near the midplane.
5. Energy loss is treated as optically thick, the opacity assumed has a special radial, temperature, or density dependence (detailed below).

Let (u, v) be the radial and azimuthal velocities in cylindrical (r, ϕ) coordinates, ρ, T, p the density temperature and gas pressure (no radiation pressure included), S the entropy, H the disk thickness, and $\Sigma = H\rho$ the surface density. Then the model is summarized by the equations:

$$\rho\left(\frac{du}{dt} - \frac{v^2}{r}\right) = -\frac{\partial p}{\partial r} - \rho\frac{GM}{r^2}, \tag{9}$$

$$\rho\left(\frac{dv}{dt} + \frac{uv}{r}\right) = -\frac{1}{r}\frac{\partial p}{\partial \phi}, \tag{10}$$

$$\frac{\partial \Sigma}{\partial t} + \frac{1}{r}\frac{\partial}{\partial r}(\Sigma u r) + \frac{1}{r}\frac{\partial}{\partial \phi}(\Sigma v) = 0, \tag{11}$$

$$\Sigma T\frac{dS}{dt} + \frac{4}{3}\frac{\sigma T^4}{\Sigma \kappa} = 0, \tag{12}$$

where σ is the Stefan-Boltzmann constant and κ the opacity. Two possible dependences are allowed:

$$\kappa = \kappa_0 r^{-1/2} \quad or \quad \kappa = \kappa_0' \rho^n T^{-(3n-1)/2}, \tag{13}$$

where n is arbitrary. Neither of these is physically very realistic, but this restriction is necessary to allow the existence of selfsimilar solutions. In the following we use the form

$\kappa \sim r^{-1/2}$. The disk thickness H is taken to be a function of r only, and identified with the pressure scaleheight at some value of ϕ.

We look for stationary solutions of equations 9-12. With $\partial/\partial t = 0$, these equations have two symmetries: invariance for rotation over an arbitrary angle in azimuth ϕ, and a scale invariance in r. As a result of this, there exist special solutions of the form

$$u = r^{-1/2}U(\psi), \qquad (14)$$
$$v = r^{-1/2}V(\psi), \qquad (15)$$
$$p = r^{-5/2}P(\psi), \qquad (16)$$
$$\rho = r^{-3/2}R(\psi), \qquad (17)$$
$$H = rs, \qquad (18)$$

where

$$\psi = \phi + B \ln r \qquad (19)$$

and B, s are constants. Thus the solutions have a logarithmic spiral symmetry; the angle between the spiral and the radial direction is given by $\tan\theta = B$. By virtue of the assumed hydrostatic balance in the vertical,

$$s = [P(0)/R(0)]^{1/2}(GM)^{-1/2}. \qquad (20)$$

For such solutions, equations 9-12 reduce to a set of four nonlinear first order ordinary differential equations for the variables U, V, P, R. The boundary conditions are jump conditions at the shock positions (located at fixed values of ψ). In addition the solutions have to pass through a sonic point type of singularity, which adds another constraint.

As a result (for details see Spruit, 1987) the free parameters of the problem are $\tan\theta$ (or equivalently the ratio of the disk temperature to the virial temperature), the ratio of specific heats γ and n_s, the number of shocks. Examples of the solutions are shown in figure 2.

4. Properties of spiral-shocked accretion

The discussion in this section is based on the self similar solutions introduced above, but the qualitative aspects will also hold for more general, non-selfsimilar cases. Some conclusions are not even restricted to spiral-shocked accretion, but should hold for any process of accretion with angular momentum.

The opening angle of the spirals is related directly to the temperature of the disk. Since the shocks turn out to be of only moderate strength, they propagate roughly at the sound speed. The condition that they yield a stationary pattern in an inertial frame then requires that the angle between the shock surface and the direction of the orbital motion be of the order $c/(\Omega r)$ if c is the sound speed. This angle is therefore of the order of the inverse Mach number of the Kepler velocity, or the square root of the ratio of the disk temperature to the virial temperature.

Not all combinations of disk temperature (measured by θ) and γ are possible. Which ones are allowed depends mostly on whether radiative losses are important. If radiative

a

b

c

Figure 2: Examples of disks formed by spiral shocked accretion flow. The pictures show the gas density, on a logarithmic scale. Hot flows (a) have more open spirals than cool ones (b). Flows with two arms have stronger shocks and accrete more efficiently than those with more arms (c).

332

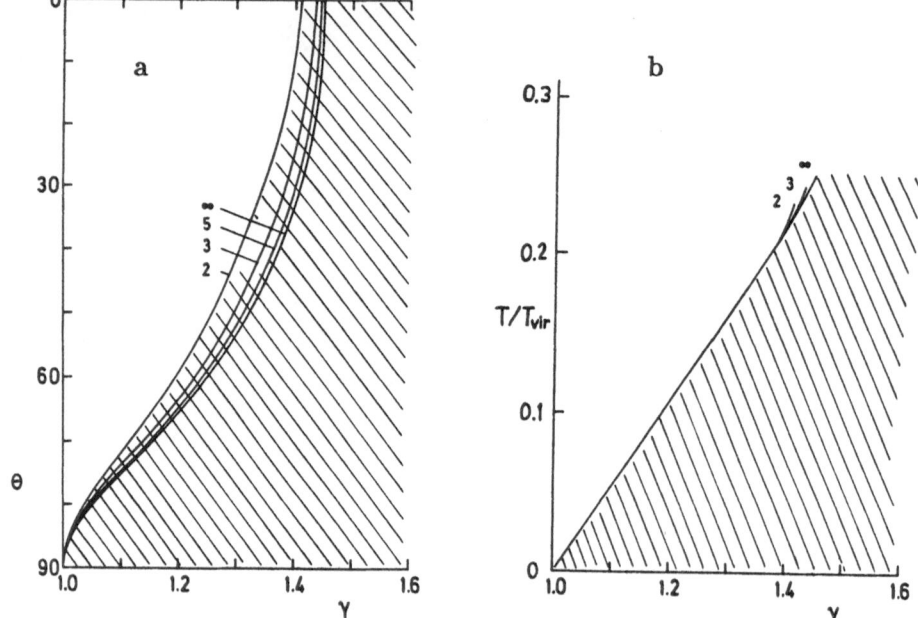

Figure 3: Spiral angle (a) and temperature (b) as functions of the ratio of specific heats for selfsimilar solutions. The shaded area shows the combinations allowed if radiative loss is included. Solutions without radiation loss are also possible; they occupy the left boundary of the shaded area. The position of this boundary depends on the number of shocks; the cases $n_s = 2, 3, 5$ and $n_s \to \infty$ are shown.

losses are ignored, equations 9-12 are invariant to multiplication of ρ and p by a common constant. We shall call the case without losses 'adiabatic' (though it is not isentropic, due to the shock dissipation). On the one hand this means that in this case the pressures and densities are determined by the equations only up to a common constant; on the other hand it implies that there is one less free parameter in the problem. Hence in this case the opening angle θ is a unique function of γ, for given n_s. This is shown in figure 3a. Solutions including radiative loss exist to the right of this relation. Figure 3b shows the same in terms of the disk temperature instead of the opening angle θ. The adiabatic shock relations for different numbers of shocks (n_s) are nearly identical in this representation, and close to the relation $T/T_{vir} = \frac{1}{2}(\gamma - 1)$.

For cases with radiative loss, the accretion rate is the natural parameter. As the accretion rate is increased, the disk temperature increases (moving up along a vertical line in figure 3). What happens at high rates depends on the ratio of specific heats. For $\gamma < \gamma_c$ ($\gamma_c \approx 1.45$ for $n_s = 2$), the disk temperature approaches a *finite* value as $\dot{M} \to \infty$. That is, for low values of γ the accretion can proceed *without energy loss*, and this condition is approached at high accretion rates, where the high surface density makes the disk optically thick. In such cases all the gravitational energy, liberated through the shocks, stays in the gas and is accreted with the flow onto the central object. To understand the conditions under which this is possible, consider the internal energy per unit mass, $e = p/[\rho(\gamma - 1)]$. If no energy was lost during accretion from infinity to a Kepler orbit at distance r, the internal energy equals the energy dissipated in the process, $\frac{1}{2}GM/r$, so that

$$T/T_{vir} = \frac{p}{\rho}\frac{r}{GM} = \frac{1}{2}(\gamma - 1). \tag{21}$$

For γ close to 1, the number of internal degrees of freedom of the gas is large, so that the temperature can stay low, and the gas bound to the central object.

For $\gamma > \gamma_c$ on the other hand radiative loss is essential for the accretion process. The maximum temperature at which accretion can proceed is now reached at a finite *maximum* accretion rate. This maximum depends on the details of how radiative losses is included.

The presence of a critical value of γ below which accretion without energy loss is possible is not a peculiarity of the model presented here. The 'slim' α-disk model also has this property and indeed the physics involved shows that it should be a very general property of accretion with angular momentum (Spruit *et al.* 1987). γ_c Is to a large degree the equivalent of the critical value $\gamma = 5/3$ that occurs in spherical accretion (Bondi, 1952).

The high accretion rates at which the role of the critical value of γ becomes apparent are not necessarily relevant for actual astrophysical objects. The limiting factor in most cases is the radiation pressure, not included in the above model. It turns out (Spruit *et al.* 1987, Spruit, 1989) that the Eddington limit becomes relevant, in most kinds of accreters, before these high rates are reached. Exceptions however may be (proto-) planets in the process of accreting gas; the effects discussed above may well be relevant there. In particular, very fast accretion not limited by radiation loss may perhaps be involved in the formation of the giant planets.

The efficiency by which the spiral shocks accrete gas can be compared with α-disk models. From the radial drift speed one can define an effective value of α (the value that an α-disk must have to get the same accretion time scale, for a given accretion rate). For the selfsimilar models this value is, in the limit of very cool models:

$$\alpha_{eff} = 0.013(H/r)^{3/2} \tag{22}$$

(Spruit, 1987; Larson, 1988). The efficiency of accretion by spiral shocks therefore decreases with disk temperature. This is because the strength of the shocks decreases with temperature (somewhat unexpected perhaps, because the Mach number of the Kepler flow increases). It must be remembered that α_{eff} is not a local quantity, since the accretion process involves a global shock wave in contrast to the α-disk model, where α can be introduced as a local quantity.

5. Applications

The most interesting question is of course whether accretion by spiral shocks is effective enough in practical situations. We look at this briefly for a few kinds of disks.

334

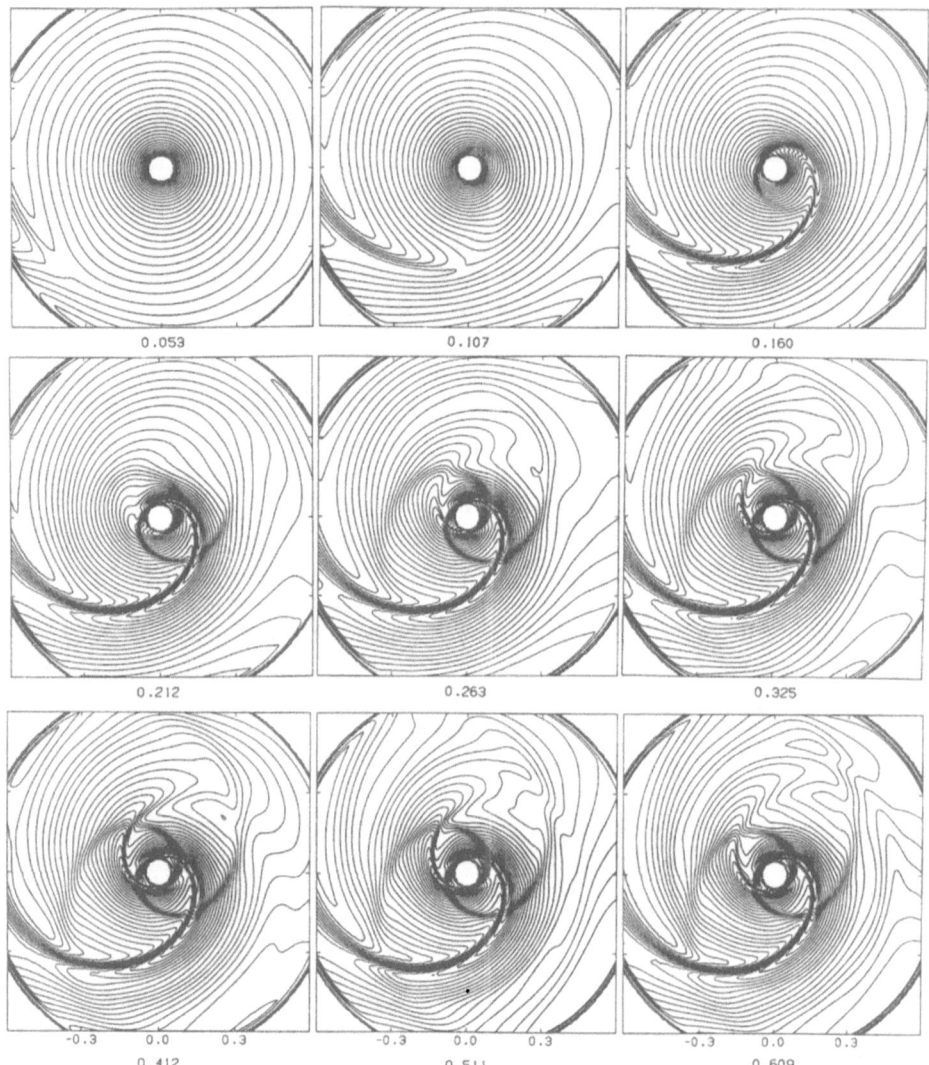

Figure 4: Development of spiral shock pattern from tidal perturbation. The disk has a temperature $T/T_{vir} = 0.1$, mass ratio is 0.1, companion is at the left at $R = -1$. Time sequence (time in units of the orbital period) shows inward propagation of the disturbance at the sound speed. The second arm forms from the weaker tidal perturbation on the opposite side and becomes visible only near the center. Reflection of the two waves at the center produces two additional leading spirals whose strength decreases outward.

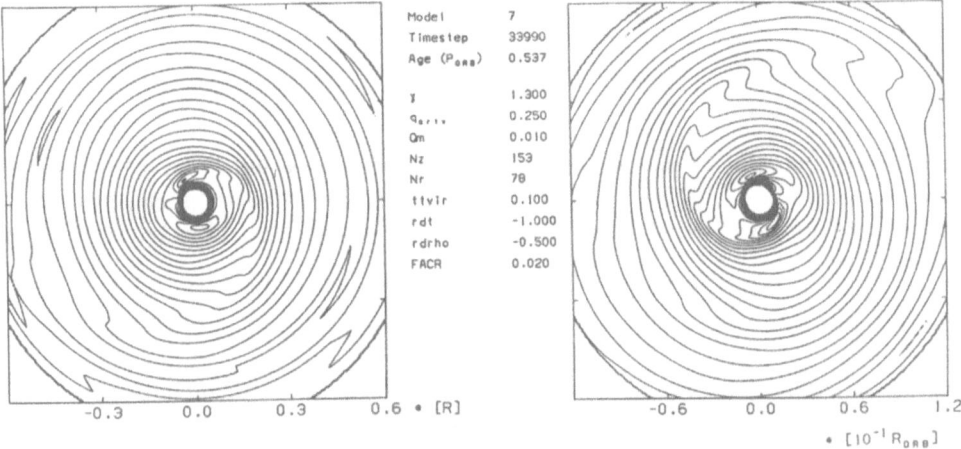

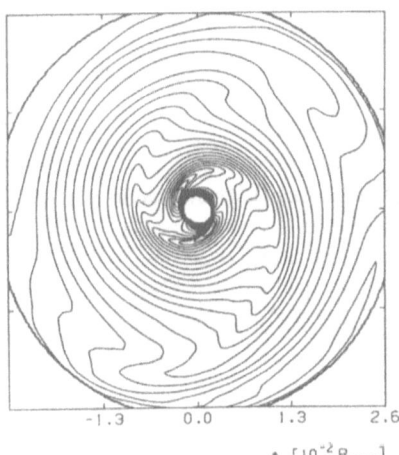

Figure 5: Amplification of a weak external perturbation into a stationary shock pattern. The figure shows (top left) a disk with temperature $T/T_{vir} = 0.1$, perturbed by a companion with mass ratio $q = 0.01$, after the pattern has settled into a steady state. The inner regions are recomputed on a 5 times smaller scale (top right), with the previous large scale result providing the outer boundary condition. This process is repeated once more to yield the picture at the left. The strength of the shocks in the last picture is close to that predicted by the selfsimilar model.

CATACLYSMIC VARIABLES

For inferred disk thicknesses, H/r of the order 0.1, the selfsimilar value of α_{eff} is (eq. 22) $\sim 4\,10^{-4}$, which is too low to be interesting for CV's. For the most common mass ratio's, $0.1 < q < 1$, the forcing by the companion is so strong however that the resulting spiral shocks have a strength much above the selfsimilar value over a large part of the disk. Only in the innermost parts of the disk do the waves approach the predicted selfsimilar value (Spruit et al. 1987). This is also demonstrated by the recent numerical simulations by Matsuda et al. (this volume), which show values of α up to 0.1. These calculations, and those reported by Różycka and Spruit (this volume) still contain certain idealizations, so that it is not entirely clear at present if spiral shocks can yield the required values of α in CV disks (see also the discussion in the next section).

An interesting aspect of spiral shocks applied to CV disks is the prediction that α would be significantly smaller in the inner than in the outer parts of the disk. This might have effects on the character of the outbursts in disk instability models.

PROTOSTELLAR DISKS

Assuming $H/r = 0.1$ for the protostellar disk the selfsimilar α-value of $4\,10^{-4}$ yields an accretion time scale of 10^6yr or shorter for distances up to 8 AU. Spiral shocked accretion might therefore play a role in the inner parts of protostellar disks (cf. also the discussion in Larson, 1988).

JUPITER-GENERATED SPIRAL SHOCKS

At the mass ratio appropriate for Jupiter $q \sim 10^{-3}$, the spiral waves are initially so weak that they steepen into shocks only in the very innermost parts of the disk. This is illustrated in figure 5, for the somewhat larger mass ratio $q = 0.01$. Jupiter-generated spiral shock waves are therefore probably not very important for the evolution of protostellar disks.

6. Numerical simulations

Numerical simulations are necessary to determine the effect of spiral shocks in the more realistic situations, such as those in cataclysmic variables. The first simulations showing spiral shocks (Sawada *et al.* 1985) were set up with this kind of system in mind.

A significant difficulty with all numerical simulations is the presence of several different time scales. It is relatively easy to follow the evolution of the disk on an orbital time scale. The accretion (radial drift) time scale τ_a however is much longer:

$$\tau_a = \frac{1}{\alpha_{eff}\Omega}\left(\frac{v_k}{c_s}\right)^2. \tag{23}$$

This time scale can be followed only for rather hot disks with high values of α_{eff}. For cool disks the evolution can usually be followed until the wave pattern is stationary, but not long enough for the accretion process to reach a steady state. This is unfortunate since it makes it hard to derive realistic estimates of α_{eff} from the simulations. Nevertheless, a true steady state has been reached for some cases by Matsuda et al. (this volume), and values of α_{eff} have been derived from these.

FORMATION OF THE SHOCK PATTERN

First, consider a few numerical calculations that illustrate the physical scenario for spiral shock formation developed in section 2.

The first example (figure 4) shows a disk perturbed by a companion with mass ratio 0.1, for a fairly hot disk ($T/T_{vir} = 0.1$). This shows how the tidal perturbation propagates into the disk, and after reflecting at the computational center settles into a two-fold shock pattern. This development takes place on a sound travel time through the disk, and a stable stationary pattern is established in only a few sound crossing times. In the outer parts of the disk the arm on the side of the companion is stronger than the opposite arm because the tidal force exciting the waves is much stronger on that side. As predicted by the scenario of figure 1 however the two arms become of equal strength near the center. Also in agreement with this prediction, the two leading spirals produced by reflection at the center decrease in strength away from the center.

The second example (figure 5) illustrates that even a very weak external perturbation saturates into the selfsimilar spiral pattern, as it propagates towards the center. In this calculation, like in the previous one, the initial radial pressure and density distributions were those of a selfsimilar disk ($\rho \sim r^{-3/2}, p \sim r^{-5/2}$).

STABILITY

The stability of a selfsimilar shock can be tested by a numerical simulation which uses it to provide initial and outer boundary conditions. Such a calculation is reported in Różycka and Spruit (this volume). This test is also a useful check on the numerical accuray of the simulation code. The two-fold pattern always appears stable. Sawada et al. (1985) report cases of unstable spiral patterns; these cases appeared mainly under conditions where the selfsimilar theory predicts that only patterns with $n_s > 2$ should exist. It is not clear therefore if patterns with multiplicity larger than two are also stable.

RELATION TO PARTICLE SIMULATIONS

In the simulations discussed so far the disk extended to the outer edge of the computational grid, where conditions were kept constant in time. More realistically, one would like to to give the disk a free outer boundary. Disks with free outer boundaries are reported in this volume by Osaki and by Whitehurst, calculated with particle simulations. Whitehurst's calculations include only a viscosity but no gas pressure. These assumptions are just the opposite of those involved in the spiral shock model, and it would be interesting to see which of the phenomena observed in the particle disks carry over in inviscid gas pressure disks; in particular the disk precessions observed by Whitehurst and by Osaki.

Preliminary results are reported in Różycka and Spruit (this volume). A significant degree of unsteadiness was observed in many cases, but most calculations have not been evolved long enough to obtain the asymptotic behavior. A case with mass ratio 0.2, initial temperature $T/T_{vir} = 0.01$ and disk edge at $r = 0.3$ (in units of the orbital separation) is shown in figure 6. The tidal distortion of the disk is only mild, a stable pair of spiral shocks forms within one orbit and is maintained for at least three orbits. For larger disk size (figures 7, 8) there still exists an initial phase during which the same spiral pattern is evident, but it is subsequently destroyed by a global rearrangement in the disk. The end result of this rearrangement depends strongly on how the energy equation is treated (i.e. where the energy dissipated in the rearrangment process goes). These results show that in order to calculate the steady state of a disk accreting by spiral shocks a large number of orbits must be followed. This has thus far been done for only a few cases (*cf.* Matsuda *et al.*, this volume).

338

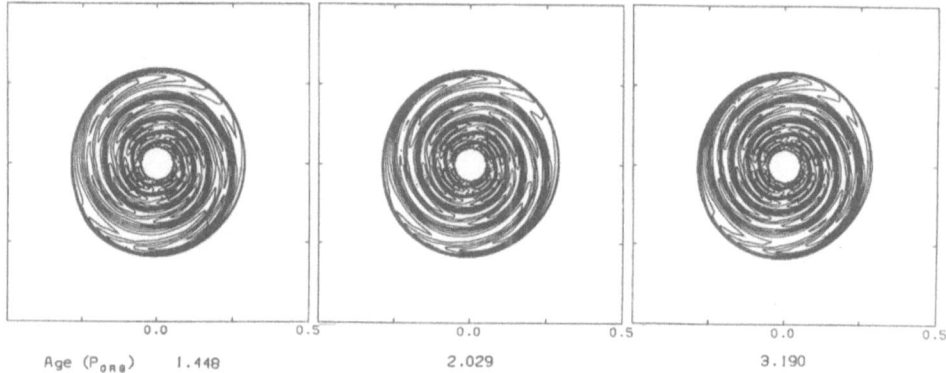

Age (P_{ORB})   1.448   2.029   3.190

Figure 6: Spiral shocks formed by tidal interaction in a disk of radius $0.3a$ (a is the orbital separation, the companion is at the left). No energy loss included, ratio of specific heats $\gamma = 1.02$. Grid size 50 (azimuthal) by 50 (radial) points. The pattern is stable on a time scale of at least 3 orbits. A small $m = 4$ distortion is present, due to the wave reflected from the center (*cf.* fig. 5).

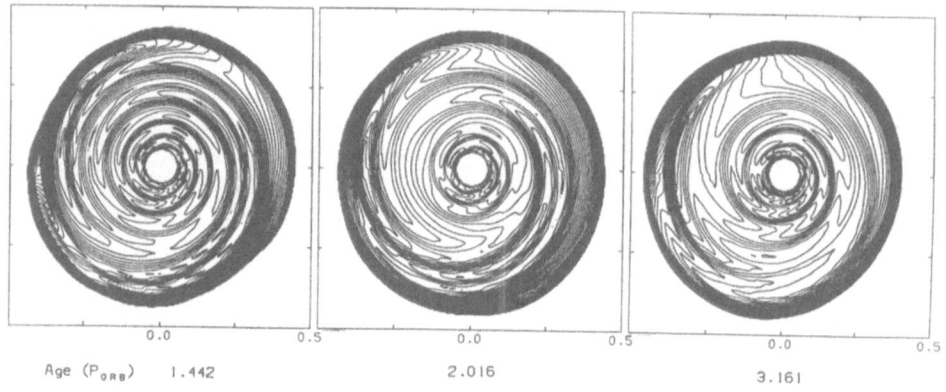

Age (P_{ORB})   1.442   2.016   3.161

Figure 7: As figure 6, initial disk size $0.4a$. Initially an apparently stable spiral pattern appears, but then the tidal force changes the shape of the disk, shearing off the outer parts of the disk, and destroying the spiral pattern in the interior.

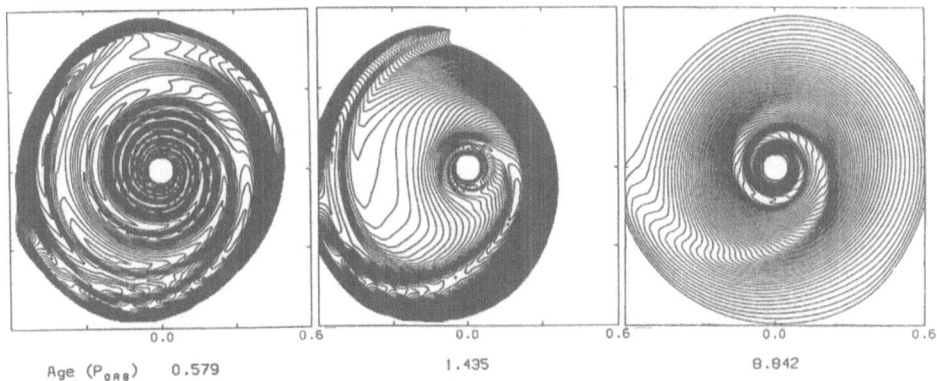

Figure 8: As figure 7 but for initial disk size 0.5 a. The same development, but more dramatic and on a shorter time scale. After the second frame a violent rearrangement takes place in the disk (not shown). The new state of the disk after this process is in the rightmost frame.

Acknowledgement

The calculations reported in figures 4, 5, and 6-8 used Michael Różycka's code, and were carried out on the Cray-XMP at the Institute for Plasma Physics in Garching.

References

Binney, J. and Tremaine, S.: 1987, *Galactic Dynamics*, Princeton University Press, Princeton

Bondi, H.: 1952, *Mon. Not. R. astron. Soc.*, **112**, 195

Donner, K.J.: 1979, PhD thesis, Cambridge University, U.K.

Drury, L.O'C.: 1977, PhD thesis, Cambridge University, U.K.

Drury, L.O'C.: 1980, *Mon. Not. R. astron. Soc.*, **193**, 337

Drury, L.O'C.: 1985, *Mon. Not. R. astron. Soc.*, **217**, 821

Glatzel, W.: 1987, *Mon. Not. R. astron. Soc.*, **228**, 77

Glatzel, W.: 1989, *J. Fluid Mech.*, **202**, 515

Goldreich, P. and Tremaine, S.: 1980, *Ap. J.*, **241**, 425

Hawley, J.F. and Blaes, O.M.: 1988, *Ap. J.*, **326**, 277

Kafka, P. and Meyer, F.: 1984, in *High Energy Transients in Astrophysics*, (Proceedings of a meeting in Santa Cruz, July 11-22 1983), S.E. Woosley, ed., ISBN 0-88318-314-5

Kojima, Y.: 1986, *Prog. Theor. Phys.* **75**, 251

Larson, R.B.: 1988, in *The Evolution of Planetary Systems* (Proceedings of a conference at the Space Telescope Institute, Baltimore, May 1988), eds. H.A. Weaver, F. Paresce and L. Danly, Cambridge University Press, in press

Lighthill, M.J.: 1978, *Waves in Fluids*, Cambridge University Press

Lin, D.N.C. and Papaloizou, J.C.B.: 1979, *Mon. Not. R. astron. Soc.*, **186**, 799

Lin, D.N.C. and Papaloizou, J.C.B.: 1986, *Ap. J.*, **309**, 846

Matsuda, T., Inoue, M., Sawada, K., Shima, E. & Wakamatsu, K.: 1987, *Mon. Not. R. astron. Soc.*, **229** 295

Michel, F.C.: 1984, *Ap. J.*, **279** 807

Narayan, R., Goldreich, P. and Goodman, J.: 1987, *Mon. Not. R. astron. Soc.*, **228**, 1

Papaloizou, J.C.B. and Pringle, J.E.: 1984, *Mon. Not. R. astron. Soc.*, **208**, 721

Papaloizou, J.C.B. and Pringle, J.E.: 1985, *Mon. Not. R. astron. Soc.*, **225**, 267

Sawada, K., Matsuda, T. and Hachisu, I.: 1986, *Mon. Not. R. astron. Soc.*, **219** 75

Sekiya, M. and Miyama, S.M.: 1988, *Mon. Not. R. astron. Soc.*, **234** 107

Spruit, H.C.: 1987, *Astr. Ap.*, **184**, 173

Spruit, H.C.: 1989, in *Magnetic fields and accretion disks in astrophysics*, ed. G. Belvedere, Kluwer Academic Publishing, Dordrecht.

Spruit, H.C., Matsuda, T., Inoue, M. and Sawada, K., 1987, *Mon. Not. R. astron. Soc.*, **229**, 517

SPIRAL SHOCKS IN ACCRETION DISKS:
A PRELIMINARY NUMERICAL STUDY.

Michał RÓŻYCZKA[1,2] and Henk C. SPRUIT[2]

[1] Warsaw University Observatory, Al.Ujazdowskie 4,
PL-00-478 Warszawa, Poland

[2] Max-Planck-Institut für Physik und Astrophysik, Karl-Schwarzschild-Str.1,
D-8046 Garching, FRG

ABSTRACT. The response of accretion disks to tidal forces is studied numerically with the help of a 2-D, second order hydrodynamical code. The code is tested on analytical self-similar models of spiral shocks and proven capable of maintaining them stationary in the grid. The disks are assumed to reside in potential wells of primary components of binary systems. The models are not structured perpendicularly to the orbital plane, and only the flow in the orbital plane is studied. A uniform inflow condition applied to the outer boundary of the disk invariably leads to the formation of two-armed spiral waves resembling the analytical similarity solutions. Provided the disk approaches the Roche lobe, fairly strong shocks are observed even for mass ratios (secondary to primary) as small as 0.01. Stopping the inflow at the outer boundary modifies the spiral pattern and causes the disk to shrink rapidly.

1. Introduction

The key problem in understanding the behaviour and evolution of accretion disks lies in the processes that redistribute angular momentum in them, thus enabling the accretion flow. Based on the standard 'α-disk' formalism, it has widely been assumed, until fairly recently, that the accretion is driven by an *ad hoc* viscosity originating from some kind of turbulence present throughout the disk. As mechanisms capable of generating strong turbulent motions, gas infall, convection and global dynamical instabilities of the disks have been suggested. Along with the viscosity, other, somewhat less popular, factors enabling the angular momentum transport like gravitational and magnetic torques have been discussed. A review covering these topics has been given by Larson (1988).

Yet another mechanism of angular momentum transport was pointed out by Shu (1976), who ascribed the major rôle in the accretion process to the "internal shock wave". His "impact-driven accretion disk" was orbiting the primary com-

F. Meyer et al. (eds.), Theory of Accretion Disks, 341–354.

ponent of a semi-detached binary system, and the wave was generated in it by the impact of the stream flowing from the secondary through the inner Lagrangian point. Soon afterwards it was demonstrated (Paczyński 1977, Papaloizou and Pringle 1977) that shock waves should spontaneously develop due to tidal effects when the outer edge of the disk approaches the Roche lobe, so that in some sense shock forcing by the stream was superfluous. Unfortunately, mainly due to a lack of numerical tools, virtually no progress was made in this field within the next decade and research on wave transport of angular momentum begun only a couple of years ago. In the mean time, Michel (1984) pointed out that if the accreting object is asymmetric (a neutron star magnetosphere for example), shocks will be generated at the inner edge of the disk which can provide the angular momentum transport needed to feed the object, at least up to some distance.

The first disk models with tidally induced shock waves were obtained by Lin and Papaloizou (1986; hereafter LP), and Sawada et al. (1986a,b; hereafter SMH). LP assumed a nonnegligible viscosity, so that even in the absence of shocks the angular momentum was transported in their models from the inner disk outwards by means of viscuos coupling. As a result, more viscous disks extended further towards the secondary than less viscous ones. However, independently of the assumed viscosity, the shocks were only found in the outermost parts of the disk, while the inner disk was rather weakly perturbed by sound waves. The number of shocks was closely related to the azimuthal mode number of the Lindblad resonance dominating at the edge of the disk, and in one case four shocks of various strength were obtained.

SMH followed the flow of the gas lost by the secondary, and the early phases of disk formation around the compact primary. No viscosity was present in their models apart from a residual numerical one, and two or three spiral shock waves extending down to $r = 0.03A$ (the orbital separation) from the primary were largely responsible for the accretion. In a subsequent paper (Sawada et al. 1987; hereafter SMIH) the shocks were found to extend down to $r = 0.01A$, i.e. essentially to the very surface of the compact star. High-γ (the adiabatic exponent) disks were substantially larger than low-γ ones, but regardless of the size of the disk no more than two shocks were found to persist in models followed by several orbital periods.

Stationary self-similar solutions for accretion disks with spiral shocks presented by Spruit (1987, 1989) were compared with time-dependent numerical simulations by Spruit et al. (1987). The same numerical techniques as in SMH and SMIH were used, and a rough agreement was found between numerical and analytical results as far as the opening angles of the shocks were concerned. In the outer parts of the disk the numerical shocks were significantly stronger than the self-similar ones; nevertheless, given the differences in physical assumptions, the agreement between the two sets of results was satisfactory.

The long-range aim of the calculations whose preliminary stage is reported in the present communication is to obtain a broad sample of shock-driven accretion disks in binary systems under various physical assumptions, and to estimate the effective values of α (the standard viscosity parameter). A brief discussion of numerical techniques and boundary conditions employed is given in sect.2. In the same section the tests of the code used for disk modelling are reported, and the models themselves are presented in sect.3. The most important results are discussed in sect.4 together with the prospects of the future research.

2. Numerical Techniques and Test Calculations.

We employ a modified version of the code used by Schwarzenberg-Czerny and Różyczka (1988). The Navier-Stokes equations are solved together with the energy equation on an Eulerian grid in a frame corotating with the binary and centered on the accreting component. Polar coordinates (r, φ) are used and a second-order, alternating direction scheme with monotonicity constraints on advected quantities is applied. The general numerical strategy follows the one outlined by Różyczka (1985), except that no more than one type of gas is allowed to reside in the grid. Shock waves are stabilized with the help of an artificial viscosity whose kinematic coefficient is proportional to the divergence of the velocity field. All components of the viscosity tensor are consistently taken into account.

At the present stage the physics of the problem is kept as simple as possible. A $\gamma-$equation of state is used, and no radiative energy loss from the disk is allowed for. The models are not structured perpendicularly to the orbital plane and only the flow in the orbital plane is considered. Thus, the problem is made strictly two-dimensional.

At the beginning of each run the disk is Keplerian and uniform. The gas can flow freely between the grid proper and the outermost ring (two radial zones broad), where the initial conditions are maintained throughout the calculations (for short, our outer boundary condition is hereafter referred to as a 'uniform inflow condition' or UIC). In the rest of the grid the gas is allowed to evolve until a clear spiral pattern emerges. In order to save the computational time, the inner boundary of the grid is placed at $r_{in} = 0.12 r_{out}$, where r_{out} (the outer radius of the disk) varies between $0.4A$ and $0.9A$. Following SMIH, we place a very low pressure gas around the primary, thus applying a 'vacuum' boundary condition (hereafter referred to as VC).

Typically, the grid contains 50×100 points in (r, φ), respectively, and about 10-15 min of the Cray X-MP CPU time is needed to complete one run.

In a test calculation the similarity solution for γ equal to 1.358 was read into the grid and allowed to evolve for 1.3 orbital periods of the outer rim of the disk. In this case no secondary was present, and we kept the similarity solution in the outermost ring of the grid instead of applying UIC. Upon comparing the evolved pattern to the original one (Fig.1) two conclusions can be drawn: that the similarity solution is stable on a time scale comparable to the orbital period of the binary P_{orb}, and that the code is capable of following the similarity flow with pronounced shock waves (the distortions of the flow that appear at the inner boundary of the grid are due to VC).

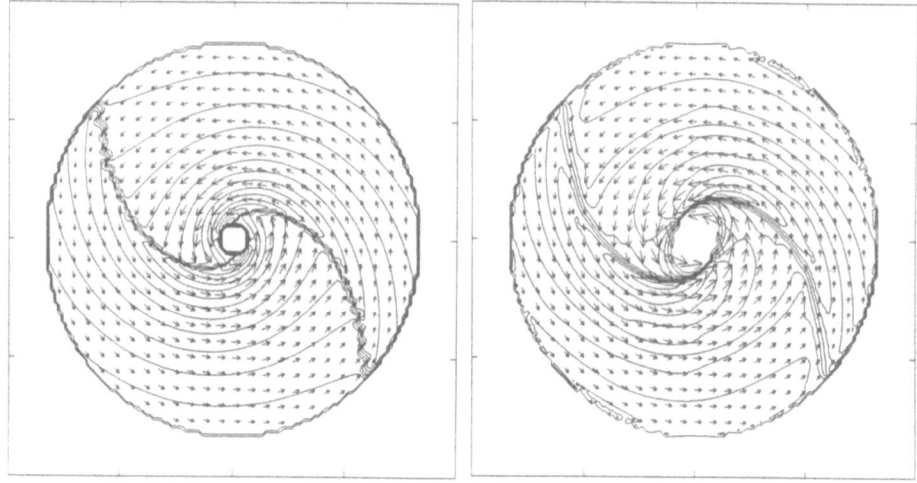

Figure 1: A self-similar wave for the adiabatic exponent γ equal to 1.358 is represented numerically on a grid of 50×100 points in (r, φ) coordinates (left) and allowed to evolve for 1.3 orbital periods of the outer rim of the disk (right). In all figures constant density lines are plotted in a logarithmic scale with $\Delta \log \varrho = 0.05$.

In the next test the influence of the artificial viscosity parameter q_{av} on the morphology of spiral patterns was examined. From Fig.2 it is evident that increasing q_{av} does not change the opening angle of the spiral. While the shocks are progressively smeared out over more and more grid points, their strength (as measured by the number of iso-density lines) remains almost unaffected. Różyczka (1985) applied $q_{av} = 1.4$, and a somewhat surprising result is that $q_{av} = 0.0$ is tolerated by the scheme. Apparently, the internal viscosity of the code, while low in terms of q_{av}, is high enough to guarantee the stability of shocks. We should also stress that the internal viscosity does not appear in keplerian disks, and thus it must originate almost entirely from non-keplerian motions.

The aim of the last test was to find the minimum time scale on which a clear spiral pattern is formed. As it can be seen in Fig.3, in the case of smaller disks $(r_{out} = 0.4A)$ that scale is as short as $0.2P_{orb}$. This time scale is determined by the sound travel time through the disk, as discussed further below.

3. Results of Calculations.

In the present section the effects of disk temperature T_d, sound velocity c_s, mass ratio $q = M_2/M_1$ (secondary to primary) and disk size in units of the orbital separation are investigated.

For $q = 0.5$ and $\gamma = 1.2$, T_d is varied between 0.05 and $0.5T_v$ (the virial temperature at the outer edge of the disk, $T_v = GM_1/r_{out}\mathcal{R}$ where $\mathcal{R}$ is the gas

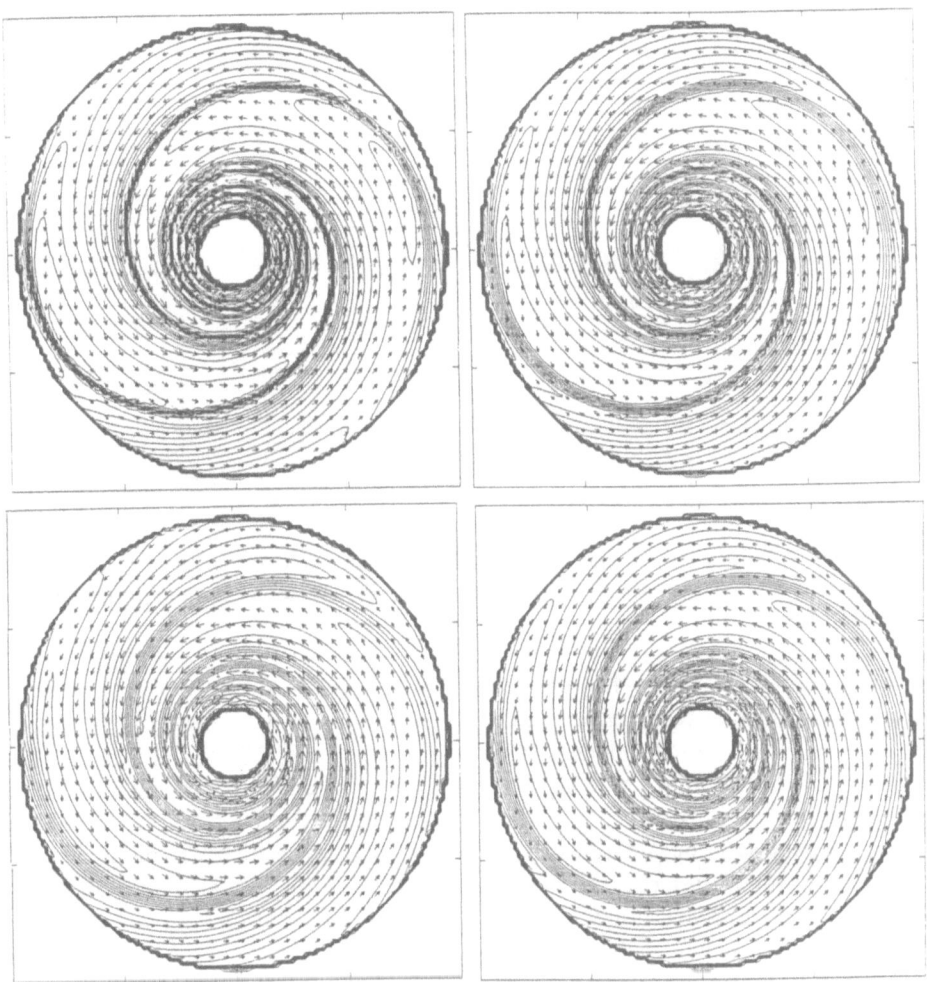

Figure 2: Tidally induced patterns in a disk evolved for $0.2P_{orb}$ (the orbital period of the binary). The artificial viscosity parameter q_{av} varies clockwise from 0.0 (upper left) through 0.25 and 0.5 to 1.0.

constant) in a series of disks with $r_{out} = 0.4A$. As shown in Fig.4, the opening angle of the spiral increases with increasing temperature, while the strength of the shock decreases (in the hottest disk the shocks practically disappear). On the other hand, the strength of the shock increases as one moves from the outer edge of the disk towards the primary. This is the effect foreseen by Spruit et al. (1987).

Varying T_d at constant γ is equivalent to varying c_s. To clarify the rôle of

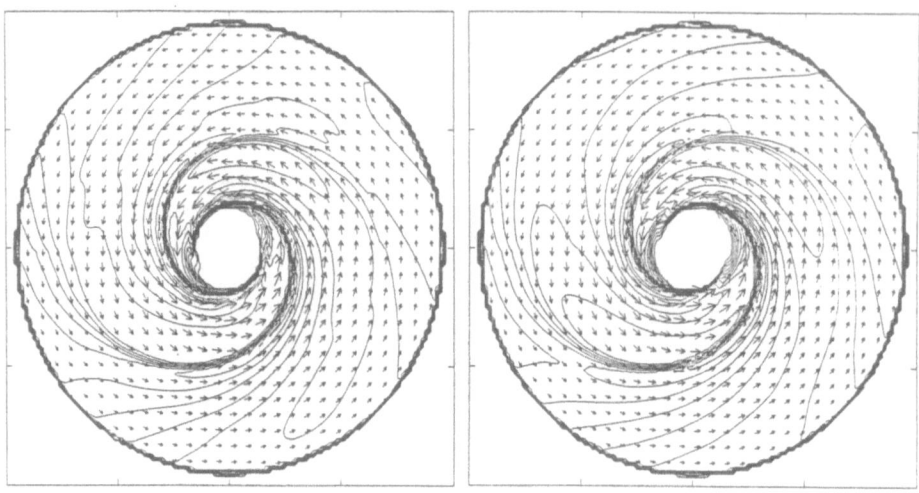

Figure 3: Tidally induced pattern in a disk evolved for $0.2P_{orb}$ (left), and $04.P_{orb}$. The disk is initially 5 times hotter than that in Fig.2, and all remaining parameters are unchanged ($q_{av} = 0.0$).

c_s, we obtained another series of models for $q = 0.5$ and $r_{out} = 0.4A$, in which c_s was constant due to simultaneous variations of T_d and γ (Fig.5). It is evident that the constant sound velocity leads to the same spiral pattern, with a slight tendency of the opening angle to increase with increasing γ. This result, and the time scale for reaching a steady pattern found above, can be understood in terms of the propagation characteristics of sound waves in the disk. Though the wave becomes stationary in a corotating frame, it is felt by the fluid as a periodic inward propagating wave with a frequency $m\Omega(r)$ where m is the number of spirals (2) and Ω the orbital frequency at distance r from the central point mass. Since the shocks are relatively weak their radial propagation speed is nearly the radial component of the group speed for sound waves in the disk. For not too open spirals, this is close to the adiabatic sound speed c_s. Thus a localized disturbance which is stationary in a corotating frame will set up a sound wave which will also become stationary on the sound travel time in the disk. The wave pattern will be independent of γ, keeping the sound speed constant. These are the observed properties of the spiral shocks. Thus, we may identify our shocks as inertially (centrifugal and Coriolis forces) and gravitationally modified sound waves excited by periodic compressions of the outer disk in the tidal field. Except for this excitation, the tidal force is unimportant for the resulting shape of the wave.

The dependence of the spiral pattern on the amplitude of the compression was examined by varying q and r_{out}. As it can be seen (Fig.6), the shock strength decreases rapidly with q decreasing from 0.5 to 0.1, while the opening angle remains essentially unchanged. The shock regains its original strength (characteristic of

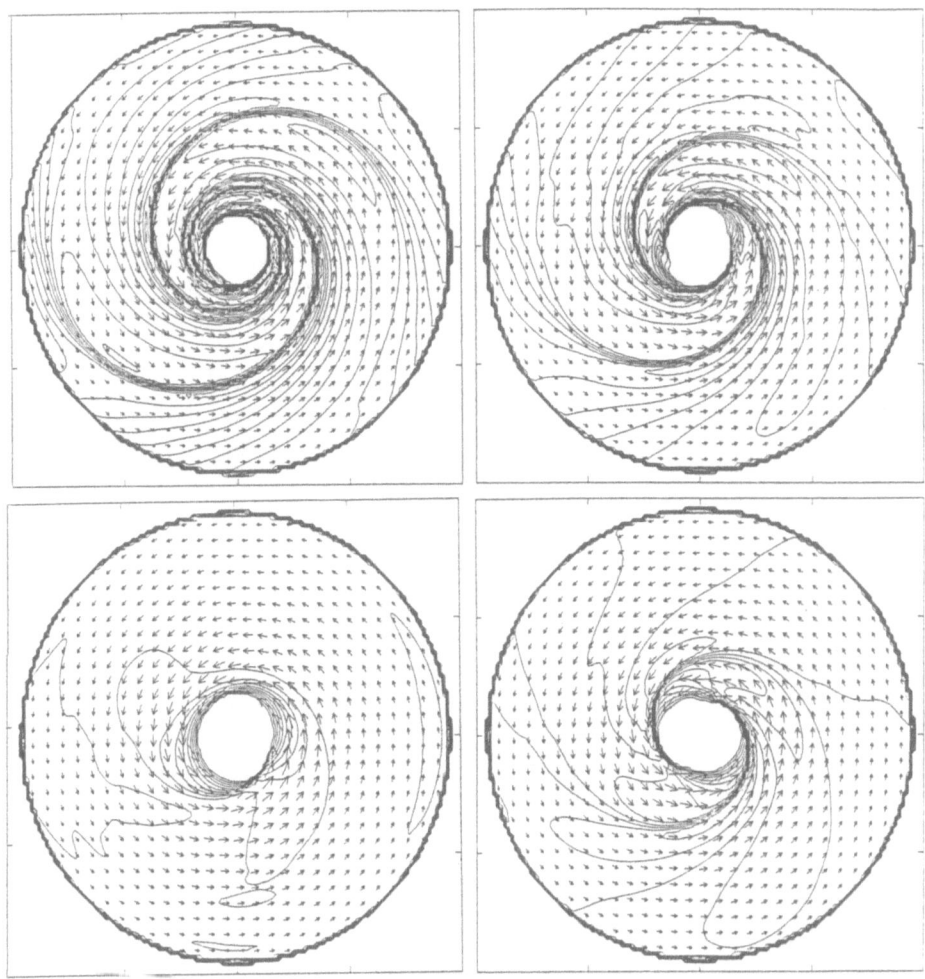

Figure 4: Tidally induced patterns in disks evolved for $0.2P_{orb}$. The temperature of the initial model varies clockwise from 0.05 (upper left) through 0.1 and 0.2 to $0.5T_v$ (the virial temperature at the outer rim of the disk). Mass ratio q and the outer radius of the disk r_{out} are equal to 0.5 and 0.4 respectively.

$q = 0.5$) when r_{out} is increased from $0.4A$ to $0.6A$, so that the disk touches the Roche lobe. At the same time, the opening angle of the spiral tends to increase in response to modified potential (at least in the outer disk), thus supporting the identification of waves as inertially and gravitationally modified .

Further decreasing q by a factor of 10 at $r_{out} = 0.6A$ causes the spiral pattern

348

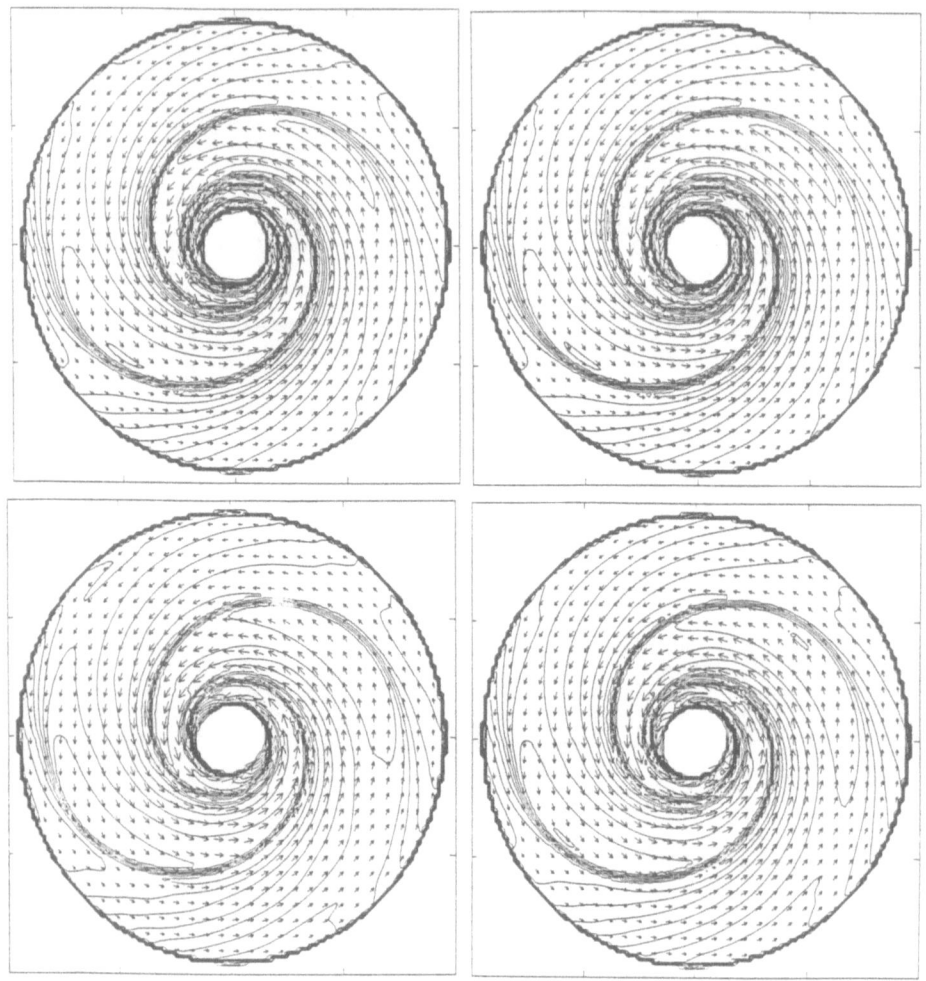

Figure 5: Tidally induced patterns in disks evolved for $0.2P_{orb}$. In all initial models the adiabatic sound speed is $0.2v_{out}$ (the Keplerian velocity at the outer rim of the disk). The adiabatic exponent increases (while the temperature of the initial model correspondingly decreases) clockwise from 1.1 (upper left) through 1.2 and 1.67 to 2.0. The initial temperature of the hottest disk is $0.11T_v$.

to almost vanish (compare appropriate frames in Figs. 6 and 7). A subsequent increase of r_{out} to $0.8A$ (so that it touches the Roche lobe again) revives the pattern. Now, however, the arms are of unequal strength, with the one excited away from the secondary being weaker. The same story repeats when q drops by

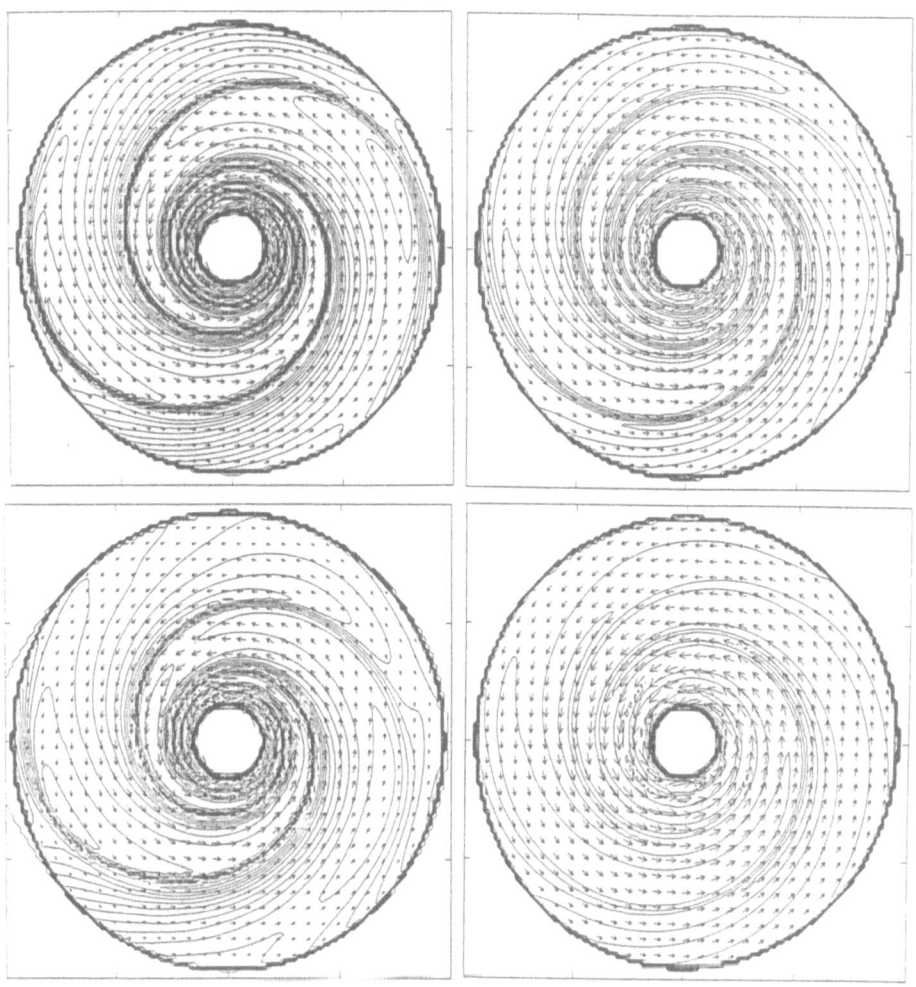

Figure 6: Tidally induced patterns obtained for various q and r_{out}. q decreases clockwise from 0.5 (upper left) through 0.2 to 0.1 while r_{out} is unchanged and equal to $0.4A$ (the orbital separation of the binary). In the last frame $q{=}0.1$, and $r_{out} = 0.6A$, so that the disk touches the Roche lobe (thin line).

another factor of 10 (the pattern almost disappears), and r_{out} increases to $0.9A$ (the pattern emerges again). However, this time, due to an insufficient amplitude of the exciting impulse the wave does not steepen into a shock and propagates inwards as a weak, almost one-armed perturbation. Note that the outer Roche lobe is visible in the corners of the last frame of Fig.7, while the inner lobe is

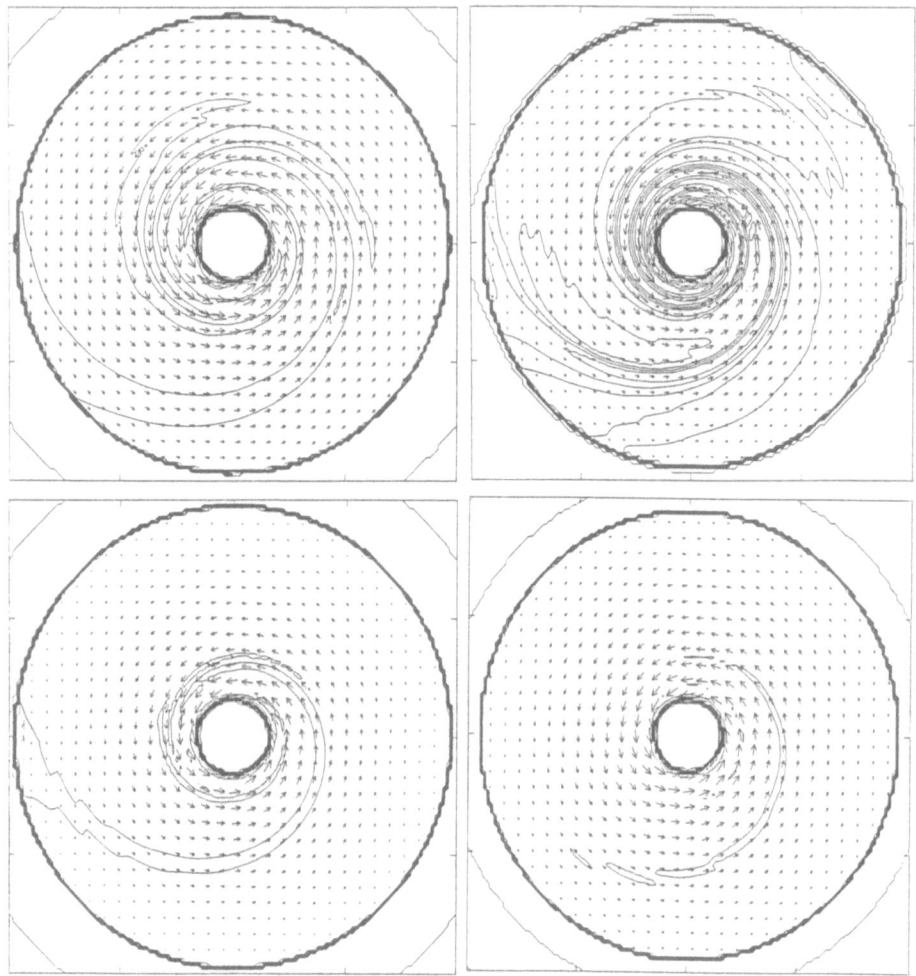

Figure 7: Tidally induced patterns obtained for $q=0.01$(upper row) and $q=0.001$ (lower row). Clockwise from upper left, r_{out} is equal to 0.6, 0.8, 0.8 and 0.9A. In the last frame the outer Roche lobe is clearly seen, while the inner one is almost entirely filled by the disk.

almost entirely filled by the disk.

To sum up, we find that strength and shape of the pattern depend on the amplitude and location of the exciting impulse - i.e. on mass ratio of the binary and outer radius of the disk. One may expect the pattern to converge to the self-similar one in the inner disk (Spruit et al. 1987). For this to occur in our models, the accretion time scale has to be covered by the calculations rather than the dynamical one, which means an at least hundredfold increase of the CPU time if the highest effective shock-induced viscosity indeed corresponds to $\alpha = 0.01$ in the standard theory (Spruit 1987). Matsuda et al. (this volume) report results for a mass ratio of 1, where the calculations have indeed been followed far enough to reach a stationary state on this time scale.

4. Discussion.

According to our results, uniform Keplerian disks evolving in binary systems rapidly develop very regular spiral waves and/or shocks. If UIC is applied, the spiral pattern is stable on a dynamical time scale approximately equal to the orbital period of the binary P_{orb}. The opening angle of the spiral depends mainly on c_s, and to a lesser degree on the size of the disk (the latter effect is probably limited to the outer disk). The amplitude of the wave (or the strength of the shock) grows larger throughout the disk when T_d decreases or when either q or r_{out} increases. Also, the amplitude tends to increase as one moves from the outer edge of the disk inwards. Two-armed spirals are always formed, but the arms can be of unequal strength when $q \leq 0.01$ or the disk resides fairly deep inside the Roche lobe. The shocks easily penetrate the disk down to the minimum inner radius of our grid ($0.05A$ from the center of the primary).

The latter two results differ from those published by LP, who found as many as four spiral arms, but restricted to the outermost disk. We interpret this as due to the higher companion masses used here. A wave action conservation argument (Spruit et al. 1988) shows that the two-armed spiral wave set up by the weak tidal force for LP's low mass ratios are should become visible only close to the primary mass. The range in radius, $r = 0.25A$ to $1.0A$ covered by LP is not sufficient to see the effect for mass ratios less than about 0.01. As far as the predominance of two-armed patterns and deep penetration are considered, our results corroborate those of SMH and SMIH. The fact that our patterns are in all cases stable whereas their high-γ disks oscillate strongly may be attributed to extremely different outer boundary conditions (UIC versus a strongly nonuniform inflow and free boundary).

Obviously, our models are highly simplified and can be made more realistic in many ways. The main objections can be raised to entirely unphysical UIC and excessively high temperatures of the disks (as an example - in a binary with $M_1 = 0.5M_\odot$ and $A = R_\odot$ the coolest of our disks would have T_d approaching $2 \times 10^5 K$).

Replacing UIC with a free outflow (and no inflow) condition leads to the situation shown in Fig.8: one of the arms becomes much stronger than the other, and a wake begins to form downstream of it. At the same time, the who le outer disk shrinks fairly rapidly while developing radial density gradients (one may expect that both processes will eventually lead to the formation of a gap discussed by LP). We note that a similar behaviour of disks with free outer boundaries was

352

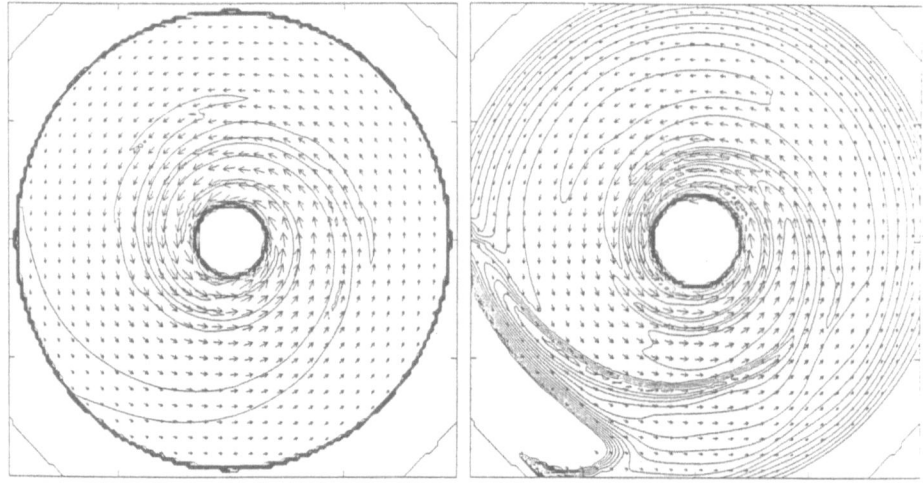

Figure 8: Tidally induced patterns obtained for q=0.01 with UIC (a uniform Keplerian inflow, left) and with no inflow outer boundary condition.

observed by Schwarzenberg-Czerny and Różyczka (1988). Their results, however, cannot be directly compared to ours because of an entirely different treatment of the interface between the disk and the ambient medium.

Coming back to UIC and lowering the temperature of the disk (to about $6 \times 10^4 K$ in our example) produces the patterns shown in Fig.9. It is evident that the waves propagate too slowly to reach r_{in} within one P_{orb} and that the opening angle of the spiral decreases markedly upon penetrating the disk. In fact, the spiral is wound so tightly in the inner half of the disk, that it is not properly resolved by our standard grid. It is also clearly visible that UIC does not perform well in the sense that the edge of the disk separates from the edge of the grid. The decrease of the opening angle is a consequence of the lower sound the speed. In a still cooler disk $(T_d = 6 \times 10^3 K$ in our example) similar effects can be seen, with the exception that an even larger central area remains free from perturbations after one P_{orb}.

From above discussion and exploratory calculations it is evident that much additional computational effort will be needed before our ultimate goal, the estimate of the efficiency of wave-driven accretion in realistic disk models, is reached. In particular, the input physics should be improved (variable thickness of the disk, better equation of state, allowance for nonadiabaticity of the flow) and various boundary conditions should be examined. It should be noted that the physically most interesting disks (that are cool in terms of T_v) will require additional special treatment in order to resolve tightly wound shocks. Special attention should also be paid to the interface between the disk and the ambient medium: it may turn out that the second-order methods are too diffusive to model it in a reliable way.

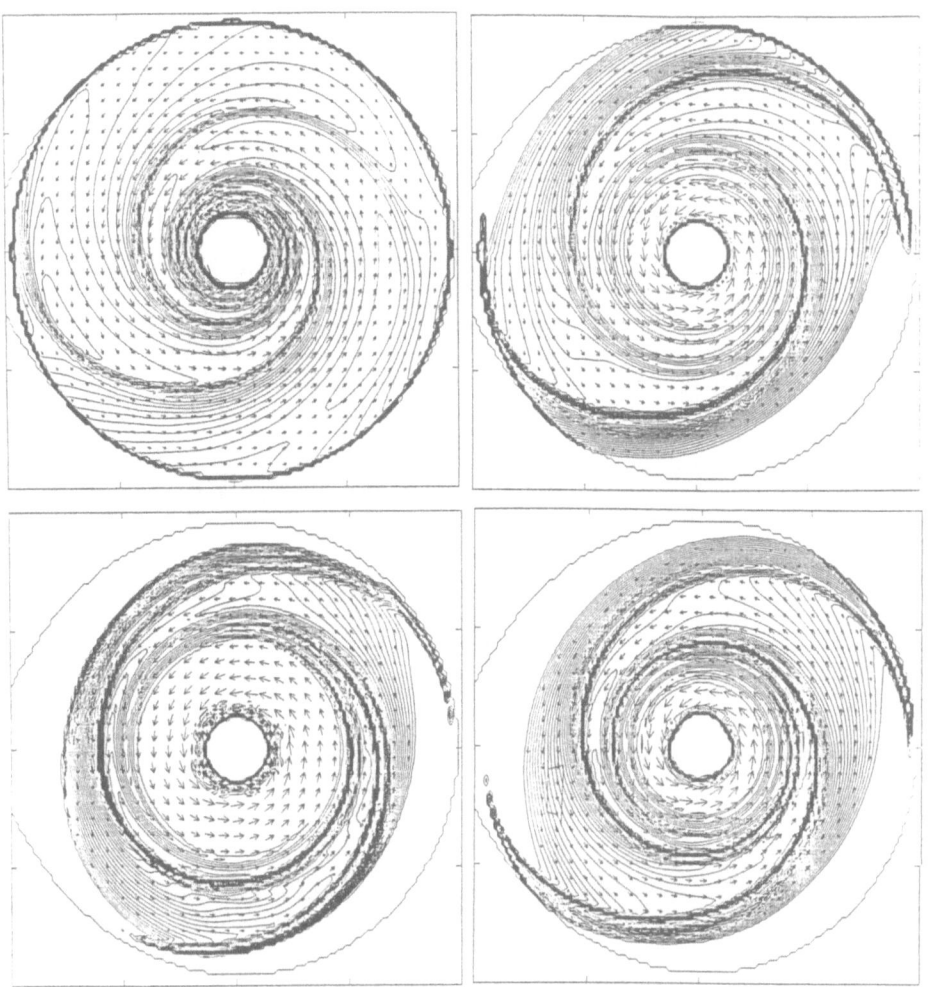

Figure 9: Tidally induced patterns obtained for $q=0.01$. Clockwise from upper left, the temperature of the initial model is equal to 0.03, 0.003, 0.003 and $0.0003T_v$. The disks are evolved for $0.33P_{orb}$ in the upper row, and for $1.0P_{orb}$ in the lower one.

ACKNOWLEDGEMENTS. While in Poland, MR was supported by the grant CPBP 01.11 from the Polish Academy of Sciences. The code used in this research was developed and tested with the help of an IBM PC computer on loan from Princeton University Observatory.

REFERENCES.

Larson, R. 1989, paper presented at a workshop on *The Formation and Evolution of Planetary Systems* held at the Space Telecope Science Institute, Baltimore, May 9-11.

Lin, D.N.C. and Papaloizou, J. 1986, *Astrophys.J.* **307**, 395.

Michel, F.C. 1984, *Astrophys. J.* **279**, 807.

Paczyński, B. 1977, *Astrophys.J.* **216**, 822.

Papaloizou, J. and Pringle, J.E. 1977, *M.N.R.A.S.* **181**, 441.

Różyczka, M. 1985, *Astron. Astrophys.* **143**, 59.

Sawada, K., Matsuda, T. and Hachisu, I. 1986a, *M.N.R.A.S.* **219**, 75.

Sawada, K., Matsuda, T. and Hachisu, I. 1986b, *M.N.R.A.S.* **221**, 679.

Sawada, K., Matsuda, T., Inoue, M., and Hachisu, I. 1987, *M.N.R.A.S.* **224**, 307.

Schwarzenberg-Czerny, A. and Różyczka, M. 1988, *Acta Astron.* **38**, 189.

Shu, F. 1976, in *IAU Symposium No 73, Structure and Evolution of Close Binary Systems*, ed. P.Eggleton, S.Mitton and J.Whelan (Dordrecht: Reidel), p.253.

Spruit, H. 1987, *Astron. Astrophys.* **184**, 173.

Spruit, H. 1989, *this volume.*

Spruit, H., Matsuda, T., Inoue, M. and Sawada, K. 1987, *M.N.R.A.S.* **229**, 517.

MASS TRANSFER BY TIDALLY INDUCED SPIRAL SHOCKS IN AN ACCRETION DISC.

Takuya MATSUDA[1], Nobuhiro SEKINO[1], Eiji SHIMA[2], Keisuke SAWADA[2], and Henk SPRUIT[3]

[1] Department of Aeronautical Engineering, Kyoto University, Kyoto 606, Japan

[2] Aircraft Engineering Division, Kawasaki Heavy Industries, Kakamigahara 504, Japan

[3] Max-Planck Institute für Physik und Astrophysik, Institute für Astrophysik, Karl-Schwarzschild Str. 1, D-8046 Garching bei München, FRG

ABSTRACT. Numerical simulations of two-dimensional adiabatic inviscid flow in an accretion disc in a binary system or a proto-planetary system are performed using a second-order accurate implicit Roe upwind scheme. A purpose of the work is to estimate quantitatively the amount of mass accretion rate due to spiral shocks produced by a less massive component. The mass ratio of the binary system, q, is varied to see the tidal effect of the less-massive component on the formation of spiral shocks. Five cases, namely $q = 0, 10^{-3}, 10^{-2}, 10^{-1}, 1$, are examined. In the case of $q \geq 10^{-2}$, steady spiral shocks persist until the end of the calculations, i.e. about 15-20 rotation periods. If q is as small as 10^{-3}, i.e. the case of the Jupiter, the amplitude of the spiral waves is not very strong. On the basis of calculated mass accretion rate, we can determine an effective α parameter in the standard accretion disc model. It depends on q, and we found that $4 \cdot 10^{-4} \leq \alpha_{eff} \leq 4 \cdot 10^{-2}$ for $10^{-3} \leq q \leq 1$.

1. Introduction.

1.1 α-DISC MODEL.

In the standard accretion disc model proposed by Pringle & Rees (1972) and Shakura & Sunyaev (1973), it has been commonly assumed that the accretion is driven by an effective viscosity produced by some kind of turbulence, of which magnitude is parametrized by the dimensionless viscosity α. In the α-disc model, the angular momentum of the gas in an accretion disc is assumed to be transported by this. Thus, the gas element losing its angular momentum, spirals in towards the central object (See Pringle, 1981; Petterson, 1983; Frank, King& Raine, 1985 for

355

F. Meyer et al. (eds.), Theory of Accretion Disks, 355–371.

reviews of the α-theory). However, the origin of the turbulence is still controversial.

A class of instabilities introduced into the astrophysical context by Drury as early as 1977 (see also Drury, 1980, 1985) has attracted considerable attention following the work of Papaloizou and Pringle (1984, 1985). These instabilities have been investigated in a variety of toroidal and disk-like geometries. Various dependences of rotation rate on distance have been used, usually in the form $\Omega \propto r^{-s}$, with s =const. (cf. the references given in Narayan et al. ,1987). For a Keplerian rotation law, and for a thin disk geometry, the growth rates turn out to be modest. Fully nonlinear 2-dimensional simulations by Blaes and Hawley (1988) of these instabilities show significant effects, but only if s is close to 2 (constant angular momentum). Nonlinear simulations of the shearing sheet (representing a small section of a thin disk) by Kaisig (1989a,c) show that for $s = 3/2$ the unstable waves saturate at a finite amplitude, *without* breaking down into turbulence. This holds both for two- and for threedimensional perturbations. Though the Mach number of the velocity perturbations reaches a maximum of about 0.5, the angular momentum transported by these nonlinear waves is very small ($\alpha \leq 10^{-3}$). Only for s near 2 do the waves become sufficiently violent to carry a significant angular momentum flux, in agreement with the findings of Blaes and Hawley.

In view of these results, it is not clear that a turbulent viscosity is the dominant mechanism for angular momentum transfer. On the other hand, any dissipation of the rotational energy coupled with a tidal effect due to a companion star in a close binary system inevitably leads to accretion of the disc gas.

1.2. SPIRAL SHOCK THEORY.

An alternative mechanism to the α-disc theory is angular momentum transfer by shock waves coupled with a tidal effect. Shu (1976) speculated that a spiral shock is generated by the stream of gas flowing from the companion star. Michel (1984) calculated the angular momentum transport due to shocks produced by nonaxisymmetries imposed at the inner edge of the disk (e.g. accretion onto a magnetosphere). Numerical simulations of a gas flow in a close binary system have been done by many workers (Prendergast & Taam, 1974; Sorensen et al., 1975; Flannery, 1975; Lin & Pringle, 1976; Hensler, 1982), although none of them found such a spiral shock.

Recently Sawada et al. (1986a,b, 1987) performed elaborate calculations on the flow using a modern numerical scheme and a super-computer, and they found spiral shocks in the accretion disc. Note that these shocks are not generated by the impact of the streaming gas from the companion star as was envisaged by Shu (1976) but generated tidally. This will be clearly shown in the present calculations, in which no stream of gas is assumed. The mechanism of the formation of spiral shocks is much the same as that occurring in barred galaxies (Sorensen et al., 1976; Matsuda et al., 1987).

Spruit (1987) obtained self-similar solutions of spiral shocks by a semi-analytic manner. He computed angular momentum transfer rate and estimated the effective α parameter to be $10^{-4} \leq \alpha_{eff} \leq 10^{-2}$. Spruit et al.(1987) compared these self-similar solutions with numerical results, and obtained reasonable agreement. They pointed out that at low values of the ratio of specific heats ($\gamma \leq 1.45$) accretion is possible without radiative losses.

1.3 JUPITER-DRIVEN WAVES.

There is another important field of application of the theory of accretion discs apart from those in close binary systems, i.e. the solar nebula or a protostellar nebula. Observations of newly formed stars show that the gaseous protostellar nebular disc disappears about 10^6 years (see reviews by Larson 1988, Ohtsuki & Nakagawa 1989).

Larson (1988) considered various mechanisms of angular momentum transport in protostellar discs, e.g. turbulent viscosity, gravitational torques, magnetic fields and wave transport. He concluded that the last possibility, i.e. wave transport which is nothing but the mechanism we are considering in the present paper, is the most promising one. He suggested that Jupiter-like planets generate waves strong enough to deplete the interior parts of a disc in a time scale of the order of 10^6 years. A value for α_{eff} of 10^{-3} suffices for depletion of the region interior to 5 AU (the orbit of Jupiter).

Jupiter-driven waves for the solar-nebula have been before considered by Lin & Papaloizou (1979, 1986) and Goldreich & Tremaine (1980). The waves are primarily excited at the Lindblad main resonance situated at 0.63 times Jupiter's distance from the Sun. Lin and Papaloizou (1986) developed a numerical method for calculating a steady disc with tidal forcing by a low mass companion. They concentrated on the processes determining the position of the disc's outer edge, and obtained detailed results for this quantity that agree with earlier analytic estimates.

Sekiya, Miyama & Hayashi (1987, 1989) computed the formation of Jupiter-driven waves using a three-dimensional smooth particle hydrodynamics technique. Like Lin and Papaloizou, they were mainly concerned with the removal of gas from the region close to Jupiter.

In this context, we consider in the following some quantitative aspects of spiral shock theory. The objective of the present work is to compute mass accretion rate in accretion discs in binary systems with mass ratio ranging from unity down to 10^{-3} (the case of Jupiter).

2. Assumptions and Method of Calculations.

2.1 BASIC ASSUMPTIONS.

We consider a binary system composed by a main star with mass M_1 and a purturber with mass M_2. The mass ratio is defined as $q = M_2/M_1$. We consider cases of $10^{-3} \leq q < 1$. As a special case, we also consider a single star with $q = 0$. If the mass ratio is as small as 10^{-3}, it is a planetary system rather than a binary system.

Let us consider a gaseous accretion disc surrounding the main star. We assume that the disc has a constant thickness so that two-dimensional approximation can be applied. We neglect the effects of radiation, magnetic fields, and dust. We calculate two-dimensional inviscid flow about the main star. Apart from numerical dissipation, the calculations contain no viscosity. The gas is adiabatic except for the presence of shocks and is characterized by the ratio of specific heats, γ. In the present work, we fixed $\gamma = 1.4$.

We consider the region surrounding the main star, and no gas supply from the purturber is assumed. This simplification is made to show that spiral shocks can form from the tidal perturbation and do not require the presence of a gas stream. Self-gravity of the disc is neglected, and the gravitational field in the system is given by the Roche potential.

2.2 METHOD OF CALCULATIONS.

2.2.1. Numerical scheme. The basic equations are the two dimensional time-dependent Euler equations written in a frame rotating with the perturber. The equations are made dimensionless by the rotation period of the system, Ω_0, the separation of two stars, r_p, and the initial uniform density of gas, ρ_0.

The basic equations are discretized by a finite volume method or a cell method, in which physical variables are defined at the cell centre. Numerical fluxes on the cell faces are obtained by solving a Riemann problem between two states in the neighbouring cells. The Riemann problem is solved approximately by Roe's method (Roe, 1981; Matsuda et al. 1989). The time integration is done implicitly using LU-ADI method (Matsuda et al. 1987). The CFL number is about 10. The numerical accuracy of the method is second-order both in space and in time.

2.2.2. Numerical grid. We construct a polar grid with 50 radial and 100 azimuthal grid points about the main star. We call this grid the coarser grid. In one case, the number of grid points is increased to 100×100 in order to see the effect of resolution. This grid is called the finer grid. We place a hole with radius of 0.05 at the centre. This value corresponds to $0.25AU$ in the case of the solar nebula, and is well within the orbit of Mercury.

The outer boundary is at $r = 0.75$ for $q \leq 10^{-1}$ and at $r = 0.45$ for $q = 1$. In one case we enlarged it to $r = 1.5$ in order to see the effect of an outer boundary. Note that the Lindblad main resonance is at $r = 0.63$.

2.2.3. Initial condition. In the present work we do not intend to make a very concrete model for a particular object, and we assume rather simple initial configuration of gas. At $t = 0$ gas of uniform density rotating with the Keplerian velocity about the main star is assumed. The temperature of gas is assumed to be $T \sim r^{-1}$ so that the initial Mach number of the gas, M, is constant. We set $M = 6$ in the present calculations. We have also tested the case in which initial velocity distribution is determined by the balance of forces including the pressure force. We found that the final result does not depend on the initial choice of velocity.

2.2.4. Boundary conditions. The choice of boundary conditions on the inner and the outer numerical boundary is rather important. We treat boundary conditions as follows. We place fictitious cells just outside of a boundary, and we prescribe physical variables in the fictitious cells. Numerical fluxes on the boundary wall are computed by solving a Riemann problems between states. We tested five types of boundary conditions described bellow.

  a Vacuum condition: The density and the pressure in the fictitious cells is set at a fixed very low value, equivalent to the disk having a free inner edge.

  b Reflective condition: The boundary wall is a reflecting wall.

  c Ambient condition: The value of physical variables in the fictitious cells are the same as those in the neighbouring interior cells at $t = 0$, and these are fixed in the calculations.

  Other conditions such as a pressure condition and an extrapolation condition are also tested.

We applied only the ambient condition at the outer boundary, while we tested all the conditions stated above on the inner wall. We will show that the vacuum condition gives us the most stable results.

The computations were performed on the Fujitsu VP400E or VP200 vector processors. The CPU time per step per grid point is 2.7×10^{-5} sec and 3.9×10^{-5} sec, respectively. We computed up to $t \sim 120$ corresponding to 20 rotation periods, and it takes two to three CPU hours.

3. Results.

3.1 SPIRAL SHOCKS FOR VARIOUS MASS RATIO.

Figures 1a-d show density contours for the models with $q = 1, 10^{-1}, 10^{-2}, 10^{-3}$, respectively. Vacuum condition was used at the inner boundary for these cases. Figures 2a, b show the Mach number contours for the models with $q = 1, 10^{-2}$, respectively. In these calculations the coarser grid was used except the case of $q = 10^{-3}$, in which the finer grid was used in order to overcome too much numerical diffusion.

As can be seen in these figures, we have clear two-armed spiral waves. They are actually spiral shocks for $q \geq 10^{-2}$. In the case of $q = 10^{-3}$, it is difficult to say that the spiral waves are shocks or simply sound waves in this resolution. As was

360

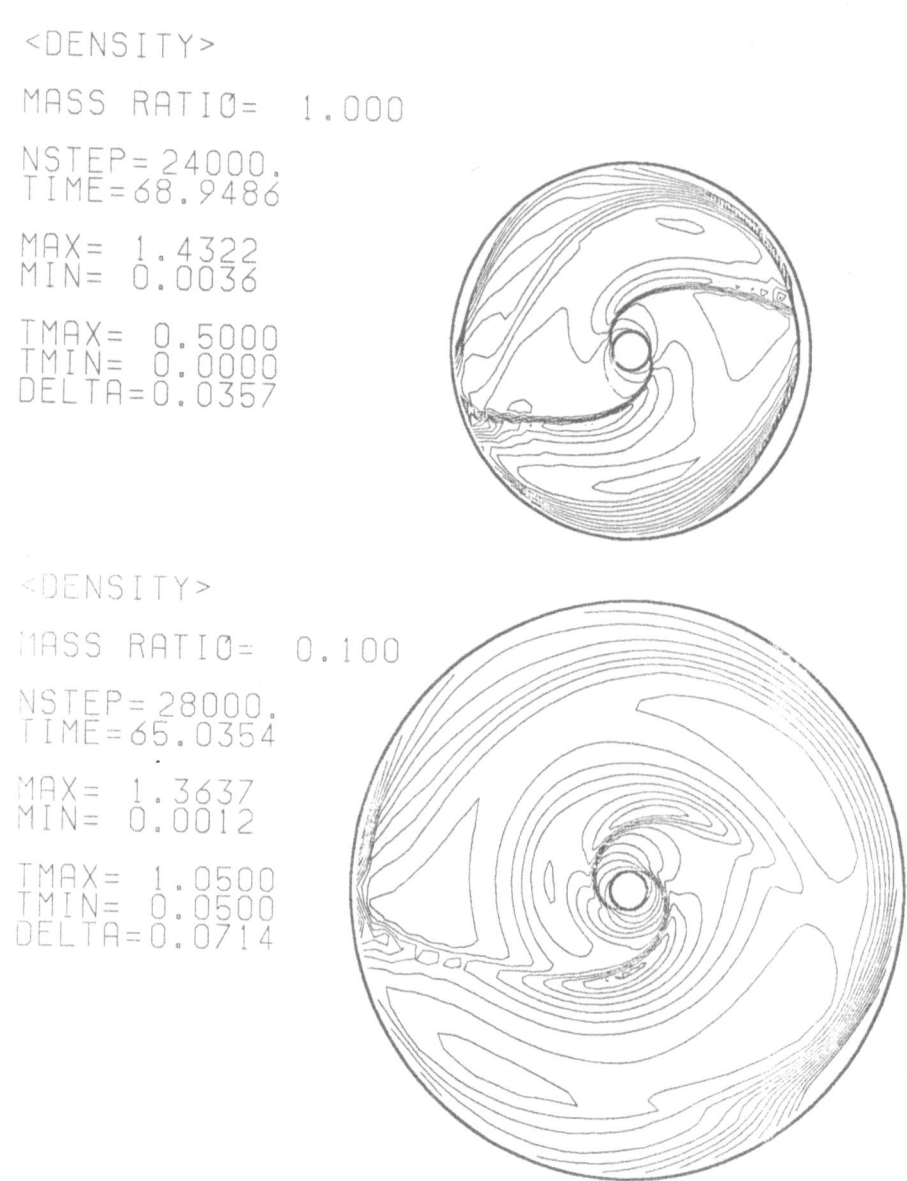

Figure 1: a) Density contours for the mass ratio, $q = 1$, computed on the coarser grid having 50×100 resolution. The vacuum condition was applied at the inner boundary. The pattern at $t = 68.9$ corresponding to the numerical time step 24000 is shown. MAX and MIN show the maximum and the minimum values occurring in the figure, while TMAX and TMIN show the maximum and minimum contour lines, respectively. DELTA is the contour spacing; b) $zeta = 10^{-1}$; c) $zeta = 10^{-2}$; d) $q = 10^{-3}$. The finer grid with 100×100 grid points was used in this case.

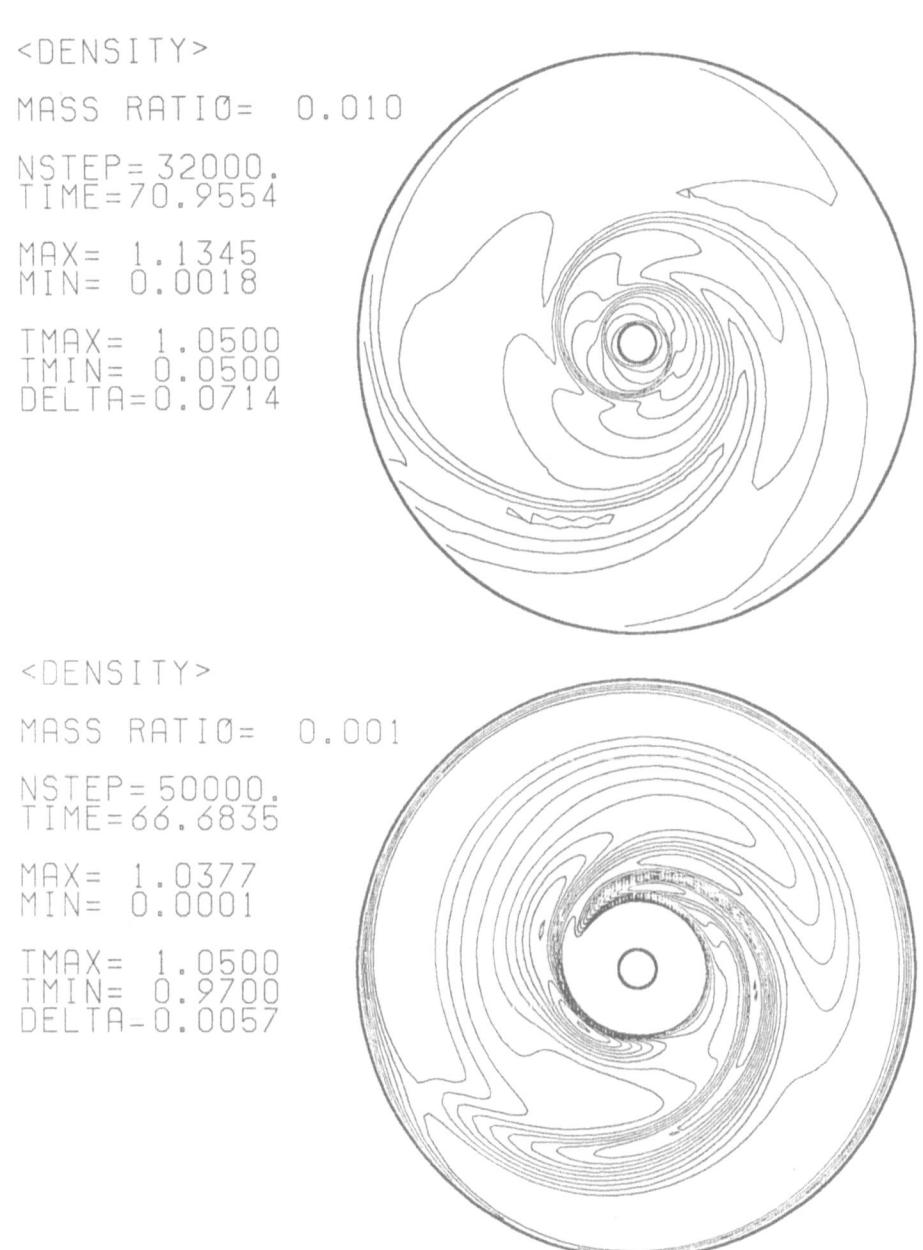

<DENSITY>
MASS RATIO= 0.010
NSTEP=32000.
TIME=70.9554
MAX= 1.1345
MIN= 0.0018
TMAX= 1.0500
TMIN= 0.0500
DELTA=0.0714

<DENSITY>
MASS RATIO= 0.001
NSTEP=50000.
TIME=66.6835
MAX= 1.0377
MIN= 0.0001
TMAX= 1.0500
TMIN= 0.9700
DELTA-0.0057

Figure 1: continued

pointed out in the introduction, it is clear that these spiral waves are not generated
by the impact of a stream from the companion but by the tidal force. The spiral
waves seem to originate at the Lindblad resonance at $r = 0.63$ (see Fig 1b).

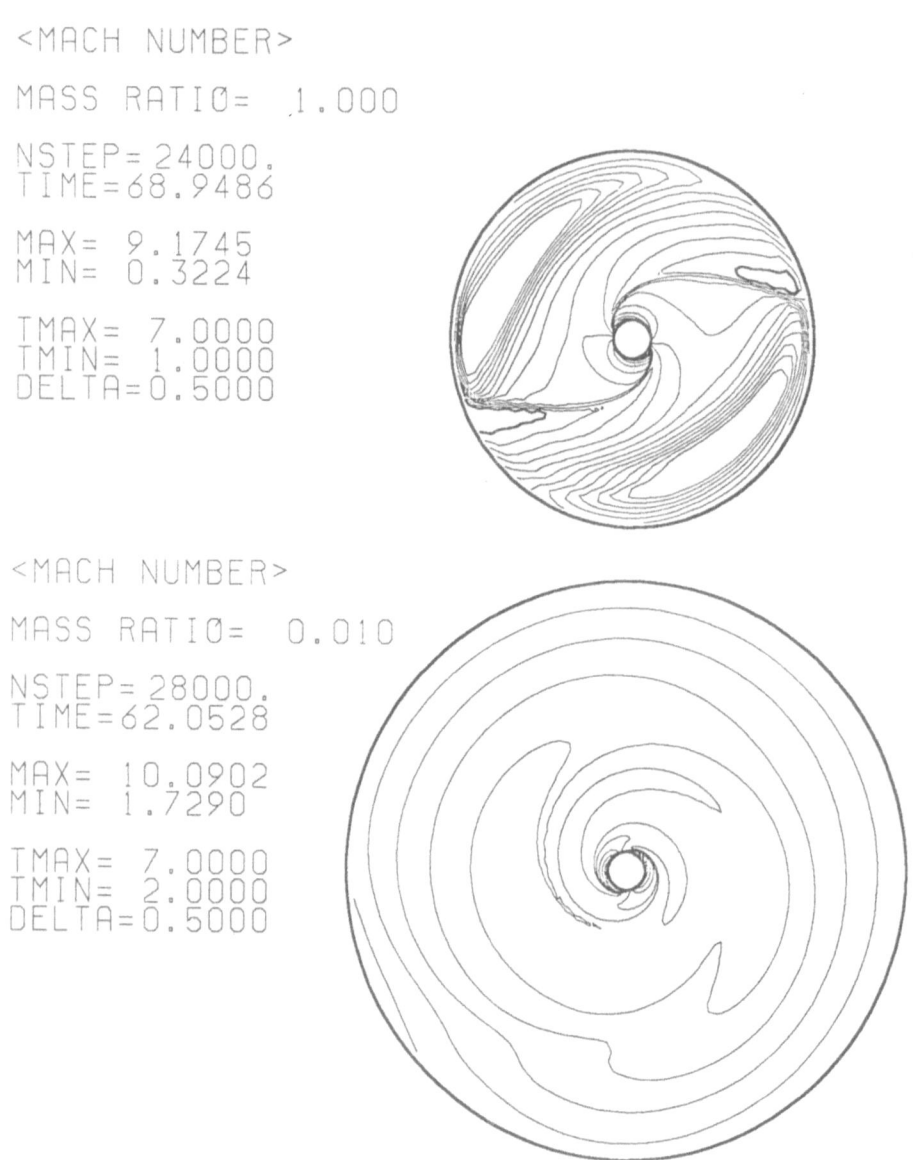

Figure 2: a) Mach number contours for $q = 1$. The thick lines show sonic lines. b) $q = 10^{-2}$.

3.2 EFFECTS OF BOUNDARY CONDITIONS.

In order to see the effect of boundary conditions on the result, we tested conditions other than the vacuum condition. Figure 3 shows the density contours for $q = 10^{-1}$,

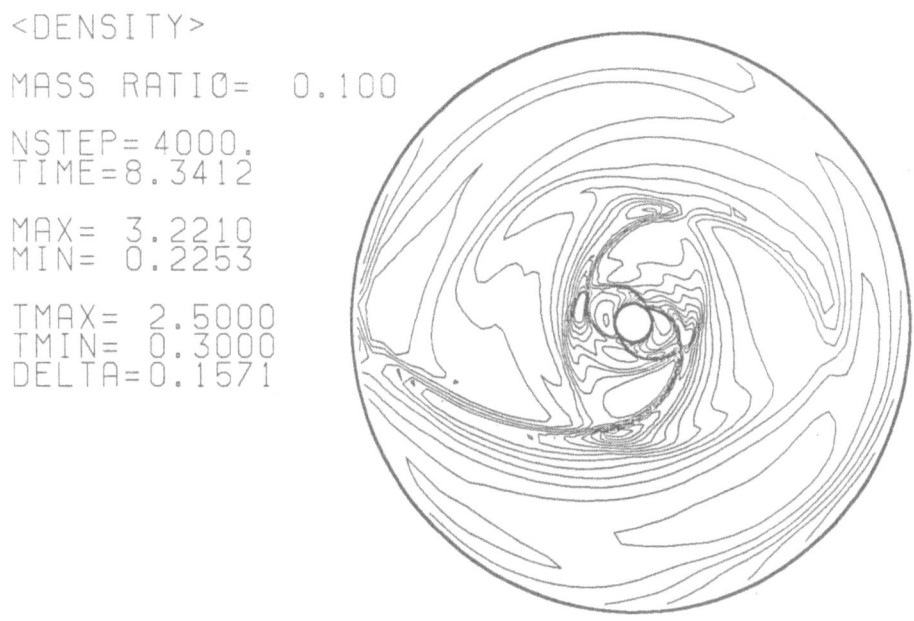

```
<DENSITY>

MASS RATIO=  0.100

NSTEP= 4000.
TIME=8.3412

MAX=  3.2210
MIN=  0.2253

TMAX=  2.5000
TMIN=  0.3000
DELTA=0.1571
```

Figure 3: Same as Fig.1b except the reflective condition assumed on the inner boundary.

in which the reflective condition was used at the inner boundary.

Spiral waves originating at Lindblad resonance propagate inward to become spiral shocks. They are reflected at the inner wall and become leading spiral shocks. At the intersections of two shocks we can see density maximum. The leading spiral shocks become weaker when they propagate outward.

In order to see the effect of outer boundary, we enlarged the outer boundary. Figure 4 shows the density contours for the same case as Fig. 1b except that the radius of the outer boundary was doubled. The perturber seen at the left of the main star was represented by a point mass. Comparing Fig. 1b and Fig. 4, we may say that the spiral shocks are not affected by the outer boundary very much.

To see the effect of the inner condition further, we computed the axisymmetric case, i.e. $q = 0$. In this case, there should be no radial mass flow. However, because of numerical error, there exists a small (artificial) mass flow. We monitored the mass flow rate across a circle with various radius. Figure 5a and b show the time evolution of the mass flow rates across the inner boundary. Fig.5a shows the case of the vacuum condition, while Fig. 5b does the case of the ambient condition. In the latter case, sound waves generated somehow seem to be reflected at the inner boundary, and this causes very oscillatory mass flow rate. It is interesting to note that the time averaged absolute mass flow rate is larger for the latter case. This

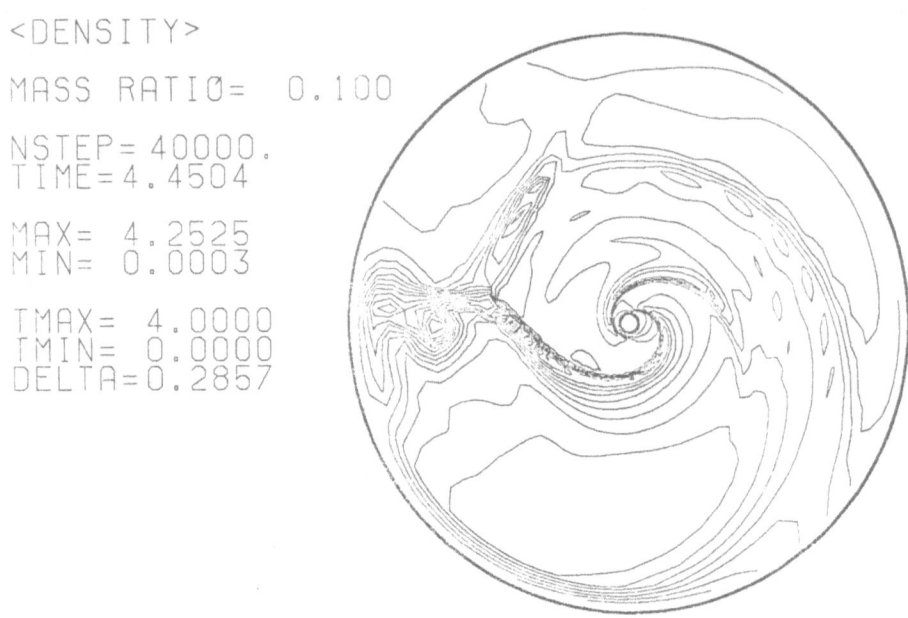

```
<DENSITY>

MASS RATIO=  0.100

NSTEP=40000.
TIME=4.4504

MAX=  4.2525
MIN=  0.0003

TMAX=  4.0000
TMIN=  0.0000
DELTA=0.2857
```

Figure 4: Same as Fig.1b except the radius of the outer boundary is doubled.

may be due to that these waves in the latter case carry angular momentum outward to enhance mass accretion.

3.3 MASS TRANSFER BY SPIRAL SHOCKS.

Figure 6a-d show the radial dependence of the mass flow rate for cases with $q = 1 - 10^{-3}$, and Fig. 6e shows this for the axisymmetric case. In these calculations the coarser grid was used. Note that only the region of $r \leq 0.45$ was computed in Fig. 6a.

One can see that there exist significant mass inflow in the inner region and mass outflow in the outer region for the cases with $q \geq 10^{-2}$. In the case of $q = 10^{-3}$, it is difficult to say that the computed mass flow is real or not, since Fig. 6d is essentially the same as Fig. 6e, i.e. the axisymmetric case. The computed mass flow may be fictitious due to numerical errors. In order to test this possibility, we computed the axisymmetric case using the finer grid. Fig. 6f shows the mass flow rate computed using the finer grid. The mass outflow in the outer region seen in Fig. 6e is reduced by an order of magnitude. Nevertheless, the mass inflow rate at the inner wall was unchanged. This is probably due to our vacuum inner condition

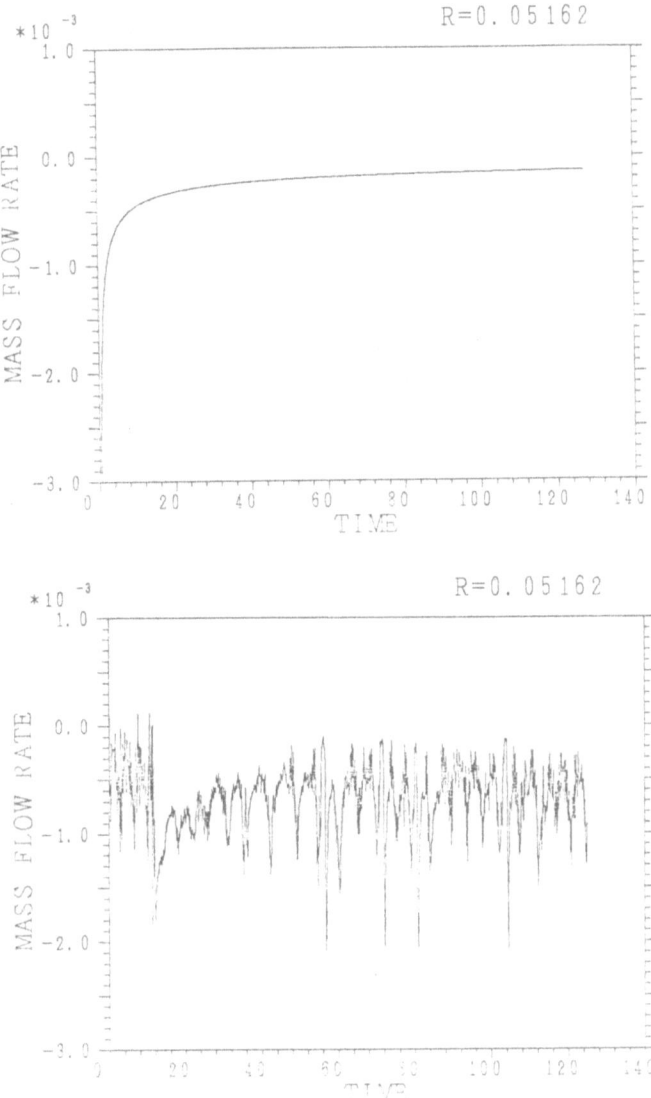

Figure 5: a) Time history of mass flow rate across the inner boundary on which the vacuum condition was applied. Axisymmetry is assumed in this case. b) Same as *a* except the ambient condition was used.

4. Conclusions and Discussions.

Figure 7 shows the summary of our calculations. It shows the mass inflow rate across the inner boundary at a fixed time, $t = 68$, for various mass ratios, q. The

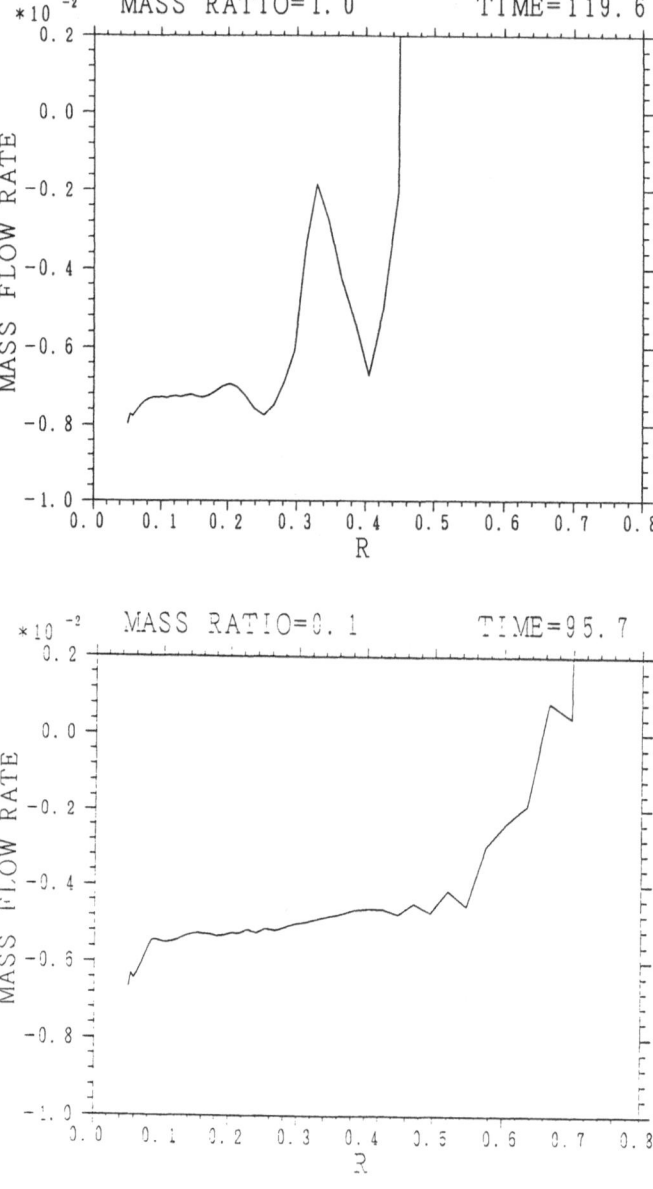

Figure 6: a) Mass flow rate as a function of radial position for the case of $q = 1$. b) $q = 10^{-1}$; c) $q = 10^{-2}$; d) $q = 10^{-3}$; e) $q = 0$. In these cases the coarser grid was used. f) The case $q = 0$ was computed on the finer grid.

horizontal lines in the figure show the mass inflow rate for the axisymmetric case, due to the numerical dissipation. The dotted line represents the result computed on the coarser grid, the solid line shows it on the finer grid. These show the errors

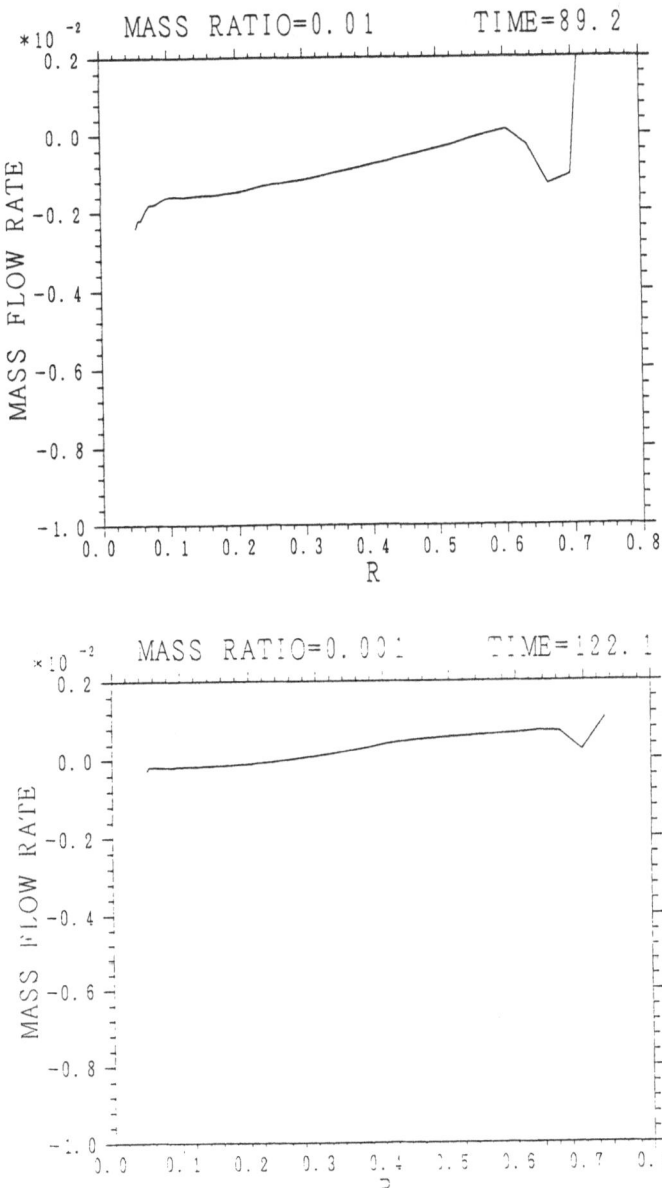

Figure 6: continued

in our calculations. The circles represent the calculations with the coarser grid, the triangles those with the finer grid. To the extent that the calculated mass inflow rates are well above the horizontal lines, these results are reliable. The results for $q \geq 4\,10^{-2}$ are an order of magnitude larger than the errors. The accretion rates for $q = 10^{-3}$ are only a factor two larger than the errors, and they are marginal.

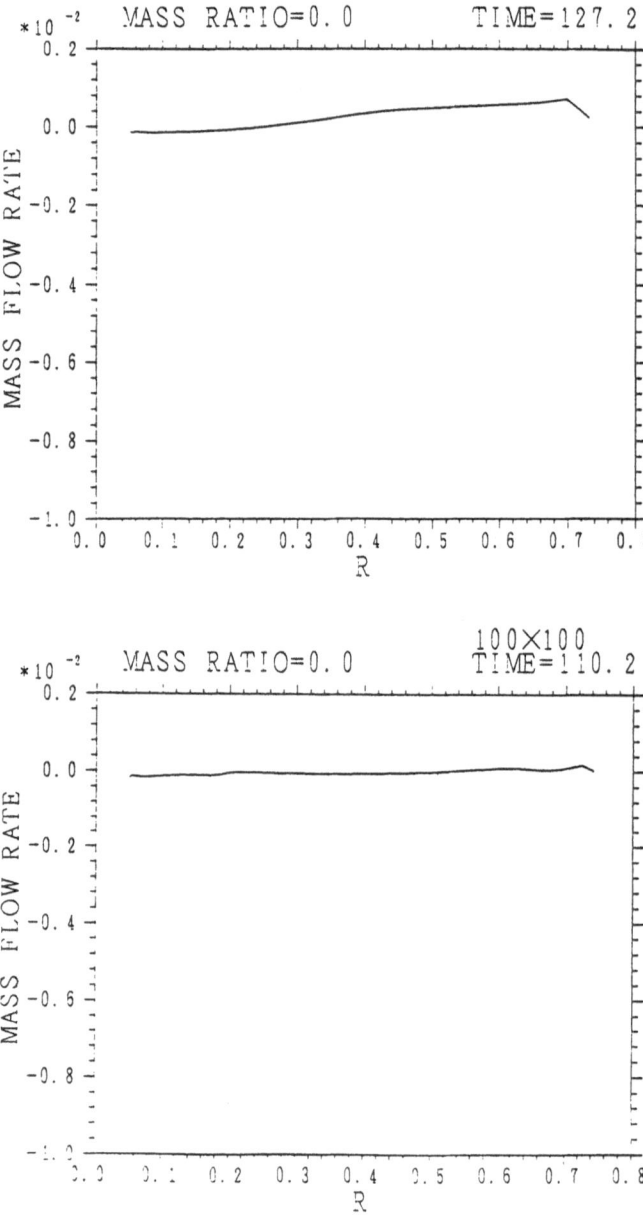

Figure 6: continued

From the mass inflow rate, $\dot{M}$, we can estimate a timescale τ, over which a significant amount of gas is accreted, as

$$\tau = M/\dot{M}.$$

Neglecting factors of order unity, $\dot{M} \sim 10^{-2}$ for $q \geq 10^{-1}$ gives us $\tau \sim 10^2$, which is about 10 rotation periods. Assuming $\dot{M} = 10^{-4}$ in the case of $q = 10^{-3}$, τ is about

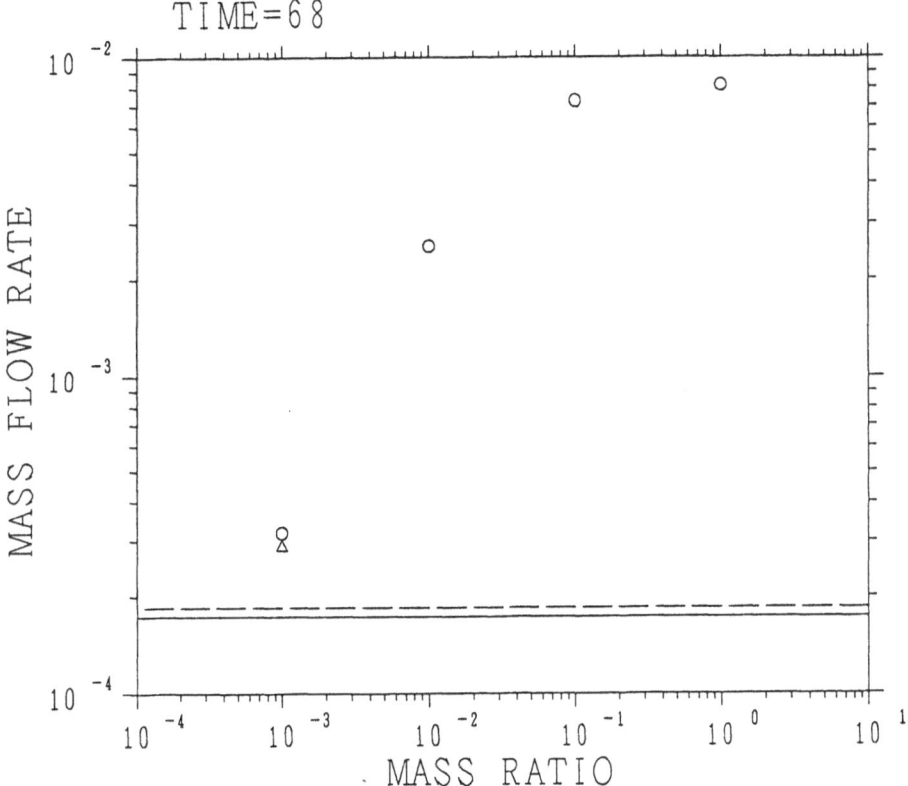

Figure 7: Computed mass inflow rate across the inner boundary against mass ratio q. The triangle shows results computed on the finer grid. The horizontal lines in the figure show mass inflow rate for the axisymmetric case, due to numerical viscosity. The dashed line: coarser grid, solid line: finer grid.

10^3 rotations. This is only 10^4 yr for the Jupiter driven waves, and it is probably too short. This means that the present mechanism is very efficient for the accretion of disc material.

In the standard accretion disc model, the mass inflow rate is given by

$$\dot{M} = 3\pi\mu\nu,$$

where μ and ν are surface density and viscosity of gas. In the α-model,

$$\nu = \alpha c^2/\Omega,$$

where c and Ω are the sound speed and the angular velocity. In our units μ and Ω are of order unity, and c is the inverse of the Mach number. Therefore, we may set

$$\alpha_{eff} \sim 4\dot{M}.$$

We have $\dot{M} = 10^{-4} - 10^{-2}$ for $q = 10^{-3} - 1$, and we may say that $\alpha_{eff} = 4 \cdot 10^{-4} - 4 \cdot 10^{-2}$. Since this estimate is very crude and $\dot{M}$ is measured at a rather late time, $t = 68$, it is not unreasonable to conclude that $\alpha_{eff} \sim 0.1$ for an accretion disc in a close binary system.

In the present calculations a very important effect, i.e. radiation loss, is not included. If this is taken into account, the disc is cooled down and the Mach number will be increased. This makes the shocks stronger, although the decrease of pitch angle of the spirals reduces the Mach number normal to the shocks. We need further investigation to clarify the cooling effect.

ACKNOWLEDGEMENTS. This work was supported by the Grant-in-Aid for Scientific Research (63540195) of the Ministry of Education and Culture in Japan, and also by the Institute of Space and Astronautical Science, Japan. Computations were performed at the Data Processing Center of Kyoto University. T. M. would like to thank Inamori foundation for financial support.

REFERENCES.

Drury, L.O'C.: 1977, PhD Thesis, Cambridge University, UK

Drury, L.O'C.: 1980, *Mon. Not. R. astron. Soc.*, **193** 337

Drury, L.O'C.: 1985, *Mon. Not. R. astron. Soc.*, **217** 821

Flannery,B.P.: 1975, *Ap. J.*, **201** 661

Frank,J., King,A.R. & Raine,D.J.: 1985, *Accretion power in astrophysics*, Cambridge University Press

Goldreich, P. & Tremaine,S.: 1980, *Ap. J.*, **241** 425

Hensler,G.: 1982, *Astr. Ap.*, **114** 309; **114**, 319

Kaisig,M.: 1989a,b, *Astr. Ap.*, **in press**

Kaisig, M.: 1989c, in preparation

Larson, R.B.: 1988, in *The formation and evolution of planetary systems* eds. H.A. Weaver et al., Cambridge University Press

Lin, D.N.C. & Pringle, J.E.: 1976, in *Structure and evolution of close binary systems* eds. P. Eggleton et al., Reidel, Dordrecht

Lin, D.N.C. & Papaloizou, J.: 1979, *Mon. Not. R. astron. Soc.*, **186** 799

Lin, D.N.C. & Papaloizou, J.: 1986, *Ap. J.*, **307** 395

Matsuda, T., Inoue, M., Sawada, K., Shima, E. & Wakamatsu, K.: 1987, *Mon. Not. R. astron. Soc.*, **229** 295

Michel, F.C.: 1984, *Ap. J.*, **279** 807

Narayan, R., Goldreich, P. & Goodman, J.: 1987, *Mon. Not. R. astron. Soc.*, **228** 1

Ohtsuki, K. & Nakagawa, Y.: 1989, *Prog. Theor. Phys. Suppl.* No. 96 *Origin of solar system*, Chap.17

Papaloizou, J.C.B. & Pringle, J.E.: 1984, *Mon. Not. R. astron. Soc.*, **208** 721

Papaloizou, J.C.B. & Pringle, J.E.: 1985, *Mon. Not. R. astron. Soc.*, **225** 267

Petterson, J.A.: 1983, in *Accretion-driven stellar X-ray sources* eds Lewin, W.H.G.& van den Heuvel, E.P.J., Cambridge University Press

Prendergast, K.H.& Taam, R.E.: 1974, *Ap. J.*, **189** 125

Pringle, J.E.: 1981, *Ann. Rev. Astr. Ap.*, **19** 137

Pringle, J.E. & Rees, M.: 1972, *Astr. Ap.*, **21** 1

Sawada, K., Matsuda, T.& Hachisu, I.: 1986a, *Mon. Not. R. astron. Soc.*, **219** 75

Sawada, K., Matsuda, T.& Hachisu, I.: 1986b, *Mon. Not. R. astron. Soc.*, **221** 679

Sawada,K., Matsuda,T., Inoue,M.& Hachisu,I.: 1987, *Mon. Not. R. astron. Soc.*, **224** 307

Sekiya, M., Miyama, S.M.& Hayashi, C.: 1987, *Earth, Moon, Planets* **39**, 1

Sekiya, M., Miyama, S.M.& Hayashi, C.: 1989, *Prog. Theor. Phys. Suppl. No. 96*, Origin of solar system, Chap.27

Shakura, N.I. & Sunyaev, R.A.: 1973, *Astr. Ap.*, **24** 337

Shu, F.H.: 1976, in *Structure and evolution of close binary systems* eds. P. Eggleton et al., D.Reidel, Dordrecht

Sorensen, S.A., Matsuda,T.& Sakurai, T.: 1975, *Astrophys. Space Sci.* **33**, 465

Sorensen, S.A., Matsuda, T.& Fujimoto,M.: 1976 *Astrophys. Space Sci.* **43**, 491

Spruit, H.C.: 1987, *Astr. Ap.*, **184** 173

Spruit, H.C., Matsuda, T., Inoue, M.& Sawada, K.: 1987, *Mon. Not. R. astron. Soc.*, **229** 517

ACCRETION DISKS AND THE LINK BETWEEN AN AGN AND ITS HOST GALAXY.

Mitchell C. BEGELMAN[1], Juhan FRANK[2], and Isaac SHLOSMAN[3]

[1] JILA, University of Colorado and NIST, Boulder, CO 80309, USA

[2] Max-Planck-Institut für Astrophysik, Garching bei München, FRG

[3] Theoretical Astrophysics, Caltech, Pasadena, CA 91125, USA

ABSTRACT. We examine the possible role of accretion disks in bringing matter from the interstellar medium into the central regions of an active galactic nucleus (AGN). Thin axisymmetric accretion disks are *not* a viable means of transporting fuel to luminous AGNs on scales larger than a fraction of a parsec, because 1) the inflow time scale is too long; and 2) disks carrying the required fuel supply would become self-gravitating, leading to fragmentation and probably star formation. We consider the effects of star formation, concluding that energy input from stellar winds and supernovae is probably inadequate to increase the disk thickness to the point where disk accretion becomes viable. There are also serious obstacles to maintaining and regulating the energetics of geometrically thick, hot accretion flows, although the possibility of fuelling some AGNs through cooling flows cannot be ruled out. In this paper and the companion paper by Frank, Shlosman and Begelman (these proceedings), we outline a unified model for fuelling AGNs, in which the inflow is driven by global nonaxisymmetric gravitational instabilities on large scales, and on small scales forms a mildly self-gravitating disk of clouds in which angular momentum is transported by cloud-cloud collisions.

1. Introduction.

Even if they convert mass to energy with a high efficiency $\epsilon \sim 0.1$, active galactic nuclei (AGNs) require fuel at a rate

$$\dot{M} \sim 0.2 \left(\frac{\epsilon}{0.1} \right)^{-1} L_{45} \ M_\odot \ yr^{-1},$$

(1)

where $L_{45} \equiv L/10^{45}$ ergs s^{-1} is the luminosity. Lifetimes of AGNs are uncertain, but arguments based on statistics and the energy contents of giant radio sources suggest that this rate must be maintained for as long as 10^7–10^9 yr. Thus the question of long-term fuel supply is a fundamental problem which must be addressed by the theory of active galaxies.

F. Meyer et al. (eds.), Theory of Accretion Disks, 373–386.

Within the context of black hole accretion models, there have been two approaches to the fuelling problem. In one class of models, the AGN is regarded as a self-contained "machine" whose activity bears little immediate connection to phenomena in its host galaxy. Models in this category contain a dense star cluster which feeds the black hole through a variety of mechanisms, including tidal disruption of stars which stray too close to the hole (Hills 1975, 1978), star-star collisions (Begelman and Rees 1978, Frank 1978), and stellar ablation (Voit and Shull 1988, Shlosman 1988). In the second class of models, the black hole is fed by interstellar gas, or by gas which has been captured from the intergalactic medium. Whereas the main problem with the star cluster models hinges on the adequacy of the mass supply, there is no question that enough fuel is available on kiloparsec scales in the host galaxy. Instead, the critical problem facing models with external fuel supply is: how to get rid of the angular momentum on a short enough time scale. Over the years, a variety of mechanisms have been proposed (e.g., Bailey and Clube 1978, Gunn 1979), but no consensus has emerged as to their plausibility.

In this article, we focus on the fuelling of AGNs by inflow from kpc scales in the host galaxy. The large angular momentum content of the accreting gas suggests that the flow will form a disk, but we will argue that the way in which it loses angular momentum must be very different from the viscous angular momentum transfer which is thought to occur in standard thin accretion disks. Ulrich and Malkan (these proceedings) have reviewed the observational evidence for accretion disks in the "central engines" of AGNs. Indeed it is plausible (but by no means essential) that a structure resembling a standard thin accretion disk could exist in the inner regions of the AGN, say, within $\sim 10^3$ gravitational radii of the black hole. In this article we will not consider such small scales, but will concentrate on the region between ~ 1 parsec and ~ 1 kiloparsec from the center. This region bridges the gap between the interstellar medium (ISM) of the host galaxy and the AGN.

2. Can Thin Accretion Disks Feed AGNs?

There are two major problems in using standard thin accretion disks to fuel AGNs. First, there is a problem with the viscous inflow time scale:

$$t_{visc} \sim 10^9 \ \alpha^{-1} T_{100}^{-1} v_{\phi,100} r_{10} \ yr, \tag{2}$$

which tends to be too long at radii larger than a few parsecs. Here α is the ratio of shear-induced stress to pressure in the disk, usually assumed to be < 1; the internal disk temperature T_{100} is in units of 100 K; the orbital speed $v_{\phi,100}$ is in units of 100 km s^{-1}; and r_{10} is the radius in units of 10 pc. If t_{visc} is longer than the lifetime of the AGN (let alone the Hubble time) then it is impossible for the disk to fuel the AGN steadily from large radii.

Second, thin accretion disks cannot carry the requisite mass flux without becoming gravitationally unstable. Gaseous disks become *locally* Jeans unstable when the mean density in the disk becomes so large that the self-gravity exceeds the tidal gravitational field (Toomre 1964, Goldreich and Lynden-Bell 1965), *i.e.*, when

$$n \gtrsim n_{crit} \sim 10^6 \, v_{\phi,100}^2 r_{10}^{-2} \; cm^{-3}. \tag{3}$$

The critical density defines a critical accretion rate,

$$\dot{M}_{crit} \sim \frac{3\alpha c_s^3}{G} \sim 5 \times 10^{-4} \alpha T_{100}^{3/2} \; M_\odot \; yr^{-1}, \tag{4}$$

above which the disk is locally unstable to clumping. In eq. (4), c_s is the thermal sound speed, but a similar criterion would apply in a disk supported by turbulent pressure with the velocity dispersion replacing c_s. Note that $\dot{M}_{crit}$ is independent of both r and v_ϕ!

Global gravitational instability sets in when the mass of the disk exceeds a fraction of the external gravitating mass, *i.e.*, for accretion rates

$$\dot{M} \gtrsim \left(\frac{v_\phi}{c_s}\right)\dot{M}_{crit} \sim 5 \times 10^{-2} \alpha T_{100} v_{\phi,100} \; M_\odot \; yr^{-1}. \tag{5}$$

The global instability would probably lead to the development of a gaseous bar, which would transport angular momentum rapidly and lead to inflow on a few orbital timescales. Indeed, we argue in the companion paper by Frank, Shlosman and Begelman (these proceedings) that a nested sequence of such bars is responsible for most of the mass transport in AGNs on scales ~ 10 pc – 1 kpc.

On the basis of the arguments presented above, we conclude that thin accretion disks cannot carry enough mass to fuel luminous AGNs, nor transport it inward on a short enough timescale (from distances greater than a few parsecs), unless $\alpha \gg 1$ and/or $T \gtrsim 10^4$ K. Can either or both of these conditions be met?

2.1. DISK TEMPERATURE.

The effective temperature associated with viscous dissipation,

$$T_{eff} \sim 11 \left(\frac{\dot{M}}{1 M_\odot \; yr^{-1}}\right)^{1/4} v_{\phi,100}^{1/2} r_{10}^{-1/2} \; K, \tag{6}$$

is a lower limit on the disk temperature. The actual temperature will be higher, partly because of finite optical depth effects and partly because internal viscous dissipation is unlikely to be the main heating mechanism in the disk at the radii we are considering. Additional sources of heating may include radiation directly incident on the disk from the central engine, radiation scattered down onto the disk by gas and dust at high latitudes, radiation emitted by jets, and starlight. Whereas the surface flux due to viscous dissipation decreases with radius $\propto r^{-3}$, the incident flux (if it is present at all) is likely to decrease less rapidly than $\propto r^{-3}$. Thus, external sources of heating should dominate at large r, even if the total incident power is a small fraction of the AGN luminosity.

Shlosman and Begelman (1987, 1989) calculated central and surface disk temperatures as functions of radius, under various assumptions about heating and cooling mechanisms. The highest temperatures are obtained when the gas in the disk is dust-free, and must cool via atomic and molecular excitation. Even with the incident flux as high as 1–10% of the AGN luminosity, the temperature drops below $\sim 10^3$ K at $r \gtrsim 1$ pc, and below ~ 100 K at $r \gtrsim 10$–300 parsecs (depending on the accretion rate and other parameters). When dust is present at normal interstellar abundance (as it is likely to be in an AGN being fed by large-scale inflow), the thermal properties of the disk are completely dominated by the radiative heating and cooling of dust grains, with thermal coupling to the gas provided by grain-gas collisions. In this case, the disk temperature will drop below 100 K at radii beyond 1–10 pc in a luminous AGN.

We therefore conclude that radiative processes cannot keep a thin disk warm enough to carry the required mass flux without becoming gravitationally unstable, unless $\alpha \gg 1$. If the disk can be maintained in a highly turbulent state, then the velocity dispersion associated with the turbulence substitutes for the thermal sound speed in the disk structure equations, and larger mass flow rates are possible. In § 3 we consider whether star formation in the disk could drive the highly supersonic turbulence which would be required, and in § 5 we propose that the self-consistent outcome of gravitational instability is a mildly self-gravitating "disk" of weakly interacting clouds. First, however, we consider whether local gravitational instabilities themselves might produce an α exceeding 1.

2.2. CONSEQUENCES OF LOCAL GRAVITATIONAL INSTABILITY.

There are two divergent views as to the likely outcome of local gravitational instability in a disk. Instability will initially lead to the growth of density and pressure inhomogeneities with increasing peculiar velocities in the plane of the disk, or perhaps to the development of local spiral "fragments" (D. Lin, private communication). If these inhomogeneities interact with one another and with the mean flow, they presumably transport angular momentum outward and could provide the α for the disk. The important question is whether this process can be maintained as a steady means of angular momentum transport in a self-gravitating disk. Paczyński (1978) proposed a model for self-regulating viscosity in which the energy dissipated by collisions between clumps is enough to prevent their collapse and to keep the disk at marginal stability. The α obtainable in this model is $\ll 1$. Lin and Pringle (1987) extended this idea to strongly unstable disks, and argued that continuing gravitational instability could lead to $\alpha \gg 1$. Alternatively, if the self-gravitating clumps contract to the point where they hardly interact at all, then we would expect the disk to resemble more a stellar disk (or a disk of individual clouds:

§ 5) than a turbulent gaseous disk. In this case, we would expect the effective α to *decrease* with the development of instability, although the concomitant increase in velocity dispersion may ultimately lead to faster accretion (*cf.* § 5). Which of these outcomes is most likely in a disk feeding an AGN?

A strongly self-gravitating disk, *i.e.*, one with $n \gg n_{crit}$, is unstable over a range of wavelengths, $h \lesssim \lambda \lesssim (n/n_{crit})h$, where h is the disk thickness and λ is the scale of the unstable mode in the plane of the disk. To reconcile our notation with that of Lin and Pringle, we note that $n/n_{crit} \sim Q^{-2}$, where Q is the Toomre (1964) self-gravity parameter. Note that h must be calculated taking self-gravity into account, *i.e.*, $h \sim (n/n_{crit})^{-1/2}(c_s/v_\phi)r$. $\alpha \sim 1$ corresponds to the radial transport of angular momentum over a scale h at the local sound speed. $\alpha > 1$ would result if the typical scale of angular momentum transport (at the sound speed) were larger than h. Lin and Pringle suggested that this would happen in a strongly self-gravitating disk if the modes with $\lambda \gg h$ dominate the transport of angular momentum.

In order for the Lin-Pringle mechanism to work, the disk must be prevented from fragmenting into weakly interacting clumps. If the cooling time scale of the gas is longer than the time required for contraction (basically the gravitational free-fall time in the clump), then the pressure could build up and limit the degree of contraction. In a steady-state accretion disk, the time required to generate the thermal energy in the disk by viscous dissipation is $t_{diss} \sim \alpha^{-1}t_{orb}$, where t_{orb} is the orbital time. Since the thermal balance in the disk is probably dominated by external heating, the cooling time scale satisfies $t_{cool} < t_{diss}$. To prevent fragmentation on scales $\sim \lambda$, the cooling time should be longer than the collapse time for a clump, $t_J(\lambda) \sim (n/n_{crit})^{-1/2}(\lambda/h)^{1/2}t_{orb}$. This imposes an upper limit on α:

$$\alpha < \left(\frac{n}{n_{crit}}\right)^{1/2}\left(\frac{t_{cool}}{t_{diss}}\right)\left(\frac{h}{\lambda}\right)^{1/2}, \tag{7}$$

where λ is the scale which dominates the transport of angular momentum. Since $\alpha \sim \lambda/h$ in the Lin-Pringle picture, the maximum value of α consistent with the non-fragmentation condition is of order $(n/n_{crit})^{1/3}(t_{cool}/t_{diss})^{2/3}$.

It therefore seems more likely that a strongly self-gravitating thin disk would fragment than that it would develop an α large enough to feed a luminous AGN. We note that *nonlocal* forms of angular momentum transfer, such as large-scale spiral waves or magnetic fields, may lead to inflow with an effective $\alpha > 1$. Frank, Shlosman and Begelman (these proceedings; see also Shlosman, Frank and Begelman 1989) argue that large-scale gravitational disturbances are likely to cause inflow on scales $\gg 1$ pc.

3. Effects of Star Formation.

The likelihood of fragmentation suggests that we should examine the consequences of star formation in a self-gravitating disk. What are the masses of stars likely to form? What are the effects of star formation on the disk — specifically, will heating by stellar winds and supernovae lead to a higher disk temperature and/or greater disk turbulence, resulting in faster inflow? How long will it take to turn the gaseous disk into stars?

The factors which determine the initial mass function (IMF) of stars are as mysterious in the disk as they are everywhere else in the Universe. A crude benchmark for the upper mass limit is the mass contained in a cube of dimension h on a side:

$$M_J \sim 19 \left(\frac{n}{n_{crit}} \right)^{-1/2} T_{100}^{3/2} v_{\phi,100}^{-1} r_{10} \ M_\odot. \qquad (8)$$

This at least indicates that massive stars can form at radii $\gtrsim$ a few parsecs. If we take into account the fact that clumps with $\lambda < \lambda_{frag} \sim [1 + (n/n_{crit})(t_{cool}/t_{orb})]h$ may not be able to fragment, then the maximum stellar mass is increased by a factor $\sim (\lambda_{frag}/h)^2$.

The cumulative effects of star formation depend on the star formation rate or "efficiency" ξ, defined as the fraction of mass going into stars per local free-fall time. The total star formation rate per decade in radius is given by

$$\dot{M}_* \sim 3 \ \xi \left(\frac{n}{n_{crit}} \right) T_{100}^{1/2} v_{\phi,100}^2 \ M_\odot \ yr^{-1}. \qquad (9)$$

Star formation in the Milky Way is observed to be rather inefficient, with $\xi \sim 0.05$ for low-mass stars ($\lesssim 1 M_\odot$) and $\sim 10^{-3}$ for high-mass stars ($\gtrsim 10 M_\odot$) (Shlosman and Begelman 1989 and references therein). Adopting these as fiducial values, we find that the time scale for turning the disk into stars is

$$t_{g \to *} \sim 4 \times 10^6 \left(\frac{\xi}{0.05} \right)^{-1} \left(\frac{n}{n_{crit}} \right)^{-1/2} v_{\phi,100}^{-1} r_{10} \ yr. \qquad (10)$$

$t_{g \to *}$ is shorter than the inflow time t_{visc} (if estimated by eq. [2]), even if ξ is as small as 10^{-3}. We therefore would expect a self-gravitating accretion disk to be significantly depleted by star formation.

Both high-mass and low-mass stars inject energy into the disk: high-mass stars through radiation-driven winds and supernova explosions, and low-mass stars through winds produced during the T Tauri phase. Shlosman and Begelman (1989) present a detailed discussion of the energetic effects of star formation, the results of which we summarize here. Even if the IMF is heavily weighted towards low-mass stars, the high-mass stars probably dominate the energetics (by a factor ~ 100). The total rate of energy input,

$$\dot{E}_* \sim 10^{41} \left(\frac{\xi}{10^{-3}}\right)\left(\frac{n}{n_{crit}}\right) T_{100}^{1/2} v_{\phi,100}^2 \ ergs \ s^{-1}, \tag{11}$$

is marginally enough to keep the disk warm and/or turbulent. However, it is likely that only a small fraction (of order 0.01) of this energy is actually available for heating the disk. The rest goes into radiation through the cooling of shocked gas, or is lost through the "breakout" of blast waves or wind-driven bubbles from the disk (McCray and Kafatos 1987). A side effect of the breakout phenomenon is that star formation in one region of the disk is likely neither to inhibit nor to trigger star formation in neighboring regions, since the disturbance associated with young stars is relatively localized. Breakout may also lead to the formation of a disk corona and the ejection of cool filaments into the region above and below the disk.

Star formation in a self-gravitating disk will leave behind a flattened disk of stars in the nucleus of the galaxy. Although stellar dynamical collective effects will thicken the disk until it reaches a state of marginal self-gravity, significant rotational flattening should persist to the present time, since two-body relaxation operates too slowly to completely sphericalize such a system. Recent observational papers suggesting the presence of massive black holes in the nuclei of nearby galaxies (Dressler and Richstone 1988, Kormendy 1988) stress the importance of rotation, although the degree of flattening is uncertain. We conjecture that these stellar systems could have been formed via the fragmentation of self-gravitating disks.

4. A "Hot" Accretion Flow?

Up to this point, we have argued that a cool, thermally supported disk will remain at a temperature well below 10^4 K. We have implicitly assumed that the disk *starts out* cool. We have not fully examined the fate of an accretion flow which *starts out* hot, *i.e.*, with $T > 10^4$ K. There is good reason to consider such a flow. First, it is well known that accretion disks can exhibit a thermal bistability, *i.e.*, two thermal states can exist for the same accretion rate (Shapiro, Lightman and Eardley 1976): this result leads to the prediction of ion tori (Rees *et al.* 1982), albeit on much smaller scales than we are considering here. Second, there is strong observational evidence that cooling flows carry large quantities of hot gas into the central regions of many early-type galaxies, some of which contain AGNs (Sarazin 1986, Arnaud 1988). Third, if we make the *ansatz* that the inflow speed is limited by some effective α ($\lesssim 1$), then hot flows can support much faster inflow than cool flows with the same α, and can carry more mass without becoming gravitational unstable.

Consider a restricted subset of such flows, *i.e.*, those which are have significant rotational support, are in vertical hydrostatic equilibrium, and rely on angular momentum transport with some characteristic α ($\lesssim 1$). Normalizing the pressure to 10^{-12} dyne cm^{-2} (a few times the pressure in the ISM of the Milky Way), the temperature to 10^6 K, and the radius to 1 kpc, we find that the accretion rate is given by

$$\dot{M} \sim 0.2 \; \alpha p_{-12} T_6^{1/2} v_{\phi,100}^{-2} r_{kpc}^2 \; M_\odot \; yr^{-1}. \tag{12}$$

Figure 1 shows curves of constant $\dot{M} = 1$ and 0.1 $M_\odot$ yr^{-1} on the p–T plane for $\alpha = v_{\phi,100} = r_{kpc} = 1$. The horizontal line labeled "$T = T_{vir}$" indicates the (virial) temperature above which the gas will tend to escape from the gravitational potential, rather than accrete. Pressures 2–3 orders of magnitude above typical ISM pressure are required in order to obtain the fuelling rate for a luminous AGN without exceeding the virial temperature. This is not implausible in itself, since such pressures are inferred to exist in many early-type galaxies (Forman, Jones and Tucker 1985; Thomas et al. 1986; Canizares, Fabbiano and Trinchieri 1987). Note, however, that much higher pressures are required to carry the same mass flux at smaller radii. Also plotted on Fig. 1 is the curve "$t_{cool} = t_{ad}$", below which radiative cooling dominates over the compressional (and dissipative) heating associated with the accretion flow. Apparently, some *external source of heating is required to prevent the flow from cooling and collapsing into a thin disk.* The requirement of extra heating is even more pronounced at smaller radii.

What form could this heating take? One possibility is heating by supernovae and winds from massive stars. If the heating efficiency is high, then a massive star formation rate of one every few years is required. A high heating efficiency requires that the energy be distributed throughout the hot gas before localized regions have a chance to cool. Ideally, the supernova blast waves or stellar wind bubbles should come into pressure balance with the background medium in less than the cooling time, a condition which applies if $T \gtrsim 8 \times 10^5 p_{-12}^{1/5}$ K. Unless there is some feedback regulating the massive star formation rate, this heating mechanism appears to be unstable: if T decreases, the heating becomes less efficient and the hot gas will collapse onto the central plane; if T increases, the cooling becomes less efficient and the gas will begin to flow out in a wind. It is not absolutely clear, however, that the outflowing gas would escape from the galaxy altogether, particularly if a massive halo is present. Instead, it is possible that the adiabatic cooling of the outflowing gas could lead to the condensation of clouds which fall back into the galaxy and ultimately join the accretion flow (as in "galactic fountain" models for the ISM: Shapiro and Field 1976, Wang and Cowie 1988).

A second possibility is that the gas is heated by X-rays from the central source. If the ionization parameter Ξ (= [ionizing radiation pressure]/[gas pressure]) is larger than a few, then the gas will begin to heat towards the inverse Compton temperature of the radiation field. Fig. 1 shows schematically the threshold for runaway heating, assuming an ionizing luminosity of 10^{45} ergs s^{-1} and an AGN spectrum with an inverse Compton temperature of 10^7 K. Although recent evidence suggests that the inverse Compton temperature may be even lower in some objects

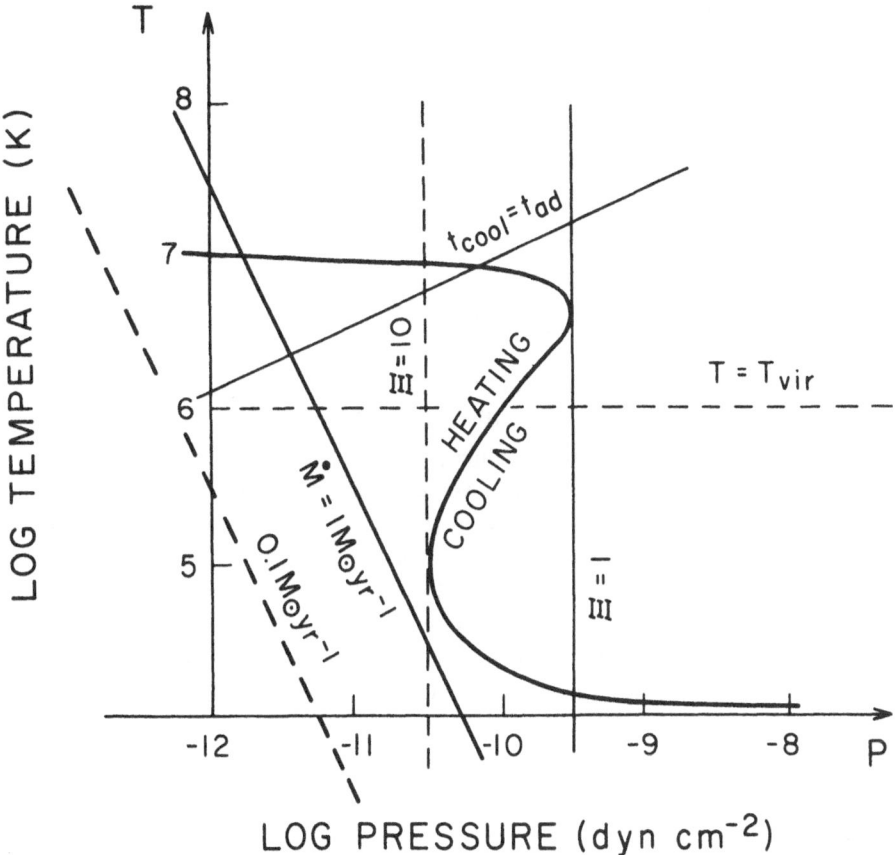

Figure 1: Parameter space in pressure–temperature plane for "hot", angular momentum-dominated accretion. Details are described in text.

(Mathews and Ferland 1987), it probably exceeds the virial temperature in most cases. Calculations by Begelman (1985) show that the gas would heat to T_{vir} and beyond in less than the inflow time. Thus, this mechanism also tends to produce a wind, rather than promoting steady accretion.

Despite the pessimistic theoretical considerations above, observations do seem to indicate that cooling flows transport large quantities of gas into the central regions of elliptical galaxies. Perhaps the assumptions adopted above are too restrictive: angular momentum transport could be more efficient than we have assumed (e.g., due to magnetic fields), or the accreting gas could be prevented from blowing away by the pressure of the intergalactic medium. Thus, one cannot rule out the possibility that cooling flows fuel some luminous AGNs (Nulsen, Stewart and Fabian 1984, Begelman 1986).

5. A Disk of Clouds.

The arguments presented in §§ 2–4 deal with an idealization of the accretion disk as a quasi-continuous flow, *i.e.*, a flow in which the accreting matter occupies most of the volume in the disk. We have shown that simple models of such flows run into serious difficulties. In this section we describe an alternative type of flow in which the "disk" is composed of randomly moving clouds, with small volume filling factor, embedded in a low density medium. The viscosity is provided by collisions between clouds. Such a "disk of clouds" could carry enough mass flux to power a luminous AGN, without running into the difficulties faced by a quasi-continuous disk.

One can draw a straightforward analogy between standard accretion disk theory and the theory applicable to a disk of discrete particles. Such a theory has been described in connection with Saturn's rings (Goldreich and Tremaine 1978). (kT/μ) is replaced by the square of the cloud velocity dispersion $(\Delta v)^2$. The coefficient of kinematic viscosity can be written $\nu \sim (\text{m.f.p.})^2/t_{coll}$, where (m.f.p.) is the appropriately defined mean free path for momentum transport and t_{coll} is the cloud collision time. A useful parameter is the cloud covering factor C, which is defined as the fraction of disk area covered by clouds and is roughly equal to t_{orb}/t_{coll}. When $C \gg 1$, clouds travel on roughly straight paths between collisions, the m.f.p. $\sim (\Delta v)t_{coll}$ and $\nu \sim (\Delta v)^2 t_{coll}$, as in an ideal gas. When $C \ll 1$, however, clouds execute epicyclic motion between collisions, the appropriate m.f.p. $\sim (\Delta v)t_{orb}$, and ν is a factor $\sim C^2$ smaller than in the previous case (Goldreich and Tremaine 1978). These results can be summarized in terms of an effective α:

$$\alpha \sim \begin{cases} C & C \ll 1 \\ C^{-1} & C \gg 1 \end{cases}. \tag{13}$$

The source of the clouds' velocity dispersion is a major uncertainty. Transport of angular momentum through cloud-cloud collisions entails viscous dissipation, which could increase Δv at the expense of the free energy associated with shear, provided that the collisions are sufficiently elastic. Normally, cloud-cloud collisions are thought to be rather inelastic — the necessary elasticity would probably require clouds in which the gas pressure is strongly dominated by magnetic energy (Krolik and Begelman 1988). As before, stellar winds and supernova explosions do not seem very promising sources of energy, because they are used inefficiently. A third possible source of energy is the free energy associated with local self-gravity in the disk. A disk of clouds with $C \ll 1$ will be subject to the stellar dynamical analogues of local gravitational instabilities which afflict self-gravitating gaseous disks. The ultimate outcome of these instabilities will be to thicken the disk (which initially has a mean density $\bar{n} > n_{crit}$) until $\bar{n} \sim n_{crit}$. If $C < 1$, the cloud collision time is longer than the growth time of instability and the disk can be kept at $\bar{n} \sim n_{crit}$

through continuous mild instability. The resulting accretion rate,

$$\dot{M} \sim \frac{3\alpha(\Delta v)^3}{G} \sim 700 \, C \left(\frac{\Delta v}{v_\phi} \right)^3 v_{\phi,100}^3 \, M_\odot \, yr^{-1}, \tag{14}$$

is adequate to fuel luminous AGNs if C and Δv take on plausible values. The values of C and Δv probably depend on the properties of the clouds and their collisions; predicting them is beyond the scope of this discussion.

6. Conclusion: A Unified Scenario for Fuelling AGNs.

In this article we have discussed several apparent dead ends in the search for a means of fuelling AGNs, and one possible resolution of part of the puzzle. While thin accretion disks may be present and may produce observable side-effects (e.g., resulting from star formation in the disk), they are unlikely to be responsible for carrying the bulk of the accreting matter. A marginally self-gravitating disk of clouds, with "viscosity" provided by cloud-cloud collisions, is an attractive alternative.

How can such a flow be established? Although the ISM in the Milky Way and other nearby spirals has a cloudy structure, the inflow rate due to cloud-cloud collisions is too small to fuel luminous AGNs. At large distances ($\gtrsim \, few \times 100$ pc) this is primarily because the velocity dispersion is too small. (The column density through the disk is not large enough to force it to thicken through gravitational instabilities.) In the molecular torus at the Galactic Center ($\sim$ few pc) the velocity dispersion and covering factor of clouds are moderately high, but $\bar{n}$ is well below n_{crit} (see, e.g., Phinney 1989 and references therein). A plausible way to create a cloud disk with a large velocity dispersion is to start with a gaseous system which is strongly gravitationally unstable, i.e., with $\bar{n} \gg n_{crit}$. The outer boundary of such a flow would have to be established in less than a few orbital timescales, but this is just what one would expect to happen in the "bars within bars" scenario described by Frank, Shlosman and Begelman (these proceedings) and Shlosman, Frank and Begelman (1989).

According to the "bars within bars" scenario, a dynamically unstable sequence of nested gaseous bars drives inflow on a few dynamical time scales. At each radius, the mass in gas is some significant fraction of the total mass interior to that radius. This implies that the gas is also strongly unstable to local gravitational instabilities, but it is the global gravitational disturbance, not the cloud-cloud collisions, which drives inflow on these outer scales. Since not all of the gas is swept inward by the bar, bar-driven inflow is likely to slow down when the flow reaches radii where the mass interior to r ceases to vary strongly with radius. In a galactic nucleus with a black hole at the center, this is probably the radius at which the black hole begins to dominate the gravitational potential, $r_h \sim 40(M_{BH}/10^8 \, M_\odot)v_{\phi,100}^{-2}$ pc. As soon as the bar-driven inflow becomes inefficient, the flow is likely to fragment (if it has not already done so) and to thicken due to local gravitational instability. We speculate that the transition in the behavior of the potential causes a transition from bar-driven inflow to inflow driven by cloud-cloud collisions.

Thus, our analysis leads us to propose a specific description of the accretion flows which fuel luminous AGNs. The main components are bar-driven inflow at large radii and a marginally self-gravitating disk of clouds at smaller radii. The relationship of these flow regions is shown schematically in Fig. 2, which also indicates that a standard thin accretion disk or torus may still form in the innermost regions. Although our model involves a specific set of processes, in its current form it does include all of the physics necessary to produce a detailed picture of the conditions in an AGN. Much work needs to be done on the cloud physics and on the transition from bar-driven to cloud-collision-driven inflow. We regard the model's flexibility as a virtue, since many of the details are likely to differ from one object to another. For example, one can envisage a case in which material flows from the bar-driven region, through the cloud disk and into the black hole at a steady rate, or a case in which the disk of clouds stores the incoming material and feeds it into the central engine over a much longer time scale. In the former case, the episode of luminous nuclear activity would last for several galactic dynamical times, and evidence for bar-driven inflow should be observed simultaneously with the nuclear activity. In the latter case, activity could proceed seemingly independently of the presence of inflow on large scales, and for an undetermined period of time.

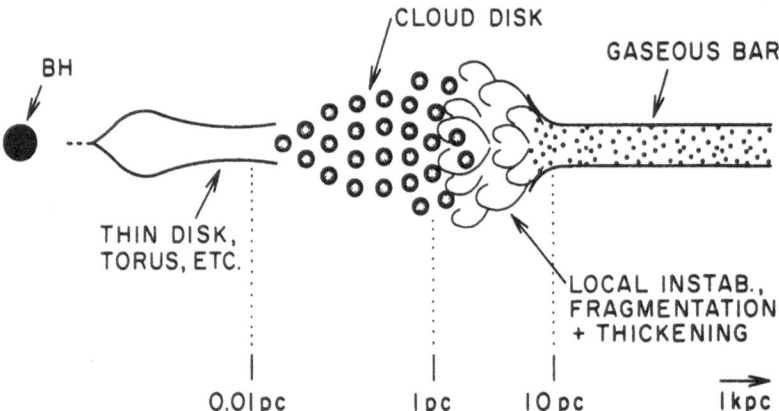

Figure 2: Schematic illustration of the radial flow regions associated with accretion of interstellar matter by an AGN.

The "disk of clouds" model has several features which may bear on important observational aspects of AGNs. If the clouds are optically thick and have a small covering factor, then they may be resistant to the rapid X-ray ablation which was pointed out and studied by Begelman (1985) and by Krolik and Begelman (1986,1988). The ability of the inflowing gas to avoid dispersal by X-ray heating is necessary for the survival of the accretion flow, since the ablation rate can exceed the accretion rate. The initial thickening and fragmentation of the disk, during which the covering factor may still be large, could produce the observational signature of the opaque molecular tori inferred to exist in Type 2 Seyferts (Antonucci and Miller 1985; Krolik and Begelman 1986, 1988) and may contribute to a thermal infrared spectral component in Type 1 Seyferts and QSOs (Sanders *et al.* 1989, Phinney [these proceedings]). The clouds may contribute to the production of the observed emission lines. Finally, the clouds themselves may be self-gravitating, and (along with a thin disk component) could serve as star formation sites in a nuclear starburst.

ACKNOWLEDGEMENTS. Thoughtful comments by M. de Kool, P. Goldreich, M. Loewenstein and S. Phinney helped us in the preparation of this manuscript. This work was supported in part by NSF grant AST83-51997, NASA Astrophysical Theory Center grant NAGW-766, and grants from Rockwell International Corporation and the Alfred P. Sloan Foundation (at JILA); and NSF grant AST84-51725 (at Caltech).

REFERENCES.

Antonucci, R. R. J., and Miller, J. S. 1985, *Ap. J.*, **297**, 621
Arnaud, K. A. 1988, in *Cooling Flows in Clusters and Galaxies*, ed. A. C. Fabian (Dordrecht: Kluwer Academic Publishers), p. 31
Bailey, M. E., and Clube, S. V. M. 1978, *Nature*, **275**, 278
Begelman, M. C. 1985, *Ap. J.*, **297**, 492
Begelman, M. C. 1986, *Nature*, **322**, 614
Begelman, M. C., and Rees, M. J. 1978, *M.N.R.A.S.*, **188**, 847
Canizares, C. R., Fabbiano, G., and Trinchieri, G. 1987, *Ap. J.*, **312**, 503
Dressler, A., and Richstone, D. O. 1988, *Ap. J.*, **324**, 701
Forman, W., Jones, C., and Tucker, W. 1985, *Ap. J.*, **293**, 102
Frank, J. 1978, *M.N.R.A.S.*, **184**, 87
Goldreich, P., and Lynden-Bell, D. 1965, *M.N.R.A.S.*, **130**, 97
Goldreich, P., and Tremaine, S. 1978, *Icarus*, **34**, 227
Gunn, J. E. 1979, in *Active Galactic Nuclei*, ed. C. Hazard and S. Mitton (Cambridge: Cambridge University Press), p. 213
Hills, J. G. 1975, *Nature*, **254**, 295
Hills, J. G. 1978, *M.N.R.A.S.*, **182**, 517
Kormendy, J. 1988, *Ap. J.*, **325**, 128
Krolik, J. H., and Begelman, M. C. 1986, *Ap. J. (Letters)*, **308**, L55
Krolik, J. H., and Begelman, M. C. 1988, *Ap. J.*, **329**, 702
Lin, D. N. C., and Pringle, J. E. 1987, *M.N.R.A.S.*, **225**, 607
Mathews, W. G., and Ferland, G. J. 1987, *Ap. J.*, **323**, 456
McCray, R., and Kafatos, M. 1987, *Ap. J.*, **317**, 190
Nulsen, P. E. J., Stewart, G. C., and Fabian, A. C. 1984, *M.N.R.A.S.*, **208**, 185
Paczyński, B. 1978, *Acta Astr.*, **28**, 91
Phinney, E. S. 1989, in *The Galactic Center, Proc. IAU Symposium 136*, ed. M. Morris (Dordrecht: Kluwer Academic Publishers).
Rees, M. J., Begelman, M. C., Blandford, R. D., and Phinney, E. S. 1982, *Nature*, **295**, 17

386

Sanders, D. B., Phinney, E. S., Neugebauer, G., Soifer, B. T., and Matthews, K. 1989, submitted
Sarazin, C. L. 1986, *Rev. Mod. Phys.*, **58**, 1
Shapiro, P. R., and Field, G. B. 1976, *Ap. J.*, **205**, 762
Shapiro, S. L., Lightman, A. P., and Eardley, D. M. 1976, *Ap. J.*, **204**, 187
Shlosman, I. 1988, in *Supermassive Black Holes*, ed. M. Kafatos (Cambridge: Cambridge University Press), p. 231
Shlosman, I., and Begelman, M. C. 1987, *Nature*, **329**, 810
Shlosman, I., and Begelman, M. C. 1989, *Ap. J.*, in press
Shlosman, I., Frank, J., and Begelman, M. C. 1989, *Nature*, **338**, 45
Thomas, P. A., Fabian, A. C., Arnaud, K. A., Forman, W., and Jones, C. 1986, *M.N.R.A.S.*, **222**, 655
Toomre, A. 1964, *Ap. J.*, **139**, 1217
Voit, G. M., and Shull, J. M. 1988, *Ap. J.*, **331**, 197
Wang, Z., and Cowie, L. L. 1988, *Ap. J.*, **335**, 168

LARGE-SCALE ACCRETION FLOWS IN AGN.

Juhan FRANK[1], Isaac SHLOSMAN[2], and Mitchell C. BEGELMAN[3]

[1] Max-Planck-Institut für Astrophysik, Garching bei München, FRG

[2] Theoretical Astrophysics, Caltech, Pasadena, CA 91125, USA

[3] JILA, University of Colorado and NIST, Boulder, CO 80309, USA

ABSTRACT. A two stage mechanism for fueling AGN is proposed which makes use of stellar dynamical and gas dynamical instabilities. The first stage adopts different forms depending on the type of host galaxy. In spiral hosts a stellar bar sweeps the interstellar material inward as a consequence of the gas losing angular momentum to the bar. In elliptical galaxies the first stage is likely to be infall. In a second stage common to both types of host, the gaseous disk accumulated in the nuclear region of the galaxy goes bar unstable again and the material flows further in. Some observational and theoretical implications of this picture are discussed.

1. Introduction.

In this talk we discuss the flow of gas in the main body of the host galaxy, from galactic scales on the order of ~ 10 kpc down to $r_h \sim 10$ pc radii where the potential of the supermassive black hole (SBH) dominates. The radius r_h is essentially the accretion radius, and for a SBH of mass M_h in a stellar environment with velocity dispersion v_c, it is given by $r_h = GM_h/v_c^2$. The flow in the SBH-dominated domain is addressed in the accompanying paper by Begelman, Frank & Shlosman (these proceedings).

We envisage the large-scale flow as a two stage process in which the first stage assumes different forms depending on the host morphology, but the second stage is common to elliptical and spiral hosts. In elliptical hosts the gas falls in from ~ 10 kpc until it becomes rotationally supported at ~ 1 kpc. In spiral hosts the first stage of inflow is accomplished sweeping the gas inwards with a stellar bar. As a result of the first stage just described a $\sim$kpc-scale, rotationally supported, nuclear gaseous disc forms which may go dynamically unstable when massive enough (Shlosman, Frank & Begelman 1989). This second stage occurs in both elliptical and spiral hosts and results in angular momentum transport and further inflow.

We discuss first the situation in elliptical galaxies, their gas content and degree

387

F. Meyer et al. (eds.), Theory of Accretion Disks, 387–395.

of rotational support. Then we consider the situation in spiral galaxies and argue that a large scale stellar bar causes the gas to lose angular momentum and to flow in. Some complications arise as a result of the presence of resonances which may have observable consequences. We then discuss the stability of the gas collected in the innermost kpc using a simple semi-empirical criterion. Finally, we summarize our picture and discuss future steps in the pursuit of this scenario.

2. Gas Inflow in Elliptical Hosts.

The gas content of elliptical galaxies has been reviewed by Schweizer (1987) and we quote here the main points. The observed gas in ellipticals comes in three components:

a) the "hot" component, with $T \gtrsim 10^6$ K and observed masses in the range $5 \times 10^8 M_\odot - 5 \times 10^{10} M_\odot$. This gas is detected as an X-ray corona which gradually sinks to the centre as a cooling flow (Biermann & Kronberg 1983; Nulsen et $al.$ 1984). The correlation between X-ray and optical luminosity in early type galaxies suggests that the hot gas originates from mass-loss from stars (Canizares, Fabbiano & Trinchieri 1987, Knapp 1987). The fate of the inflow is not clear at this point: a significant fraction of it may turn into stars on the way in, and what is left may in some cases become rotationally supported and be the origin of the "warm" gas described below.

b) the "warm" HII component at $T \sim 10^4$, detectable through optical line emission, with masses in the range $10^3 M_\odot - 10^6 M_\odot$. An overview of this component and of its main properties is given by Sadler (1987). Depending on the luminosity of the galaxy two different morphologies and dynamics are possible. In low luminosity ellipticals (with $M_B \gtrsim -19$) the HII gas appears to be clumpy, it is not rotationally supported, and the observed spectral line ratios are typical of HII regions and can be understood as the result of ionization by hot stars. In bright ellipticals, in contrast, the gas is mostly found in smooth discs of ~kpc radius, rotationally supported, and with an AGN-like spectrum requiring power-law ionization. In the case of bright centrally dominant galaxies in clusters, the ionized gas appears filamentary and not supported by rotation.

In powerful radio galaxies, the extended emission line gas has a median scale of ~ 10 kpc and the detected masses range up to $5 \times 10^9 M_\odot$ (Fosbury 1986, Baum & Heckman 1989a,b). The evidence suggests that this gas is of external origin, probably captured during an interaction. This gas appears to be ionized by the central source as the observed emission line luminosity correlates with the total radio luminosity (Baum and Heckman 1989a). This gas is therefore more intimately related to the "cold" component described below than to the nuclear emission line regions mentioned above.

c) the "cold" component with $T \lesssim 10^2$, consisting of HI and dust, with masses in the range $10^5 M_\odot - 5 \times 10^9 M_\odot$. This gas is in many cases found outside the main body of the galaxy, it is rotationally supported and has too much angular momentum to have originated in the galaxy by some mass ejection process. As the plane of symmetry of this component is frequently not aligned with any of the principal planes of the galaxy, it is thought that it could be relict from galaxy formation or it arises from mergers or captures.

An important clue to the origin of these gaseous components and their mutual relationships is provided by the kinematics of the stars in elliptical galaxies, in particular the observed degree rotational support in these systems. A very concise and useful summary of the situation is obtained by plotting the observed ratio of rotational velocity to velocity dispersion, normalized to the value expected for oblate spheroids, as a function of galactic magnitude (Davies et al. 1983; Davies 1987). The most important result of such a plot is that again ellipticals fall into two classes according to their luminosity. Bright galaxies with $M_B \lesssim -20$ have relatively low ratios, i.e. a low degree of rotational support. Bright ellipticals and powerful radio galaxies are associated and appear to have similar degrees of rotational support (Heckman et al. 1985, Fig. 4). In quite a few cases the angular momentum required for support is ten times greater than the observed value implying that material released at ~ 10 kpc can fall in to $\sim$ kpc radii before it becomes rotationally supported. Low luminosity ellipticals, on the other hand, are fully supported by rotation.

The simplest interpretation of these data is that the degree of rotational support determines how far the material released by stars can fall in during the first stage. As this stage is a prerequisite for the second stage of dynamical instability, bright ellipticals are more likely to become AGN than low luminosity ellipticals. The gas in the latter type of host is rotationally supported at galactic scales and in general will not fall in. In the absence of dissipative processes only local condensations can arise in which eventually stars form and which develop into HII regions. Hence the ionized gas is clumpy and does not possess an organized pattern of rotation. In galaxies with cooling flows the gas condensing out forms filaments and may join at smaller radii a rotationally supported disc. Observed examples of the different cases are given by Sadler (1987). The interpretation proposed here accounts for the morphology of the gas and the statistical association of AGN activity with bright ellipticals.

3. Gas Inflow in Spiral Hosts.

Adams(1977) and later Simkin, Su and Schwarz (1980) noted an excess of barred spirals among Seyfert galaxies. This suggested that the stellar bar was somehow instrumental in bringing the gas in and/or favouring the formation of the central accreting object. Numerous simulations of the flow of gas in an imposed bar-like potential have shown that the gas does indeed flow towards the centre as it loses angular momentum to the stellar bar (see e.g. Matsuda et al. 1987 and references therein, and van Albada & Roberts 1981).

We shall give here some qualitative arguments which help to understand and summarize the main properties of the inflow obtained in the numerical simulations. These arguments do not do justice to the complexity of the existing studies but will be sufficient for the purposes of this paper. The bar can be pictured as a valley in the otherwise axisymmetric potential of the host galaxy. The potential is assumed to be dominated by the stellar component and the gas is in most cases considered a non-gravitating test fluid. This valley turns at the pattern angular speed of the bar Ω_b which is slower than the local circular equilibrium velocity of the gas inside the corotation resonance (where $\Omega(R) = \Omega_b$). The gas orbiting around the centre of the galaxy inside corotation first falls into the potential trough and is accelerated and decompressed. As it flows out of the trough it is compressed and shocks. If these shocks are trailing, the gas flows inwards, and if they are leading, the gas flows outwards. If the pattern speed is sufficiently high so that no inner Lindblad resonances (ILRs) are present, then the gas flows in until the shocks become weak or the barred disturbance is relatively weak. Numerical simulations show that, if resonances occur, the shock tends to be leading in the domain between the two ILRs and a ring of gas results from the combined effect of inflow outside the ILRs and outflow in between the ILRs. Other resonances may also play a role in the formation of rings (Schwarz 1985).

To summarize we may say that, in the absence of ILRs, the gas flows inwards and accumulates in a nuclear disc of radial scale $\lesssim 1$ kpc. Further inflow is slow because the large-scale stellar bar with a radial extent of several kpc to over 15 kpc becomes inefficient at radii of the order of a tenth or so of the bar radius. If ILRs are present a ring-like gas distribution may result whose radius will depend on the position of the resonances involved. Thus the large-scale stellar bar in barred spirals may take care of the first stage of inflow in disc-like galaxies bringing in the interstellar gas from galactic scales down to $\lesssim 1$ kpc. In the next section we shall see how this nuclear disc, which forms in different ways depending on the type of host galaxy, may itself be unstable and drive the inflow further.

4. The Nuclear Disc.

The gas accumulated in the innermost kpc or so is rotationally supported and cool in the sense that $T_{gas} << T_{vir}$. In the absence of a detailed knowledge of the mass and angular momentum distributions in the disc only the crudest discussion of stability is possible. In a recent paper we applied the semi-empirical rule suggested by Ostriker and Peebles (1973) to discuss the stability of the nuclear gaseous disc (Shlosman, Frank and Begelman 1989). The O/P criterion is a generalization of the classical results on the stabilty of rotating fluids to stellar and gaseous discs (see discussion in Binney & Tremaine 1987 pp. 328 and 374) in terms of the parameter $t = T_{rot}/|W|$, where T_{rot} is the kinetic energy of rotation and W is the gravitational potential energy of the system. It is easy to see from the Virial Theorem that $0 \leq t \leq 1/2$. The O/P rule states that a rotating system will be dynamically unstable if $t > t_{crit}$, where $t_{crit} \approx 0.14$ for a stellar system (Ostriker & Peebles 1973), and $t_{crit} \approx 0.26$ for a fluid system (Bardeen 1975).

Let us assume that the nuclear gaseous disc of radial scale a contains a fraction g of the total mass of the host galaxy, and represent the "fixed" stellar component by the sum of a disc and a spherical halo with fractional mass h, both with radial scale b. Then it is possible to calculate the virial parameter t as a function of g, h and the ratio of scales a/b. By requiring $t > t_{crit}$ we can obtain domains of stability/instability in parameter space. In Shlosman, Frank & Begelman 1989, we take Kuzmin/Toomre discs and a Plummer halo in order to calculate an example designed to illustrate the kind of situation envisaged here. It turns out that t is only weakly dependent on h and Fig. 1 shows the domains of stability/instability obtained for the above example. As the nuclear disc is likely to be inhomogeneous probably the stellar $t_{crit} = 0.14$ is relevant.

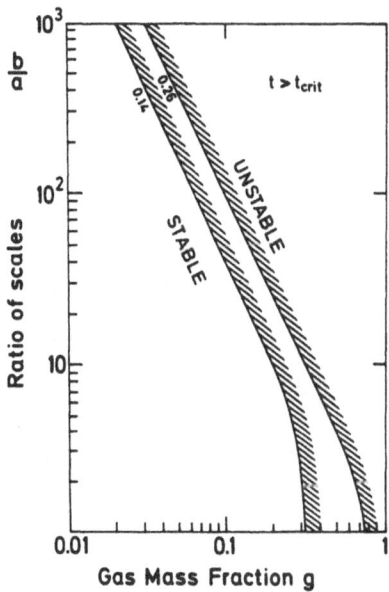

Figure 1: The domains of dynamical stability and instability for two values of t_{crit} as a function of the gas mass fraction and the ratio of the scale b of the "fixed" stellar potential to the scale a of the gaseous disc.

Although this is a particular example chosen for computational ease, we expect that at least qualitatively our conclusions will not depend strongly on the details. Quite generally one expects that for a given disc mass fraction there will be a minimal disc scale for stability. If the disc is smaller than some critical scale it will decouple dynamically from the host potential and be dynamically unstable if cool. Conversely, if the nuclear disc formed by infall or bar sweeping is such that

$b \sim 10a$, then Fig. 1, taken at face value, would imply that $g \gtrsim 20$ % is required for the nuclear disc to go unstable. Presumably this dynamical instability will result in something like a bar-driven spiral which transports angular momentum outwards and generates further inflow.

If the instability is efficient in removing angular momentum from the disc without leaving much mass behind, this process may result in a runaway instability which can only stop when the potential of the black hole becomes dominant at $r \lesssim 10$ pc (see Begelman *et al.* in these proceedings). If no black hole is present initially, it is conceivable that this process might lead to its formation.

5. Conclusions.

The best way to summarize our results is in the form of a "flow diagram" in which the possible pathways leading to the formation of an AGN or a starburst galaxy are indicated (see Fig. 2). The dominant processes operating and the radial scales on which they are relevant are also schematically shown (See also Hernquist 1989).

The essential step in the path leading to an AGN is the dynamical instability of the nuclear gaseous disc at scales $\lesssim 1$ kpc. This process provides the "missing link" by bridging the gap between the black hole dominated central region and the inflow at galactic scales. If this step is missing because the disc is not massive enough then the most likely outcome is a nuclear or ring-like starburst. In this sense a starburst galaxy is a *failed* AGN but further evolution might in some cases lead to an AGN at a later stage. In spiral hosts the first stage of inflow is induced by a stellar bar. As the second stage required for fuelling an AGN is a gaseous bar, we expect Seyfert galaxies to be "bars within bars" and to display evidence of current or past bar instabilities (Shlosman, Frank & Begelman 1989).

Numerical simulations of the tidal interaction and merging picture by Hernquist (1989) show that an encounter triggers the inflow of the gaseous component and that (at least in some cases) the gas goes bar unstable generating further inflow. Although this does confirm qualitatively our ideas, there are difficulties in trying to make more quantitative checks using these simulations: they are already expensive and further improvements in the spatial resolution would probably prove prohibitively expensive for the time being.

The gaseous bar is most likely in gas rich hosts linking perhaps the activity to gas content and hence to morphological type. For example we note the preference of both spiral starbursts and Seyferts for intermediate Hubble types Sb–Sbc which are richest in gas (Verter 1987, 1988; Haynes & Giovanelli 1984).

In powerful radio galaxies there is a tendency for the rotation axis of the very extended emission line gas to align with the radio axis (Heckman *et al.* 1985). While the stellar rotation axis of ellipticals associated with radio sources tends to be aligned with the minor axis, there is no clear tendency for the radio axis to align with the minor axis (Davies & Birkinshaw 1988). This suggests that either

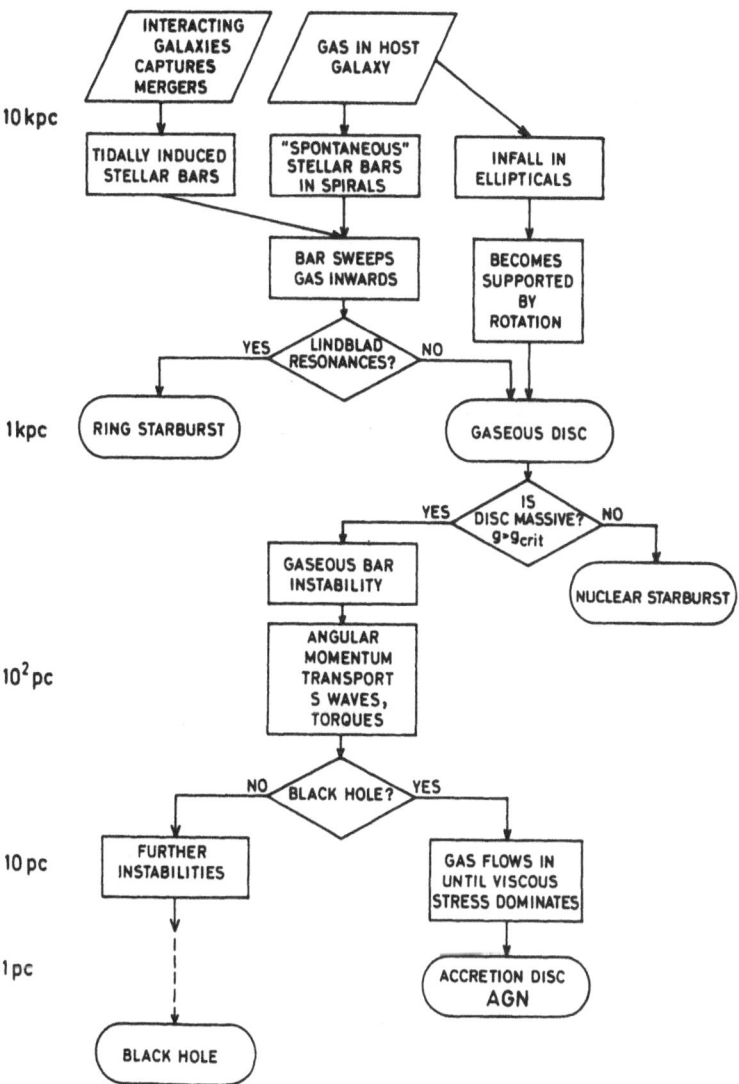

Figure 2: Different pathways by which host galaxies can acquire a massive black hole and/or become an AGN or a starburst. The approximate radial scales at which these processes take place are indicated on the left.

the gas has mostly an extragalactic origin or that triaxiality of the host is involved (Heckman *et al.* 1985). Further studies with high spatial resolution of the central regions of elliptical galaxies covering a wide range of radio power are necessary to answer the most important questions about the origin and angular momentum content the gas fuelling the activity.

The efficiency of sweeping by the large scale stellar bar in spiral hosts depends on its strength and pattern speed, and on the mass distribution of the host galaxy (presence and location of Lindblad resonances) but we do not have a quantitative model for these processes. Also our present understanding of angular momentum transfer in self-gravitating discs and the resultant mass inflow is far from complete. Further theoretical work on these areas is required. Probably the most likely area of progress in the near future lies in observational high resolution studies of the innermost regions of active and normal galaxies. For example the morphology of the innermost few hundred pc may reveal evidence of dynamical instabilities. High resolution and sensitivity studies in narrow band optical wavelengths, HI and CO observations, and IR photometry may reveal disturbances in the gaseous and stellar components which support the idea of a gaseous bar.

ACKNOWLEDGEMENTS. This work was supported in part by NSF grant 83-51997, NASA Astrophysical Theory Center Grant NAGW-766, grants from Rockwell International Corporation and the Alfred P. Sloan Foundation (at JILA), and NSF grant AST84-51725 (at Caltech).

REFERENCES.

Adams, T. F. 1977. *Astrophys. J. Supp. Ser.* **33**, 19.

Bardeen, J. M. 1975. *Dynamics of Stellar Systems*, ed. A. Hayli, Dordrecht: Reidel, p. 297

Baum, S. A. & Heckman, T. 1989a. *Astrophys. J.* **336**, 681.

Baum, S. A. & Heckman, T. 1989b. *Astrophys. J.* **336**, 702.

Biermann, P. & Kronberg, P. P. 1983. *Astrophys. J. (Letters)* **268**, L69.

Binney, J. & Tremaine, S. 1987. *Galactic Dynamics*, Princeton: P. Univ. Press

Canizares, C. R., Fabbiano, G. & Trinchieri, G. 1987. *Astrophys. J.* **312**, 503.

Davies, R. L. 1987. *Structure and Dynamics of Elliptical Galaxies*, IAU Symp. No. 127, ed. Tim de Zeeuw, Dordrecht: Reidel, p. 63.

Davies, R. L. & Birkinshaw, M. 1988. *Astrophys. J. Supp. Ser.* **68**, 409.

Davies, R. L., Efstathiou, G., Fall, S. M., Illingworth, G. & Schechter, P. L. 1983. *Astrophys. J.* **266**, 41.

Fosbury, R. A. E. 1986. *Structure and Evolution of Active Galactic Nuclei*, ed. G. Givricin, F. Mardirossian, M. Mezzetti, and M. Ramonella, Dordrecht: Reidel, p. 297.

Haynes, M. P. & Giovanelli, R. 1984. *Astron. J.* **89**, 758.

Heckman, T. M., Illingworth, G. D., Miley, G. K. & van Breugel, W. J. M. 1985. *Astrophys. J.* **299**, 41.

Hernquist, L. 1989. Presented at the Texas Symposium 1988

Knapp, G. R. 1987. *Cooling Flows in Clusters and Galaxies*, ed. A.C. Fabian, Dordrecht: Kluwer Acad. Pub., p 93.

Matsuda, T., Inoue, M., Sawada, K., Shima, E. & Wakamatsu, K. 1987. *Mon. Not. R. astr. Soc.* **229**, 295.

Nulsen, P. E. J., Stewart, G. C. and Fabian, A. C. 1984. *Mon. Not. R. astr. Soc.* **208**, 185.

Ostriker, J. P. & Peebles, P. J. E. 1973. *Astrophys. J.* **186**, 467.

Sadler, E. M. 1987. *Structure and Dynamics of Elliptical Galaxies*, IAU Symp. No. 127, ed. Tim de Zeeuw, Dordrecht: Reidel, p. 125.

Schwarz, M. P. 1985. *Mon. Not. R. astr. Soc.* **212**, 677.

Schweizer, F. 1987. *Structure and Dynamics of Elliptical Galaxies*, IAU Symp. No. 127, ed. Tim de Zeeuw, Dordrecht: Reidel, p. 109.

Shlosman, I., Frank, J. & Begelman, M. C. 1989. *Nature* **338**, 45.

Simkin, S., Su, H., & Schwarz, M. P. 1980. *Astrophys. J.* **237**, 404.

van Albada, G. D. & Roberts, W. W. 1981. *Astrophys. J.* **246**, 740.

Verter, F. 1987. *Astrophys. J. Supp. Ser.* **65**, 555.

Verter, F. 1987. *Astrophys. J. Supp. Ser.* **68**, 129.

THE DISK ACCRETION OF A TIDALLY DISRUPTED STAR ONTO A MASSIVE BLACK HOLE.

John K. CANNIZZO[1], Hyung Mok LEE[2], and Jeremy GOODMAN[3]*

[1]Department of Physics, McMaster University
[2]Canadian Institute for Theoretical Astrophysics
[3]Princeton University Observatory

ABSTRACT. We consider the tidal disruption of a star by a moderately massive black hole such as might be found in the nucleus of a galaxy like M31. The initial eccentric orbit of the stellar debris will circularize near the tidal radius after experiencing strong shocks. We study the evolution of the accretion disk produced by this torus using a time dependent α-disk model. We find that the light-to-mass ratio of the disk-plus-black-hole exceeds unity for several thousand years after disruption. Some fraction of galaxies should be extremely bright at far UV wavelengths if they contain black holes of mass $10^{6-8} M_\odot$.

1. Introduction.

Supermassive black holes have been one of the favored models for powering the observed high luminosities of active galactic nuclei (AGN) and quasars (e.g. Rees 1984). On this hypothesis, some nearby galaxies should harbor black holes of mass $\gtrsim 10^8 M_\odot$ (cf. Phinncy 1983), and many more might have black holes of lesser mass (too small for the relic of a bright quasar). Such black holes might announce their presence in ways more dramatic than high stellar velocities. Given a supply of gas to accrete, the black hole would be expected to revert to quasar-like activity, at least temporarily. Tidal disruption appears to be the most important fueling mechanism for the conditions expected in the nuclei of nearby galaxies. Reasonable estimates of the tidal disruption rate for M32 and our galaxy are about $10^{-4} M_\odot \ yr^{-1}$.

Our approach is very similar to that taken by Gurzadyan and Ozernoy (1979, 1980). However, we argue that they greatly overestimated the initial radius of the disk, which had the effect of exaggerating the timescale for accreting the debris. The surprise is that the timescale remains very long after these initial conditions are corrected. We also explain both our results and theirs by reference to exact self-

---

* Sloan Foundation Fellow and David and Lucille Packard Foundation Fellow

F. Meyer et al. (eds.), Theory of Accretion Disks, 397–401.
© *1989 by Kluwer Academic Publishers.*

similar solutions of the α-disk equations, and we use these solutions to demonstrate the insensitivity of our results to our assumptions about the disk.

2. Tidal disruption and Circularization.

2.1. THE TIDAL DEBRIS.

If a Newtonian approximation for the gravitational field is used, the tidal radius is given by

$$R_T \approx R_* \left(\frac{M_b}{M_*}\right)^{1/3} = 1.5 \times 10^{13} \; cm \; \left(\frac{M_*}{M_\odot}\right)^{-1/3} \left(\frac{M_b}{10^7 \, M_\odot}\right)^{1/3} \left(\frac{R_*}{R_\odot}\right), \quad (1)$$

where M is the mass of the black hole and R_* and M_* are the radius and mass of the incoming star.

Stars fated for disruption approach the black hole on nearly parabolic orbits. The *mean* orbital energy per unit mass of the debris will be of order $v_*^2/2$ but, unless the material leaves the star in directions exactly orthogonal to its orbital motion, there will be a spread in energies of order $v_* v_p$ (e.g. Rees 1988), which is much larger than the mean since the pericentral velocity $v_p \sim 0.5c(M_b/10^7 M_\odot)^{1/3}$. Thus about half of the debris will be unbound immediately. The remaining half will lie on orbits with periods ranging upwards from

$$P_{min} \approx 2\pi \frac{GM_b}{(v_p v_*)^{3/2}} \approx 0.7 \; yr \left(\frac{M_b}{10^7 M_\odot}\right)^{1/2}. \quad (2)$$

Since the bound material is roughly uniformly distributed in energy, it will be distributed as $P^{-5/3}$ in orbital period, so that most of it will have orbital periods shorter than a few times P_{min}.

So for a few years, there will be a steady stream of material returning for the first time to the vicinity of R_T. The relativistic precession of the latus rectum is substantial ($1.5\pi R_s/R_p$ radians per orbit) and the ingoing gas stream will intersect the outgoing stream near R_T in strong shocks. (If the black hole rotates, these intersections may be considerably delayed by precession of the orbital plane due to the Lense-Thirring effect.) 2.2. THE ACCRETION DISK.

Once a near equilibrium torus is formed as a result of the processes described above, the subsequent evolution is slower and relatively predictable. The material will spread out into a viscous accretion disk. The evolution during most of this phase is remarkably insensitive to the details or even the overall magnitude of the viscosity. This is fortunate, since the source of the viscosity is not understood.

The evolution of the surface density distribution Σ is given by a diffusion equation (Bath and Pringle 1981). We adopt the so-called α disk model (Shakura and

Sunyaev 1973) and take the viscosity to be proportional to the gas pressure rather than to the total pressure in order to avoid the Lightman and Eardley (1974) instability.

3. Time Dependent Results.

We have made several computations with a number of different initial conditions. For the purpose of illustration we present results for a run with $R_p = 2 \times 10^{13}$ cm, $\alpha = 1$, and $m_d = 0.5 M_\odot$. The grid contains 360 radial points. Drawing from observational constraints on α obtained by comparing theoretical and observed time scales of dwarf nova eruptions we adopt $\alpha = 1$.

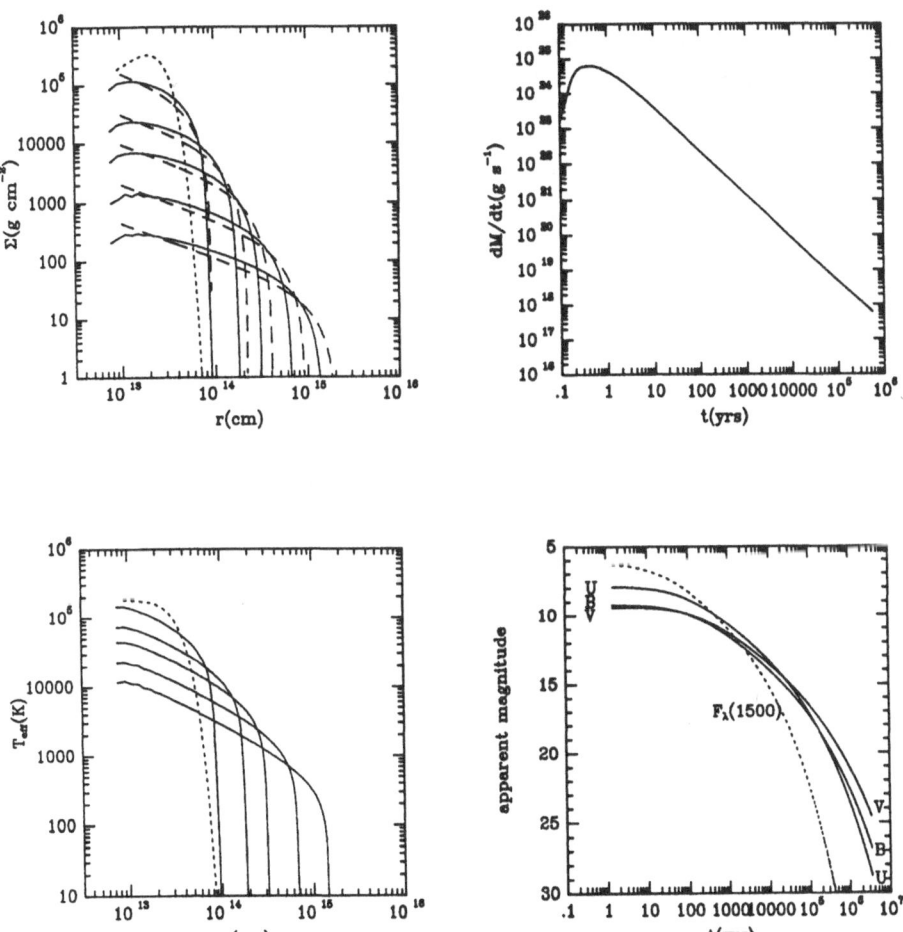

Figure 1 - 4

Figure 1 portrays the evolution of the surface density at the times $t(\text{yrs}) = 0$, 5.87, 62.1, 323, 2880, 23600. The dashed curves show the similarity solution given by equation (5) below. Figure 2 shows the evolution of the rate of mass loss from the inner grid point onto the black hole. The bolometric luminosity L_B produced by this accretion is $\sim \epsilon \dot{M} c^2 \simeq 10^{20} \dot{M}(g/s)$. Thus initially L_B exceeds 10^{44} erg/s and for later times L_B and $\dot{M}$ decline as $t^{-1.2}$. The effective temperature distribution at various epochs is depicted in Figure 3. Figure 4 shows the temporal evolution of the UBV apparent magnitudes and the monochromatic flux density at 1500 A. The monochromatic flux is scaled by $-2.5F - 20$. We take $D = 1$ Mpc, as would be roughly appropriate for M31 or M32.

4. Exact Results.

The power-law behavior of $\dot{M}$ after the first 10 years or so strongly suggests that the evolution has become self-similar. In fact, Pringle (1974) gives exact self-similar solutions to the nonlinear diffusion for Σ when $\nu \propto r^p \Sigma^q$, and at late times, when radiation pressure is neglegible compared to gas pressure

$$\nu = C r \Sigma^{2/3},$$

$$C \equiv \alpha^{4/3} \left(\frac{k_B}{\mu m_p} \right)^{4/3} \left(\frac{\kappa}{12 a c G M_b} \right)^{1/3}, \tag{4}$$

where μ is the molecular weight, m_p the proton mass, and k_B is Boltzmann's constant. This is of Pringle's form, and the explicit solution is

$$\frac{\Sigma(r,t)}{\Sigma_0} = \left(\frac{t}{t_0} \right)^{-15/16} f \left[\frac{r}{r_0} \left(\frac{t}{t_0} \right)^{-3/8} \right], \tag{5}$$

$$\text{where} \quad f(u) \equiv (28)^{-3/2} u^{-3/5} \left(1 - u^{7/5} \right)^{3/2}.$$

The constant dimensional scales Σ_0, r_0, and t_0 are arbitrary except that they satisfy

$$t_0^{-1} = \frac{C \Sigma_0^{2/3}}{r_0}. \tag{6}$$

The solution (5) extends in radius from the origin out to $u = 1$.

5. Discussion.

The luminosities we compute are much greater than those seen in nuclear regions of M31 or M32. Although the initial e-folding time for accreting the post-circularization torus is a few years, the mass-to-light ratio stays below 1 for about 2000 years. In contrast with Gurzadyan and Ozernoy (1979, 1980), we find that the peak luminosity is close to the Eddington limit for the black hole, whereas their peak luminosity actually declines with increasing M_b. We also find a rather shorter duty cycle for $L_d/M_b > 1$; they find almost continous emission if M_b is larger than a critical value of order $10^5 M_\odot$. These differences can be traced primarily to our assumption that the initial disk is circular, whereas they took it to be highly eccentric with a mean radius much larger than the disruption radius.

6. Conclusion.

We have constructed a simple model for the gaseous disk formed from the tidal disruption of a star by a black hole in the core of a galaxy. The remnants from each star will quickly circularize into a torus of matter close to the black hole. This material will form into an accretion disk of a few tenths of a solar mass. Although the initial characteristic e-folding time scale for accretion onto the black hole is short, the disk luminosity remains high for a time comparable to the time scale between successive tidal disruptions. Thus one could not argue that the outburst duty cycle is short and that galaxies like M32 have massive black holes which we are now seeing in a (high probability) quiescent stage.

ACKNOWLEDGEMENTS. We acknowledge useful conversations with G. Ferland, B. Peterson, M. Rees, and S. Tremaine. Financial support was provided by the NSERC of Canada at CITA and McMaster University, and by Fellowships from the Alfred P. Sloan Foundation and the David and Lucille Packard Foundation (to JG).

REFERENCES.

Bath, G. T., and Pringle, J. E. 1981, *M. N. R. A. S.*, **194**, 967
Gurzadyan, V. G., and Ozernoy, L. M. 1979, *Nature*, **280**, 214
Gurzadyan, V. G., and Ozernoy, L. M. 1980, *Astr. Ap.*, **86**, 315
Lightman, A. P., and Eardley, D. M. 1974, *Ap. J. (Letters)*, **187**, L1
Phinney, E. S. 1983, *Ph.D. Thesis*, Cambridge University
Pringle, J. E. 1974, *Ph.D. Thesis*, Cambridge University
Rees, M. J. 1984, *Ann. Rev. Astr. Ap.*, **22**, 471
Rees, M. J. 1988, *Nature*, **333**, 523
Shakura, N. I., and Sunyaev, R. A. 1973, *Astr. Ap.*, **24**, 337

LINE RADIATION FROM STATIONARY ACCRETION DISKS.

Johannes ADAM[1], Davina E. INNES[1], Giora SHAVIV[2],
Herbert STÖRZER[1], Rainer WEHRSE[1]

[1] Institut für Theoretische Astrophysik, Im Neuenheimer Feld 561,
D-6900 Heidelberg, Fed. Rep. of Germany

[2] Physics Department TECHNION, IL-32000 Haifa, Israel

ABSTRACT. We apply a finite difference method to the 3D radiative line transfer in accretion disks. The two-level-atom approximation with complete re-distribution is used. As a first step we assume disks which are homogeneous in temperature and pressure. We give monochromatic images of the intensity distribution on the surface of CV disks for several assumptions on the scattering and disk thickness. In the LTE case earlier results (e.g. by Smak, 1969) are confirmed. However, for scattering conditions in thin disks we get a completely different intensity pattern. In a parameter study for geometrically thick disks we show a variety of line profile shapes.

The effect of a corona overlying a disk atmosphere on the Balmer line profiles is studied with a ray tracing technique. Using Stark profile functions and a parametrization of the coronal temperature, we show that the shape of the total profile strongly depends on the structure of the transition zone.

1. Introduction.

Stationary accretion disks are characterized by a relatively large number of parameters (for example M_*, $\dot{M}$, r_{in}, r_{out}, α and element abundance cf. Frank, King and Raine, 1985, for a review) compared to effective temperature, gravity and composition for normal ("compact") stellar atmospheres. Even if solar composition is assumed, it seems impossible with present day observations to determine these disk parameters from the continuum energy distribution alone. In addition the continuum, in the accessible wavelength range, contains little information about the disks' coronae whose existence is established (Mason, 1986; van der Woerd, 1987) but whose heating mechanism is still under discussion (Galeev, Rosner and Vaiana, 1979; Begelman, McKee and Shields, 1977; Shaviv and Wehrse, 1986).

It is expected from our knowledge of stellar atmospheres that the modeling of spectral lines, and in particular their profiles, can considerably constrain the large parameter space and give empirical evidence about the spatial distribution of

F. Meyer et al. (eds.), Theory of Accretion Disks, 403–418.

temperature, pressure and velocity in the disks.

The transfer of line radiation in accretion disks cannot be adequately described with the classical approximations used for static plane-parallel stellar atmospheres. In a static atmosphere, photons prefer to escape along the normal with respect to the surface. However in rotating disks photon escape is impeded, due to vanishing Doppler shifts, along the normal, and in the center and the anticenter direction as well as in the direction of the moving particles. Thus photons will preferentially leave the disk in some oblique direction. In addition, if the disk has a Keplerian rotation law, its velocity gradient is steeper closer to the star, so that photon escape is easier here than in the outer parts of the disk. Since the photon mean free path is often long compared to the length scales for substantial change in the disk structure the numerical treatment of line transfer must account for these anisotropic propagation characteristics and the temperature and density distribution.

Up to now these features have not been taken into account consistently. Instead, as in the calculation of continuum radiation (*cf.* G.Shaviv and R.Wehrse, this volume), the disk was divided into a series of independent rings. Each ring was regarded as a static plane parallel atmosphere. Then the surface emissivity of each ring was evaluated for the inclination of the binary and the total profile was just the sum of the spectra of the rings (Smak, 1969; Smak, 1981; Mayo, Wickramasinghe and Whelan, 1980; van Groningen, 1983; La Dous, 1989). Horne und Marsh (1986) took the anisotropic emission in the rings into account and calculated line profiles for isothermal disks in LTE. Deviations from LTE were considered by Williams and Shipman (1988). They calculated local escape probabilities from a single annulus of a differentially rotating disk for a multi-level H atom. These calculations can qualitatively explain many observed features, but their use for a quantitative understanding seems to be limited.

Since we think that the next step for realistic modeling of line profiles from accretion disks is the inclusion of multidimensional radiative transfer, we first (Sect. 2) describe a new numerical method for the solution of the 3D radiative transfer equation. We apply the method to study, in Sect. 3, the formation of resonance lines (including the effect of scattering) in disks which have homogeneous density and temperature distributions. We restrict our investigation to such simple cases in order to demonstrate the complexity of the transfer and to show that even here a large variety of different profile types can occur.

In Sect. 4 we consider α-disks with vertical structure and calculate the profiles of $H\alpha$ and $H\delta$ with a ray tracing technique in which the formal solution of the 1D radiative transfer equation is integrated along a large number of rays (*cf.* Störzer, 1987; Baschek *et al.*, 1988; Adam, Störzer and Duschl, 1989). It is seen that Stark broadening results in very broad absorption wings, whereas the corona temperature is reflected in the central emission component. The results are summarized in Sect. 5.

2. Radiation Transfer in Spectral Lines.

2.1. ASSUMPTIONS AND BASIC EQUATIONS.

With **n** the direction of the radiation, I_λ the specific intensity at wavelength λ, κ_λ^L and κ^c the absorption coefficients in line and continuum and S^L and S^c the source functions of line and continuum, the standard form of the equation of radiative transfer reads

$$n \cdot \nabla I_\lambda = \kappa_\lambda^L S^L + \kappa^c S^c - (\kappa_\lambda^L + \kappa^c) I_\lambda. \tag{2.1}$$

Because we shall test a new method of solving Eq. (2.1), we use very simple atomic physics: a *two level atom* and *complete re-distribution* (e.g. Athay, 1972; Cannon, 1985). Then the source function S^L is given by

$$S^L = (1 - \epsilon)\bar{J} + \epsilon B \tag{2.2}$$

with the scattering parameter ϵ

$$\epsilon = \frac{C_{ul}}{C_{ul} + A_{ul} + B_{ul}B} \tag{2.3}$$

and the average intensity $\bar{J}$

$$\bar{J} = \frac{1}{4\pi} \int_{4\pi} \int \phi_\lambda(\lambda, n) I_\lambda d\lambda d\Omega. \tag{2.4}$$

In the usual nomenclature C_{ul} is the collisional de-excitation rate, B_{ul} and A_{ul} are the Einstein rate coefficients for induced and spontaneous de-excitation, B is the Planck function and ϕ_λ is the profile function. In the limit $\epsilon \to 0$ there is scattering only ($S^L \to \bar{J}$). The case $\epsilon = 1$ corresponds to LTE ($S^L \equiv B$). For S^c we shall use B throughout.

We treat the problem in the *observers frame* and write the absorption coefficient at position $s(x, y, z)^T$ as

$$\kappa_\lambda^L(s) := \kappa_0^L \Delta\lambda_{th}\sqrt{\pi} \; \phi_\lambda((\lambda - \lambda_0) + \frac{\lambda}{c}v(s) \cdot n) \tag{2.4}$$

with v the velocity of the gas elements and n the direction vector of the absorbed photon. Aberration and relativistic effects are not included in this study. For a Maxwellian distribution of the thermal velocities the profile function ϕ_λ is a Gaussian

$$\phi_\lambda(\lambda) = \frac{1}{\Delta\lambda_{th}\sqrt{\pi}} \; exp-[(\lambda - \lambda_0)/\Delta\lambda_{th}]^2 \tag{2.5}$$

with a width $\Delta\lambda_{th}$. The values of κ_0^L and of κ^c are taken to be parameters.

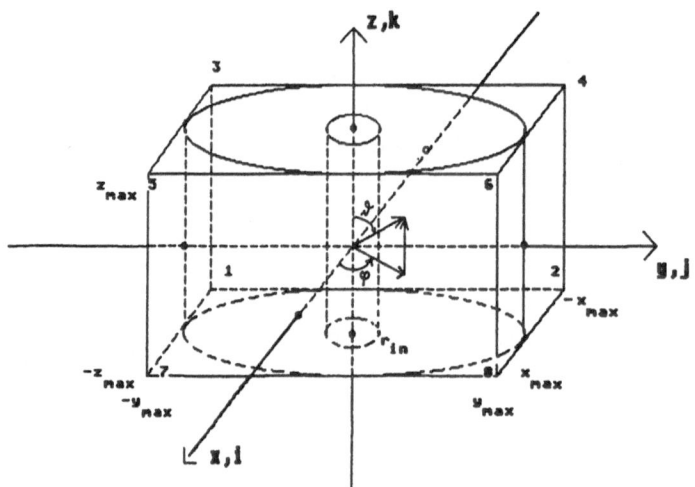

Figure 1: Geometry of the model and definition of the system of coordinates.

Our disk model is also simple. The disk is assumed to be a *homogeneous cylinder* ($B = const.$, $\kappa_0^L = const.$, $\kappa^c = const.$, $\epsilon = const.$) with an central void. All gas elements move around the z-axis on circular orbits. Then the velocity $\mathbf{v}$ is given by

$$v(s) = \begin{pmatrix} -v_\varphi(r)\,sin\,\varphi \\ v_\varphi(r)\,cos\,\varphi \\ 0 \end{pmatrix}, \qquad (2.6)$$

with the Keplerian velocity $v_\varphi = (GM_*/r)^{1/2}$, the distance from the rotation axis $r = \sqrt{x^2 + y^2}$, $sin\,\varphi = y/r$ and $cos\,\varphi = x/r$. With the direction angles ϑ, φ (*cf.* Fig.1) the direction vector has the components $n = (cos\,\varphi\,sin\,\vartheta, sin\,\varphi\,sin\,\vartheta, cos\,\vartheta)^T$.

2.2 NUMERICAL METHOD.

Although the disk is cylindrically symmetric we write the equation of radiation transfer in cartesian co-ordinates because the five directional derivatives that occur in a cylindrical co-ordinate system (*cf.* Mihalas, 1978, §2.2) make cylindrical co-ordinates disadvantageous. Thus Eq. (2.1) becomes

$$n_x \frac{\partial I_\lambda}{\partial x} + n_y \frac{\partial I_\lambda}{\partial y} + n_z \frac{\partial I_\lambda}{\partial z} = \kappa_\lambda^L S^L + \kappa^c S^c - (\kappa_\lambda^L + \kappa^c)I_\lambda \qquad (2.7)$$

where S^L, S^c and κ^c are, in general, functions of the spatial coordinates x, y and z, and I_λ and κ_λ^L depend on both position and direction $\mathbf{n}$. Because S^L contains the averaged intensity $\bar{J}$ Eqs.(2.7) are a system of partial integro-differential equations for I_λ.

We define an equidistant three-dimensional rectangular mesh with gridpoints (x_i, y_j, z_k) $(i = 1, \ldots, N_x, j = 1, \ldots, N_y, k = 1, \ldots, N_z)$

$$x_i = x_{min} + \frac{x_{max} - x_{min}}{N_x - 1} i = -R + 2\frac{R}{N_x - 1} i$$

$$y_j = y_{min} + \frac{y_{max} - y_{min}}{N_y - 1} j = -R + 2\frac{R}{N_y - 1} j \qquad (2.8)$$

$$z_k = z_{min} + \frac{z_{max} - z_{min}}{N_z - 1} k = -H + 2\frac{H}{N_z - 1} k$$

(*cf.* Fig.1). r_{in} and R are the inner and outer radius of the cylinder and H its (half) height. We use a first order implicit scheme to integrate Eq.(2.7) (Stenholm, Störzer and Wehrse, 1989). By implicit we mean that the r.h.s. is evaluated at the new gridpoint (x_i, y_j, z_k), so that Eq.(2.7) then reads (we are now omitting the subscript λ)

$$n_x \frac{I_{ijk} - I_{i-\alpha,j,k}}{x_i - x_{i-\alpha}} + n_y \frac{I_{ijk} - I_{i,j-\beta,k}}{y_j - y_{j-\beta}} + n_z \frac{I_{ijk} - I_{i,j,k-\gamma}}{z_k - z_{k-\gamma}} = \kappa^L_{ijk} S^L_{ijk} + \kappa^c_{ijk} S^c_{ijk}$$

$$- (\kappa^L_{ijk} + \kappa^c_{ijk}) I_{ijk}. \qquad (2.9)$$

α, β und γ are increments $+1$ or -1 depending on the choice of n. $\kappa^L_0 = 0$ and $\kappa^c = 0$ on gridpoints lying outside the disk volume.

$\bar{J}$ can be written as a quadrature sum over $I_\lambda(n)$

$$\bar{J}_{ijk} \approx \frac{1}{4\pi} \sum_{i_\lambda=1}^{N_\lambda} A^{(\lambda)}_{i_\lambda} \sum_{i_\varphi=1}^{N_\varphi} A^{(\varphi)}_{i_\varphi} \sum_{i_\vartheta=1}^{N_\vartheta} A^{(\vartheta)}_{i_\vartheta} \phi_\lambda(\Delta\lambda_{i_\lambda} + \frac{\Delta\lambda_{i_\lambda} + \lambda_0}{c} v_{ijk} \cdot n_{i_\varphi,i_\vartheta}) \times \qquad (2.10)$$

$$sin\,\vartheta_{i_\vartheta} I_{ijk}(\Delta\lambda_{i_\lambda}, \varphi_{i_\varphi}, \vartheta_{i_\vartheta})$$

with $\Delta\lambda = \lambda - \lambda_0$ and integration points at $\Delta\lambda_{i_\lambda}$, φ_{i_φ}, ϑ_{i_ϑ} and weights $A^{(\lambda)}_{i_\lambda}$, $A^{(\varphi)}_{i_\varphi}$ and $A^{(\vartheta)}_{i_\vartheta}$. Eq.(2.9) corresponds to a system of linear equations. The dimension of the matrix is $N_x \times N_y \times N_z \times N_\vartheta \times N_\varphi \times N_\lambda$. With the values that we used in applications, there are $63 \times 63 \times 63 \times 12 \times 4 \times 80 = 9.6 \; 10^8$ matrix elements. Problems with this dimension have to be tackled with an iterative procedure. This means that we need to store $N_x \times N_y \times N_z = 2.5 \; 10^5$ elements of both I_{ijk} and $\bar{J}_{ijk}$.

The system of Eqs.(2.9) is solved recursively. To demonstrate this, we show in Fig.2 the analogous two-dimensional mesh. The intensity is zero on the *boundary* for a given direction **n**. Likewise, the boundary faces themselves are determined by **n** uniquely. Beginning at the corner, where the known boundaries meet, we compute I_{ijk} with the procedure sketched in Fig.1. We solve Eq.(2.9) for I_{ijk}:

$$I_{ijk} = \frac{\kappa^L_{ijk} S^L_{ijk} + \kappa^c_{ijk} S^c_{ijk} + n_x \frac{I_{i-\alpha,j,k}}{x_i - x_{i-\alpha}} + n_y \frac{I_{i,j-\beta,k}}{y_j - y_{j-\beta}} + n_z \frac{I_{i,j,k-\gamma}}{z_k - z_{k-\gamma}}}{(\kappa^L_{ijk} + \kappa^c_{ijk}) + \frac{n_x}{x_i - x_{i-\alpha}} + \frac{n_y}{y_j - y_{j-\beta}} + \frac{n_z}{z_k - z_{k-\gamma}}}. \qquad (2.11)$$

408

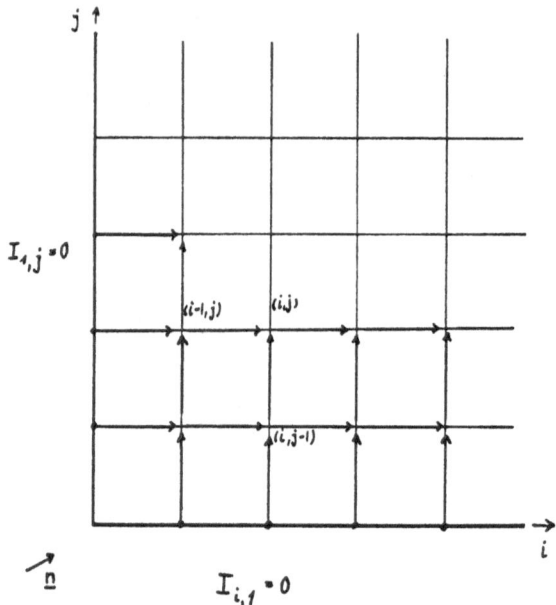

Figure 2: Two-dimensional analog for solving the discretized equation of radiative transfer. Starting with two gridpoints with known intensity, we calculate the solution for a new gridpoint.

As a consequence of the implicit discretization this method is unconditionally stable because only positive terms occur in Eq.(2.11). For Eq.(2.11) we need S_{ijk}^L. Because S^L contains $\bar{J}$ which is an average of I for all directions $\mathbf{n}$ and wavelengths λ, we proceed iteratively ("Λ-iteration"):

$$\text{step 0:} \quad \bar{J}_{ijk}^{(0)} \equiv 0$$

$$S_{ijk}^{(0)} = \epsilon_{ijk} B_{ijk}$$

$$I_{ijk}^{(0)} = \quad \text{Eq.(2.11) for many } \mathbf{n} \text{ and } \lambda$$

$$n = 1$$

$$\text{n-th correction step:} \quad \bar{J}_{ijk}^{(n)} = \quad \text{Eq.(2.10)} \quad (*)$$

$$S^{L(n)} = (1 - \epsilon_{ijk})\bar{J}_{ijk}^{(n)} + \epsilon_{ijk} B_{ijk}$$

$$I_{ijk}^{(n)} = \quad \text{Eq.(2.11)}$$

$$n = n + 1$$

$$\text{go to } (*)$$

It is well known that convergence is very slow for small values of ϵ. However, it can be substantially accelerated by using the extrapolation method of Ng and the approximate operator method, both described in Olson, Auer and Buchler (1986).

Additionally, for cylindrical configurations with the prescribed velocity field, we can use symmetries in angles and wavelength to reduce CPU time by a factor of 16. Acceleration and symmetries are given by Adam (1989).

If the source functions are known, it is easy to calulate the flux F_λ of the line for any inclination $i(\hat{=}\vartheta)$. We use $\varphi = 0$ and get like in Eq.(2.11)

$$I_{ijk} = \frac{\kappa^L_{ijk}S^L_{ijk} + \kappa^c_{ijk}S^c_{ijk} + n_x\frac{I_{i-1,j,k}}{x_i-x_{i-1}} + n_z\frac{I_{i,j,k-1}}{z_k-z_{k-1}}}{(\kappa^L_{ijk} + \kappa^c_{ijk}) + \frac{n_x}{x_i-x_{i-1}} + \frac{n_z}{z_k-z_{k-1}}}. \tag{2.12}$$

for $(i = 2,\ldots,N_x,\ j = 1,\ldots,N_y,\ k = 2,\ldots,N_z)$. The flux $F_\lambda := \int I\,n\,dA$, integrated over the visible faces A of the grid 3-4-6-5 and 5-6-8-7 (see Fig.1) is

$$F_\lambda \approx \sum_{i=1}^{N_x-1}\sum_{j=1}^{N_y-1}\frac{1}{4}(I_{i,j,N_z} + I_{i+1,j,N_z} + I_{i,j+1,N_z} + I_{i+1,j+1,N_z})\Delta x\,\Delta y\ sin\,i+$$

$$\sum_{j=1}^{N_y-1}\sum_{k=1}^{N_z-1}\frac{1}{4}(I_{N_x,j,k} + I_{N_x,j+1,k} + I_{N_x,j,k+1} + I_{N_x,j+1,k+1})\Delta y\,\Delta z\ cos\,i.$$

$$\tag{2.13}$$

3. Applications of 3D line transfer.

One advantage of using such simple homogeneous disks and atomic physics is that a complete parameter study can be carried out. We investigate disk models that have been proposed for disks in *Cataclysmic Variables* (CV) and in *Active Galactic Nuclei* (AGN). For CV's, we start with a "base model" where the geometrical parameters r_{in}, R and H are fixed and the radiation parameters ϵ, κ^L_0 and κ^c are varied. For AGN's we fix R and the vertical total optical depths, but change the aspect ratio H/R.

For the simplified disks, we have to specify the width of the profile function $\Lambda\lambda_{th}$, which is composed of a thermal and a turbulent contribution. It can be estimated using scale arguments. For a disk in hydrostatic equilibrium, its height, temperature and Keplerian velocity at R are related by (e.g. Treves, Maraschi and Abramowicz, 1988)

$$\frac{H}{R} \approx \frac{c_s(R)}{v_\varphi(R)}, \tag{3.1}$$

with c_s the sound speed. Because the thermal velocity v_{th} is of the order of c_s and $v_{turb} = \alpha c_s \approx \alpha v_\varphi(H/R)$ we get a total broadening of

$$\Delta\lambda_{th} = \lambda_0\frac{\sqrt{c_s^2 + v_{turb}^2}}{c} \approx \lambda_0 d\sqrt{1 + \alpha^2}\ \frac{v_\varphi(R)}{c}, \tag{3.2}$$

where we introduced the aspect ratio $d := H/R$. $\Delta\lambda_{th}$ is assumed constant in the disk.

Figure 3: Surface intensity maps of a very thin disk (a), a thin disk with constant line source function (b), and a thin disk with a variable source function due to scattering (c). High intensity is bright grey. The total optical depth in the line is equal in all cases.

3.1. LINES IN CATACLYSMIC SYSTEMS.

For CV disks we use as fixed parameters $r_{in} = 10^9$cm, $R = 2 \; 10^{10}$cm, $H = 10^9$cm, $d = 1/20$ and $M_* = 0.5 \; M_\odot$ and study variations for $\epsilon = 1$, 10^{-2} and 10^{-4}, the vertical optical depths at midplane in the line center $\tau_0^L = 1$, 10 and 100, and in the continuum $\tau^c = 10^{-2}$ and 1.

3.1.1. Monochromatic Images. The position of radiating surface elements at a given wavelength in the line profile has been discused for infinitely thin disks ($d \to 0$) by Horne and Marsh (1986). In Fig.3a we show as a corresponding case the monochromatic image of the disk for $\Delta\lambda = \lambda - \lambda_0 = 4$Å. The white arc is the position of constant radial velocity $v_x \; sin \, i$ according to the given $\Delta\lambda$. The dashed arcs mark the extent of the radiating area, which is determined by the width of the profile function. In Fig.3b the disk has $d = 1/20$, $\tau_0^L = 100$ and $\tau^c = 0.01$. Here the finite disk thickness allows radiation to escape from the "front side" of the emitting volume. Therefore, the (projected) radiating area spreads over the dashed arcs towards the observer (*cf.* Baschek *et al.*, 1988). For this case we assumed $\epsilon = 1$, so that the line source function is constant at each height. Fig.3c shows the situation for the scattering case with $\epsilon = 10^{-2}$, where $S^L(z)$ has a parabola-like shape with a maximum at the midplane. Whenever radiation escapes close to the surface, like between the dashed arcs, the source function is small and so the intensity is low. However, the "front sides" of the emitting volume mainly have a high source function, so that their intensity is high. This is the cause of the high intensity bows. Note, that the former bright regions of Fig.4a are now dark! This is a common property of all our $\epsilon < 1$ calculations.

3.1.2. Line Profiles. In Fig.4 we show the line profiles for inclination $i = 80°$. The double peaks are a characteristic of rotating geometrically thin disks if viewed at i greater than about 15° (Smak, 1969, 1981; Horne and Marsh, 1986). The wavelength offset of the peaks $\Delta\lambda_p$ is given by the projected velocity at R (Smak, 1981)

$$\Delta\lambda_p \approx \pm\lambda_0 \frac{v_\varphi(R)}{c} \, sin \, i. \tag{3.3}$$

Ripples in the profiles can be understood as a consequences of the finite optical and geometric extension of the disk.

3.2. LINES FROM GEOMETRICALLY THICK DISKS.

We expand the parameter space to geometrically thick disks, which may represent conditions in AGNs. Here we adopt as the base parameters $M_* = 10^{10} M_\odot$, $r_{in} = 9 \; 10^{15}$cm, $R = 2 \; 10^{17}$cm, $\epsilon = 10^{-2}$, $\tau_0^L = 1$, and $\tau^c = 10^{-2}$. In Fig.5 we show the variation of the line profiles for two inclinations and three aspect ratios H/R.

412

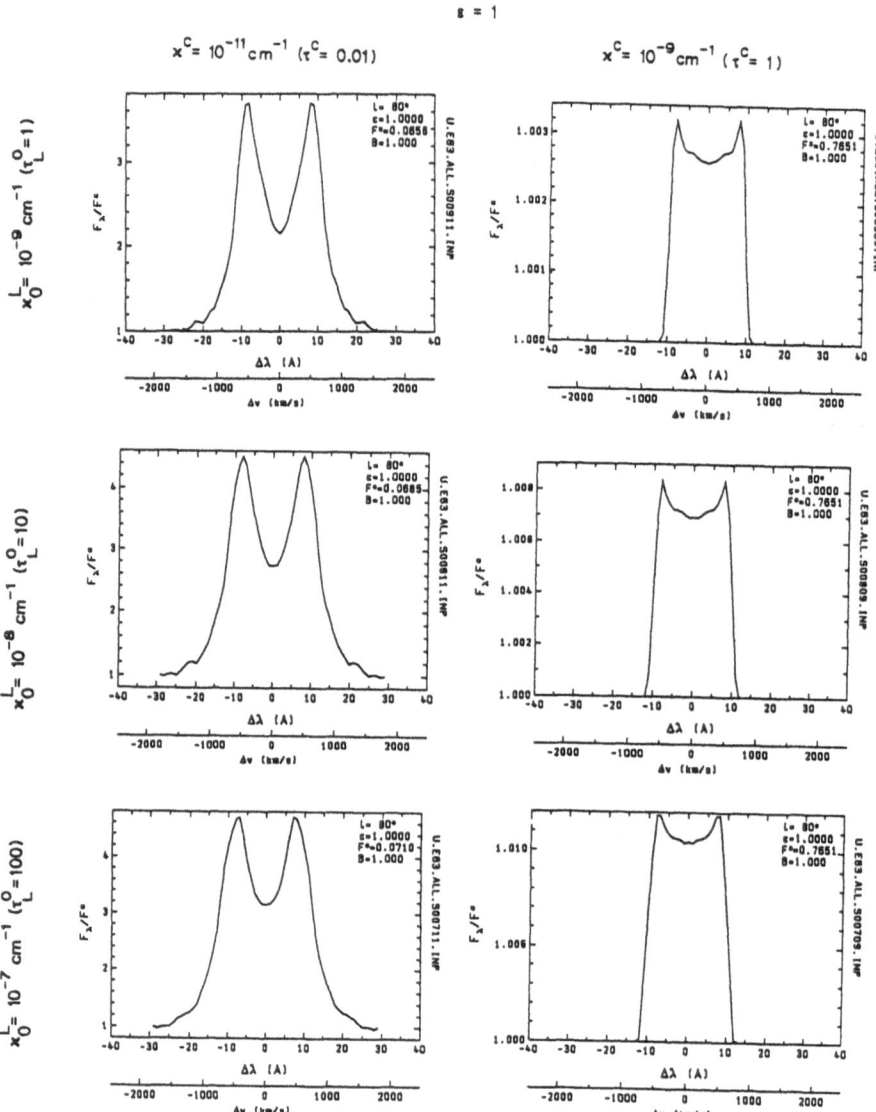

Figure 4: Normalized theoretical line profiles for homogeneous geometrically thin disks. Parameters are the vertical optical depths in line center τ_0^L and in continuum τ^c, and the scattering parameter ϵ.

According to Eq.(3.2) the profile function is broader for a thicker disk. Because the λ-separation of the peaks is smaller than the broadening $\Delta\lambda_{th}$, the profile for the $H/R = 1$ disk is roughly Gaussian, whereas for the thinnest disk the profile may show double peaks. The criterion for double peaks simply is

$$\frac{\Delta\lambda_{th}}{\Delta\lambda_p} = \frac{(H/R)\sqrt{1+\alpha^2}}{\sin i} < 1. \tag{3.4}$$

In fact, it is dangerous to use the approximation Eq.(3.1) which is valid for thin disks. For example the temperatures corresponding to c_s would be too high. However, we want to show that it is possible to produce a variety of line signatures.

4. Line profiles from disks with coronae.

In this section we study how the presence of a corona effects Balmer line profiles. These lines are formed at the disk surface, where much of the heating may occur, so they may prove useful in distinguishing between various heating mechanisms.

The disk structure is computed self-consistently with the radiation field in the Grey approximation (Adam et al., 1988; Adam, 1989). Naively one might expect that because a corona only changes the emission, not the absorption, from a single surface element, any optically thin disk model could be used. However, since the observed profile is the sum of profiles from each surface element of the disk, it is the overall balance between emission and absorption that counts. Therefore we use a more realistic optically thick disk structure.

Typical disk structures have been described by Shaviv and Wehrse (1986), Adam et al. (1988) and Adam (1989). The structure consists of a geometrically thin, optically thick disk, temperature decreasing with distance from the midplane, and optically thin higher temperature ($\sim 10^5$ K) corona. With this sort of structure hydrogen emission lines may be formed in the transition region between the disk and corona (Baschek et al., 1988) and if electron densities in the disk are greater then 10^{12} cm^{-3} the emission may be flanked by broad absorption wings (Cheng and Lin, 1989).

As an example we compute line profiles from a steady α-disk with mass accretion rate 10^{-10} M$_\odot$ yr^{-1}, a Keplerian rotation law and gas in LTE. Since the validity of LTE and the simple parameterization of viscous dissipation in the corona and transition layer are highly questionable, their kinetic temperatures are rather uncertain. We therefore assume that above the disk there is a corona with a radial temperature dependence (Adam, Störzer and Duschl, 1989)

$$T(r) = T_{in} \left(\frac{r}{r_{in}}\right)^{-1/2} \tag{4.1}$$

where T_{in} is the temperature of the corona at the inner radius r_{in}, and that this is connected to the disk by a transition zone whose height is a small fraction (≈ 0.05)

414

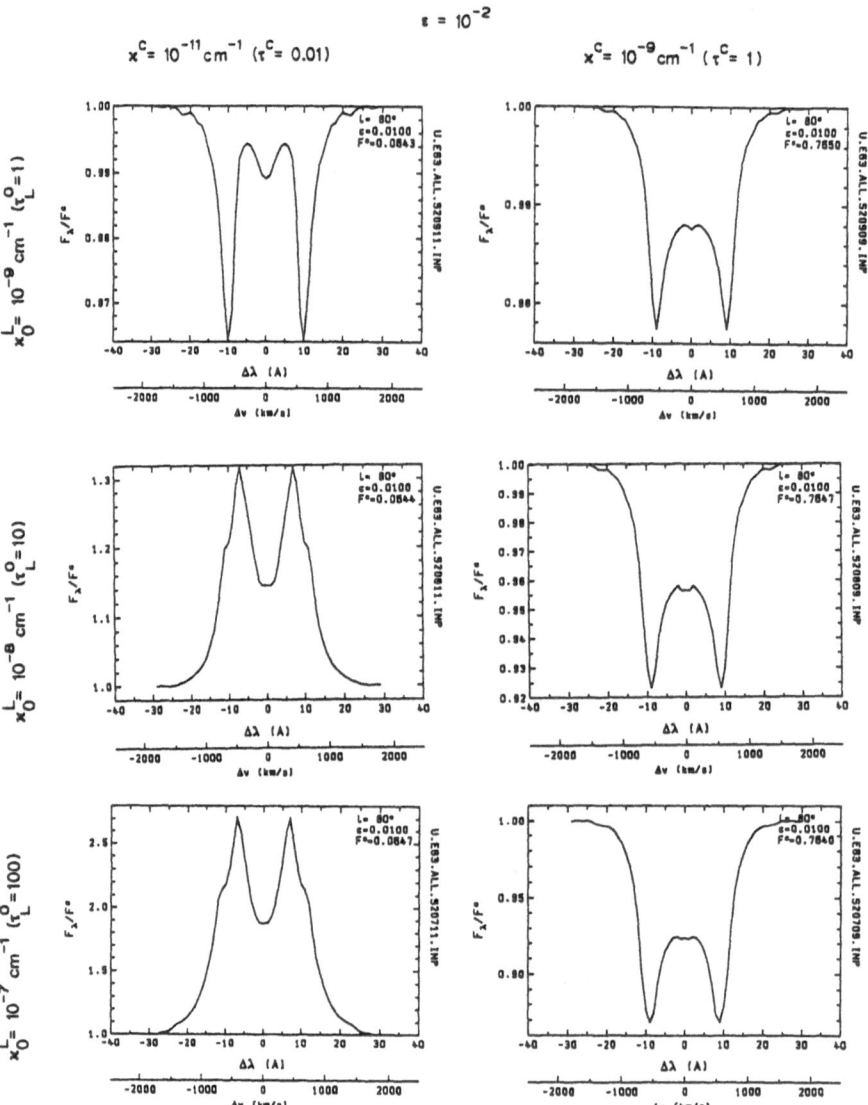

Figure 4: continued

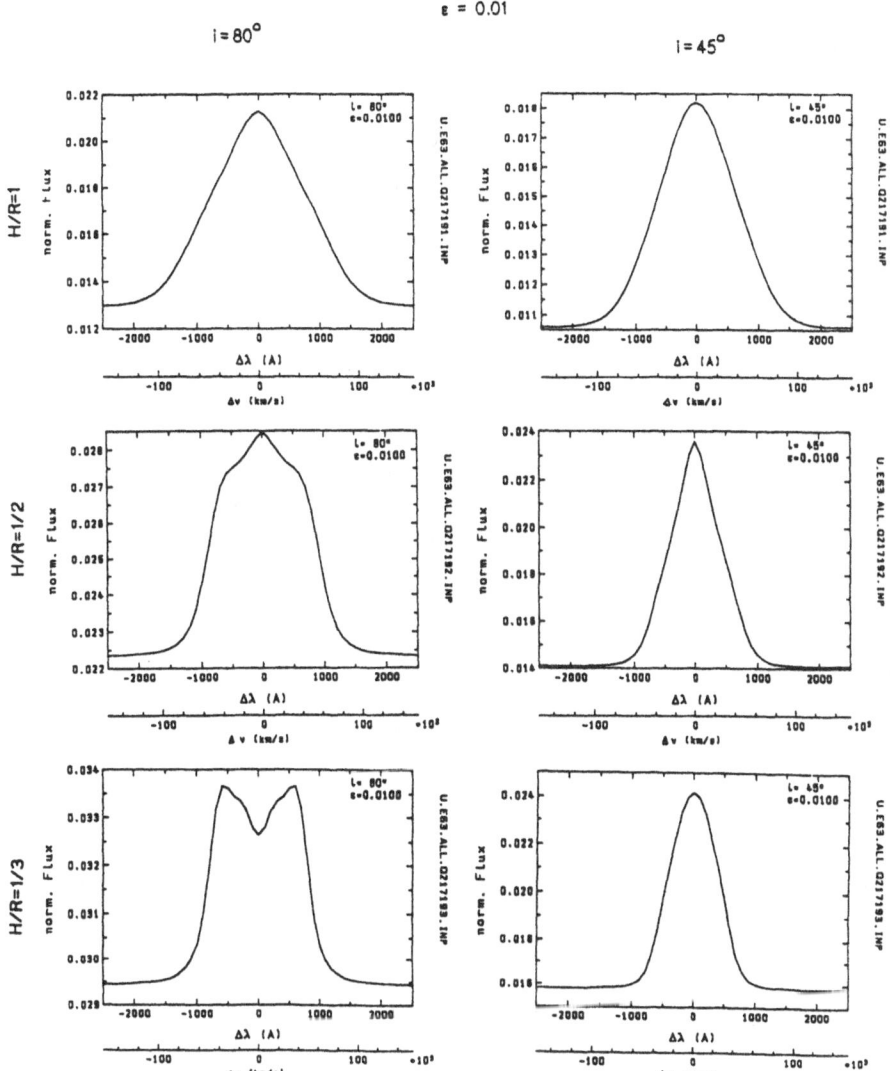

Figure 5: Theoretical line profiles for homogeneous geometrically thick disks. The parameters are the aspect ratio H/R and the inclination.

of the disk height. The temperature variation with height through the transition zone is assumed to be logarithmic, the pressure is assumed constant and, as for the disk, the gas is assumed to be in LTE.

Detailed line profiles are computed using a ray tracing technique (Baschek *et al.*, 1988; Adam, Störzer and Duschl, 1989). The intensity emerging from the disk

416

at wavelength λ from a point at radius r, azimuthal angle φ and inclination i is

$$I_\lambda(r,\varphi,i) = \int_0^{\tau_\lambda^{max}} S_\lambda(\tau_\lambda)e^{-\tau_\lambda}\,d\tau_\lambda \qquad (4.2)$$

where the τ_λ is the sum of the line and continuum optical depths, S_λ is the source function and the integral is along a ray through the disk. At distance d the flux from the disk, outer radius R, is then given by

$$F_\lambda = \frac{1}{d^2}\int_0^{2\pi}\int_0^R I_\lambda(r,\varphi,i)(\cos i - \cos\varphi\sin i\tan\eta(r))r\,dr\,d\varphi \qquad (4.3)$$

where $\tan\eta(r)$ is the disk gradient at r.

Line profiles of Hα and Hδ at an inclination angle 30° for different coronal temperatures are shown in Fig. 6. We assume that the source function is equal to the local Planck function. For the (microscopic) profile functions Stark profiles of the unified theory are used (Vidal, Cooper and Smith, 1973).

All profiles show broad absorption lines extending over 5000 km s^{-1}. Line profiles from models with T_{in} greater than $2\,10^5$ K have emission cores. These emission profiles are all double peaked because there is significant emission from gas in the transition zone where the electron densities are less than 10^{10} cm^{-3}. As the temperature of the corona is reduced the emission component at a particular (r,φ) decreases and becomes swamped by the effects of absorption from other lines-of-sight.

In recent papers Lin, Williams and Stover (1988) and Cheng and Lin (1989) have demonstrated that the broad single peaked Balmer line profiles may be produced by high electron densities in disk. However, as shown here, if there is a corona there will be a transition zone with lower electron densities so that the width of lines formed here will be of the order 10's rather than 1000's of km s^{-1}. Depending on the structure of the transition zone these narrow lines may contribute significantly to the overall emission line profile. In this case the disk profile would show double peaked structure.

5. Conclusions.

In this paper we have shown that even disks that are homogeneous in temperature and density can give rise to essentially all types of line profiles which are observed from systems which supposedly contain disks. The spatially resolved brightness distribution is complex and in most cases cannot be derived from surface emissivities with a simple radial dependence. For more realistic disk models with vertical temperature structures we have learnt that the line profiles are sensitive to the structure of the region connecting the photosphere and the corona (transition zone). Therefore, careful observations and modeling of line profiles should enable us to determine

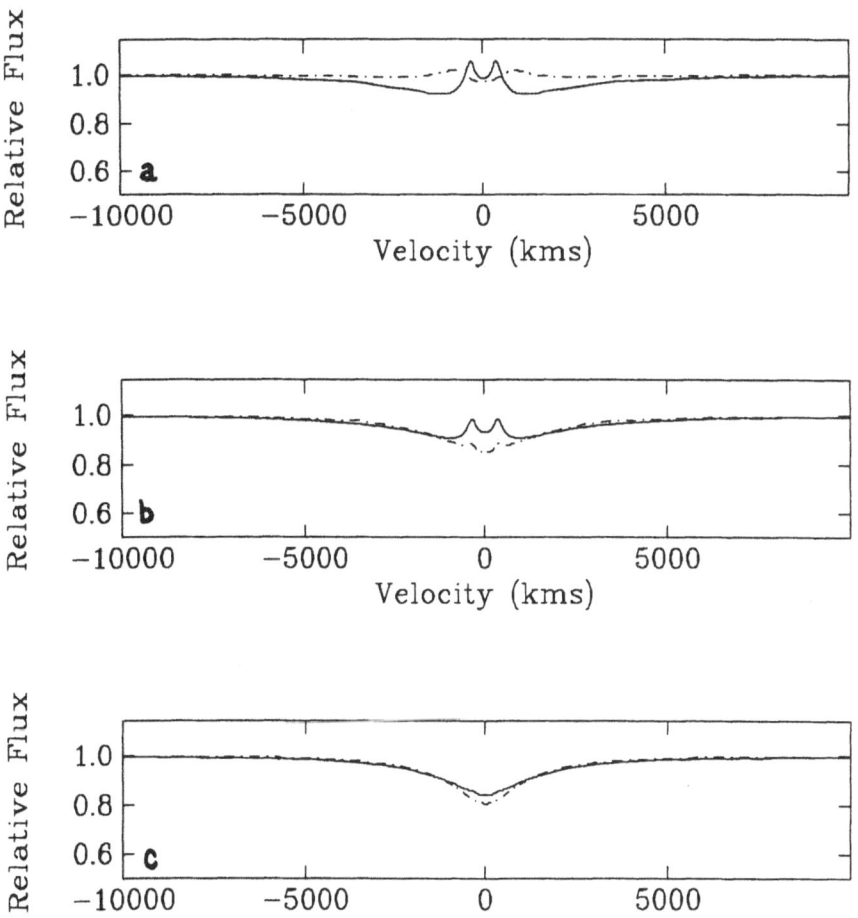

Figure 6: Hα (solid line) and Hδ (dashed line) line profiles from the disk with different corona temperatures: a) $T_{in} = 4\,10^5$ K; b) $T_{in} = 2\,10^5$ K; c) $T_{in} = 8\,10^4$ K.

empirically the structure of disk atmospheres and coronae and to constrain the local rate of energy dissipation (i.e. constrain α). Our models can also be used to simulate eclipses, which – when compared with time resolved observations – could provide even more detailed information on the spatial distribution of matter in accretion disks.

ACKNOWLEDGEMENTS. Financial support from the Deutsche Forschungsgemeinschaft (SFB 328) is acknowledged.

REFERENCES.

Adam, J.: 1989, *Ph. D. thesis*, Univ. Heidelberg

418

Adam, J., Störzer, H., Duschl, W.J.: 1989, *Astron. Astrophys.* in press

Adam, J., Störzer, H., Shaviv, G., Wehrse, R.: 1988, *Astron. Astrophys.* **193**, L1

Athay, R. G.: 1972, *Radiation transport in spectral lines*, Reidel. Dortrecht

Baschek, B., Adam, J., Plate, R., Störzer, H., Wehrse, R.: 1988, *Adv. Space Res.* **8**, (2)315

Begelman, M. C., McKee, C., Shields, G. A.: 1983, *Astrophys. J.* **271**, 70

Cannon, C. J.: 1985, *The transfer of spectral line radiation*, Cambridge University Press. Cambridge

Cheng, F. H., Lin, D. N. C.: 1989, *Astrophys. J.* **337**, 432

Frank, J., King, A. R., Raine, D. J.: 1985, *Accretion power in astrophysics*, Cambridge University Press. Cambridge

Galeev, A. A, Rosner, R., Vaiana, G. S.: 1979, **229**, 318

Horne, K., Marsh, T. R.: 1986, *Mon. Not. R. astr. Soc.* **218**, 761

LaDous, C.: 1989, *Astron. Astrophys.* **211**, 131

Lin, D. N. C., Williams, R. E., Stover, R. J.: 1988, *Astrophys. J.* **327**, 234

Mason, K. O.: 1986, *The Physics of Accretion onto Compact Objects*. p. 29. eds Mason, K. O., Watson, M. G. and White, N. E., Springer-Verlag, Heidelberg.

Mayo, S. K., Wickramasinghe, D. T., Whelan, J. A. J.: 1980, *Mon. Not. R. astr. Soc.* **193**, 793

Mihalas, D.: 1978, *Stellar Atmospheres*, W.H.Freeman. San Francisco

Olson, G. L., Auer, L. H., Buchler, J. R.: 1986, *J. Quant. Spectroscop. Transfer* **35**, 431

Shaviv, G., Wehrse, R.: 1986, *Astron. Astrophys.* **159**, L5

Smak, J.: 1969, *Acta Astron.* **19**, 155

Smak, J.: 1981, *Acta Astron.* **31**, 395

Stenholm, L., Störzer, H., Wehrse, R.: 1989, *in preparation*

Störzer, H.: 1987, *Diplomarbeit*, Univ. Heidelberg

Treves, A., Maraaschi, L., Abramowicz, M.A.: 1988, *Publ. Astr. Soc. Pacific* **100**, 427

van der Woerd, H.: 1987, *Astrophys. Space Sci.* **130**, 225

van Groningen, E.: 1983, *Astron. Astrophys.* **126**, 363

Vidal, C. R., Cooper, J., Smith, E. W.: 1973 *Astrophys. J. Suppl.* **25**, 37

Williams, G. A., Shipman, H. L.: 1988, *Astrophys. J.* **326**, 738

CONTINUUM SPECTRA OF ACCRETION DISCS.

Giora SHAVIV[1] and Rainer WEHRSE[2]

[1] Dept. of Physics and Space Research Institute,
Israel Institute of Technology, Haifa, Israel.

[2] Institut für Theoretsche Astrophysik, Heidelberg University,
Heidelberg, FRG.

1. Introduction.

Accretion discs have come to be known as a frequent phenomenon in astrophysics (Frank, King and Raine 1985). While accretion discs are of various dimensions, scales, temperatures and so on, we limit the discussion here to accretion discs around white dwarfs. Extension of the present discussion to other conditions, in particular accretion discs in premain sequence stars and self gravitating accretion discs, will be made in subsequent communications.

Accretion discs around white dwarfs are mostly found in cataclysmic variables of various types. Here the accretion disc is formed by mass transferred from the companion, usually a late type star, onto the primary which is a white dwarf. The system of two stars and an accretion disc has several sources of radiation on top of the accretion disc. These are the white dwarf, the red dwarf, the hot spot and sometimes the gas stream from the red dwarf to the disc. Any effort to analyze these systems and find their parameters requires the unfolding of the total radiation into the different components. Obviously a trustful theory of the accretion disc is essential for such a procedure. In many cases both stars are faint and the accretion disc dominates over all other radiation sources in the system. Consequently, most of the information about the properties of the system must be extracted from the radiation emerging from the disc. It is therefore again important to understand how the radiation is formed and to be able to model it trustfully so as to be able to use theoretical models for observational data analysis and testing theoretical models and understanding of the physical processes which go on in such objects.

Many of the systems show frequent light variations on many time scales. However, we expect that the discs return to a quiescent steady state phase on a relatively short time scale. In any case, our discussion here is restricted to steady state discs.

F. Meyer et al. (eds.), Theory of Accretion Disks, 419–444.

The disc is assumed to possess axial symmetry out to its outer radius. Furthermore, the disc is considered to be geometrically thin. This assumption implies a Keplerian velocity. The matter in the disc losses gradually its angular momentum and moves towards the accreting star. At the same time the angular momentum flows outwards.

One of the most important problems in the accretion disc theory is the source of the viscosity which operates in the disc and drives the matter onto the star. It is not the purpose of this paper to discuss this elusive viscosity. We merely point out that: a) We do not know the exact nature of the viscosity b) Once certain assumptions are made about the nature of the viscosity law, models can be constructed and compared with observations of accretion discs. These observations, when examined on the background of theoretical models, should enable us to eliminate inadequate theories and restrict possible range of parameters for the discs. On the other hand, it is important to produce theoretical models under different assumptions about the viscosity and find out how the viscosity affects the radiation that emerges from the accretion discs. c) There are global parameters, like the total luminosity emerging from the disc and the effective temperature at a given radius which do not depend on the detailed structure of the disc, for example the assumed viscosity etc. Furthermore, these global parameters are frequently found by analyzing only a small part of the spectrum of the object. Without a trustful theory such determinations can be very biased.

The matter composing the disc must loose angular momentum but also releases energy as it sinks into the gravitational potential of the star. Hence the total luminosity of the disc is:

$$L_{disc} = \frac{GM_{wd}\dot{m}}{2}\left(\frac{1}{R_{star}} - \frac{1}{R_{out}}\right) \tag{1-1}$$

where R_{out} is the outer radius of the disc, R_{star} is the radius of the star, M_{wd} the mass of the white dwarf and G the constant of gravitation. The accretion rate is $\dot{m}$. Frequently $R_{out} >> R_{star}$ and hence the second term in the bracket is neglected. This is not the general case.

Since the disc is assumed to be geometrically thin a corollary is the assumption that the largest gradients are in the vertical direction and all the energy released by the matter when crossing radial distance ΔR flows in the vertical direction only. The radial velocity V_r is usually very small compared with the Keplerian velocity at R. The energy carried in the radial direction can therefore be neglected in this approximation.

The energy released from a ring ΔR at R is $\Delta r dL_{disc}/dr$ and it defines an effective temperature. We have therefore a well defined $T_{eff}(R)$ relation. The simplest way to model the emerging radiation is to assume that the radiation emitted by each ring can be described by a black body. The radiation by the collection of black bodies is then summed up to form the radiation from the disc (Lynden-Bell 1969,

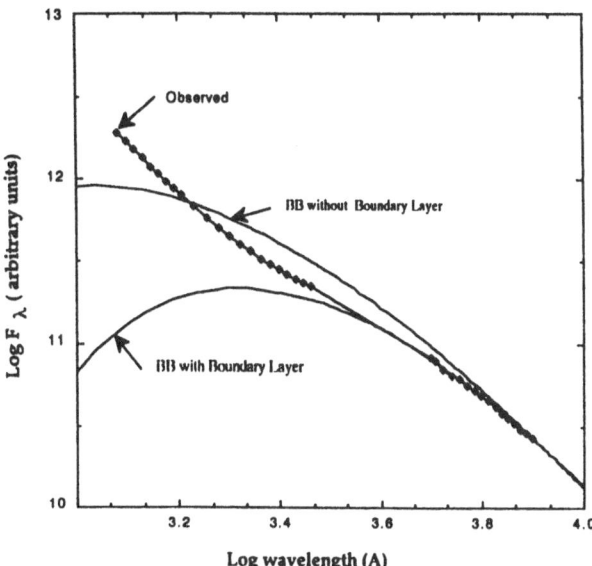

Figure 1: A comparison of the observations of RW Sex with the energy distribution predicted by black bodies. The observations are from Haug (1987). The comparison is made with and without a boundary layer. In the first case it is assumed that the temperature in the boundary layer tends to zero at the surface of the star. In the second case thge temperature is allowed to rise according to the general law for the temperature in the disc. The assumed accretion disc is 10^{18} g/s and the mass of the white dwarf is $0.8 M_{sun}$.

Pringle and Rees 1972, Shakura and Sunyaev 1973, Bath et al 1974, Tylenda 1977, Schwarzenberg-Czerny and Rozyczka 1977 and Pacharintanakul and Katz 1980. See also the monograph by Frank,King and Raine 1985). We provide in Fig. 1 a typical example of how poor black body models can be in describing the observed results. Two $T_{eff}(R)$ relations are shown in the example. In the first case a boundary layer between the Keplerian disc and a non rotating star is assumed to exist and the temperature of the boundary layer approaches that of the stellar surface which in turn is assumed to be sufficiently small to have no effect on the spectra. So long as the real temperature of the WD is small relative to the maximum temperature in the disc, the larger surface area of the white dwarf is not sufficient to compensate for the lower effective temperature and the exact value of the surface temperature has no effect on the spectral energy distribution. In the second case we assume that there is no boundary layer and that T_{eff} is allowed to increase to

$$T_{surface} = \left(\frac{3G}{8\pi\sigma} \frac{M_{wd}\dot{m}}{R_{wd}^3} \right)^{0.25} \tag{2-1}$$

The black body curves were shifted up and down (rescaling of the absolute flux) so as to fit the flux in the long wavelength regime. It is obvious that non of these models agrees with the observations of RW Sex or similar systems described by Haug (1987).

Regev (1983) argues that the exact boundary condition between the disc and the star may depend on the rotation of the white dwarf. King and Shaviv (1984) argue that if the boundary layer is optically thin it is thermally unstable and expands to form a corona around the WD. While the exact value of T_{surf} depends therefore on the state of the WD and the nature of the boundary layer, the value for T_{surf} given by the above expression is the maximum one for a disc without any boundary layer. The boundary layer is assumed to be a transition layer in which the temperature changes from that of the disc to that of the surface of the white dwarf. The temperature can be higher for a corona emitting predominantly X rays. It is clear from Fig. 1 that the extra UV and the IR flux seen in the observations cannot be reproduced by black body models with or without a boundary layer. On the other hand, this Fig. demonstrates how the boundary layer and its temperature distribution affects the spectra in the UV.

The next step in modeling the radi ation emerging from accretion discs is to fit a stellar atmosphere. The first to do so, Schwarzenberg-Czerny and Rozyczka (1977), used a grid of published stellar atmospheres for $T_{eff} < 60,000K$ and black bodies for higher temperatures. The reason was the non-availability of stellar models for such high temperatures. However, as will be shown later, the discrepancy between the detailed models and black bodies is larger at high temperatures. Stellar atmospheres were used for the first time by Kiplinger (1979,1980) and by Mayo, Wickramasinghe and Whelan (1980). Mayo et al (1980) assumed a very simple disc model with for example a constant (and prescribed) height to radial distance ratio, namely, the location of the photosphere was assumed a priori to be at a certain height above the plane of symmetry. They then integrated a stellar atmosphere rather then interpolate between a grid for the following reasons (which are valid also in our case).

i) The general range of (T_{eff}, g) parameters of a disc was not well covered by published grids of stellar atmospheres;

ii) They wanted to obtain the angular distribution of the emergent intensities $I_\nu(\mu)$;

iii) They calculated profiles of absorption lines.

They further assumed the Eddington temperature distribution to be valid and in addition took the gravitational acceleration $g = g_z$ to be constant. These authors integrated the hydrostatic equation in the z direction to obtain the density and pressure. For the emerging flux the plane-parallel transfer equation was solved with the Planck function as the source function. Since the line profiles depend on the stratification it is not to be expected that constant g models will yield agreement with the observations.

Williams (1980) and Williams and Ferguson (1982) investigated the emission lines from optically thin accretion discs of cataclysmic variables. They assumed an isothermal temperature profile in the vertical direction and solved for the temperature assuming that the cooling rate equals the heating rate. This treatment is good only for optically thin discs. Yet they applied it to very high accretion rates. Even at the low accretion rate of $10^{15}g/s$ only the outer parts of the disc are optically thin while the inner parts are thick. Williams and Williams and Ferguson derived extremely high He abundances for the accretion disc. Such a huge He abundance poses severe problems to the theory of stellar structure as well as to the theory of the formation of close binary systems. The main purpose of the authors was to explain the emission lines seen in various discs. As will be seen later, our model predicts the existence of a corona with emission lines as a consequence of the self consistent model.

Wade (1984) assumed the usual run of $T_{eff}(R)$ (with a boundary layer). He used a constant gravitational acceleration and derived the exact value for it from some global considerations (Herter et al 1979). Moreover, the disc was assumed to be isothermal in the vertical direction at each radius. On top of the disc a Kurucz's (1979) model was then interpolated and fitted at each radial point. Hence the run of g in the atmosphere is the one appropriate to stars. The agreement between these models and observed accretion discs is not good (Wade 1987).

LaDous (1987,1988) calculated synthetic optical and UV of stationary accretion discs. The assumptions are essentially those of Wade (1984) except for the fact that all atmospheres were calculated (assuming the assumptions usually made in stellar atmosphere calculations) and not interpolated from a grid.

Several factors combine to make the disc atmosphere so different from that of stars:

(i) The run of the gravitational acceleration with height in the atmosphere is different. In most stars g stays effectively constant while in red giants it decreases somewhat with height. In discs we find that the gravitational acceleration increases significantly in the region in which the spectrum is formed. The changes of g in the optically thin part may be very large (recall that large disc regions are often scattering dominated);

(ii) The optical depth of the disc is finite so that the boundary condition at the symmetry plane can seriously affect the outgoing radiation field;

(iii) The geometry of the disc is neither plane parallel nor spherical.

As a consequence, the problem of radiative transfer and in particular, the temperature optical depth relation must be derived from a set of assumptions that pertains to the disc.

A large number of observers and theoreticians is interested only in the value of the accretion rate $\dot{m}$ and not in the details of the model. Consequently, a common wisdom is that knowing the details of the disc is not mandatory if only $\dot{m}$ is desired. We comment that (a) even for a black body model disc the size of the disc (inner and outer radius) cannot be ignored and (b) since frequently only a small part of the spectrum is observed, bolometric corrections are required and they can be found only from good and trustful theoretical models. Hence, even the most fundamental parameter for the disc - the accretion rate - cannot be determined without specifying the structure of the disc. Methods like fitting a single temperature black body or a stellar atmosphere (cf. Wade 1984) are bound to yield dubious results. The number of parameters in any disc model is relatively large and fitting the theory to a small number of data , say 2-3 monochromatic magnitudes, may result in large errors in the good case or in a meaningless large range of parameters in the worst case. Since the number of parameters is high, a fit over a range of wavelengths as wide as possible is imperative if results with high degree of confidence are desired. Obviously the more data is fitted (lines, Balmer jump etc.) the better.

In view of the above discussed situation we determined to solve the radiative transfer problem from a suitable (basic) set of assumptions and under the conditions appropriate to the disc and coupled in a self consistent manner to the hydrostatics. The method used is a general one and can be used to check the effect of various assumptions on the energy mechanism, viscous energy dissipation, structure of the boundary layer etc. Features ignored in this work and delayed to later publications are the effects of the two stars, the hot spot, the stream of gas and inhomogeneities on the disc.

The structure of the paper is as follows: We discuss in Sect. 2 the basic assumptions applied for the construction of the model. The method for solution and the iteration scheme are described in Sect. 3. In Sect. 4 we present results for typical disc models and show how the various assumptions affect the emergent flux. Finally, a brief comparison with observation is carried out in Sect. 6.

2. The Physical Model and Basic Assumptions.

The following assumptions were made in the models presented in this paper.

a) Steady state accretion. We assume that the disc is in a steady state. The integration of the flow of matter in the z direction is the same for all radii. Thus, if at radius R the disc has a height Z_o, defined to be the height where the optical depth at 5000A vanished and $P_g = 10^{-6} dyne/cm^2$, then the rate of accretion, or mass flow through the disc is constant, namely

$$\dot{m} = 2\pi R \int_0^{z_0} \rho v_r \, dz = const. \tag{1-2}$$

where v_r is the velocity of the matter in the radial direction and ρ the density. The coordinate perpendicular to the plane of the disc is denoted by z.

b) Axial symmetry. The outer radius of the disc is sufficiently close to the accreting object that the disc can safely be assumed to possess axial symmetry.

c) The disc is geometrically thin. At any radial point R the width Z_0 of (half) the disc satisfies

$$Z_0 << R \tag{2-2}$$

d) Following assumption (c) we assume that the radial gradients are small relative to the vertical ones so that

$$\frac{\partial P}{\partial R} << \frac{\partial P}{\partial Z}, \frac{\partial T}{\partial R} << \frac{\partial T}{\partial Z}. \tag{3-2}$$

and

$$\frac{\partial F}{\partial R} << \frac{\partial F}{\partial Z}, v_r << v_{kep} \tag{4-2}$$

where v_{kep} is the Keplerian velocity at R. P and F are the gas pressure and the total energy flux respectively.

e) The velocity of the matter in the disc is essentially the Keplerian one except for a narrow neighborhood near the star, where a boundary layer forms between the fast (Keplerian) disc and the slow (or even stationary) star. The existence of a boundary layer affects the run of the effective temperature with radius (see above). In the models described here, $T_{eff}(R)$ is an input function. The model is valid for any $T_{eff}(R)$ law and starts to fail as soon as the assumption (d) breaks down.

f) The disc is in hydrostatic equilibrium in the vertical direction namely,

$$\frac{dP}{dz} = -g\rho = \frac{GM_{wd}}{R^3} z\rho \tag{5-2}$$

g) The energy generated in the disc can be expressed as a function of the location in the disc.

h) The radiative transport equation is

$$n.\nabla I = \kappa(S - I) \tag{6-2}$$

where S is the source function for the radiative field and κ the opacity at wavelength λ. n is a unit vector in the ray direction.

i) Whenever the temperature gradient in the vertical direction is greater than the adiabatic one we assume that convection sets in. Convection is treated in the mixing length approximation and we take for the mixing length one pressure scale height. Frequently the convection is very efficient. We then limit the convective flux at each point to the energy generation by viscous dissipation integrated over z up to this point minus the radiative flux carried by the adiabatic temperature gradient.

i) The above system of equations is supplemented by the equation of state and expressions for the opacity . In all results reported in this paper we used solar composition of hydrogen, helium and metals.

j) Since in this paper we are mainly interested in the continuum emission, LTE was assumed in the calculation of the population of electrons in the various atomic levels.

3. Method of Solution.

The disc problem differs from the problem of stellar structure or stellar atmosphere in the following fundamental senses. In the problem of stellar interior, one as a rule uses the mass as an independent Lagrangian coordinate and imposes the outer boundary condition at essentially fixed Lagrangian mass. The condition at the center is obviously the vanishing of the flux. In stellar atmospheres with slab geometry a given radiation flux is assumed at the bottom of the atmosphere which is placed at sufficiently high optical depth. The location of the photosphere or the place where g is known, has absolutely no effect on the results.

In the problem of the disc the exact height of the photosphere is not known a priori and so is the gravitational acceleration. A change in the height of the photosphere means a change in the gravitational acceleration and with it a change in the entire structure. Hence, we have first to find the location of the photosphere. Next, unlike stellar atmosphere, the flux is not constant throughout the height of the the the disc. On one hand the energy produced in the optically thin layers is (probably) small because the density and pressure are low. But on the other hand, since these layers are poor radiators (and absorbers) a small energy production in these layers has a large effect on the final temperature these layers will eventually settle to in the equilibrium state (Shaviv and Wehrse 1986). Here we only mention that in some theories for the viscosity in the disc and angular momentum transfer, the energy dissipation is by means of a wind and then the dissipation and energy source is in the very outer layers (Begelman et al 1983) It is not at all clear how the stresses are transferred in the vertical direction. In any case, the radiative transfer must be solved under the boundary condition at the symmetry axis z = 0 namely, the vanishing of the energy flux F. Last but not least, a Lagrangian coordinate in the usual meaning cannot be defined since the mass at a given R and z is not known a priori. The volume density, the surface density and even the total surface density are not known before the start of the calculations.

In view of the above described situation we approached the problem in the following way. A temperature - optical depth relation is assumed and a height Z_0 where the optical depth is 10^{-n} where n = 4-6, is guessed. The exact value of n is so chosen as to prevent the formation of a corona in the models for the continuum described below. A large value for n gives rise to a corona which is optically thin and does not affect the continuum. On the other hand it slows a bit the convergence. At this point we know the total luminosity from the ring and the effective temperature. With the aid of $T = T(\tau)$ we can integrate the hydrostatic equilibrium from Z_0 downward to z = 0. If the height Z_0 is the correct one, then the total energy generation in the disc (calculated on the basis of the present structure) is equal to the total luminosity that we assumed to emerge from this ring. If this condition is not satisfied, we iterate for Z_0 keeping the $T = T(\tau)$ law fixed until the total energy generation in the disc is equal to the luminosity from the ring . When this loop converges, the hydrostatic model is consistent with the energy generation condition, but not the radiation field. We turn now to the iteration for the radiation field. The integration of the hydrostatic equation is carried out using a 4th order Runge-Kutta integration scheme. The relative accuracy in the integration is 10^{-4}.

The iteration for the radiation field is carried out in the following way: the run of the density, pressure, monochromatic opacities etc, with height are taken from the hydrostatic calculation and are kept constant during the iteration for the radiative field. Also, the energy produced in each height Dz is kept constant. The radiative transport equation is inverted now into an equation for the temperature at every height in the disc. When the iterations for the temperature converge we calculate a new $T = T(\tau)$ relation whereby the temperatures are the new ones and the optical depths are still the old ones. If the new $T = T(\tau)$ relation agrees well with the previously used relation, the iteration is stopped and the hydrostatic structure is consistent with the radiative transfer. However, if the two temperature relations do not agree, the whole process of solving for the hydrostatics and then for the radiative field starts again but now with the new $T = T(\tau)$ relation.

There are therefore three basic iteration loops in the calculations:

1) Solve for the hydrostatic structure (say pressure and density height distribution) for a given $T = T(\tau)$ relation. At the end of the iteration we know the height Z_0 and the location of the photosphere. The hydrostatic structure 'produces the required total luminosity'.

2) Solve for the radiative field and the temperature distribution for given R(z), q(z) and the optical depths $\tau(z)$ distributions. At the end of the iteration we obtain a new $T = T(\tau)$ relation.

3) Repeat the previous two iterations until the temperature distribution converges.

The basic problem we encountered with the application of this iteration scheme is the communication between the two iterations: the one for the hydrostatics and the one for the radiative field. Suppose for example that the independent variable is the optical depth and the information from the hydrostatic solution is transferred to the radiative solution in the form of functions which depend on optical depths. As soon as the radiative iteration starts and the temperatures change, the optical depth changes and the new temperature does not refer to the same physical point (height) in the disc as the original temperature was. In other words, since there is no Lagrangian coordinate there is no 'absolute reference frame' and the iteration pendulates over a perpetually changing independent coordinate.

To overcome the problem we define a semi-Lagrangian coordinate defined as

$$\eta = \frac{z}{z_0} \tag{1-3}$$

where Z_0 changes during the iterations.

The new independent coordinate η is discretized in a universal way, namely the points η_i, i=1,2,... I where the solution is provided are kept fixed during the iterations. The total number of points in the independent variable (semi-Lagrangian coordinate) and the values η_i are input to the problem. The distribution of η_i is based on the analysis of previous models and what we found is required to provide the most accurate solution with a fixed total number of points. Previous attempts in which the independent variables were z, the optical depth τ or the mass, had a variable number of integration points and the points were distributed by the integration routine according to the accuracy requirements as determined by the integration routine. These methods are superior to the one used here from the point of view of accuracy. However, the η method is much more robust and poses much less problems in convergence. The previous methods failed to converge quite often. One can say that in this method we keep the shape of the distribution between two successive iterations and the change is only in the scale. Consequently, previous models with different parameters can serve as a reasonable initial model and save many iterations.

The hydrostatic equations are now written as:

$$\frac{d\tau}{d\eta} = z_0 \kappa$$

$$\frac{dP}{d\eta} = z_0 g \rho \tag{2-3}$$

The optical depth used in these equations and in the temperature is a monochromatic opacity calculated at a representative wavelength. It serves essentially as a scale.

The temperature distribution in the vertical direction of the disc is determined from the energy equation:

$$\int_0^\infty (J(z, R, \lambda) - B(T(z, R), \lambda)) \kappa(\lambda), d\lambda = -E(z, R) \tag{3-3}$$

where J is the mean intensity, B the Planck function and κ the absorption coefficient. E is the energy input per unit time and mass. The energy input is given by:

$$E(z, R) = \bar{E}(z, R) + \frac{1}{\rho} \frac{dE_{conv}}{dz} \tag{4-3}$$

where $\bar{E}$ is the energy generation due to viscous forces and E_{conv} is the convective energy flux. Here we assume that E(z,R) has been calculated during the integration of the hydrostatic equation and is therefore a given function in the radiative transfer calculations.

The solution of equation for T(z,R) is carried out by expressing J in terms of B by means of the radiative transfer equation. Under the assumptions made here the radiative flux is only in the vertical direction and we assume that the various radial shells are independent of each other. The neglect of the radial radiative flux is at most few percent and in most cases much less. The radial temperature gradient is extremely small compared to the vertical one so long as the assumption of a geometrically thin disc is valid. We check this assumption a posteriori and find that it is satisfied in all cases presented here. The radiative transport equation is solved in the two stream approximation, namely we have a flux I^+ in the outward direction (positive z) and a flux I^- in the inward direction. The intensity vector is now written as

$$I = \begin{pmatrix} I^+ \\ I^- \end{pmatrix} \tag{5-3}$$

and the radiative transfer equation becomes:

$$\pm\frac{dI^\pm(z,R,\lambda)}{dz} = -(\kappa(\lambda) + \sigma(\lambda)I^\pm(z,R,\lambda) + \sigma(\lambda)J(z,R,\lambda) + \kappa(\lambda)B(T(z,R),\lambda) \tag{6-3}$$

where $I^\pm$ are the specific intensities in the positive and the negative direction and σ_λ and κ_λ are the scattering and absorption coefficients respectively. The mean intensity J is given by

$$J(z,R,\lambda) = \frac{1}{2}\left(I^+(z,R,\lambda) + I^-(z,R,\lambda)\right) \tag{7-3}$$

The solution of the radiative equation is carried out over a discrete grid. We divide the disc at every radial point into 2N layers. Since the mid-plane is a symmetry plane, it is sufficient to calculate the radiation and temperature field for the upper (or lower) part of the disc. We employ therefore only N points in the actual calculation and impose the symmetry condition on the radiation field. The points are identical to the points used in the hydrostatic equation.

The radiative transfer is described by the specific intensities at the boundaries of the layers. The transmission and reflection coefficient are evaluated for each layer and the matrix (t^+, t^-, r^+, r^-) as well as the vector of the internal sources (S^+, S^-) are formed. The radiative equation is now written in the following form:

$$\begin{pmatrix} I^+_{j,n+1} \\ I^-_{j,n} \end{pmatrix} = \begin{pmatrix} t^+_{j,n} & r^+_{j,n} \\ r^-_{j,n} & t^-_{j,n} \end{pmatrix} \begin{pmatrix} I^+_{j,n} \\ I^-_{j,n+1} \end{pmatrix} + \begin{pmatrix} S^+_{j,n} \\ S^-_{j,n} \end{pmatrix} \tag{8-3}$$

(compare with Peraiah 1984). The subscript j refers to wavelength and the index n to the depth. In order to simplify the notation we have omitted the reference to the radial point at which the radiative transfer is solved.

The boundary conditions are the following: For the outer boundary condition we assume at the moment no incident flux and hence:

$$I^-_{j,1} = 0 \tag{9-3}$$

Our formalism is a general one and will be later applied to cases in which there is an incident flux on the disc. At the symmetry plane the specific intensities in the positive and negative directions must be equal and hence:

$$I^+_{j,N+1} = I^-_{j,N+1} \qquad (10\text{-}3)$$

The system can be written in a compact matrix notation in the following way (Wehrse 1981)

$$A_j I_j = S_j \qquad (11\text{-}3)$$

where the vector I_j contains all the specific intensities at wavelength λ_j, the matrix A_j consists of all the corresponding coefficients and S_j comprises all the source terms. Since the source terms are proportional the the Planck function B_j we can write

$$S_j = Q_j B_j. \qquad (12\text{-}3)$$

The solution for the specific intensities is now given by:

$$I_j = A_j^{-1} Q_j B_j \qquad (13\text{-}3)$$

and

$$J_j = G A_j^{-1} Q_j B_j \qquad (14\text{-}3)$$

where the matrix G contains the summation weights (see eq. 6-3) . In terms of these quantities the energy equation (2-3) can be written in the following way:

$$\sum_j a_j \kappa_j (1 - G A_j^{-1} Q_j) B_j = E, \qquad (15\text{-}3)$$

where κ_j is a matrix comprising all absorption coefficients at wavelength λ_j and a_j are the weights for the wavelength integration.

The determination of the temperature structure as expressed by the vector of temperatures is carried out by means of a Newton-Raphson method. The iterations are performed by means of the following corrections for the temperature:

$$\Delta T = \left\{ \sum_j a_j \kappa_j (1 - G A_j^{-1} Q_j) \frac{\partial B_j}{\partial T} \right\}^{-1} \left\{ E - \sum_j a_j \kappa_j (1 - G A_j^{-1} Q_j) B_j \right\}.$$
$$(16\text{-}3)$$

In the actual calculation we compute $A_j^{-1} Q_j$ as described by Kalkofen and Wehrse (1984). Since the matrices $a_j \kappa_j (1 - G A_j^{-1} Q_j)$ do not change during the iteration they need to be evaluated only once and stored.

Furthermore we assume that the Planck function varies linearly with depth within any given shell (of the optical path). The linear function is different in each optical depth shell. Thus the Planck function is 'given' at the grid point of the optical depth and the points are connected in a linear way. The coefficients ϵ_λ are approximated by a suitable average value. At this point the transfer equation can be solved analytically and we find for a layer of optical depth τ:

$$t^+ = t^- = \frac{4\sqrt{\epsilon}e^{-\sqrt{\epsilon}\tau}}{(1+\sqrt{\epsilon})^2 - (1-\sqrt{\epsilon})e^{-2\sqrt{\epsilon}\tau}}, \tag{17-3}$$

$$r^+ = r^- = \frac{(1-\epsilon)(1-e^{-\sqrt{\epsilon}\tau})}{(1+\sqrt{\epsilon})^2 - (1-\sqrt{\epsilon})^2 e^{-2\sqrt{\epsilon}\tau}}, \tag{18-3}$$

$$\begin{pmatrix} Q^+ \\ Q^- \end{pmatrix} = \begin{pmatrix} (1-1/\tau) + t^+/\tau - r^+(1+1/\tau) & 1/\tau - t^+(1+1/\tau) + r^+/\tau \\ 1/\tau + r^+/\tau - t^+(1+1/\tau) & (1-1/\tau) - r^+(1+1/\tau) + t^+/\tau \end{pmatrix}. \tag{19-3}$$

with ϵ being the ratio of the absorption to the extinction. In the numerical calculations we used 75 z points (75 layers) and up to 300 frequencies distributed in such a way as to give the best coverage at each temperature. Since the effective temperature changes along the radial distance it is necessary to change the relevant frequencies with the effective temperature, thus the frequencies change with the effective temperature. The entire disc was divided into 50 radial rings and the above procedure repeated for each one. The flux emerging from each ring is integrated to yield the total flux according to

$$F_{total} = 2\pi \int_{R_{wd}}^{R_{out}} F(R)R dR. \tag{20-3}$$

4. Some General Results.

The model and the method of solution presented so far are general and can be applied to various models for discs. The results that we presents here and use later in the comparison with the observations assume the following specific assumptions about the effective temperature as a function of the radius and the viscous energy generation. The effective temperature at each R follows

$$T_{eff}^4 = \left(\frac{3}{8\pi\sigma}\right)\left(\frac{GM_{wd}\dot{m}}{R^3}\right)\left(1 - \left(\frac{R_{wd}}{R}\right)^{1/2}\right). \tag{1-4}$$

Clearly this assumption means that the white dwarf does not rotate.

For the viscous energy generation we assume here the a model, namely,

$$E = (3/2)\alpha\Omega P/\rho, \tag{2-4}$$

where α is a dimensionless parameter, Ω the angular frequency, P the gas pressure and ρ the density.

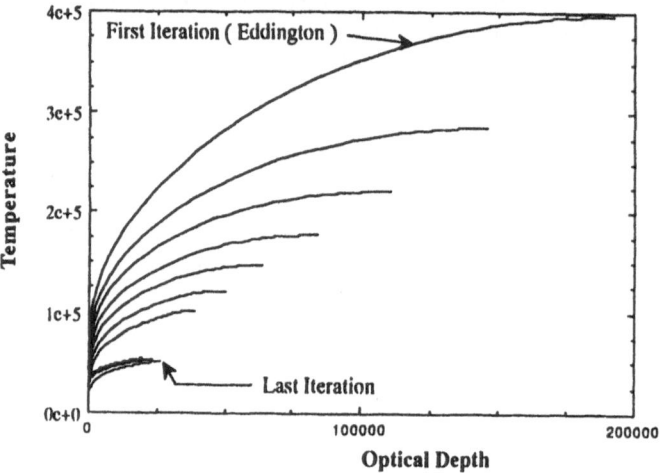

Figure 2a: The variations of the temperature profile during the iteration to the converged structure.

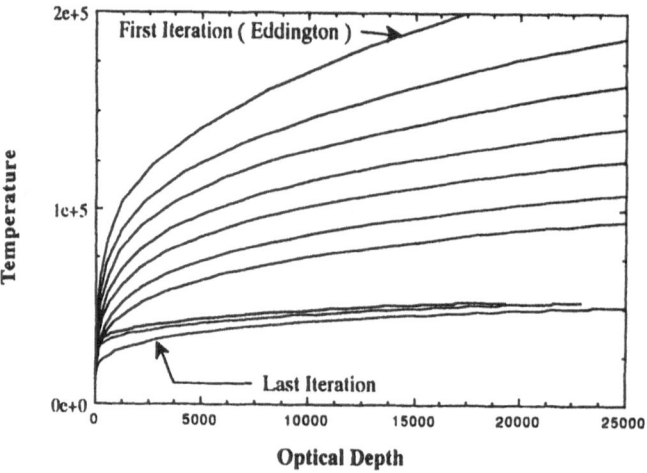

Figure 2b: The temperature profile during the iterations for the structure. It is the same as shown in Fig. 2a but for a smaller range of optical depths.

The first question that comes to mind is: are the iterations for the temperature profile necessary? A significant part in the complication of the calculation is associated with this iteration and hence the question whether we could simply assume some $T(\tau)$ law. In Figs. 2a and 2b we show how the temperature profile changes during the iterations. It should be recalled that the density profile and the entire structure change with the iterations. Moreover, since we iterate for the location of the photosphere (so as to satisfy the condition on the energy generation) the location of the photosphere changes during the iterations. The initial temperature profile assumed in the case shown in the figures is the Eddington grey body approximation. As can be seen from the figure, the changes in the temperature profile are enormous. Note that the total optical depth changes by at least an order of magnitude during the iterations. Our results show that the actual temperature distribution in the disc is significantly flatter than the grey body approximation.

The next question is: what happens to the temperature in the optically thin region where the radiation emerges from the star? For that reason we show in Fig. 2b the same temperature distribution as is shown in Fig. 2a but this time for a smaller range of optical depths (rather than the entire range as was shown before.) The general conclusion about the flatness of the temperature distribution is demonstrated again here. The inapplicability of the Eddington law is obvious.

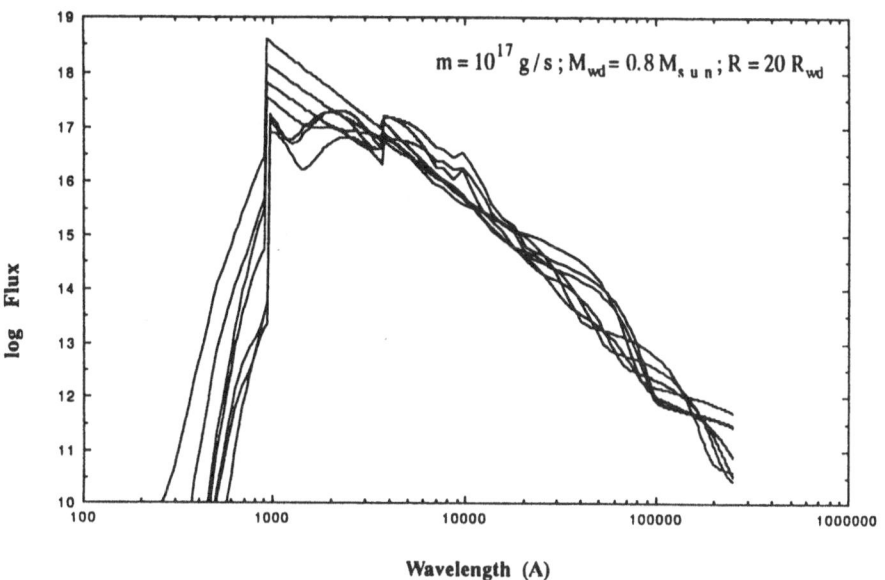

Figure 3a: The variation of the energy distribution during the iterations.

The most important, however, is the behavior of the radiative flux and how it varies with the iteration. The variation of the flux with the iteration is shown in Figs. 3a and 3b. The first figure shows the entire range of wavelengths while the second figure concentrates on the region of the Balmer jump. We stress that all the energy distributions satisfy the the energy condition, namely, the total output of all distributions is the same. The only difference between the different energy distributions is the location of the photosphere! When the location of the photosphere changes so does the run of g and $T(\tau)$. We think that these figures demonstrate the necessity to iterate for the temperature distribution and the entire structure is reli-

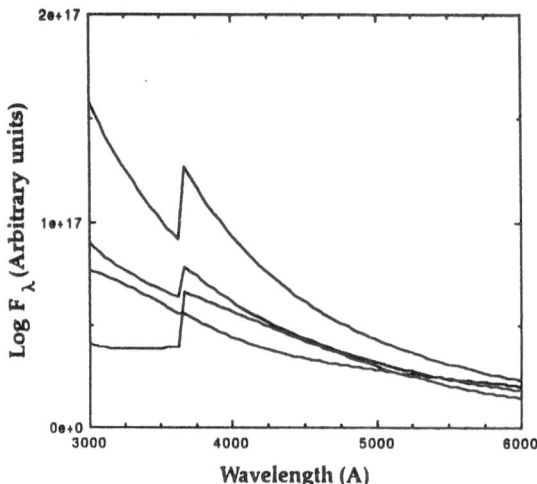

Figure 3b: The variation of the energy distribution during the iterations in the optical range as taken every few iterations.

able energy distribution are attempted. The Balmer jump for example changes by at least an order of magnitude during the iterations. There are several reasons for the large changes in the flux during the iterations. We think that among the most important ones is the redistribution of the energy below and above the Lyman limit. Small changes below the Lyman limit have large effects at longer wavelengths. Next in importance are the variations of the temperature-optical depth relation (which in turn depends on the location of the photosphere).

Many of the researchers who fitted atmospheres with different effective temperatures and gravitational accelerations assumed that the location of the photosphere can be approximated by a straight line running at some constant angle to the symmetry axis. Obviously, the angle this line makes with the z plane is a free parameter in the model. As explained before, we iterate for the location of the photosphere and hence we do not require this assumption. We show in Fig. 4a how the height of the photosphere varies with radial distance. The case shown in Fig. 4a corresponds to a low accretion rate, namely $10_{15}g/s$. The location of the photosphere is most sensitive to the accretion rate and the assumed α. The mass of the white dwarf has a secondary effect, at least far away from the accreting object. First we note that the linear law is valid only as far as $0.2R_{wd}$. Beyond this point this is not true anymore. Typical accretion discs are assumed to extend radially out to about 0.7 of the L_1 or equivalently to about 50 to 60 R_{wd} . It is therefore conspicuous that the assumption of linear relation for the location of the photosphere is not valid in this case. The figure also demonstrates that the assumption of constant g is quite poor in this regime.

In Fig. 4b we show the variation of the height of the photosphere with radial

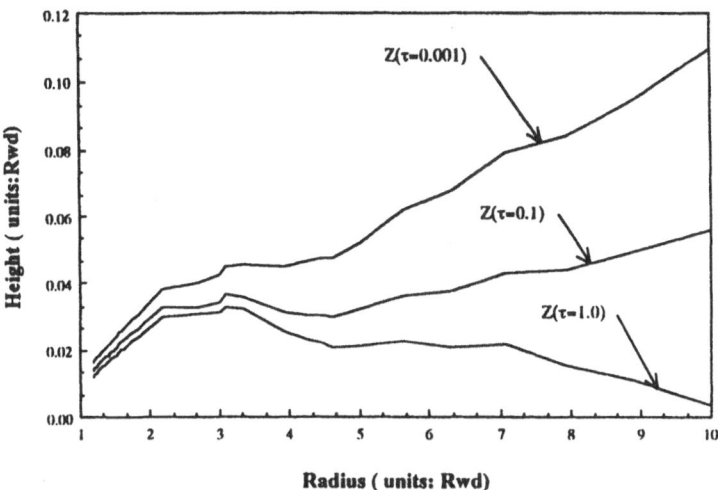

Figure 4a: The height of the photosphere as a function of the radial distance. The model is calculated for $\dot{m} = 10^{15}$ g/s, $\alpha = 1$ and a white dwarf mass 0.8 M_{sun}

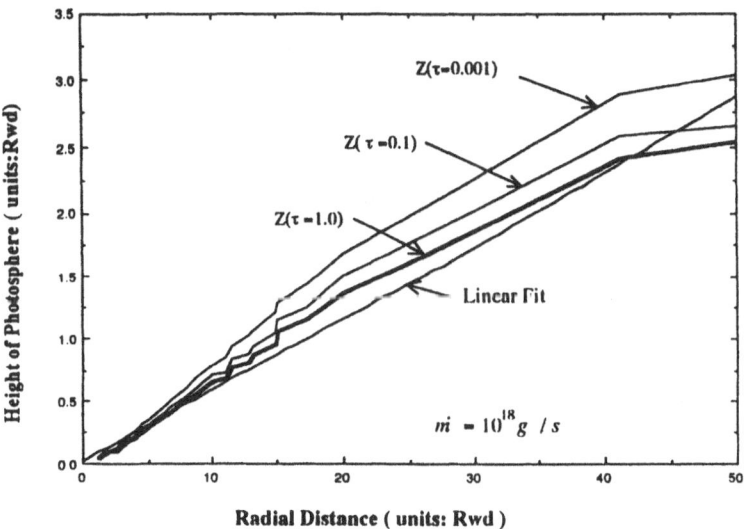

Figure 4b: The variation of the height of the photosphere as a function of the radial distance. The data are for $\dot{m} = 10^{18}$ g/s, $\alpha = 0.5$ and a white dwarf mass of $0.8 M_{sun}$. The linear fit is to the curve $\tau(5000A) = 1$.

distance for a high accretion rate $(10^{18} g/s)$. While the deviation from a straight line is smaller then in the previous case, the linear fit is not very good either. The particular fit shown is the figure corresponds to a straight line inclined at 70 to

436

the symmetry axis. Our method does not permit us to abandon the iteration for the height of the photosphere and asses the effect of this assumption on the final calculated energy distribution.

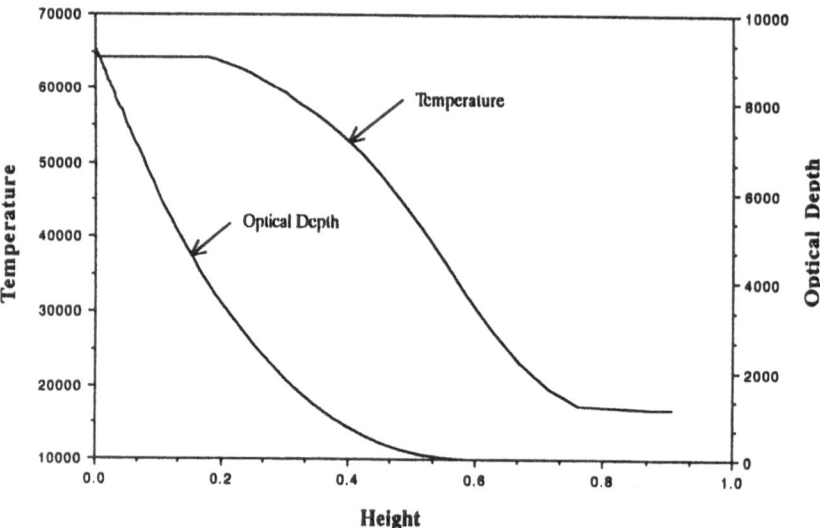

Figure 5a: A typical run of the temperature and optical depth as a function of the height in the disc. The height is given in units of the white dwarf radius.

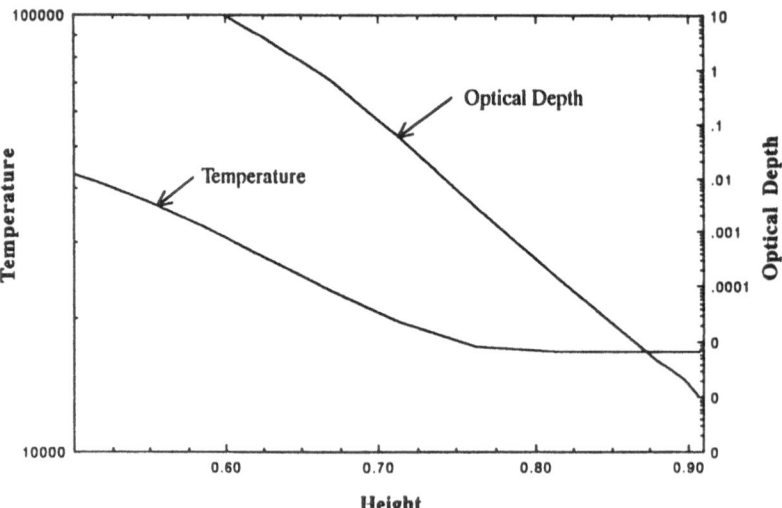

Figure 5b: The run of the temperature and the optical depth with height above the disc plane. The height is given in units of the white dwarf radius.

The typical run of the temperature with optical depth was shown before. Next we show in Figs. 5a and 5b how the optical depth and temperature vary with height above the symmetry plane. First we note the flat temperature distribution. The reason is convection. In this example convection sets inside the disc close to the symmetry plane. Since convection is very effective the temperature distribution is flat. Consequently, the temperatures at the symmetry axis are not very high. Fig. 5b shows that the emitting region is very wide and again (about a third of the height) and the simple Eddington law for $T(\tau)$ just does not exist here.

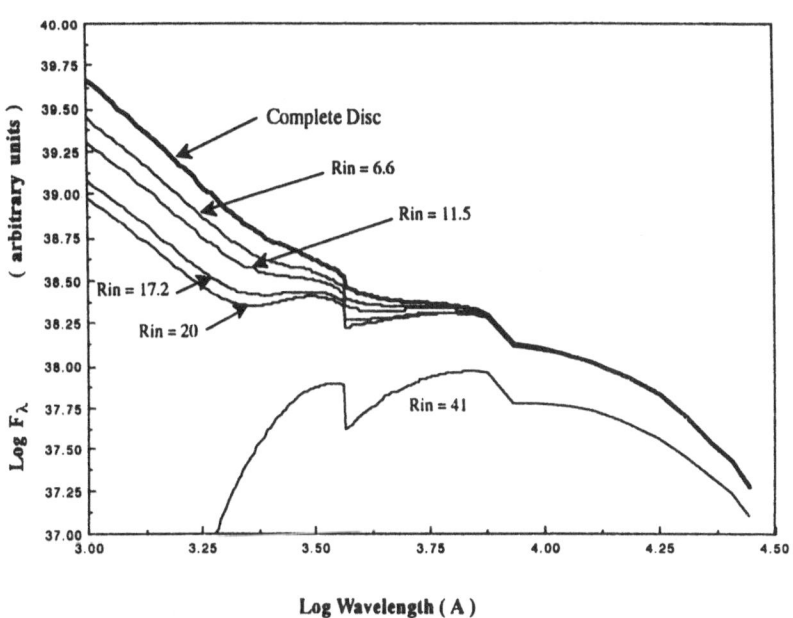

Figure 6: The effect of the inner raius on the energy distribution. The model is for $\dot{m} = 10^{18}$g/s,$\alpha = 0.5$ and a white dwarf mass of $0.8M_{sun}$.

Interesting information is obtained from analyzing discs with different outer or inner radii. This analysis can inform us about the physical size of the disc, the possible existence of a magnetic field on the white dwarf which is sufficiently strong to prevent the existence of the inner parts of the disc etc. The effect of the inner radius on the total energy distribution is shown in Fig. 6, namely, the figure compares discs with different inner radii. The disc itself is assumed to extend out to $50R_{wd}$ (about 0.7 of L1 for typical cataclysmic variables). The effect of the inner radius (in this particular model) on the final distribution is quite large in the range of wavelengths corresponding to the UV and the visual. Actually the slope in the visible changes sign as the inner radius changes. In Fig. 7 we demonstrate the effect of the outer radius. We see that the slope at about $log \lambda > 3.5$ takes a different value once the outer radius is above about $15R_{wd}$.

Valuable insight into the structure of the spectra is obtained when plotting the energy distribution as a function of wavelength and radial distance. Such a figure is shown in Fig. 8. One can easily see why the Balmer jump, which may be seen in

438

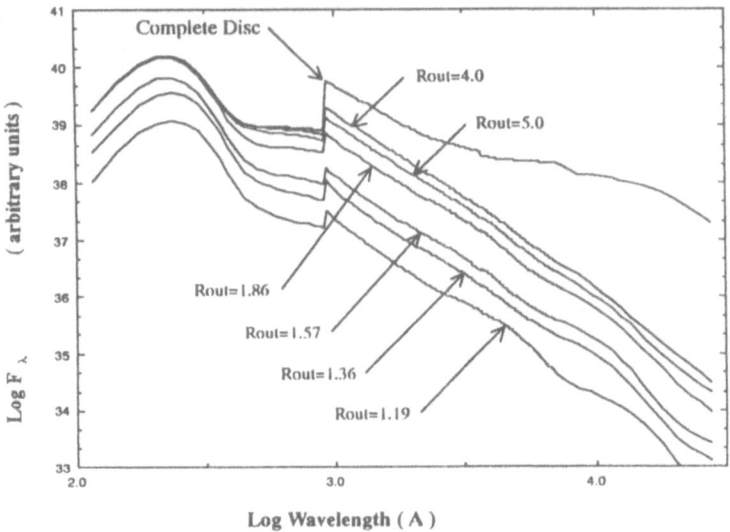

Figure 7: The effect of the outer radius of the disc on the predicted energy distribution. The model is for $\dot{m} = 10^{18}$g/s,$\alpha = 0.5$ and a white dwarf mass of $0.8M_{sun}$.

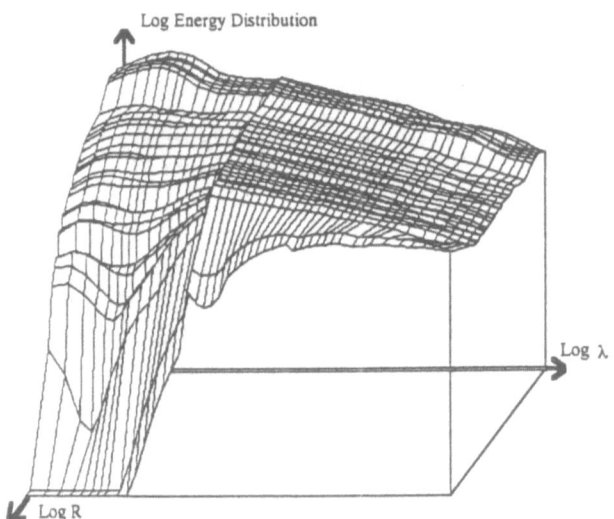

Figure 8: Isometric plot of the emergent flux as a fuction of wavelength and radial distance. All scales are logarithmic.

the energy distribution emerging from some radial rings does not appear in the final and total energy distribution. We find that the Balmer jump is strong in the outer parts of the disc and decreases as we move inward. The absolute value of the flux increases as the radial distance decreases. The outcome is that for this accretion rate the inner parts dominate and give rise to almost no Balmer jump when the entire disc is taken into account. The UV bump at 400A is contributed only by the inner most parts. Note that while the total luminosity per radial distance increases as the radius decreases, the flux in the IR decreases (most of the rise is in the UV) for a while and increases again for the inner most rings. We expect that detailed information obtained in eclipsing systems will enable the identification of the various regions discussed here.

5. Instabilities in the Radiative Transfer.

Shaviv and Wehrse (1986) and Adam, Stoerzer, Wehrse and Shaviv (1987) discussed the instabilities that arise from the energy generation in the optically thin region and the particular behavior of the opacity. The first instability occurs whenever E/κ is a decreasing function of the pressure. Since the opacity dependence is usually higher than the first power and since in the a model the energy generation rate is proportional to the pressure it is found that the temperature rides strongly in the outer parts of the atmosphere leading to the formation of a corona. Shaviv and Wehrse demonstrated the instability in the radiative transfer numerically by introducing a parameter which cuts the energy generation in the optically thin regions in an artificial way. The understanding of these instabilities is important in the construction of the continuum (to which the corona does not contribute in any significant way) in general and understand how we chose the initial τ to start the integration inwards.

If the region under discussion is optically thin then we can ignore the radiative transfer equation completely (it will not describe the main energy losses). We will show later that is neglect is not crucial for the corona and when we do include the radiative transfer we run into the second instability. The energy balance in the optically thin region is

$$2/3\alpha\Omega P/\rho = C\rho^2 T^\beta \tag{1-5}$$

where on the LHS we have the α model energy losses and on the RHS the radiative cooling. C and β are two constants. Assume now an ideal gas equation of state $P \propto T$ and get for the temperature

$$T \propto P^{-2/(3-\beta)} \tag{2-5}$$

and hence the temperature increases indefinitely as the pressure decreases. Shaviv and Wehrse overcame the problem of infinite temperatures by eliminating the tiny region of $\tau < 10^{-6}$ from the calculations. However, if thermal conduction is included in the calculations then a natural temperature for the corona arises and the assumption of small optical depth can be checked a posteriori (Rosner, private

440

communication). When thermal conduction is included in the energy balance we have:

$$2/3\alpha\Omega P/\rho - C\rho^2 T^\beta = -\nabla.F_{cond} \qquad (3-5)$$

where F_{cond} is the conduction flux from the corona back into the photosphere of the disc. (We remark here that in the original demonstration of the instability the radiative interaction between the corona and the photosphere was included in the calculations). Let Z_{cond} be the height at which the temperature attains its maximum value. Then the divergence of the conduction flux is approximately given by $\kappa_{cond}T^{7/2}/Z_{cond}$ and hence to the extent that the transition region from the photosphere to the corona is all in the optically thin region and takes place over a distance much smaller than a pressure scale height (so that the pressure can be assumed constant) we have

$$C\left(\frac{\mu m_H}{k}\right)\rho^2 T^{\beta-11/2} - \frac{\kappa_{cond}}{Z_{cond}^2} \approx 4/3\alpha\Omega T^{-5/2} \qquad (4-5)$$

This equation provides an estimate for the maximum temperature of the corona. The purpose of this derivation is only to show that the instability exists with and without the radiative transfer equation (as used by Shaviv and Wehrse). Only a proper treatment of the conduction and the radiative transfer will provide the exact value for the temperature of the corona.

The instability found by Adam et al. has to do with the solution of equation (2-3) for the temperature stratification. The particular behavior of the opacity leads to a situation where the equation has several solutions for the temperature. Furthermore, the solutions for the temperature in not a continuous function of the variables (i.e. the height). We find that the solutions for the temperature in two neighboring layers can be very different and the gradient in the temperature very large. Obviously, electron conduction must be included in this case (which we did not include in the calculation reported here) in order to derive the consistent structure of the corona along with the photosphere of the disc. Detailed calculations show that the maximum temperature of the corona is few $10^5 K$ and the optical depth is less then 10^{-4}. In view of these results we ignore the contribution of the corona to the continuum radiation calculation reported here. This is the reason for the particular choice for the initial value of the optical depth taken in the integration for the hydrostatic equation.

6. Comparison with Observations.

Cataclysmic systems with an accretion disc have several radiation sources on top of the accretion disc itself. The white dwarf, the red dwarf, the hot spot, the boundary layer and even the stream of mass itself all contribute to the total radiation coming from these objects. The separation between the different contributions requires the use of the theory and hence it is complicated to use such systems to check the prediction of an accretion disc model. The situation is better is systems with high accretion rate where the disc dominates the emerging radiation from the system. For this reason we compare in this paper the theoretical model with high accretion rate systems. Nova-like variables are known to have a high accretion rate. Of particular importance in validating the theory and fitting the observed spectra is the use of a wide range of wavelengths. The contribution of certain parts of the disc to a particular wavelength range may be more important than other parts. However, all the parts of the disc contribute to all wavelengths. Hence, a fit over a wide wavelength range is necessary, if a good agreement between the theory and observation is desired.

The following the system RW Sex has been observed over a wide frequency range and the results assembled by Haug (1987). Hence we examine the continuum energy distribution from this system assuming that all the observed flux is due to the accretion disc. There is good reason to suppose that the energy emitted by some of the 'other' sources in the system may depend on the total accretion rate. For example,the total emissivity of the hot spot is probably related to the total accretion rate. However, the hot spot is likely to contribute in the IR if at all. Thus one should be prepared to see deviations in the IR. Fortunately, these are not seen in this systems and we find a good agreement over the entire observed range of frequencies. The analysis of lines and other details will be carried out elsewhere.

Haug (1987) presented data of UX-UMa systems over a wide range of wavelengths. The data so collected refers to the stationary state and hence can serve as test for the theory of accretion discs. The nova-like systems were studied in great detail since 1965 and reviews are given by Wargau et al (1983), Haug and Drechsel (1985), Greenstein and Oke (1982).

The data presented by Haug was composed of observations in the UV (IUE low dispersion spectra (1200-2900A), optical spectroscopy (4000 -10,000A) and the near IR (JHK(L) data). Typical error of the optical and the IR photometry is estimated at $0^m.02$ and $0.^m05$ respectively.

Figs. 9a,b present the comparison of the present model with the observed energy distribution from the system RW Sex. The comparison is carried out with two accretion rates. As can be seen, an assumed accretion rate of $10^{17}g/s$ produces a good agreement between the theory and the observations over a wide range of wavelengths. The high accretion rate $10^{18}g/s$ does not reproduce the observed data as well.

The particular shape of the energy distribution and the maximum near 400A are due to several effects, the most important ones being the decrease in the opacity for wavelengths below the Lyman jump and the peak in the black body distribution corresponding to the high effective temperatures (40,000 to 70,000K) in the inner parts of the discs.

The good fit in the UV range of $\lambda = 1000-3000A$ is amazing. What is the source of this good fit? As the effective temperature in the disc increases the maximum

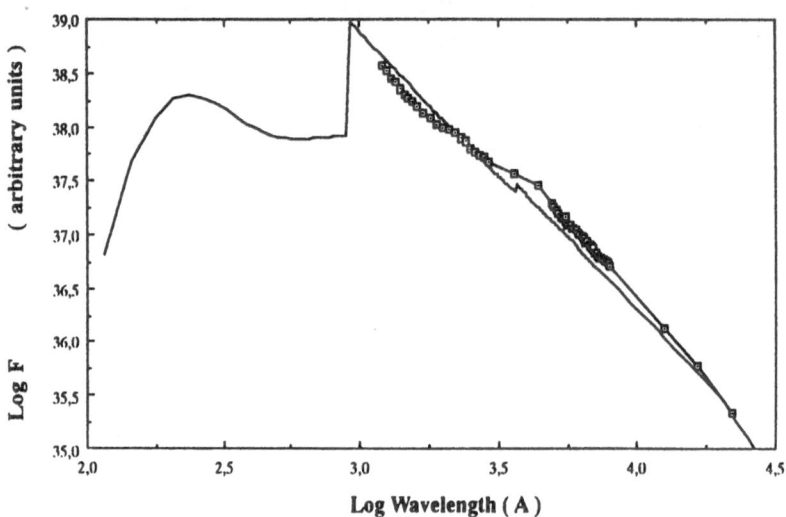

Figure 9a: Comparison between the predicted energy distribution (full curve) and the observed one (squares, according to Haug, 1987) for RW Sex. The theoretical model assumes $\dot{m} = 10^{18}$g/s, $\alpha = 0.5$, a white dwarf mass of $0.8 M_{sun}$ and an outer radius of $15 R_{wd}$.

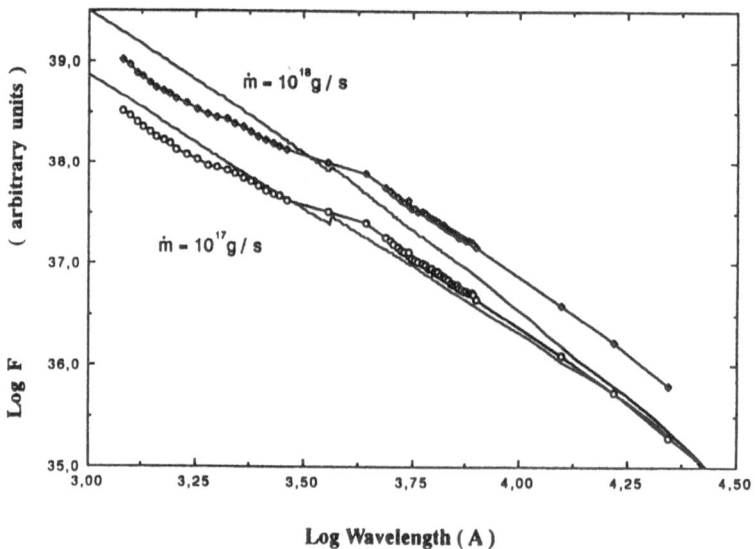

Figure 9b: Comparison of the observed energy distribution of RW Sex (after Haug) with two theoretical calculations of different accretion rates.

of the energy distribution shifts towards lower wavelengths. The opacity below the Lyman edge is very high. The only way to conserve energy is to shift photons from the high frequencies above the Lyman edge to low frequencies below it and thus fill the range above these strong absorption edge. Note also the carbon edge which for some radii is very important. We gave in Fig. 6 and 7 a set of energy distributions for single rings. We note the very sharp cutoff at the absorption edges and the flatness of the individual distributions. The details of the redistribution depend on the structure of the photosphere and cannot be estimated in a simple way.

Of interest is the invisibility of the Balmer jump. When the radiation field is inconsistent with the hydrostatics frequently the Balmer jump is seen in the theoretical energy distribution (cf LaDous 1988, Kriz and Hubeny 1987). The observations do not seem to show such a jump.

The model for the continuum radiation from accretion discs presented here succeeds to predict the continuum energy distributions observed from three nova like variables. The fit is obtained over a large range of wavelength and actually covers all the observed range. Haug (1987) used a somewhat higher rate ($3.910^{17} g/s$) and hence gets a disc luminosity which is also higher by the same factor.

ACKNOWLEDGEMENTS. This work was supported by the Deutsche Forschungsgemeinschaft (SFB 328).

REFERENCES.

Adam, S., Stoerzer, H., Wehrse, R. and Shaviv, G. 1987, Astron. Astrophys. 193, L1
Bath, G.T., Evans, W.D., Papaloizou, J. and Pringle, J.E. 1974, M.N.R.A.S. 169, 447
Begelman,M.,C., McKee, C.,F., and Shields, G., A. 1983 Astrophys. J. 271,70
Begelman,M.,C., McKee, C.,F. 1983 ,Astrophys. J. 271,89
Frank, J., King,A.R. and Raine,D.J. 1985 Accretion power in astrophysics, Cambridge University Press
Greenstein, J.L. and Oke,J.B. 1982, Astrophys. J. 258, 209
Haug, K. 1987, Astrophys. Space Science 130, 91
Haug, K. and Drechsel, H. 1985, Astron. Astrophys. 151, 157
Herter, T. Lacasse, M.G., Wesemael, F. and Winget, D.E. 1979 Astrophys. J. Suppl. 39, 513
Kalkofen, W., Wehrse, R. 1984 in Methods in radiative transfer, W. Kalkofen,ed., Cambridge University Press, p. 307
King, A., R., and Shaviv, G. 1984, Nature 308, 519
Kiplinger, A.L. 1979, Astrophys. J. 234, 997
Kiplinger, A.L. 1980, Astrophys. J. 236, 839
Kriz, S. and Hubeny, I. 1987, Astrophys. Space Science 130, 341
LaDous, C. 1986, in Cataclysmic Variables,NASA/CNRS Monographs Series on Nonthermal Phenomena in Stellar Atmospheres, M.Hack, ed.,
LaDous, C., 1989, Astron. Astrophys. 211, 131
Liang,E.P.T. and Price, R.H. 1977, Astrophys. J. 218, 247
Lynden Bell, D. 1969, Nature, 233, 690
Mayo, S.K., Wickramasinghe, D.T. and Whelan, J.A.J. 1980, M.N.R.A.S. 193, 793
Pacharintanakul, P., and Katz, J.I. 1980, Astrophys. J. 238, 985
Peraiah, A. 1984 in Methods in radiative transfer, W. Kalkofen,ed., Cambridge University Press, p. 281

444

Pringle, J.E. and Rees, M.J. 1972, Astron. Astrophys. 21, 1
Regev, O. 1983, Astron. Astrophys. 126, 146
Rudiger 1987, Acta Astronomica 37,223
Schwarzenberg-Czerny, A. and Rozyczka, M. 1977, Acta Astron. 27, 429
Schwarzenberg-Czerny A. 1981, Acta Astronomica 31,3
Shakura, N.I., and Sunyaev,R.A. 1973, Astron. Astrophys. 24, 337
Shaviv, G. and Wehrse, R. 1986, Astron. Astrophys. 159, L5
Tylenda, R. 1977, Acta Astron. 31, 267
Wade, R. 1984, M.N.R.A.S. 208, 381
Wade, R. 1987 IAU Colloq. 95, Faint Blue Stars
Wade, R. 1987, M.N.R.A.S. 208,381
Wehrse, R. 1981, M.N.R.A.S. 195, 553
Wehrse, R. and Shaviv, G, 1988 in preparation
Warner, B. 1987, M.N.R.A.S. 227,23
Wargau, W., Drechsel, H., Rahe,J., and Bruch,A. 1983, M.N.R.A.S. 204, 35
Williams, R.E.,1980, Astrophys. J. 235, 939
Williams, R.E., and Ferguson, D.H. 1982, Astrophys.J. 257, 672

INFLUENCE OF RADIATIVE TRANSFER ON THE VERTICAL STRUCTURE OF ACCRETION DISKS.

Ivan HUBENY

High Altitude Observatory, National Center for Atmospheric Research[1]
Boulder, CO 80307-3000, U.S.A.

ABSTRACT. A brief review of various theoretical approaches to model accretion disks is presented. Emphasis is given to models that determine self-consistently the structure of a disk together with the radiation field. It is argued that a proper treatment of the vertical structure is essential for calculating theoretical spectra to be compared with observations. It is demonstrated that neither the blackbody nor model stellar atmosphere fluxes provide a satisfactory approximation of the emergent radiation. Finally, several models calculated with various degree of sophistication in treatment of interaction of radiation with matter, and with different values of viscosity, are intercompared, and possible diagnostic consequences are briefly discussed.

1. Spectroscopic Diagnostics of Accretion Disks.

The knowledge of the vertical structure of accretion disks is a crucial factor that determines the quality of corresponding spectroscopic diagnostics. Most studies assume that the radial structure is given by the canonical model (Shakura and Sunyaev 1973; Lynden-Bell and Pringle 1974; Pringle 1981; for a recent review see Frank, King, and Raine 1985), which assumes a cylindrically symmetric, stationary Keplerian disk. In contrast, the vertical structure is treated with a very varying degree of sophistication.

Two early approaches avoid calculation of vertical structure completely: The theoretical emergent spectrum is calculated either as a superposition of black body fluxes corresponding to the local effective temperature given by the canonical model (Lynden-Bell 1969); or by using model stellar atmosphere fluxes instead of black body fluxes (Mayo, Wickramasinghe, and Whelan 1980; Wade 1984). Both these approaches have recently been tested by Wade (1988), who has shown that neither one reflects the physics of disks properly. He based this conclusion on a comparison

---

[1] The National Center for Atmospheric Research is Sponsored by the National Science Foundation.

F. Meyer et al. (eds.), Theory of Accretion Disks, 445–456.

between theoretical predictions and observations. There are, however, also several fundamental objections to the above approaches, namely a) a disk does not have to be optically thick in the vertical direction, which is the implicit assumption underlying both approaches; b) the effective temperature is not the only parameter that determines the emergent radiation; and c) a disk is neither a vertically homogeneous layer in thermodynamic equilibrium (which is needed to get blackbody emergent radiation), nor does it have the same vertical structure as a classical stellar atmosphere, particularly due to dissipation of mechanical energy in the disk.

The next category of approaches comprises models constructed assuming optically thin, vertically homogeneous disks. The emergent spectrum is then calculated by solving a simplified radiative transfer; the optically thin approximation enables one to write analytical expressions for the emergent flux (Williams 1980; Tylenda 1981; Williams and Ferguson 1982; Williams and Shipman 1988). The obvious drawback of these methods is that a disk need not generally be optically thin.

Finally, there are several studies that calculate the vertical structure. Some of them do not treat the internal radiation field in detail, using basically the diffusion approximation for radiation transfer (Meyer and Meyer-Hofmeister 1982, 1983; Cannizzo and Wheeler 1985; Cannizzo and Cameron 1988). Vertical structure of accretion disks, with self-consistent treatment of radiative transfer, has been calculated by Kříž and Hubeny (1986), Shaviv and Wehrse (1986), and by Adam et al. (1988).

2. Vertical Structure of Accretion Disks; Comparison to Stellar Atmospheres.

The general equations governing the vertical structure of individual rings are the following: Assuming that the radial component of gravity of the central star is balanced by the centrifugal force of the Keplerian rotation, and neglecting self gravity of the disk, one may write the equation of vertical hydrostatic equilibrium as

$$\frac{dP}{dz} = -g(z)\rho , \qquad (1)$$

where z is the geometrical distance from the central plane, P and ρ are the total pressure and density, respectively, and

$$g(z) = Qz , \quad Q = Gm_*/R^3 , \qquad (2)$$

is the effective vertical gravity. G is the gravitational constant, m_* is the mass of the central star, and R the distance of the given disk ring from the central star. Here the neglected terms are of the order of $(z/R)^2$. In contrast to stellar atmospheres, the gravity now depends on the vertical coordinate z.

It is advantageous to introduce the mass-depth variable $m(z) = \int_z^\infty \rho dz$. The total column mass, $M = m(0)$, is related to the usual surface density Σ through

$M = \Sigma/2$. Equation (1) may then be simply written as $dP/dm = g$, but due to the dependence of g on z no simple integral of this equation can be written.

The second basic equation is the energy balance equation,

$$D_{mech} = D_{rad} + D_{conv}, \tag{3}$$

which states that the energy dissipated in unit volume, D_{mech} is equal to the net radiation (plus, in general, convective) loss per unit volume. Assuming that energy is released through the radial shear of the Keplerian motion, the former is given by

$$D_{mech} = (9/4)Qw\rho, \tag{4}$$

where w is the (generally z-dependent) kinematic viscosity. The net radiation loss is generally given by

$$D_{rad} = 4\pi \int_0^\infty [\eta(\nu,z) - \chi(\nu,z)J(\nu,z)]d\nu, \tag{5}$$

where η and χ are the monochromatic emission and absorption coefficients, respectively, J is the mean intensity of radiation, and $D_{conv} = dF_{conv}/dz$, F_{conv} being the vertical convective flux; $F_{conv} = 0$ in convectively stable layers.

Since the radiation enters the energy balance equation explicitly, the third basic equation of the problem is the radiative transfer equation, written as

$$\mu\frac{\partial I(\nu,\mu,z)}{\partial z} = -\chi(\nu,z)I(\nu,\mu,z) + \eta(\nu,z), \tag{6}$$

where $I(\nu,\mu,z)$ is the specific intensity of radiation at frequency ν; μ is the cosine of the polar angle.

The corresponding boundary conditions are as follows: The lower boundary condition expresses the symmetry of the disk with respect to the central plane,

$$I(\nu,\mu,m = M) = I(\nu,-\mu,m = M) . \tag{7}$$

The upper boundary condition reads

$$I(\nu,-\mu,m = 0) = I^{ext}(\nu,\mu) , \quad \mu > 0. \tag{8}$$

where I^{ext} is a prescribed incident (generally angle-dependent) intensity of external radiation. One usually assumes $I^{ext} = 0$, but in some cases there may be an appreciable external irradiation of the disk by the central star; this case will be briefly discussed in Section 5. These equations are complemented by definition equations for the absorption and emission coefficients, which in turn require the corresponding equations for the relevant atomic level populations.

Summarizing, there are three basic differences from the classical stellar atmospheres case: i) z-dependent gravity; ii) the presence of the mechanical energy dissipation term D_{mech} in the energy balance equation; consequently the total radiation (plus generally convective) flux is not constant; and iii) the optical thickness is neither infinite, nor an a priori known finite quantity. Instead, its value should be determined self-consistently with other structural parameters.

On the other hand, the system of basic structural equations describing the vertical structure of a disk is rather similar to that for stellar atmospheres, so that conceptually similar numerical methods may be used in both cases. Also, the spectroscopic diagnostics of accretion disks may, and should, profit from experience gained during several decades of studying stellar atmospheres. As an illustration I mention the work of Kříž and Hubeny (1986), who solved the above set of disk structural equations by a modification of the complete linearization method devised originally for classical stellar atmospheres (Auer and Mihalas 1969; Mihalas 1978).

3. Simple Analytical Model for the Temperature Structure.

As in the case of classical stellar atmospheres, simplified models constructed using some appropriately defined frequency-averaged opacities serve both as a guide to the general nature of the results to be expected from more realistic calculations, as well as as an excellent starting approximation for a subsequent iterative method, as for instance the complete linearization. Since one usually makes an approximation of the equality of different mean opacities, these models are often referred to as *gray* models.

Gray models of accretion disks were constructed by Kříž and Hubeny (1986), who used them as a starting solutions for their method, by Shaviv and Wehrse (1986), and by Adam et al. (1988). All these authors solved the gray problem numerically. Nevertheless, it is possible to derive some simple analytic expressions that represent a generalization of corresponding classical stellar atmospheric results.

The frequency integrated moments of the transfer equation may be written, assuming LTE, as

$$\frac{dH}{dm} = \kappa_J J - \kappa_B B , \qquad \frac{dK}{dm} = \kappa_H H, \qquad (9)$$

where J, H, K are the frequency-integrated moments of the specific intensity, $B = (\sigma/\pi)T^4$ is the frequency integrated Planck function; σ being the Stefan-Boltzmann constant. Further, κ_J, κ_B, κ_H are the usual absorption mean, Planck mean, and flux mean opacity, respectively (Mihalas 1978). All mean opacities are defined here as opacities per unit mass. Notice also that while the absorption and Planck means are defined through the *true* absorption coefficient, the flux mean is defined through the *total* extinction (absorption + scattering) coefficient.

Using Eqs. (4), (5), (9), and neglecting convection, the energy equation (3) may be written

$$\frac{dH}{dm} = -E\,w(m)\,, \quad E = (9/16\pi)Q\,, \tag{10}$$

where we indicate that viscosity is generally depth-dependent. Note that in the case of classical stellar atmospheres Eq. (10) simply reads $dH/dm = 0$. In other words, the total vertical radiation flux is not conserved in the accretion disks, in contrast to classical stellar atmospheres. The solution of Eq. (10) is

$$H(m) = EM\bar{w}[1 - \theta(m)], \tag{11}$$

where

$$\theta(m) = \int_0^m w(m')dm'/(\bar{w}M)\,, \quad \bar{w} = \int_0^M w(m)\,dm/M\,. \tag{12}$$

θ is a monotonically increasing function between $\theta(0) = 0$ and $\theta(M) = 1$; in the case of depth-independent viscosity $\theta(m) = (m/M)$.

Combining Eqs. (9) and (10) one gets after some straightforward algebra

$$T^4(m) = \frac{3}{4}T^4_{eff}\left\{\gamma_J[\tau_H - \tau_\theta + \gamma_H/\sqrt{3}] + \frac{1}{3M\kappa_B}\frac{w(m)}{\bar{w}}\right\} \tag{13}$$

where the effective temperature is defined by

$$(\sigma/4\pi)T^4_{eff} \equiv H(0); \quad i.e. \quad T_{eff} = (4\pi EM\bar{w}/\sigma)^{1/4}\,, \tag{14}$$

γ_J and γ_H are factors of the order of unity, defined by

$$\gamma_J = \kappa_J/(3\kappa_B f_K)\,; \quad \gamma_H = \sqrt{3}\,f_K(0)/f_H \tag{15}$$

where $f_K = K/J$ and $f_H = H(0)/J(0)$ are the Eddington factors, τ_H is the optical depth associated with the flux mean opacity; and

$$\tau_\theta = \int_0^m \kappa_H(m')\theta(m')dm'. \tag{16}$$

may be called the "viscosity-weighted" flux-mean optical depth.

Equation (13) is an essentially exact LTE temperature stratification within a disk. However, it is quite formal because the quantities $\gamma_J, \gamma_H, \tau_H, \tau_\theta$, and κ_B are not a priori known. In order to provide a physical insight, we invoke several further approximations: i) We put $\gamma_J = \gamma_H = 1$, which follows from two separate approximations, namely the Eddington approximation $f_K = 1/3$, $f_H = 1/\sqrt{3}$; and the equality $\kappa_J = \kappa_B$, which is usually a good approximation (Mihalas 1978). Notice however that the equality $\gamma_H \simeq 1$ is a good approximation only in the case of no or weak incident radiation (see Section 5). ii) We assume that $\tau_H = \tau_R$, τ_R being the Rosseland mean opacity. Next, we write $\kappa_B = \epsilon\kappa_R$, where ϵ may roughly be understood as an average of κ_ν/χ_ν, where κ_ν represents the true absorption, while χ_ν the total extinction (absorption + scattering) coefficient. iii) Next we assume that the mean opacities and viscosity vary slowly with m, which yields $\tau_\theta \simeq (1/2)\tau_R^2/\tau_{tot}$ and $M\kappa_B \simeq \epsilon\tau_{tot}$, where τ_{tot} is the total Rosseland optical depth of the disk midplane (i.e. the total optical thickness of the disk is $2\tau_{tot}$). Equation (13) then reads (writing τ for τ_R)

$$T^4 = \frac{3}{4}T_{eff}^4 \left[\tau(1 - \tau/2\tau_{tot}) + 1/\sqrt{3} + \frac{1}{3\epsilon\tau_{tot}} \frac{w(m)}{\bar{w}} \right] \tag{17}$$

This result is a generalization of the well-known gray temperature distribution for stellar atmospheres, and reveals several interesting features.

Consider first the case $\tau_{tot} \gg 1$, i.e. the optically thick ring. Then for $1 \leq \tau \leq \tau_{tot}$, the factor ϵ is likely to be of the order of unity. The last term of (17) is then negligible, and one gets

$$T^4 = \frac{3}{4}T_{eff}^4(\tau + 1/\sqrt{3}) \quad for \quad \tau > 1, \quad \tau_{tot} \gg 1 \tag{18}$$

i.e. the temperature distribution is *identical* (within the Eddington approximation) to the stellar atmospheric distribution. This explains why the emergent spectrum from an optically thick ring of a disk may reasonably be approximated by the stellar atmospheric spectrum. However, the temperature at low optical depths $\tau \ll 1$ may still differ significantly from the stellar atmospheric temperature, basically due to the presence of the last term in Eq. (13) or (17) - see next Section.

In the limit of very low total optical depth, $\tau_{tot} \ll 1$, we get

$$T = T_{eff} \left[w(\tau)/(4\epsilon\tau_{tot}\bar{w}) \right]^{1/4}, \tag{19}$$

which reveals two interesting facts. Firstly, the local temperature may be much higher than the effective temperature and, secondly, the local temperature need not necessarily be constant, as it is usually assumed, due to the presence of the terms ϵ and $w(\tau)/\bar{w}$ which may depend strongly on depth.

Another interesting feature is the midplane temperature in the effectively thick rings. Assuming $\epsilon \simeq 1$ and $w/\bar{w} \simeq 1$, one gets for $\tau_{tot} > 1$ the midplane temperature $T = (3/8)^{1/4}T_{eff}\tau_{tot}^{1/4}$, while the atmospheric value at the same depth is roughly $T = (3/4)^{1/4}T_{eff}\tau^{1/4}$, i.e. the disk midplane temperature is lower than the corresponding atmospheric one. This is explained by the fact that the radiation from the midplane may escape equally well to both sides of the disk. In other words, the total radiation flux at the midplane is zero, in contrast to stellar atmospheres. This phenomenon is important for moderately optically thick disks, since for very large optical thickness the regions $\tau \simeq \tau_{tot} \gg 1$ do not appreciably influence the emergent radiation and hence are not very interesting from the point of view of spectroscopic diagnostics.

4. Question of High-Temperature Surface Layers.

Letting $\tau \to 0$, and assuming first a depth-independent viscosity, we get from (17)

$$T(0) = T_{eff} \left[\sqrt{3}/4 + 1/(4\epsilon\tau_{tot}) \right]^{1/4}. \tag{20}$$

Since τ_{tot} is basically given by the global properties of the disk (above all by the mass accretion rate $\dot{m}_*$ and the distance from the central star R), and is thus fixed for a given disk ring, the crucial factor that determines the surface temperature is ϵ.

At first sight, one may argue that ϵ goes to zero for $\tau \to 0$. Indeed, adopting the view that ϵ represents an average of $\kappa_\nu/(\kappa_\nu + \sigma_\nu)$; κ_ν and σ_ν being the true absorption and scattering coefficients, respectively; κ_ν decreases with τ more quickly than σ_ν (since σ_ν is proportional to n_e, the electron density, while κ_ν is roughly proportional to n_e^2). The total opacity is then more and more dominated by electron scattering, and consequently ϵ goes to zero. As a result, the temperature increases indefinitely! Physically, this temperature increase is caused by the fact that energy is dissipated at all depths (recall the assumption of depth-independent viscosity), but the gas at small optical depths only *scatters*; it does not absorb or emit radiation, i.e. it possesses no efficient mechanism to reradiate the dissipated energy.

This is precisely the argument raised by Shaviv and Wehrse (1986) and Adam *et al.* (1988), although they used somewhat different language. They also suggested a way to avoid an indefinite temperature rise, which can be easily seen in our formalism: The critical last term of Eq. (17) contains $w/\bar{w}$, so that letting $w(m)/\bar{w}$ decrease with m more rapidly than ϵ prevents the "thermal catastrophe" of the disk.

However, the argument of Shaviv and Wehrse and Adam *et al.* is not quite correct. The crucial point is that the intuitive interpretation of ϵ as an average of the ratio of pure absorption to total extinction is incorrect. The correct definition of ϵ follows from Eq. (13), namely that $\epsilon\tau_{tot} = \kappa_B M$. The Planck mean opacity is determined predominantly by frequency regions of *high opacity*, i.e. by strong resonance lines. Physically this means that regions with $\tau \ll 1$ are not poor emitters as argued by Shaviv and Wehrse. These layers may in fact emit considerable amount of radiation energy in strong resonance lines of abundant species, such as H I, Mg II, C II, Al III, Si IV, C IV, N V, O VI, etc., each line being an efficient cooler for a certain temperature range.

The above analysis was meant as a simple demonstration that the temperature need not necessarily increase to infinity for models with depth-independent viscosity. It is quite clear that in the superficial layers the LTE approximation is absolutely inadequate, and should be replaced by a more realistic NLTE approach. In any case, the question remains whether the line cooling will be efficient enough to remove the thermal instability by itself, or a decrease of viscosity with depth will still be needed. This can be solved only by future detailed calculations.

5. Influence of External Irradiation.

In case where the disk is irradiated by substantial external radiation, Eq. (8), the temperature structure given by Eq. (13) is formally correct, but the equality $\gamma_H \simeq 1$, used in Sections 3 and 4 is no longer a useful approximation. In this case, it is advantageous to consider the surface mean intensity $J(0) = J^{(0)}(0) + J^{ext}$, where $J^{ext} = \int_0^\infty \int_0^1 I^{ext}(\nu, \mu)d\nu d\mu$ is the zeroth moment of the external radiation. It can be shown (Hubeny 1989) that if the second Eddington factor is now defined as $f_H = H(0)/J^{(0)}(0)$, the useful approximate equality $\gamma_H \simeq 1$ is recovered. Putting, for illustration, $J^{ext} = WB(T_*)$, where T_* is the effective temperature of the central star, and W is the dilution factor (a function of the radius of the central star, the distance from the central star, and generally of the shape of the disk), we finally obtain an interesting expression

$$T^4(m) = \frac{3}{4}T_{eff}^4 \left\{ \gamma_J \left[\tau_H - \tau_\theta + \gamma_H/\sqrt{3} + 4f_K(0)W(T_*/T_{eff})^4 \right] + \frac{1}{3M\kappa_B} \frac{w(m)}{\bar{w}} \right\}$$
(21)

which represents a generalization of Eq. (13). In contrast to other studies that consider an irradiation of the disk by the central star (Friedjung 1985; Adams and Shu 1986; Kenyon and Hartmann 1987), the present one explicitly shows that the effect of irradiation depends not only on the radial distance from the central star, but also on the vertical distance from the central plane.

Indeed, if $W \ll 1$ and $T_{eff} \simeq T_*$, the additional term is unimportant, and the temperature distribution in the disk is unaffected by the external irradiation. If, in contrast, $(T_*/T_{eff}) \gg 1$, the irradiation is important. In the extreme case, the overall temperature distribution is dominated by external radiation, and we get from Eq. (21)

$$T \simeq W^{1/4}T_* , \quad for \quad \tau < W(T_*/T_{eff})^4,$$
(22)

i.e. the local temperature in the disk is influenced anywhere from the surface down to optical depths which depend on the dilution factor (and hence on the radial distance from the central star), and on the effective temperature of both the disk and the central star.

6. Some Numerical Results and Discussion.

Having obtained an analytic expression for the temperature - optical depth relation, one may proceed analogously as in the case of stellar atmospheres to construct simplified models of the vertical structure of accretion disks. Essentially, the problem consists in solving iteratively the hydrostatic equilibrium equation (1) together with the $T - \tau$ relation. If Eq. (17) is employed, the resulting model is called *gray* model. One may also consider the general LTE expression (13); in this case there is an extra iteration loop of solving the radiative transfer equation. Details of the numerical method are presented elsewhere (Hubeny 1989).

To compare results following from various approximations, I have calculated the following types of models for several representative disk rings:

i) LTE-gray models, with $T - \tau$ given by (17).

ii) "Improved gray" models, adopting Eq. (13) instead of (17) for temperature. Numerical procedure is basically a generalization of the Unsöld-Lucy method (Hubeny 1989).

iii) Exact LTE model, calculated by a modification of the complete linearization method, similarly as described by Kříž and Hubeny (1986). Compared to the latter study, I used here a substantially improved computer program called TLUS-DISK, which is basically a modification of the non-LTE stellar atmospheric program TLUSTY described in detail in (Hubeny 1988). Both programs are specifically devised to construct non-LTE models. Results of non-LTE accretion disk calculations will be reported in a future paper.

iv) For purposes of comparison, I have also calculated several LTE model stellar atmospheres, corresponding to the same basic input parameters as the disk models, and with same same chemical composition.

The global input parameters of the overall disk model are the radius and the mass of the central star, R_* and m_*; the mass flux through the disk, $\dot{m}_*$; and some suitable parameters that specify the viscosity. The given ring is then identified by specific R, the distance from the central star. Due to the lack of any reliable theory, the choice of the type of parametrization of viscosity is essentailly a matter of taste. I have adopted here the Reynolds number (also called β) prescription (Lynden-Bell and Pringle 1974; Williams and Ferguson 1982; Kříž and Hubeny 1986)

$$\bar{w} = (Gm_*R)^{1/2}/Re, \qquad (23)$$

where Re, the effective number of the accretion flow, is assumed to be a depth-independent, input parameter of the model. Since even less is known about the dependence of viscosity on depth, I adopt a simple power law paramatrization suggested by Kříž and Hubeny (1986),

$$w(m) = (\zeta + 1)\bar{w}(m/M)^\zeta, \qquad (24)$$

where ζ is another input parameter of the model. In all calculation reported below I adopted $\zeta = 2/3$; a comparison of models calculated with various ζ will be discussed in a separate paper.

The illustrative models are calculated for $m_* = 1M_\odot$, $R_* = 5 \times 10^8 cm$ (the corresponding radius of a white dwarf), with $\dot{m}_* = 10^{-11}$ or $10^{-10} M_\odot/year$. These are the typical values for a quiescent or near-quiescent state of a typical dwarf nova. The disk is assumed to be composed of hydrogen and helium, with $N_{He}/N_H = 0.1$

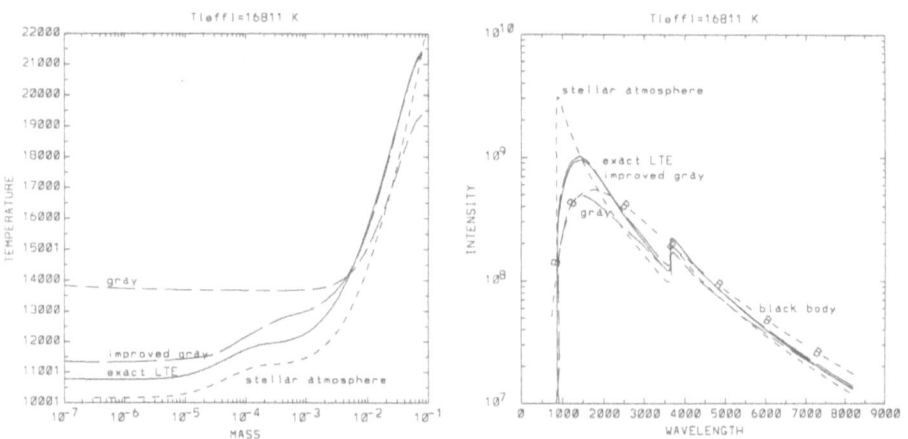

Figure 1a,b: Temperature (Fig.1a) and normally emergent intensity (Fig.1b) for various models of a ring with $R/R_* = 1.3611$ in a disk with $m_* = 1\ M_\odot$, $R_* = 5 \times 10^8 cm$, $\dot{m}_* = 10^{-11}\ M_\odot/year$, and $Re = 5000$. For comparison, the values for the LTE stellar atmosphere model with the same effective temperature are also displayed.

Figure 1a displays the temperature for models constructed with various degree of sophistication, for $\dot{m}_* = 10^{-11} M_\odot/year$, $R/R_* = 1.3611$ [$= (7/6)^2$, which is the distance where the canonical model predicts the maximum effective temperature], and $Re = 5000$. These values yield the effective temperature $T_{eff} = 16811$K, and the total Rosseland optical depth $\tau_{tot} = 2.8$, i.e. the ring is just marginally optically thick. The improved gray and exact LTE models exhibit an almost identical temperature in deeper layers, where most UV and visible continua are formed, and differ only in outer layers. The gray temperature distribution is significantly different; and the corresponding LTE stellar atmosphere temperature is systematically lower everywhere.

The corresponding normally emergent intensity is shown in Fig.1b, where also the black body intensity is displayed. It is readily seen that while the visible continuum is roughly the same for all models, the UV continuum is rather sensitive to the adopted model. In particular, the stellar atmospheric intensity is too high

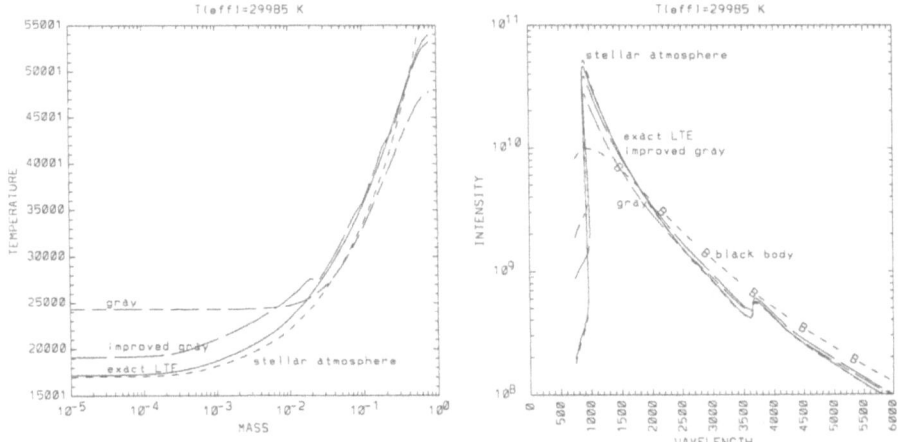

Figure 2a,b: The same as in Fig.1, but for $\dot{m}_* = 10^{-10}\ M_\odot/year$.

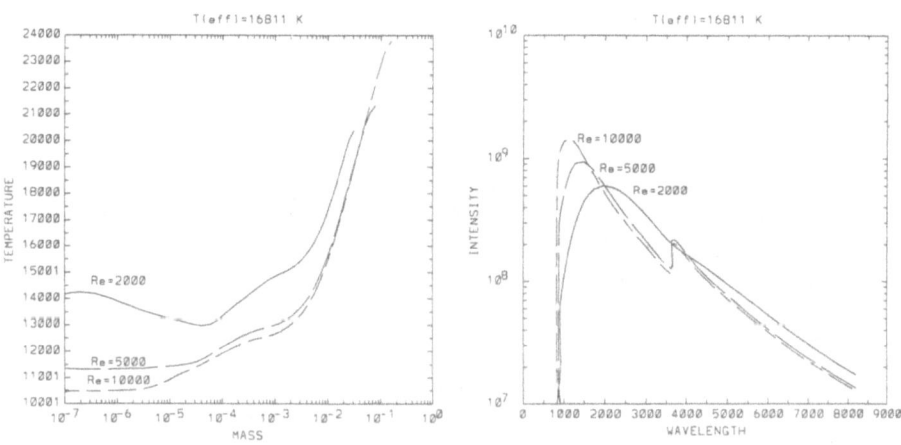

Figure 3a,b: The same as in Figure 1, but for various values of the Reynolds number.

betwen 1000 and 1500 angstroms, and too low for longer wavelengths, while the black body flux is too low shortward and too high longward of 2000 angstroms. This demonstrates quantitatively the conclusion discussed in Section 1 that both black-body and model atmosphere fluxes can be used only with great caution.

Analogous results for $\dot{m}_* = 10^{-10} M_\odot/year$, with otherwise the same input parameters as before, are displayed in Fig.2. Here, $T_{eff} = 29985$K and $\tau_{tot} = 17.3$. The ring is now reasonably optically thick; this explains why all models give similar results. However, the black body intensity again provides only a poor approximation of the emergent spectrum.

Finally, Fig.3 compares models calculated for the same parameters as in Fig.1, but with various values of the Reynolds number. All the displayed models are the improved gray models. Fig.3a demonstrates that the total column mass, and hence the total optical depth, decreases with decreasing Re, while the local temperature generally increases with decreasing Re. Perhaps more interesting is the corresponding emergent spectrum (Fig.3b), which exhibits quite a strong dependence on Re, particularly in the UV region. This result implies that there is a hope that a careful spectroscopic diagnostics of accretion disks based on self-consistent models of the vertical structure may shed some light into the long-standing problem of the nature and magnitude of viscosity in the accreton disks.

ACKNOWLEDGEMENTS. This work was supported in part by NASA grant ADP U-003-88 through the University of California, Los Angeles. I would like to thank Drs. M. Plavec, D. Hummer, and J.-P. Lasota for stimulating discussions, and I gratefully acknowledge the support allocated by NATO for the Workshop.

REFERENCES.

Adam, J., Störzer, H., Shaviv, G., and Wehrse, R.: 1988, *Astron. Astrophys.* **193**, L1
Adams, F. C., and Shu, F. H.: 1986, *Astrophys. J.* **308**, 836
Auer, L. H., and Mihalas, D.: 1969, *Astrophys. J.* **158**, 641
Cannizzo, J. K., and Cameron, A. G. W.: 1988, *Astrophys. J.* **330**, 327
Cannizzo, J. K., and Wheeler, J. C.: 1984, *Astrophys. J. Suppl.* **55**, 367
Frank, J., King, A. R., and Raine, D. J.: 1985, *Accretion Power in Astrophysics*, (Cambridge: Cambridge University Press)
Friedjung, M.: 1985, *Astron. Astrophys.* **146**, 366
Hubeny, I.: 1988, *Computer Phys. Commun.* **52**, 103
Hubeny, I.: 1989 (in preparation)
Kenyon, S. J., and Hartmann, L.: 1987, *Astrophys. J.* **323**, 714
Kříž, S., and Hubeny, I.: 1986, *Bull. Astron. Inst. Czech.* **37**, 129
Lynden-Bell, D.: 1969, *Nature* **223**, 690
Lynden-Bell, D., and Pringle, J. E.: 1974, *Mon. Not. R. A. S.* **168**, 603
Mayo, S. K., Wickramasinghe, D. T., and Whelan, J. A. J.: 1980, *Mon. Not. R. A. S.* **193**, 793
Meyer, F., and Meyer-Hofmeister, E.: 1982, *Astron. Astrophys.* **106**, 34
Meyer, F., and Meyer-Hofmeister, E.: 1983, *Astron. Astrophys.* **128**, 420
Mihalas, D.: 1978, *Stellar Atmospheres* (San Francisco: Freeman)
Pringle, J. E.: 1981, *Ann. Rev. Astron. Astrophys.* **19**, 137
Shakura, N. I., and Sunyaev, R. A.: 1973, *Astron. Astrophys.* **24**, 337
Shaviv, G., and Wehrse, R.: 1986, *Astron. Astrophys.* **159**, L5
Tylenda, R.: 1981, *Acta Astron.* **31**, 127
Wade, R. A.: 1984, *Mon. Not. R. A. S.* **208**, 381
Wade, R. A.: 1988, *Astrophys. J.* **335**, 394
Williams, R. E.: 1980, *Astrophys. J.* **235**, 939
Williams, R. E., and Ferguson, D. H.: 1982, *Astrophys. J.* **257**, 672
Williams, G. A., and Shipman, H. L.: 1988, *Astrophys. J.* **326**, 738

DUSTY DISKS AND THE INFRARED EMISSION FROM AGN.

E. Sterl PHINNEY
Theoretical Astrophysics, 130–33 Caltech
Pasadena, CA 91125 U.S.A.

ABSTRACT. The distortions inferred in the gaseous disks of active galaxies suggest that a significant, and possibly dominant fraction of the $1 - 1000\,\mu m$ radiation observed from AGN must be thermal emission from gas and dust heated by the central source. We report calculations of the growth and sublimation of dust grains in the outer parts of accretion disks appropriate to AGN. The thermal state of the gas undergoes a sudden change at the radius where the dust sublimates. The outer portion of the accretion disk radiates at $0.5 - 5\,\mu m$; free-free emission from gas whose dust has sublimated contributes to the flux at $0.5 - 2\,\mu m$. If this thermal emission dominates the flux from radio-quiet quasars, it naturally explains the frequency and depth of the universal minimum in νF_ν at $10^{14.5}\,Hz$. Free-free emission from the photoionized surface layers of the disk at larger radii produces a radio flux at $\nu \lesssim 10^{11}\,Hz$ comparable to that observed in radio-quiet quasars. The far-infrared and submillimeter emission from radio-quiet quasars and Seyfert galaxies is more naturally interpreted as reradiation by dust, than as nonthermal emission from the inner accretion disk.

1. Introduction.

It is generally accepted that emission from heated dust produces the steep far-infrared continua of Seyfert 2's and the IRAS warm galaxies. It is also generally acknowledged that the rapidly variable infrared emission in the BL Lacs and optically violently variable quasars must be produced by non-thermal processes. Both thermal and non-thermal emission evidently occurs in nature, in objects of comparable luminosities. The controversy over the nature of the emission in radio-quiet quasars is thus not over whether either emission process is physically possible or plausible. Both are, and both are probably present at a significant level in all objects. The question is simply which happens to predominate in radio-quiet quasars.

In most optically selected quasars, the $1–100\mu$ infrared luminosity comprises 10–50% of the bolometric luminosity (Sanders *et al.* 1989). Since the energy liberated per octave in radius in an accretion disk scales as r^{-1}, the high relative infrared luminosity requires that the ultimate source of energy for the infrared radiation be within a few times the inner radius of the accretion disk. If the

457

F. Meyer et al. (eds.), Theory of Accretion Disks, 457–470.

infrared radiation is emitted from a region comparable in size to that of its energy source, it must be nonthermal: blackbody limits on the source size at $1\,\mu$ and at $60\,\mu$ are respectively $> 0.1\,pc$ and $> 100\,pc$ for the most luminous sources in figure $1a$ and $> 0.003\,pc$ and $> 10\,pc$ for the least luminous. If the infrared radiation is thermal, and therefore emitted at large radii, it *must* be reprocessed energy transported to the emission radii by radiation or by mechanical means (e.g., a jet).

Dust heated locally by stars may contribute to some of the infrared emission from quasars. It is unlikely, however, that this dominates in the majority of sources. For stars to produce the typical infrared luminosity of $10^{13}L_{\odot} = 0.5\,M_{\odot}c^2\,yr^{-1}$ would require a star formation rate of *at least* $500\,M_{\odot}yr^{-1}$ (this lower limit assumes that only massive O stars are formed; a normal IMF would require a rate approximately 10 times higher). Over the $\sim 10^8\,yr$ lifetime of a quasar, $> 5 \times 10^{10} M_{\odot}$ of gas would have to be processed in massive stars. Dust would form from the metals produced. So much gas and dust, pushed to high latitudes by supernova explosions, would inevitably absorb and reradiate much of the luminosity from the central source. But the optical and ultraviolet radiation from quasars, variable on timescales $\lesssim 10\,yr$ (Usher 1978) must come from such a central relativistic source. Since its luminosity is comparable to the infrared luminosity, we conclude that reradiation from gas and dust heated by it would necessarily be *at least* comparable to anything contributed by stars. In what follows we therefore ignore the heat input from stars, except insofar as they provide a natural minimum dust temperature $\gtrsim 25\,K$.

Quasars and Seyfert galaxies appear to be located in galaxies amply supplied with interstellar medium. The disks of gas and dust in normal galaxies exhibit warps on all scales, and would intercept and re-radiate $\sim 10\%$ of the luminosity from a central source. Several lines of evidence suggest that the warps are even more severe in Seyferts and quasars, so that an even larger fraction of the central luminosity would be reradiated. Provided most of the reradiating material is located in an optically thick disk, the central source will never appear absorbed or reddened —either we have a clear line of sight, or the source is entirely occulted. This picture is thus consistent with the absence of reddening or absorption by broad emission line clouds in quasars.

We describe below the opacities and physical state of gas in the disks at distances from $10^{-3}\,pc$ to $10^4\,pc$ from the central source. Dust grains can form and grow in the disk within $1\,pc$. At much smaller radii, however, even graphite grains will sublimate. When this occurs, the gas loses its primary coolant, and heats until it reaches a new thermal equilibrium at $\gtrsim 10^4\,K$. The superposition of emission from radii inside and outside this transition point naturally explains the minimum in νL_{ν} at $\nu = 10^{14.5}\,Hz$ observed in most quasars. The characteristic scale length of dust in galaxies ($\sim 2-10\,kpc$) naturally explains both the frequency and steepness of the drop in νL_{ν} at $\nu \lesssim 10^{12}\,Hz$. The normalization of infrared luminosity relative to the UV and X-ray luminosities of quasars is consistent with expected covering factors and space-densities. Free-free emission from the photoionized zones on the illuminated surfaces of the disk naturally provides a flat-spectrum radio flux comparable to that observed in many quasars (Antonucci & Barvainis 1988), and may contribute to the optical continuum emission. It appears therefore that the emission at wavelengths $1 - 1000\,\mu m$ from Seyfert galaxies and quasars other than OVV's is naturally explained as thermal reradiation from the nuclear disk and the interstellar medium of the host galaxy. Although non-thermal emission from the central source may contribute in some objects at some times, a significant contribution from thermal emission seems unavoidable.

2. Dust and Warps.

Most previous discussions of dust and gas in the host galaxy of an active nucleus have concentrated on spherically symmetrical distributions of dust (Rees et al. 1969, Bollea & Cavaliere 1976, Barvainis 1987—the latter also considered conical sectors, or have concentrated on their effects on emission line clouds: Davidson & Netzer 1979, MacAlpine 1985 and references therein). A few have considered disk-like distributions (Begelman, McKee & Shields 1983, Begelman 1985, Shlosman & Begelman 1988), but have assumed the disks were planar and relatively thin, so that they intercepted only a small fraction ($\lesssim 1\%$) of the luminosity of the central accretion disk. Neither of these distributions is very likely. The gas and dust most probably lie in what could crudely be described as a heavily warped (and probably clumpy) disk. Beyond about a kiloparsec radius, the orbital time $t_{orb} \sim 10^8(r/3\,\text{kpc})$ yr is so long that gas captured or disturbed by a recent interaction with another galaxy will not have had time to settle into the preferred plane of the host galaxy's potential in a quasar lifetime ($\sim 10^8$ yr). Since quasars seem commonly to be involved in interactions or mergers (Hutchings 1983, MacKenty & Stockton 1984), large warps and streamers are to be expected. The warped disk of NGC 5128 (Centurus A) is a famous example. Moving inwards, warps on kiloparsec scales can be produced by counter rotating bars (Vietri 1986), by continuing infall of gas (Ostriker & Binney, 1989), and by Kelvin-Helmholtz instability as the disk rotates through a pressure-supported corona (Gunn 1979). Though such warps are difficult to detect in external galaxies which are not nearly edge-on, let alone in quasars, the gas at ~ 3 kpc in our own Galaxy is tilted by about 15° with respect to the plane defined at larger radii (Vietri 1984), and most well-studied Seyfert galaxies have strong kiloparsec-scale bars (Adams 1977).

The same warp-inducing processes can operate on scales of parsecs. The molecular torus extending from 2–8 pc from our Galactic center is tilted by $\sim 15°$ (in the opposite direction from the 3 kpc warp!) with respect to the plane defined by the stars. The parsec-scale dust torus in NGC 1068 (Antonucci & Miller 1985) has its axis at right angles to the kiloparsec-scale disk of gas and stars (Wilson & Ulvestad 1982), and the axes of similar tori inferred to exist in other Seyferts make *random* angles to the minor axes of the disks of their host galaxies (Unger et al. 1987, Haniff et al. 1988).

We conclude that warps are common and substantial enough to allow the nuclear continuum to illuminate the dust and gas on scales from 1–10^4 parsecs. The broad-line region (BLR) will be obscured by this dust over a range of viewing angles comparable to the covering factor of the dust disk. Except in the case that the narrow line gas is cospatial with the dusty disk in a symmetrical warp, it is difficult to prevent the narrow line region from being visible from some viewing angles when the BLR is obscured. This occurs for a fraction of sky of order the covering factor of the dust disk extending beyond the NLR: typically $\sim 0.05 - 0.1$. Since molecular clouds are larger than the $0.1 - 1\,pc$ scale of the BLR, the same fraction of quasars with obscured broad line regions would be expected even if the dust were in clouds at high latitudes rather than in a disk. These objects would appear as "quasar-2's" (by analogy to Seyfert 2's). Such objects have been rare in optical surveys, but appear common in infrared-selected samples (Sanders *et al.* 1988b). In radio samples they may masquerade as narrow line radio galaxies (Scheuer 1987, Barthel 1989).

We now examine the state of dust in a warped disk illuminated by the central accretion disk of a quasar, and its possible relevance to the infrared and submillimeter spectra of quasars. We postpone to section 5 a discussion of the vertical structure of an illuminated disk, and the expected optical and radio emission therefrom.

The equilibrium temperature T_g of dust grains of characteristic radius $a \gtrsim 20\text{\AA}$ at distance r from a radiation source of luminosity density L_ν ($erg\ s^{-1}Hz^{-1}$) is determined implicitly by

$$\frac{1}{16\pi r^2} \int_0^\infty L_\nu Q_{abs}(\nu)\, d\nu = \int_0^\infty \pi B_\nu(T_g) Q_{abs}(\nu)\, d\nu \tag{1}$$

where $Q_{abs}(\nu)$ is the absorption efficiency (cross section in units of πa^2), at frequency ν. Graphite grains with $a \sim 0.1\mu m$ are transparent to X-rays with $h\nu > 0.4\ keV$ and for $0.1\ keV < h\nu < 0.28\ keV$ (K-edge of Carbon), so $Q_{abs}(\nu) \propto a$ at those energies. At wavelengths $\lambda < 2\pi a$ where the grain is not transparent $Q_{abs} \simeq 1$. At longer wavelengths $\lambda > 2\pi a$ the grain becomes a weakly coupled antenna, $Q_{abs} \simeq (2\pi a/\lambda) f(\lambda)$ where

$f(\lambda)$ depends on the dielectric tensor and shape of the grains (fits for various grain compositions and shapes, and discussion of the Mie scattering range can be found in Martin 1978, Draine & Lee 1984, Wright 1987, and references therein). Observations of galactic dust indicate that outside resonances $f(\lambda) \propto \lambda^{1-\alpha}$, with $1 \lesssim \alpha \lesssim 2$ (Whittet 1988).

In quasars, most of the energy from the central source is emitted at frequencies for which $Q_{abs}(\nu) \simeq 1$, and reradiated at wavelengths $\lambda > 2\pi a$. Hence if we ignore heating and cooling of grains by collisions with atoms (generally weak), the temperature of directly illuminated grains is given approximately by

$$\frac{L}{16\pi r^2} \simeq \sigma T_g^4 \langle Q_{abs}(T_g) \rangle \tag{2}$$

where $\langle Q_{abs}(T) \rangle$ is the Planck-averaged absorption efficiency (cf. Draine 1981), $\langle Q_{abs}(T) \rangle \propto T_g^\alpha$. Crudely, therefore, $T_g \propto [L/(r^2 a)]^{1/(4+\alpha)}$. At a given distance from the source, the smallest grains in thermal equilibrium ($\sim 30\text{Å}$) will be about a factor of 2 hotter than the largest grains commonly considered ($\sim 0.3\,\mu m$).

The heat capacity of grains with $a \lesssim 20\text{Å}$ is so low that their temperature fluctuates, being significantly affected by the absorption of a single UV photon. Grains deeper in the dusty disk will not be exposed directly to UV radiation from the central source, but to longer-wavelength re-emission from shielding gas and dust. This shielded dust will have a temperature nearly independent of a, and slightly lower (by a factor $\sim (T_{re}/2500\,K)^{1/(4+\alpha)}$) than that of the directly illuminated grains of temperature T_{re}.

The emission from dust at temperature T_g at long wavelengths scales as $\epsilon_\nu \propto \nu^{2+\alpha}T_g$, peaks very sharply at $h\nu \sim (3+\alpha)kT_g$ ($\lambda \sim 30\,T_2^{-1}\mu m$, where $T_g = 100\,T_2 K$) and declines exponentially at higher frequencies. Except for the small grains of fluctuating temperature, which can contribute high frequency emission from regions where the equilibrium temperature is low, most emission at frequency ν will come from the radius where the equilibrium dust temperature $T_g \sim h\nu/(3 + \alpha)k$, i.e., from a radius $r \propto L^{1/2}\nu^{-(4+\alpha)/2}$, and the flux of radiation at that frequency will (for an isotropic central source) be proportional to the fraction of the sky at the central source covered by dust at the apppropriate radii. This radius-frequency scaling can cause curious effects. At frequencies where α is large (e.g., $1 < \lambda < 7\,\mu m$, where $\alpha \simeq 1.7$ — Whittet 1988), the temperature is nearly independent of radius, enhancing the probability of having a large warp at an appropriate radius, and hence a large sky covering factor and a large $\sim 3\,\mu m$ flux (see figure 2). This may be the cause of the "$3 - 5\,\mu m$ bumps" commonly observed in AGN (Edelson & Malkan 1986).

3. Outskirts of the Galaxy and Submillimeter Spectrum.

At low frequencies α is also large ($\alpha \to 2$ when 0-frequency isotropic conductivity dominates the dielectric tensor), so that the temperature changes only slowly with radius at the outskirts of the galaxy ($\sim 3 - 30\,kpc$) where the molecular gas distribution of galaxies rapidly becomes very patchy. The characteristic temperature at the outskirts of a typical galaxy would be $\sim 30 - 50\,K$ (see figure 1).

A dust layer will absorb UV flux incident at an angle θ to the normal of the layer if its column density $\Sigma > 10^{-2}\cos\theta$ g cm^{-2}. With a Galactic gas-to-dust ratio spread over a disk of radius r_{kpc} this column corresponds to a mass of gas $M_H \simeq 2 \times 10^8 \cos\theta r_{kpc}^2\,M_\odot$. The gas masses in nearby AGN are inferred from observations of CO (which of course depletion relates more directly to the dust than to M_H!) to be of order 10^8–$10^{10}\,M_\odot$ (Sanders, Scoville, & Soifer 1988a), and even in a young galaxy such as might surround a high redshift quasar it is unlikely that there would be more than $\sim 10^{11}\,M_\odot$ of processed gas. Consequently the disk will become optically thin beyond a few kpc (it could become thin at a smaller radius if most of the dust is clumped into clouds with $\Sigma \gg 10^{-2} \cos\theta$ g cm^{-2}, and such clouds at larger radii could preserve dust in neutral cores, but their covering factor would necessarily be very small).

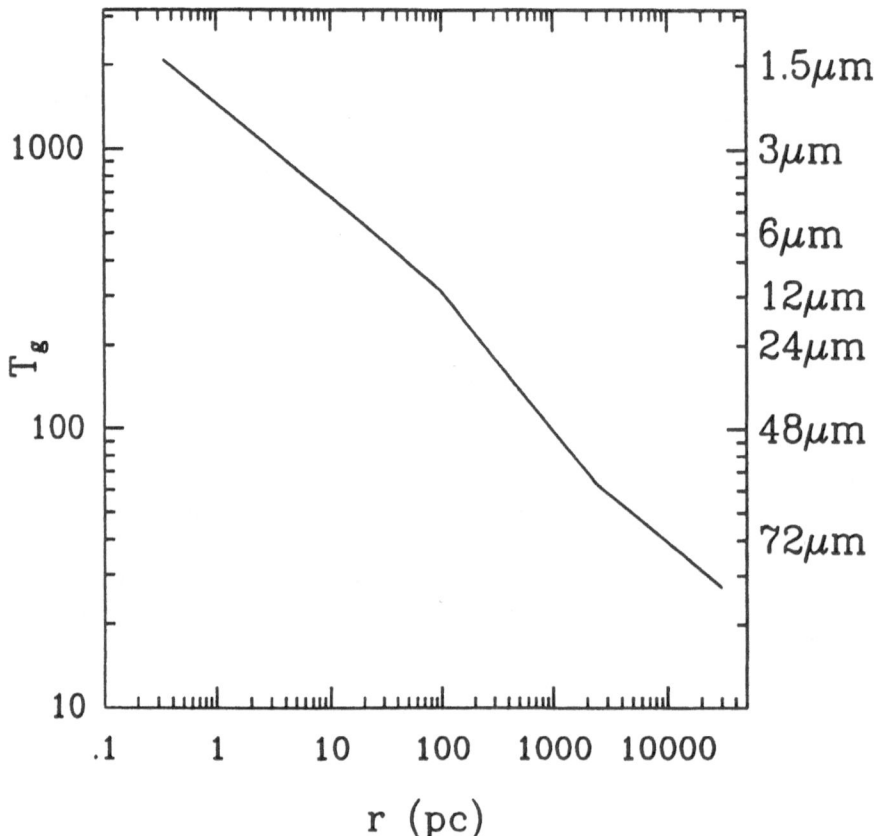

Figure 1: Temperature of directly illuminated 0.1μm graphite grains as a function of distance from a central UV source of luminosity $L_{UV} = 10^{46}$ $erg\,s^{-1}$. Planck-averaged absorption efficiencies used are from Draine & Lee 1984. For sources of other luminosities and grains of other sizes, the temperature in the flatter portions of the curve ($T > 300$ K, $T < 60$ K) scales roughly as $(L_{UV}/a)^{1/6}$, and in the steeper portion (60 $K < T < 300$ K) as $(L_{UV}/a)^{1/4}$. For a galaxy disk with a smooth logarithmic warp, dust at the given temperatures contributes predominately to the flux at wavelengths marked on the right.

Since the covering factor of dusty material at 20 K (~ 100 kpc) is expected to be very small, the spectrum of dust reradiation will be characterized by a rather well-defined minimum temperature, and should roll over sharply at wavelengths $\lambda \gtrsim 200\mu$m with $F_\nu \propto \nu^{2+\alpha}$ at longer wavelengths (as for similar reasons it is observed to do in starburst galaxies and Galactic HII regions -cf. Telesco & Harper 1980). The precise form of the spectrum at $60\mu m < \lambda < 200\mu m$ will vary depending on the covering factor at large radius, which could be enhanced by the presence of companion galaxies (whose starlight maintains a minimum dust temperature $\sim 20\ K$!), tidal tails, and the like.

As we discuss in section 5, at frequencies $\nu < 10^{11}$ Hz, free-free emission from photoionized gas at the illuminated face of the disk will dominate the spectrum. Figure 2 shows the spectrum of continuum reradiation from gas and dust in an exponential disk with a logarithmic warp (d(covering factor)$/d\ln r = const$). To illustrate how material at large radii can affect the far infrared and submillimeter spectrum, we show the effect of adding reradiation from a $2 \times 20\ kpc$ slab of dust extending from 10 to 30 kpc (which could represent a companion galaxy or a tidal tail).

4. Transition Disk.

Between $r \sim 1\ kpc$ and the black hole's accretion disk proper ($r \lesssim 10^3\ GM_h/c^2 \sim 0.03M_8\ pc$ for a black hole of mass $M_h = 10^8 M_8\ M_\odot$) lies the transition disk. Unlike the accretion disk, this disk will be heated primarily by external radiation rather than by the local release of gravitational binding energy. Beyond $r \sim 20\sigma_2^{-2}M_8\ pc$ the stars in the galaxy, of velocity dispersion $10^2\sigma_2 km\ s^{-1}$ determine the potential. A photoionized or molecular disk will be self-gravitating and Jeans unstable if its column density exceeds

$$\Sigma_{sg} \sim \Omega^2 h/G \sim 100T_3^{1/2}\sigma_2 r_{pc}^{-1}\ \text{g cm}^{-3}\ , \tag{3}$$

where h is the scale height of the disk and $T = 10^3\ T_3 K$ the gas temperature. If this gas flows in on a timescale t_{in} to supply the black hole with an accretion rate $\dot{M} = \dot{m}M_\odot\ yr^{-1}$, then $\Sigma \sim 1\dot{m}r_{pc}^{-1}v_{300}^{-1}\ (t_{in}/t_{orb})$g cm^{-2} where the orbital speed of the gas is $300v_{300}kms^{-1}$ and t_{orb} is the orbital time of gas at radius r. Purely local viscosity results in enormous inflow times, and Begelman, Frank and Shlosman (papers in these proceedings, and references therein) have argued that the angular momentum transport is determined by self-gravitation and global bar instabilities, so that the mean column density always exceeds Σ_{sg}. Dust in such a column has an enormous optical depth to UV photons: $\tau \gtrsim 10^4 T_3^{1/2}\sigma_2 r_{pc}^{-1}$. Provided the disk is warped (by any of the mechanisms described in section 2) or flared, infrared reradiation is inevitable. A disk warped through angle θ will intercept and reradiate a fraction $\sim \theta/3$ of the luminosity of the central source. For idealized dust of constant α, if $C = d$(covering factor)$/d\ln r \propto r^q$, then in the 'inertial range' ($kT_{min} \ll h\nu \ll kT_{max}$) the superposition of dust emission from all radii produces a reradiated spectrum with $L_\nu \propto \nu^{-s}$, where

$$s = 1 + q(4 + \alpha)/2\ . \tag{4}$$

464

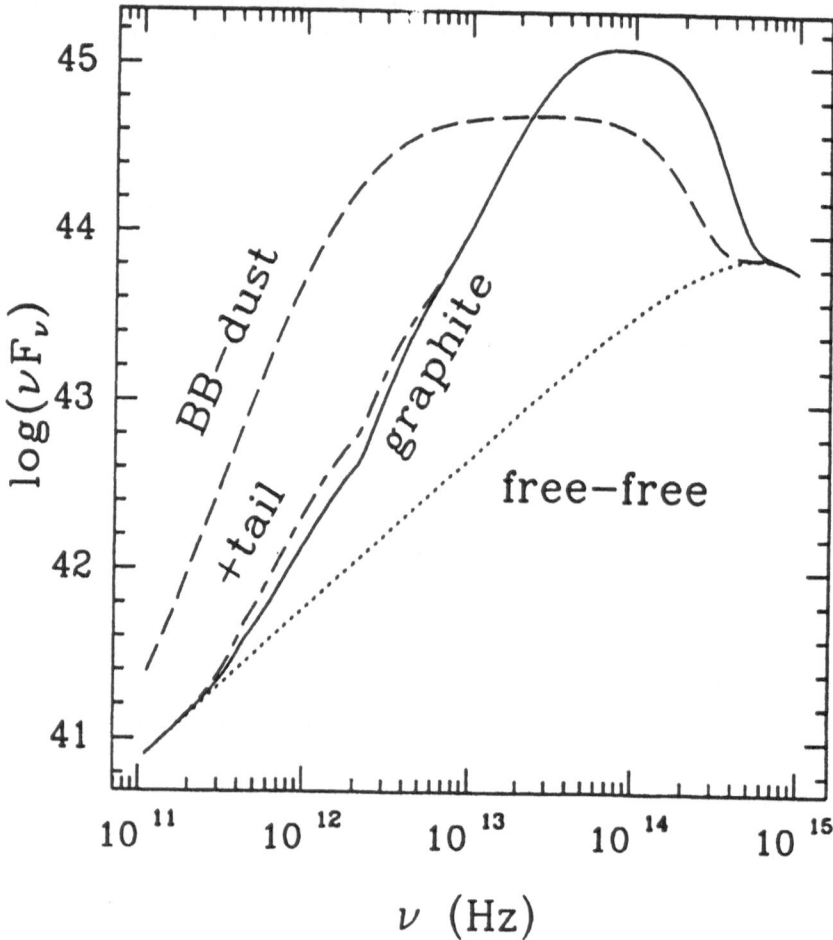

Figure 2: Spectrum of reradiation from dust and photoionized gas in a warped disk surrounding a quasar with $L_{UV} = 10^{46}$ erg s^{-1}. The solid line represents a disk containing 0.1μm graphite dust having $C = d(\text{covering factor})/d\ln r = 0.1\,exp(-r/10\,kpc)$. Note the "5$\mu$m bump." The dot-dashed line shows the result of adding 0.03 to C for $10 < r < 30\,kpc$—representing a tidal tail or companion galaxy. The dashed line shows the spectrum of reradiation from (very large) grains assumed to radiate as black bodies, with the *same* covering factor distribution as for the solid line. The dotted line shows the contribution of free-free emission from the photoionized zones above the dust. There is no freedom in its normalization relative to the dust spectra.

Since substantial warps are to be expected on all scales, we expect in some *average* sense $q \simeq 0$ and hence $s \sim 1$. A diversity of bumps and wiggles is to be expected in individual objects, depending on the radii (and hence, via figure 1, wavelength) where their warps are most pronounced, the size and composition of their dust grains, and whether they are viewed from an angle where dust at a large radius absorbs re-emission from the interior. This seems in accord with the infrared spectral energy distributions of AGN: a great variety, with a median $s \sim 1$ (Sanders *et al.* 1989, 1988b). As discussed in section 2, the properties of more realistic galactic dust modify slightly the simple result of equation (4), and by flattening the $T(r)$ relation introduces a propensity for $3-5\mu m$ "bumps" (see figure 2).

5. Surface Physics and Radio Emission.

It is fortunate that large column densities of material (equation 3) are expected close to the active nucleus, for there is an enormous radiation pressure on dusty gas directly exposed to a quasar's UV flux. The opacity of a single grain is $\sigma/m \simeq 2 \times 10^4 (0.1\mu/a) cm^2 g^{-1}$. If the grains are coupled by collisions or Lorentz forces to gas, then with a Galactic gas to dust ratio, the total opacity is $\kappa_{gr} \simeq 200\ cm^2 g^{-1}$ (roughly the cross section to mass ratio of an interstellar cloud with $A_v \sim 1$). This is nearly 10^3 times larger than the Thomson opacity $\kappa_T = 0.4\ cm^2\ g^{-1}$ which defines the Eddington limit, so radiation pressure will tend to expel unshielded dusty gas from all but the outer edges of the host galaxy. In the absence of forces other than radiation pressure and gravity, at radius r_{pc} pc gas in which dust survives will reach a terminal velocity

$$v_\infty \sim 10^4 L_{46}^{1/2} r_{pc}^{-1/2} \kappa_2^{1/2} (1 - 0.004 L_{46}^{-1} \kappa_2^{-1} M_8 - 2 \times 10^{-4} L_{46}^{-1} \kappa_2^{-1} \sigma_2^2 r_{pc})^{1/2} km\ s^{-1}$$

$$(5)$$

where $\kappa_{gr} = 10^2 \kappa_2\ cm^2\ g^{-1}$. The innermost

radius at which dust can survive (section 6) defines a characteristic velocity $v_{max} \sim 210^4 L_{46}^{1/4} \kappa_2^{1/2} km\ s^{-1}$.

The component of this radiation force normal to the disk surface, which must be balanced by vertical pressure gradients in a quasi-static situation, would destroy in a dynamical time any warp in a disk with $\Sigma < 50 r_{pc}^{-1} \sigma_2^{-2} \cos \theta$. Since this is generally less than Σ_{sg} (equation 3), warps can survive. The tangential force along the disk surface cannot be balanced, so the surface layers will be ablated. The radiation force acts primarily only on a thin layer of column density $\sim \kappa_{gr}^{-1}$. Large-scale Kelvin-Helmholtz instabilities might couple the motion of that layer to underlying material, but the resulting shocks would destroy the dust in the surface layers and thus reduce the rate of ablation. With and without coupling, we find that the ablation timescale is considerably longer than the inferred inflow timescales (see section 4), so the disk is likely to survive. The fate of the dust in the surface layers is less clear. The dust can be destroyed by sputtering if it develops a high speed Δv relative to the surrounding gas. Collisional coupling maintains $(\Delta v)^2 \sim p_{rad}/\rho_{gas}$ (proportional to the ionization parameter) small enough near the disk that grains can survive. But since accelerated dust must pass through shocks to follow a warped surface, it may well be destroyed long before it and its associated gas reach the terminal speed (5).

The column density of a Strömgren length is $\Sigma_s = 0.05 \Xi T_4\ g\ cm^{-2}$, where

Ξ is the ratio of ionizing radiation pressure to gas pressure, and $10^4 T_4 K$ is the temperature of the photoionized gas. Plausible transition-zone disks have $\Xi \simeq 10^{0 \pm 2}$ at all radii. Except in the outskirts of the galaxy $\Sigma_{sg} \gg \Sigma_c$, so only a thin surface layer will be photoionized. If dust survives in this layer (as is especially likely in the low-density outer regions of the galaxy) resonance line emission from this layer will be converted into infrared photons, while other emission lines will be heavily reddened. If dust is destroyed (as is likely in the inner regions), the surface may contribute to (and perhaps dominate!) the emission line flux from the AGN (Collin-Souffrin 1987).

Free-free emission from the photoionized surface layers of the warped disk with electron density n_e will be optically thin at frequencies $\nu > 10^{10} (n_e/10^5 cm^{-3})^{1/2}$ Hz, so radio frequency free-free emission will come predominantly from the outer parts of the disk, while free-free at millimeter wavelengths could have a comparable contribution from the inner parsecs. It is easy to show that the free-free luminosity density $L_\nu(ff)$ (the optically thin free-free emissivity, integrated over the photoionized volume), is simply related to the infrared luminosity re-radiated by the underlying (and/or cospatial, if dust survives in the photoionized surface) gas:

$$L_\nu(ff) = L_{IR}/\left[2 \times 10^{15} \; Hz \cdot (5/g(\nu, T))\right] , \tag{6}$$

where $g(\nu, T)$ is the Gaunt factor. The resulting free-free luminosities are comparable to those of the flat-spectrum components observed in quasars by Antonucci and Barvainis (1988), and also to the level of the radio detections in most PG quasars (Kellermanm reported in Sanders *et al.* 1989).

6. Sublimation and the Temperature Gap.

The rate of sublimation of a dust grain at temperature T is given (in g $cm^{-2} s^{-1}$) by

$$Q = \sum_i \left(p_{si}(T) - p_i\right) S_i(T) \sqrt{\frac{m_i}{2\pi kT}} , \tag{7}$$

where the index i runs over all equilibrium gas-phase species (e.g. C, C_2, CO, etc. for graphite grains), p_i is the partial pressure of species i, $S_i(T) \sim 1$ is the sticking fraction for molecules colliding with the grain surface, and p_{si} is the saturation vapor pressure for species i. The saturation vapor pressure is given by

$$p_{si} = \varsigma_i(T) kT \frac{(2\pi m_i kT)^{3/2}}{h^3} \; exp \left[-\frac{h_0}{kT} - \Delta\right] , \tag{8}$$

where ς_i is the product of the rotational, vibrational, and electronic partition functions for the gas-phase species i, and h_0 is the heat of sublimation at zero temperature, on a per atom basis. $\Delta(T)$ is a complicated expression involving integrals of the specific heat of the solid (Reif 1965, p. 367). Graphite ($h_0/k = 8.58 \times 10^4 \; K$

to C, $h_0/k = 9.82 \times 10^4 \, K$ to C_2, $h_0/k = 9.45 \times 10^4 \, K$ to C_3, Kelley 1973) is the most refractory substance (with the exception of Tungsten!); silicate grains have $h_0/k \simeq 6.6 \times 10^4 \, K$. We have fitted the thermodynamic data and sublimation measurements of graphite in vacuum (Kelley 1973), and find that the saturation vapor pressure of the dominant gas-phase constituent is well fitted by

$$p_s = 6 \times 10^{15} \, exp[-92200/T] \, \text{dyn cm}^{-2} \, . \tag{9}$$

The gas pressure in an α accretion disk is

$$p = \frac{GM\dot{M}}{4\pi r^3 \alpha c_s} = 0.01 M_8 \dot{m} r_{pc}^{-3} \alpha^{-1} T_3^{-1/2} \, \text{dyn cm}^{-2} \, , \tag{10}$$

This pressure is comparable to the pressure $\sim 10^{-2} \text{dyn cm}^{-2}$ in broad-line clouds. Since the solar abundance of carbon, $C/H = 4 \times 10^{-4}$, the partial pressure of carbon, were it all in the gas phase, would be

$$p_C \simeq 10^{-5} M_8 \dot{m} r_{pc}^{-3} \alpha^{-1} T_3^{-1/2} \, \text{dyn cm}^{-2} \, . \tag{11}$$

At radii $\sim 1 \, pc$, grains grow at an impressive rate: $a/\dot{a} \sim 5n_9(a/0.1\,\mu m) \, yr$; in fact the temperatures and densities are quite similar to those in red-giant winds where interstellar dust is believed to form. [But here the grain and gas temperatures need not be equal: the ratio of a grain's radiative cooling luminosity to the rate at which it exchanges energy with gas via collisions is $L/H = 10 n_{11}^{-1} T_{g,3}^5 / T_{H,3}^{3/2}$, where the grain temperature is $10^3 T_{g,3} \, K$ and the gas temperature and density are $10^3 T_{H,3}$ and $10^{11} n_{11} \, cm^{-3}$, respectively]. However, comparing with equation (7) and equation (9), we see that when the grain temperatures T_g exceed $2000 \, K$, graphite grains will certainly begin to sublimate rather than grow. For $T_g > 2100 \, K$, the timescale for sublimation of a $1 \, \mu m$ graphite grain becomes shorter than $10^3 \, yr$, the timescale on which in could be replenished by inflow. From figure 1, we see that this temperature is reached at $\sim 0.3 \, pc$ in our fiducial quasar.

When the dust sublimates, the gas loses its primary opacity and coolant. As the temperature rises above $\sim 3000 \, K$, most common molecules are destroyed, and the opacity drops precipitously by several orders of magnitude (Alexander et al. 1983). The gas in the interior of the disk is then unable to remain in thermal equilibrium at temperatures $2000 \lesssim T \lesssim 7000 \, K$, and must inevitably heat. Above $\sim 10^4 \, K$, the opacity rises abruptly to near its former level as hydrogen is ionized, providing the gas with a new thermal equilibrium state. Unless there is no warp (or down-scattering of radiation onto the disk from electrons or a jet), the transition disk at $r < 0.3 \, pc$ is thus constrained by the heating from the central source to be optically thin, with $T \sim 10^4 \, K$, until $r \lesssim 0.02 \, pc$, when the incident flux can be carried by optically thick thermal emission, and the temperature will begin to rise above $10^4 \, K$ in the accretion disk. The absence of thermalized emission from material with temperatures $2000 \lesssim T \lesssim 7000\,K$ provides a natural explanation for the minimum in νL_ν at $\nu = 10^{14.5}\,Hz$ ($\lambda = 1\,\mu$m) observed in almost all quasars (Neugebauer et al. 1987). [The reader may mentally add an accretion disk spectrum to the right of figure 2]. Since in this interpretation the frequency is a universal constant, determined (up to very slowly varying logarithms) by the heats of sublimation and dissociation of dust and molecules, and by the ionization of hydrogen, the minimum in the reradiated νL_ν will always be present, and observable unless filled in by starlight or a non-thermal contribution to the spectrum.

7. Conclusions.

Figure 3 summarizes our description of reradiation from gas and dust in the host galaxy of a quasar. Provided quasars are located in galaxies like those of their lower luminosity cousins, it is hard to imagine how thermal reradiation can fail to make a significant contribution to the infrared luminosity of quasars. As we have outlined above, assuming that this *dominates* provides attractively natural explanations for the shape of the far infrared and submillimeter spectrum, for the high-frequency radio emission, for the "$3 - 5\mu$m" bump, and for the universal minimum in νL_ν at $\nu = 10^{14.5}\,Hz$. Some support for the latter can be adduced from the elegant observations of near-infrared variability in Fairall 9 by Clavel et al. (1989). The general absence of variability at longer infrared wavelengths (Neugebauer et al. 1989) is at least consistent with a thermal interpretation. Still, objections can be raised: e.g., the general absence of emission features associated with polycyclic aromatic hydrocarbons and silicates (Moorwood 1989). Nonthermal models can be contrived (and the author must confess to some involvement!) to explain many of the same features, though perhaps less naturally.

This debate will ultimately be resolved by a measurement of the brightness temperatures. Nonthermal models of the submillimeter spectrum (de Kool & Begelman 1989) require brightness temperatures $T_B > 10^{10}\,K$, while thermal models require $T_B < 10^3\,K$. The two models thus predict angular sizes for the emitting region differing by a factor $\gtrsim 10^4$. A submillimeter interferometer with a few kilometers baseline would determine the nature of the emission. Continued monitoring of infrared variability will also discriminate between thermal and nonthermal models (though one should beware of sources like 3C273, where a weak Blazar component seems occasionally to wobble into the line of sight). The model outlined above predicts hysteresis in the near infrared flux: when the central source increases in brightness, the near infrared flux can rise with it (with only a mean light-travel time delay). But the rise will sublimate the dust in a larger area than before, and if the central luminosity subsequently falls faster than grains can reform and grow, the dust-free "hole" will produce a large dip in the near infrared νL_ν, at the frequencies corresponding to the missing temperatures. If the emitting dust grains are aligned by shear or a globally anisotropic magnetic field, their thermal radiation could be significantly polarized.

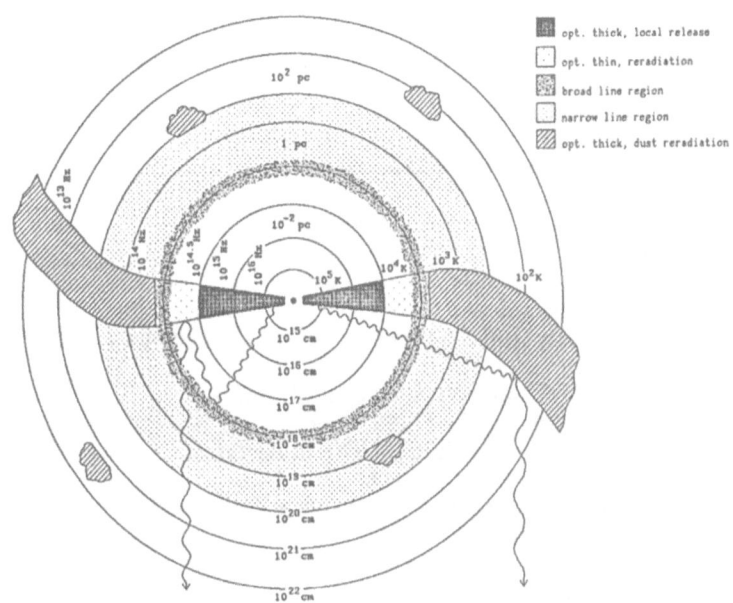

Figure 3: Cartoon illustrating the state of and frequencies of reradiation from gas in a quasar's host galaxy. Note the sudden transition in the gas temperature at $\sim 10^{18}\, cm$ caused by the drop in opacity when dust sublimates.

It is devoutly to be hoped that we will soon be freed from our embarrassing ignorance (after 20 years of observations and theoretical activity) of whether the infrared emission from quasars arises from a source $10^{14}\, cm$ across, or from one $10^{22}\, cm$ across.

ACKNOWLEDGEMENTS. I thank: Liz for typing, Gerry Neugebauer, Dave Sanders, Tom Soifer and Ski Antonucci for making this subject impossible to ignore, and the Irvine Foundation, the Boeing Corporation, and the NSF for support under Presidential Young Investigator grant AST 84-51725.

REFERENCES.

Adams, T.F. 1977 *Ap. J. Suppl.*, **33**, 19.

Alexander, D.R., Johnson, H.R., and Rypma, R.L. 1983, *Ap. J.*, **272**, 773.

Antonucci, R.R.J., and Miller, J.S. 1985 *Ap. J.*, **297**, 621.

Antonucci, R.R.J., and Barvainis, R. 1988 *Ap. J.*, **332**, L13.

Barthel, P. 1989 *Ap. J.*, **336**, 606.

Barvainis, R. 1987 *Ap. J.*, **320**, 537.

Begelman, M.C., McKee, C.F., and Shields, G.A. 1983 *Ap. J.*, **271**, 70.

Begelman, M.C. 1985 *Ap. J.*, **297**, 492.

Bollea, D. and Cavaliere, A. 1976 *Astr. Ap.*, **49**, 313.

Clavel, J., Wamsteker, W., and Glass, I.S. 1989 *Ap. J.*, **337**, 236.

Collin-Souffrin, S. 1987 *Astr. Ap.*, **179**, 60.

Davidson, K. and Netzer, H. 1979 *Rev. Mod. Phys.*, **51**, 715.

de Kool, M. & Begelman, M.C. 1989 *Nature*, **338**, 484.

Draine, B.T. 1981 *Ap. J.*, **245**, 880.

Draine, B.T., and Lee, H.M. 1984 *Ap. J.*, **285**, 89.

Edelson, R.A., and Malkan, M.A. 1986 *Ap. J.*, **308**, 59.

Gunn, J.E. 1979 in *Active Galactic Nuclei*, eds. C. Hazard and S. Mitt on, (Cambridge: Cambridge U. Pr.), 213.

Haniff, C.A., Wilson, A.S., and Ward, M.J. 1988 Preprint.

Hutchings, J.B., 1983 *Pub. A. S. P.*, **95**, 799.

Kelley, K.K. 1973 In *Selected Values of Thermodynamic Properties of the Elements*, Am. Soc. Metals, Metals Park, pp. 87 *ff*.

MacAlpine, G.M. 1985 in *Astrophysics of Active Galactic Nuclei*, ed. J.S. Miller (Mill Valley: Univ. Science Books), 259.

MacKenty, J.W., and Stockton, A. 1984 *Ap. J.*, **327**, 116.

Martin, P.G. 1978 *Cosmic Dust*, Oxford: Clarendon Press.

Moorwood, A.F.M. 1989 in *Infrared Spectroscopy in Astronomy*, ESA SP-290, eds. A.C.H. Glasse, M.F. Kessler, and R. Gonzalez-Riesta.

Neugebauer, G., Soifer, B.T., Matthews, K., and Elias, J.H. 1989 *A. J.*, **97**, 957.

Ostriker, E.C., and Binney, J.J. 1989 *M. N. R. A. S.*, **237**, 785.

Rees, M.J., Silk, J.I., Werner, M.W., and Wickramasinghe, N.C. 1969 *Nature*, **223**, 788.

Reif, F. 1965 *Fundamentals of Statistical and Thermal Physics*, (New York: McGraw-Hill).

Sanders, D.B. Scoville, N.Z., and Soifer, B.T. 1988a *Ap. J.*, **335**, L1.

Sanders, D.B., Soifer, B.T., Elias, J.H., Madore, B.F., Matthews, K., Neugebauer, G., and Scoville, N.Z. 1988b, *Ap. J.*, **325**, 74.

Sanders, D.B, Phinney, E.S., Neugebauer, G., Soifer, B. T., and Matthews, K. 1989 *Ap.J.* in press.

Scheuer, P.A.G. 1987 in *Superluminal Radio Sources*, eds. J.A. Zensus and T.J. Pearson (Cambridge: Cambridge U. Pr.), 104.

Shlosman, I., and Begelman, M.C. 1988 *Nature*, **329**, 810.

Telesco, C.M., and Harper, D.A. 1980 *Ap. J.*, **235**, 392.

Vietri, M. 1986 *Ap. J.*, **306**, 48.

Unger, S.W., Pedlar, A., Axon, D.J., Whittle, D.M., Meurs, E.J.A., and Ward, M. J. 1987 *M. N. R. A. S.*, **228**, 671.

Usher, P.D. 1978 *Ap. J.*, **222**, 40.

Whittet, D.C.B. 1988 in *Dust in the Universe*, eds. M.E. Bailey and D.A. Williams, (Cambridge: Cambridge U. Pr.) 25.

Wilson, A.S. and Ulvestad, J.S. 1982 *Ap. J.*, **263**, 576.

Wright, E.L. 1987 *Ap. J.*, **320**, 818.

INDEX.

471